SAMPLING METHODS
CENSUSES AND SUR

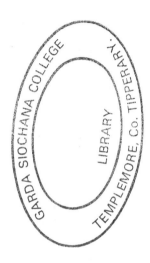

SAMPLING METHODS FOR CENSUSES AND SURVEYS

FRANK YATES, C.B.E., Sc.D., F.R.S.

Senior Scientist, Rothamsted Experimental Station
Formerly Head of the Department of Statistics
and Deputy Director

FOURTH EDITION
Revised and Enlarged

Quis separabit nos

CHARLES GRIFFIN & COMPANY LTD
London and High Wycombe

CHARLES GRIFFIN & COMPANY LIMITED
Registered Office:
Charles Griffin House, Crendon Street
High Wycombe, Bucks, HP13 6LE
England

First edition   1949
Second   ,,   1953
Third   ,,   1960
   Second impression   1965
   Third   ,,   1971
Fourth edition   1981

ISBN 0 85264 253 9

Printed in England by Henry Ling Ltd, Dorchester, Dorset

PREFACE TO THE FOURTH EDITION

THIS book has been written primarily for those who have little or no previous training in mathematical statistics, but who have some training or experience in the presentation and handling of statistical data. It was consequently not written in the form of a mathematical treatise, and mathematical proofs have not been included. On the other hand an attempt was made to cover all the modern developments of sampling theory which are of importance in census and survey work, and to give an adequate discussion of the complexities that are encountered in their practical application. This necessitated fuller treatment of the subject than is to be found in textbooks on mathematical statistics, or than is normally included in statistical courses. Indeed, the orderly development imposed by the preparation of a book revealed various theoretical gaps which had to be filled in.

The work had its origin in a request of the United Nations Sub-Commission on Statistical Sampling, at their first session held at Lake Success in September, 1947, that a manual be prepared to assist in the execution of the projected 1950 World Census of Agriculture, and the 1950 World Census of Population. The Sub-Commission were particularly impressed with the need for a wider use of sampling in the less developed areas, and it was originally intended that only the sampling problems encountered in censuses and surveys in these areas should be dealt with. On reviewing the matter, however, I came to the conclusion that conditions differed so greatly in different areas that it would be necessary to cover a wide variety of methods, which in essentials differed little from the methods appropriate to censuses and surveys in more fully developed areas. It therefore seemed best to take the opportunity of writing a more general book.

The various computational procedures have been illustrated, as far as practicable, by numerical examples. These examples in the main have an agricultural background, since this type of data was most readily accessible and was also particularly relevant to the original purpose of the book. Students without statistical knowledge are strongly advised to rework these examples for themselves. They will thereby acquire a much better grasp of the statistical techniques involved and at the same time develop their computational skills. Such tasks are greatly eased by intelligent use of modern electronic desk calculators.

Two new chapters were added to the second edition, but no changes were made in the original text. A further chapter on electronic computers, then in their infancy, was included in the third edition.

Although cards are still commonly used for the conversion of survey data to machine readable form, computers have completely replaced punched card equipment for tabulation and for more advanced types of analysis required for the investigation of causal relations, etc. Some account of these developments was therefore necessary if the book is to serve as a guide to their effective use.

This necessitated more extensive revision than was attempted in the earlier editions. The chapter on computers has been completely rewritten, and contains an outline of some of the general programs that are now widely available, their strengths and weaknesses, and directions in which we may look for further improvement. The latter part of Chapter 5 has been rewritten to take account of the extended facilities for data preparation, editing and tabulation, and the sections on punched card tabulating machinery have been deleted. Chapter 9 on critical analysis has been reorganized and extended to include a more detailed account of the uses of linear and generalized linear models in survey analysis and their relation to regression methods and the analysis of variance, with examples of the interpretation of computer output. Further treatment has been given in Sections 7.11 and 8.11 to the sampling errors of estimates from random samples in which the unit of sampling differs from the unit of analysis, and such of the miscellaneous developments included in Chapter 10 of the second edition as are still of interest have been transferred to the earlier chapters. Some changes and additions have been made in other parts of the book, but much of the text stands and this has enabled much of the earlier material to be reproduced photographically.

It did not appear feasible to preserve the old section numbers throughout. For the most part the sections in each chapter have been renumbered consecutively, but there are a few interpolations in the main sequences of Chapters 7 and 8 to avoid extensive changes in cross-references.

I was at first disposed to omit the bibliographies included in the earlier editions, but on examination it seemed they were worth retention, as the titles give an interesting indication of the problems that were of concern at that time. A further supplementary bibliography would have occupied far too much space, but I have included references to two recent specialized bibliographies by Tore Dalenius, one on non-sampling errors containing some 1500 references, the other on information privacy containing some 1200 references, the latter subdivided according to subject-matter.

I have been much helped in the preparation of this edition by Ruth Hunt, who has been associated with this book since its inception, and by my niece Jillian Bridgman while she was working at Rothamsted. Thanks are also due to R. J. Baker, of the Rothamsted Statistics Department, who ran the GLIM analyses reported in Chapter 9, and to other members of the Statistics and Computer Departments who have helped in various ways.

F. YATES

ROTHAMSTED EXPERIMENTAL STATION,
31st March, 1980

CONTENTS

CHAPTER 4

PRACTICAL PROBLEMS ARISING IN THE PLANNING
OF A SURVEY

CHAPTER 5

PROBLEMS ARISING IN THE EXECUTION AND ANALYSIS OF A SURVEY

CHAPTER 6

ESTIMATION OF THE POPULATION VALUES

CHAPTER 7

ESTIMATION OF THE SAMPLING ERROR

CHAPTER 8

EFFICIENCY

CHAPTER 9

CRITICAL ANALYSIS OF SURVEY DATA

CHAPTER 10

COMPUTER PROGRAMS FOR SURVEY ANALYSIS

CONTENTS

APPENDIX TABLES

xiii

LIST OF EXAMPLES

Bold type indicates a section number

SHORT COURSE OF READING

forming an introduction to the statistical aspects of sampling

(1)   General introduction: Chapters 1 and 2.
(2)   Random samples: 3.1, 3.2, 6.1–6.4, 7.1–7.4, 7.11, 4.32.
(3)   Stratified samples: 3.3, 3.5, 3.7, 6.5–6.7, 7.5 (para. 1–7 and example 7.5.a), 7.6, 7.7.
(4)   Efficiency (random and stratified samples): 8.1–8.3, 8.5, 8.16.
(5)   Multi-phase samples and supplementary information: 3.12, 6.8–6.12, 7.5 (remainder), 7.8–7.10, 7.12, 8.6, 8.6.1.
(6)   Multi-stage samples: 3.8, 6.18, 6.19, 7.11, 7.17.
(7)   Sampling with probability proportional to size: 3.9–3.11, 6.16, 6.17, 7.15, 7.16, 4.16.
(8)   Systematic samples: 3.6, 3.14, 3.15, 7.18.
(9)   Miscellaneous: 3.19, 4.33, 5.12, 5.19–5.21, 8.18, 8.21, 9.1–9.3.

SAMPLING METHODS FOR CENSUSES AND SURVEYS

CHAPTER 1

THE PLACE OF SAMPLING IN CENSUS WORK

1.1 The sampling process

Sampling, that is, the selection of part of an aggregate of material to represent the whole aggregate, is a long-established practice. Simple examples are provided by a handful of grain taken from a sack, or a piece of cloth cut off a roll. In these cases little attention need be paid to the selection process, since the whole of the material is similar or well-mixed, and any part of it if not too small is likely to be closely representative of the whole. When, however, the aggregate to be sampled consists of units which are somewhat dissimilar amongst themselves, and which are not well-mixed, a small sample of these units may not be representative of the whole aggregate. Even if units are selected from different parts of the aggregate, and other suitable precautions are taken, the sample is likely to a certain extent to be unrepresentative owing to the chance inclusion of an undue proportion of units of a particular type. It will clearly not be representative if units of a particular type are chosen deliberately to the exclusion of other types, or if the process of selection is such that certain types of unit are favoured at the expense of others. Thus in sampling a heap of coal by taking a few shovelfuls from the edges, too great a proportion of the large lumps will be obtained, since the large lumps tend to roll down the sides and be distributed round the edges of the heap. Similarly in the sampling of continuous material, a single portion, even if quite large, may not be adequately representative ; a piece of cloth cut off the end of a roll in which the quality of the weaving varies progressively, will not form an adequate sample of the whole roll.

Census and survey work is normally carried out on material made up of dissimilar units. Censuses of population, censuses of industrial production, and censuses of agriculture have the common feature that the aggregate of material embraces a large number of separate units which are often markedly dissimilar in various respects. In many cases the purposes for which the information is required are adequately served if a proportion only of the units are covered, but because of the dissimilarity of the different units neither haphazard nor casual selection, and still less deliberate selection, can be expected to provide a representative sample. Rigorous processes of selection have therefore to be used.

Censuses carried out on a properly selected sample will be called *sample censuses*. There has in the past been a tendency to use the term sample to refer to the results of an attempted complete census in which there has been failure to obtain information from a substantial proportion of the units. Its use in this sense is strongly to be deprecated ; instead the term *incomplete census* is suggested. The term *sample* should be reserved for a set of units or portion of an aggregate of material which has been selected in the belief that it will be representative of the whole aggregate.

1.2 Sampling errors

Whether or not a sample will give results which are sufficiently representative of the whole aggregate depends primarily on whether the errors introduced by the sampling process are sufficiently small not to invalidate the results for the purposes for which they are required. Even if a proper process of selection is employed, the sample cannot be exactly representative of the whole aggregate. The inevitable errors which then occur in the results are termed the *random sampling errors* of these results. The average magnitude of these random sampling errors will depend on the size of the sample, on the variability of the material, on the sampling procedure adopted, and on the way in which the results are calculated.

It is a fortunate fact that if a proper process of selection is adopted, the average magnitude of the random sampling errors, and indeed the expected frequency of occurrence of errors of any magnitude, can be calculated from the detailed results obtained from an actual sample. The methods by which this can be done depend on the mathematical theory of statistical sampling.

An extension of the analysis involved in the calculation of these errors enables the relative accuracy of the different sampling methods which can be employed on the same material to be assessed, and thus enables further surveys to be more efficiently planned.

It is the development of these processes that has changed sampling from a speculative and uncertain procedure to a method having definite and determinable precision. Sampling has thus become a reliable method in which full confidence can be placed. In addition, the possibility of setting ascertainable limits to the random sampling errors has served to throw into prominence those other types of error which arise from faulty selection processes or faulty methods of observation, or which exist in some other source of information with which the sampling results are being compared.

1.3 The place of sampling in census and survey work

Sampling will only be of use in census work if, as mentioned in Section 1.2, the sampling errors are sufficiently small not to affect the validity of the results for the purposes for which they are required. This will in part be a function of the degree to which the results have to be broken down. If only overall results for the whole population are required, a given degree of accuracy will be attained with a far smaller sample than will be the case if detailed results for different parts of the population (e.g. different regions, towns, etc.) are required. In certain circumstances the sample may have to be so large that there will be little point in using a sample census in place of a complete census. Obviously, in the extreme case where information on all the individual units is required, this can only be obtained by a complete census.

Another factor which influences the decision whether or not to use sampling is the relative difficulty and cost of organizing a sample census and a complete census. The amount of effort and expense required to collect information is always greater per unit for a sample than for a complete census. In addition a sample census presents its own organization problems, some of which are absent from a complete census, and it occasionally happens, if the information required is very simple, that a complete census can be carried out through the ordinary administrative channels, whereas a sample census requires the setting-up of a separate organization. Usually, however, if the size of the sample needed to give the required accuracy represents only a small fraction of the whole population, the total effort and expense required to collect the information by sampling methods will be very much less than that required for a census of the whole population.

In many cases, therefore, sampling results in great economy of effort. It has also other advantages which are not so immediately apparent. In the first place, the completeness and accuracy of the returns may be much more easily ensured if the information is collected from only a small proportion of the population. If, for example, questionnaires are sent through the post, it is frequently impossible in a complete census to bring pressure to bear on those who fail to make their returns, even where the completion of these questionnaires is compulsory, owing to the large numbers of individuals involved. In the case of a sample, the smaller number of individuals enables follow-up notices to be sent and telephone calls and visits to be made. The separate returns can also be much more carefully scrutinized, and further enquiries undertaken where there is reason to doubt their accuracy.

Secondly, it is possible to obtain more detailed information in a sample census. Although the burden on the individual of furnishing more detailed information is not lessened, except when different items of information can be obtained from different individuals, the individuals concerned are more likely to be willing to provide such information if they know that they represent a small sample of the whole population. Detailed information, when obtained, can be more easily handled, both at the stage of abstraction and coding of the original information and in the analysis of the coded results. Owing to the reduced volume of material that has to be handled the quality of the abstraction and analysis can also be improved, the former because a higher grade of clerical labour can be employed, with better supervision, and the latter because the data can be classified in many more ways with the same amount of computing or machine time.

Thirdly, in many types of census the use of sampling makes possible a very considerable increase in speed, both in the execution of the field work, and in the analysis of the results. Speed in analysis can also be obtained, in the case of a complete census, by taking a sample of the returns for abstraction and analysis. This device is frequently of value for providing preliminary results quickly, even when a final analysis of the whole of the returns is ultimately required.

The use of sampling is essential for investigations of the sociological type in which extensive and detailed information has to be collected from individuals, many of whom have neither the education nor experience required to answer detailed questionnaires without assistance. It is equally essential in investigations requiring skilled physical observations and measurements. Such investigations can only be carried out by the use of trained investigators, and complete investigations covering any large group of the population or body of material are consequently impossible, both on grounds of expense and because, even if the expense can be tolerated, a sufficient body of investigators can rarely be recruited and trained.

For an investigation of this kind involving the collection of elaborate information the term *survey* is usually employed. It seems a mistake, however, to confine the word survey to a sample survey or the word census to a complete census. Thus B. Seebohm Rowntree (1901), when he carried out an investigation into the social and economic conditions of all working-class families in York, correctly described this as a survey.

Although the use of sampling necessarily introduces certain inaccuracies, owing to sampling errors, the results obtained by sampling are frequently more accurate than those obtained in a complete census or survey. The random sampling errors are always assessable. The other errors to which a survey is subject, such as incompleteness of returns and inaccuracy of information, are liable to be very much more serious in a complete census than in a sample census, since far more effective precautions can be taken to see that the information is accurate and complete in a sample census. Furthermore, the use of sampling greatly facilitates the imposition of additional more detailed checks. Indeed, a complete census can only be properly tested for accuracy by some form of sampling check.

On the other hand, the claim that is sometimes made that the reliability and accuracy of the results of a properly planned sample census can be assessed with full objectivity from the results themselves is only partly true. The random sampling errors can be so assessed, and under certain circumstances it is possible to obtain comparisons between different investigators. If all investigators or respondents tend to make the same kind of error, however, this will not be revealed in the results, whether the census is complete or carried out on a sample.

In respect of coverage a sample census may in certain circumstances be less reliable than a complete census. It is, for example, relatively simple for an investigator to ascertain by direct question whether an individual has already been included in a population census, and simple intensive checks of certain areas, say villages, can be made in a similar manner to verify that there is no appreciable number of omissions. Similarly, in a survey of physical objects such as houses, a marking system or other suitable device can often be used to guard against duplication and omission. Such checks are impossible in the case of a sample census.

This is one of the most difficult points in the practical design of many sample surveys, particularly in undeveloped areas. To overcome it complete enumeration can sometimes be used in conjunction with sampling. Where a complete enumeration of the whole of the population or aggregate of material presents no particular difficulty, but where the collection of detailed information from all units would be a difficult or impossible undertaking, a complete enumeration can be carried out. This is then used as a basis for the selection of the sample for such sample censuses and surveys as are required to provide the detailed information.

1.4 Development of the use of sampling in censuses and surveys

Prior to the development of the appropriate methods of estimation of sampling errors and a clear recognition of the conditions governing satisfactory methods of selection of the sample, the use of sampling in census and survey work often proved unsatisfactory. There are many early examples of sample censuses and surveys which are defective in one way or another. Even when the basic principles of the simpler forms of sampling were understood, the attempted use of more complicated forms before methods of evaluating their errors and relative efficiency had been worked out gave rise to further defective surveys.

This has led to a certain mistrust of sampling, which still exists in some quarters. During recent years, however, there has been a rapid growth in the use of sampling in various countries. This development has been greatly stimulated by the war and its attendant measures of large-scale economic control. Such measures, if they were to be effective in the changing conditions met with in wartime, demanded an efficient and speedy information service which only the sampling method could supply. This has resulted in further improvements in technique through the stimulation of research into the theory of sampling methods, and the provision of basic data for practical investigations of the relative efficiency of the various methods in different fields.

It still remains true, however, that in inexperienced hands sampling may give unsatisfactory results, owing to the use of faulty methods of selection, inappropriate sampling design, or inefficient methods of estimation. The prime requirement of any large-scale sample survey is therefore that the organization of the survey should be carried out by a person who has adequate knowledge and experience of sampling methods and their application. The methods employed must be thoroughly sound, theoretically and practically, both in order that satisfactory results may be ensured, and also in order that mistrust cannot subsequently be engendered by criticism of the methods adopted. It must never be forgotten that it is not sufficient to provide results which are in fact correct. They must also be generally accepted if they are to have their full value.

It is sometimes stated that no large-scale sample census or survey should be carried out without the advice of an expert mathematical statistician with experience of such work. Unquestionably, if the services of such an expert can be secured this is all to the good, but my own experience is that no one expert can be expected to supervise adequately more than a very few surveys at any one time, since adequate supervision demands a very full knowledge both of the material that is to be surveyed and of local conditions, coupled with close attention to detail at all stages. An expert acting in an advisory capacity is therefore no substitute for the statistician on the spot, who must be prepared to accept responsibility for the planning, execution and analysis of the survey. To do this he must himself have both an adequate knowledge of sampling procedure and thorough knowledge of the material and local conditions.

Consequently, if full and effective use is to be made of sampling methods, statisticians and others who already have experience of the conduct of complete censuses but no training in sampling methods must themselves undertake a study of these methods, in order that they may decide in what ways these can be applied to their own problems. The function of the expert then becomes one of advice on exceptional problems, rather than one of detailed supervision.

Fortunately the principles underlying good sampling methods are not unduly difficult to understand, and provided a proper respect is observed for the fundamental rules of procedure I believe they can be successfully applied by those who have statistical experience but who are not primarily mathematical statisticians.

1.5 Method of presentation

The method of presentation adopted in this book is to take the various parts of the sampling process in roughly the order they are encountered in the execution of a census or survey, and discuss the various aspects of each part in turn. Thus Chapters 2 and 3 describe the various types of sample that can be used, and the general principles to be followed in the selection of a sample, Chapter 4 deals with the practical planning of a survey, and Chapter 5 with the problems encountered in its execution and in the abstraction of the results. Chapters 6 to 9 are concerned with the more strictly statistical problems. Chapter 6 deals with the various methods of estimating the population values, Chapter 7 with the estimation of sampling errors, Chapter 8 with the determination of the relative efficiency of the various sampling methods, and Chapter 9 with inferences on cause and effect. Chapter 10 discusses the current developments in the use of computers for survey analysis. This method of presentation has the advantage that the more practical aspects of sampling procedure are dealt with first. It is true that knowledge of the statistical techniques described in the later chapters is necessary before the relative merits of different methods of sampling any particular type of material can be accurately assessed. The detailed application of these techniques, however, is the province of those responsible for the numerical analysis of the results, whereas the planning is also the concern of those who require the information and those who are concerned with its collection. The planning can be undertaken much more efficiently, and with added interest, if all concerned understand in general terms the underlying problems. It is hoped that study of the first five chapters will give this understanding. If they also act as a stimulus to the study of Chapter 6 and the first few sections of Chapter 7 the understanding should be correspondingly deepened.

For those responsible for the numerical analysis of the results, and for the assessment of the relative efficiency of the different possible methods, thorough study of the whole book is necessary. This study should include the reworking of the numerical examples. Modern electronic desk and pocket calculators make this task much easier. Only by this procedure can a thorough grasp of the details of the various methods be obtained.

The separation of the discussion of the methods of estimation of the population values, of the sampling errors, and of efficiency necessarily involves a good deal of cross-reference, particularly in the numerical examples. Since this appeared inevitable, it was with some hesitation that the chosen method of presentation was adopted. On balance, however, this disadvantage appeared to be outweighed by the advantage of being able to present as a whole the relatively simple techniques involved in estimation before the more complicated techniques required for the estimation of error and the assessment of relative efficiency. It is believed that this will make the book more useful to those who do not require to go deeply into these latter techniques. For those who prefer it, there is nothing to prevent the simultaneous study of the corresponding sections of Chapters 6 and 7, or indeed of Chapters 6, 7 and 8. Chapters 6, 7 and 8 may also, if desired, be taken before Chapters 4 and 5, but the last part of Chapter 5 (Section 5.8 onwards) should be studied before embarking on Chapters 9 and 10.

1.6 Terminology and notation

The question of terminology was considered by the United Nations Sub-Commission on Statistical Sampling, at its second session held in Geneva in September, 1948. Their recommendations are included in a memorandum entitled *Recommendations concerning the Preparation of Reports on Sampling Surveys*. With a few minor exceptions the terminology adopted in this book is that recommended by the Sub-Commission.

New conventions have been adopted for the mathematical notation. The use of bold face and Gill Sans type for population values and their estimates, and of capital letters for the population totals, has enabled the formulæ to be presented in a very simple, and it is hoped easily understandable form. By the use of this notation the elaborate summation notation which has become current in much of the literature on sampling has been avoided. It is recognized that the notation is not particularly convenient for manuscript and typescript, but the difficulty can in fact be overcome by the use of single and double underlining, or wavy and straight underlining. In verbal description the words " population " and " estimate," or abbreviations of them, can be used.

CHAPTER 2

REQUIREMENTS OF A GOOD SAMPLE

2.1 Bias

The principal object of any sampling procedure is to secure a sample which, subject to limitations of size, will reproduce the characteristics of the population, especially those of immediate interest, as closely as possible.

At first sight it might appear that the most accurate results could be obtained by deliberate selection of the units to be included in the sample. In particular, if averages only are of interest, units might be selected which appear to be nearest to the average. If, for example, a quick assessment of the yield per acre of an agricultural crop is required, district officers might be asked to select some " average " fields in each district, and to determine the yields of these fields.

Such a sample is unfortunately very often of little value. Its primary fault is that it may well be *biased*, that is, the selection of all the fields may be affected by similar errors. Thus, in order to enhance the reputation of their districts, all district officers may tend to select fields which yield more heavily than the average, or, if they feel that the interests of the farmers or the country may be furthered by an underestimate, they may select fields which yield less than the average.

Even if the district officers can be trusted to be completely objective, considerable unconscious errors of judgment, all tending in the same direction, may still occur, and such errors may far outweigh any increase in accuracy resulting from deliberate selection. Nor will increase in the number of officers concerned in the selection necessarily improve matters, since all may be subject to the same type of error.

We may consequently distinguish between two types of sampling error, those arising from biases in selection, etc., and those due to chance differences between the members of the population included in the sample and those not included. The aggregate of the former in the sample will be termed the *error due to bias* and the aggregate of the latter the *random sampling error*, or when bias is known to be absent, the *sampling error*. The total sampling error will, of course, be made up of the bias, if any exists, and the random sampling error. The essence of bias is that it forms a constant component of error which does not decrease, in a large population, as the number in the sample increases, whereas the random sampling error decreases on the average as the number in the sample increases.

2.2 Methods of selection which give rise to bias

There are a number of ways in which faulty selection of the sample may give rise to bias. The main causes may be broadly classified as follows :—

(1) Deliberate selection of a " representative " sample. This is the type of bias described above.

9

(2) A procedure of selection depending on some characteristic which is correlated with properties of the unit which are of interest. Many haphazard selection processes give rise to biases of this kind.

(3) Conscious or unconscious bias in the selection of a " random " sample. If a proper random process is not strictly adhered to, the investigator, although claiming that his sample is random, may allow his desire to obtain a certain result to influence his selection. This type of bias is particularly serious, since its existence may not be immediately apparent.

(4) Substitution. Investigators often substitute another convenient member of the population when difficulties are encountered in obtaining information. Thus, in a house-to-house survey the next house may be taken when there is no reply. This will necessarily lead to a preponderance of houses of the type that are occupied all day, *e.g.* houses of people with families.

(5) Failure to cover the whole of the chosen sample. If no second visit is made to houses from which no reply is received there will still be bias even though no substitution is attempted. This fault is particularly prevalent in postal questionnaires, which are often very incompletely returned. Returns are clearly likely to be received from individuals who are specially interested in the objects of the survey, or possess other characteristics which make them unrepresentative of the whole population.

2.3   Avoidance of bias in selection

It is clear that, if possibilities of bias exist, no fully objective conclusions can be drawn from a sample. The first essential of any sampling procedure must therefore be the elimination of all important sources of bias.

The simplest, and the only universally certain way, of avoiding bias in the selection process is for the sample to be drawn either entirely at *random*, or at random subject to restrictions which, while improving the accuracy, are of such a nature that they do not introduce bias into the results. In some cases, however, certain forms of *systematic* selection, such as the selection of names at equal intervals down a list, or the use of an evenly spaced grid of points on a map, may be permissible.

Random selection does *not* mean haphazard selection. A random sample can only be obtained by adherence to some proper random process, such as the drawing of lots or the use of a table of random numbers. Sticking pins into a map will not give a random distribution of points in a map. The selection of houses by walking through the streets of a town will not give a random selection of houses in the town. The words " random " and " random sample " are, in fact, gravely abused. For this reason, if for no other, the method of selecting the sample should be specified in all accounts

10

of the results of sample surveys and censuses, and indeed in all sampling work.

In order to prevent careless or deliberately biased selection on the part of investigators it is often important in large-scale work for the selection to be done in some central office, in such a manner that no element of choice is left to the investigators, and in such a manner also that checks on the field work can be imposed if necessary. Even in cases in which a less rigorous method of selection may be judged to be satisfactory, it may be necessary to impose a rigorous method in order to prevent criticism on this ground by those not familiar with the details of the work.

2.4 Examples of biased selection

It may be well at this stage to give some actual examples of cases in which an unsatisfactory method of selection has introduced serious bias into the results.

The first example is taken from a paper by Kiser (1934, D). A sample of households was taken in Syracuse, U.S.A., in 1930 and 1931, with the object of making a study of morbidity. It was also intended to use this sample for the study of birth-rates. Before beginning this latter study, which was subsidiary to the morbidity study, a comparison was made of the sizes of households of the sample with those of the corresponding census tracts. This comparison is shown in Table 2.4.a. (Households of one were not included in the survey.)

TABLE 2.4.a—SAMPLE OF HOUSEHOLDS IN SYRACUSE: DISTRIBUTION OF HOUSEHOLDS ACCORDING TO SIZE, IN THE ORIGINAL SAMPLE, AND IN THE CENSUS TRACTS

Number in household	Original sample		Census tracts	
	Number	Per cent.	Number	Per cent.
2 . . .	254	19·4	1,762	26·8
3 . . .	338	25·9	1,745	26·5
4 . . .	307	23·5	1,438	21·9
5 . . .	201	15·4	853	13·0
6 . . .	106	8·1	388	5·9
7 . . .	46	3·5	208	3·2
8 . . .	25	1·9	96	1·5
9 and over .	29	2·2	86	1·3
TOTAL . .	1,306	99·9	6,576	100·1

It is immediately apparent from the table that the sample contains a considerably greater proportion of large households than exist in the whole population. Households of two are under-represented in the sample to the extent of 7·4 per cent. of all households, or 28 per cent. of the households of this size. This deficiency is attributed by Kiser to the failure of enumerators to revisit missed households, in which childless married women working away from home are likely to predominate. In order to provide a more satisfactory sample it was necessary to make a further survey of those families that were missed altogether at the time of the morbidity survey.

It is interesting to note that the sample was apparently considered satisfactory for the morbidity study, as is indicated by the statement that the workers " had been primarily concerned with securing a sample representative of the area in regard to prevalence of sickness rather than size of household." Actually such a biased sample can scarcely be regarded as wholly satisfactory for even a morbidity study, since sickness rates are likely to vary with the size and composition of the family.

The second example is one obtained at Rothamsted in an experimental sampling of a collection of stones (Yates, 1936, *b*, H). The stones, a number of flints of varying sizes, some 1200 in all, were spread out on a table, and twelve observers were each instructed to choose three samples of twenty stones which should represent as nearly as possible the size distribution of the whole collection. Table 2.4.b gives the mean weights per stone of these 36 samples, and also the true mean weight of the whole collection.

TABLE 2.4.b—MEAN WEIGHT PER STONE IN SAMPLES OF 20 STONES (*oz*)

Observer	1	2	3	4	5	6	7	8	9	10	11	12
Sample 1	1·9	2·4	2·4	1·9	2·2	2·8	2·4	1·6	2·2	2·6	2·4	2·4
Sample 2	1·8	3·0	2·4	2·0	2·7	2·6	2·6	2·0	2·2	2·2	2·4	3·0
Sample 3	1·7	2·4	2·1	2·0	3·1	2·8	2·5	2·0	2·2	3·1	1·8	2·4
Mean	1·8	2·6	2·3	2·0	2·7	2·7	2·5	1·9	2·2	2·6	2·2	2·6

Mean of all samples : 2·34 oz.   True mean : 1·91 oz.

It is apparent that there is a tendency, which is common to most observers, to select stones which are on the average larger than those of the whole collection. Of the twelve observers ten chose samples whose mean weight was above the mean weight, 1·91 oz., of all the stones, the mean for all samples being 2·34 oz. This tendency is consistent from sample to sample. Thus, of the thirty samples chosen by the above ten observers, all but two had mean weights greater than the mean weight of all stones, while all three samples of observer 1 were less than the correct mean.

In this example the selection was deliberate. A further example showing similar effects arising from haphazard selection (claimed by the observer to be " random ") is provided by some observations obtained in the course of a scheme of sampling observations on the growth of wheat instituted by the Agricultural Meteorological Committee (Yates, 1935, A).

In this scheme measurements on the heights of shoots of wheat were made at regular intervals on observation plots at a number of centres. A detailed procedure had been laid down for the random location on each occasion of 128 quarter-metre lengths of row in sets of 4 on contiguous rows. The height measurements were made on the 256 shoots at the ends of these lengths—test observations conducted at another time indicated that this method of selection was virtually random. At one centre a drill with fewer rows than normal had to be used, and as a result only 192 shoots were available for measurement on each occasion. In order to provide the number of observations laid down and thereby, as he thought, improve his results, the observer selected " at random " two additional shoots from each set of three quarter-metre lengths. Fortunately he booked the observations on these additional shoots separately.

Figures 2.4.a and 2.4.b show the distribution of the regular and additional measurements taken on the 31st May and on the 28th June respectively. The deviations from the set means of the regular measurements are shown. Suitable adjustments, details of which are given in the original paper, are made to the additional measurements to give fair representation of the variability as well as the bias in the mean.

Examination of Figure 2.4.a indicates that on this date the additional measurements show a considerable preponderance of positive deviations with a corresponding deficiency of negative deviations. There is, in fact, a tendency to select shoots which are higher on the average than those of a truly random sample, the difference in the average height being $+ 3 \cdot 3$ cm. This difference is clearly in the nature of a bias, and cannot be attributed to random sampling errors.

The situation was entirely different on the 28th June, as is shown in Figure 2.4.b. At this date the deviations of the additional measurements, both positive and negative, are smaller on the average than are those of the regular observations ; in other words, there is a tendency to select shoots which are nearer the mean height than they would be on the average in a truly random sample. In spite of this, there is again a considerable bias, this time negative, the mean difference being $- 2 \cdot 7$ cm. In this case, therefore, a single additional shoot will give a value which on the average is closer to the true mean value than is the value given by a single randomly located shoot, but as the number of shoots is increased the relative accuracy of the random sample progressively increases, and with the numbers of shoots actually taken, the random sample is considerably more accurate.

This example provides an illustration of a case where the biases on the two occasions, though arising from similar defects in selection, are of very different magnitude, and indeed of opposite sign. Consequently the difference of the two sets of measurements will also be seriously affected by bias. In

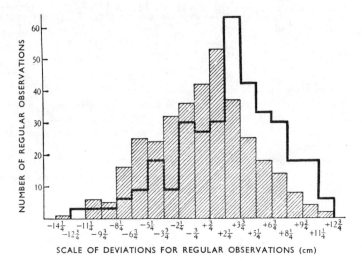

FIG. 2.4.a—DISTRIBUTION OF REGULAR OBSERVATIONS (*shaded*) AND ADDITIONAL
OBSERVATIONS (*unshaded*) OF HEIGHTS OF WHEAT SHOOTS ON 31st MAY

(By courtesy of the editor of the *Annals of Eugenics*.)

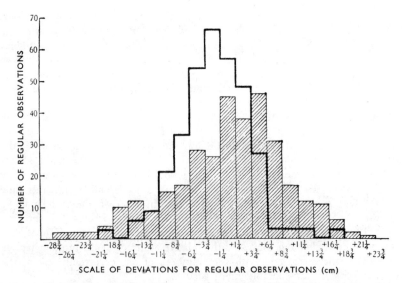

FIG. 2.4.b—DISTRIBUTION OF REGULAR OBSERVATIONS (*shaded*) AND ADDITIONAL
OBSERVATIONS (*unshaded*) OF HEIGHTS OF WHEAT SHOOTS ON 28th JUNE

(By courtesy of the editor of the *Annals of Eugenics*.)

14

this case the growth rate of the wheat would have been underestimated by nearly 10 per cent. had only the additional measurements been available.

These biases are, of course, of the type that might be expected. When the shoots are only half-grown and there is nothing much to be seen except the top leaves there will be a tendency to pick the longer shoots, but when the crop has come into ear the observer can see shoots of all lengths, and is more likely to select shoots somewhere near the average, omitting both very long and very short shoots. The strong negative bias of the last set of measurements shows that this selection was not particularly effective in improving the accuracy of the sample.

2.5 Bias arising from faulty demarcation of the sampling units

Any consistent errors in measurement will clearly give rise to bias, whether the measurements are carried out on a sample or on all the units of the population. The danger of such errors is, however, likely to be greater in sampling work, since the units measured are often smaller. Furthermore, the knowledge that had another sampling unit by chance been selected a very different value might have been obtained, may lead the inexperienced worker to believe that accuracy in the measurement of the selected units is of little importance.

When the sampling units are not natural units of the population, the selected units usually have to be demarcated at the time the measurements are taken. In crop sampling work in particular, where small areas are selected in order to obtain an estimate of the yield or other characteristics of the crop, location of the areas by means of randomly selected co-ordinates, though theoretically ensuring a random sample, will only in practice do so if the field work is carried out with complete objectivity. Since it is impossible in practice to locate the areas according to their co-ordinates by means of exact measurements, pacing or some similar approximate method must be used.

In this type of work the areas themselves should not be too small, both because errors in the demarcation of the boundaries become of increasing importance as the size of the unit is decreased, and also because the possibility of influencing the results by small changes in location, e.g. so as to include a particularly good plant, is greater the smaller the unit areas. Very small areas are capable of giving completely reliable results with experienced and well trained field-men, but may be very unreliable when used by inexperienced workers, particularly if the need for complete objectivity is not appreciated.

Sukhatme (1946a, H), for example, has reported the biases shown in Table 2.5 in some trial crop-sampling work on wheat. He himself expresses the opinion that the biases of the very small areas are due to the inclusion of border plants. This, however, would imply that the effective radius of the smallest areas, which were nominally circles of 2 ft. radius, would have to be increased by nearly 5 inches. Errors of this magnitude appear improbable, unless the observers were very careless in their work, and it seems likely that part at least of the bias has been caused by faulty location.

Eye estimates are themselves a form of measurement, but such estimates are always subject to bias, which is likely to vary from observer to observer, and is often very substantial. If eye estimates are used, steps must therefore be taken to eliminate the resultant biases by carrying out proper measurements on a sub-sample of the material. A simple example of this is provided by the 1938–9 Census of Woodlands described in Section 4.25 and Examples 6.12.b and 7.15.b. A more complicated example is discussed in Sections 6.15 and 7.17.

TABLE 2.5—BIAS IN THE USE OF SMALL-SIZE AREAS IN SAMPLE SURVEYS FOR YIELD (*Sukhatme*)

Size of area in sq. ft.	No. of areas	Average yield in maunds per acre	Percentage overestimation
Irrigated			
471·5	78	10·10	—
117·9	78	10·58	4·8
29·5	78	11·69	15·7
28·3	117	11·60	14·9
12·6	117	14·38	42·4
Unirrigated			
471·5	107	6·55	—
117·9	107	7·27	11·0
29·5	107	8·08	23·4
28·3	162	7·52	14·8
12·6	161	9·33	42·4

2.6 Bias in estimation

In addition to biases which arise from faulty processes of selection and faulty work during the collection of the information, faulty methods of analysing the results may also introduce bias. A simple example occurs in the estimation of ratios. If, for instance, an agricultural crop is grown on types of land with different levels of fertility, and if the fields on the different types of land are of different average size, the mean yield per acre estimated from the mean of the yields per acre of all the fields may be markedly different from the mean yield per acre of all the land growing the crop. To take a numerical example, if there are three types of land having average yields of 20 cwt., 15 cwt. and 10 cwt. per acre respectively, and fields of an average size of 5 acres, 10 acres and 15 acres respectively, the number of fields on each

16

type of land being the same, the mean yield per acre over the whole of the land will be given by the *weighted* mean

$$\frac{5 \times 20 + 10 \times 15 + 15 \times 10}{5 + 10 + 15} = 13\tfrac{1}{3} \text{ cwt. per acre}$$

whereas the mean of the yields per acre of all fields will be 15 cwt. Consequently the bias in the estimate by the latter process will be about 12 per cent.

Biased estimates can be avoided by using the proper processes of estimation. This matter will be dealt with more fully in Chapter 6.

2.7 Circumstances in which bias is permissible

Although avoidance of any substantial bias is usually of the utmost importance, particularly in censuses on which administrative action has to be based, absence of bias is not always essential. In some types of investigation a certain amount of bias, provided it is reasonably constant, can be accepted. In censuses which are repeated at frequent intervals with a view to determining the changes rather than absolute values, for instance, a small overall bias may be of little consequence, provided it is constant in time. Similarly in surveys which have as their main objective the comparison of different groups of the population a bias which is approximately constant from group to group will be of little importance. The investigator must also avoid attaching exaggerated importance to minor sources of bias which, in fact, can only produce errors which are trivial relative to the random sampling error.

2.8 Methods of reducing the random sampling error

Once the absence of any important bias has been ensured, attention can be turned to the random sampling errors. These must clearly be sufficiently small to achieve the accuracy required.

Apart from errors due to bias, the simplest way of increasing the accuracy of the sample is to increase its size. Other things being equal, the random sampling error is approximately inversely proportional to the square root of the number of units included in the sample.

The accuracy attained will, however, depend not only on the number of units included in the sample, but also on the variability per unit ; or, more strictly, on that part of the variability per unit which contributes to the sampling error. It is here that the complications of sampling procedure, both of design and of subsequent analysis, arise. By suitable processes of selection, which while imposing restrictions on fully random selection do not introduce bias into the results, the part of the variability per unit which contributes to the sampling error can often be substantially reduced, and the size of the sample required for a given accuracy thereby diminished.

The simplest type of restriction is that known as *stratification*. The population is " stratified " or divided into blocks of units in such a manner

that the units in each stratum or block are as similar as possible.   Each of
the strata is then sampled at random.   If the same proportion is taken from
each stratum, it is clear that each stratum will be represented in the correct
proportion in the sample, and consequently differences between different strata
are eliminated from the sampling error. ⟀

In addition to stratification there are a number of other devices, which
will be discussed in more detail later, by which the accuracy of the sampling
procedure can be increased, often very substantially.   The three most important
are : utilization of supplementary information, use of a variable sampling
fraction (sometimes called " optimal allocation ") and multi-stage sampling.

Utilization of *supplementary information,* that is information which is derived
from sources other than the sampling scheme, or from a more extended sample
than that on which information on the main characters is collected, takes a
number of forms.   A simple example will illustrate the general principle.
Suppose that an estimate of the wheat yield of a country is required, and that
a random sample of wheat fields has been taken and the total yield of each
field determined.   We can then estimate the total wheat yield of the country,
either (*a*) by multiplying the total yield of the sample by the reciprocal of the
proportion of the fields included in the sample, or (*b*) by calculating the mean
yield per acre of the sampled fields (by dividing the total yield of all the sampled
fields by their total area, so as to avoid bias) and multiplying this mean yield
by the total acreage of wheat in the country.   The latter estimate can only be
made if the total acreage of wheat in the country is already known with sufficient
accuracy, e.g. from returns made by the farmers or from a larger sample.
If this information is available the second estimate is likely to be considerably
more precise than is the first, since the variability of the total yields, which
in so far as the yield per acre is constant will be proportional to the areas of
the individual fields, is likely to be considerably greater than the variability
of the yields per acre of the individual fields.

The use of a *variable sampling fraction,* i.e. the inclusion of different
proportions of the different strata in the sample, enables the more important,
or more variable, parts of the population to be sampled more intensively.
If this is done it will of course be necessary to weight the contributions of the
different strata to the total in the correct proportions.

The optimal sampling fractions depend on the relative variability of the
different strata into which the population is divided for the purpose of taking
a sample.   Thus, if it is required to determine the number of workers in a
given industry, it will be better to take a much larger fraction, possibly all,
of the large factories than of the smaller factories.

In *multi-stage sampling* the population is divided into a number of first-stage
sampling units, which are sampled in the ordinary manner, the selected
first-stage units being subdivided into smaller second-stage units, which are
also sampled.   Further stages can be added if required.   Thus, for example,
in a population survey, a sample of all towns and villages may be taken, and
in each of the selected towns and villages a sub-sample of all households may

18

be taken, with, possibly, for certain purposes, a further sub-sample of individuals from the selected households.

2.9   Choice of unit

In some classes of material there is considerable choice in the type and size of sampling units, and this gives further scope for increase in the efficiency of the sampling procedure. In general, when a given proportion of the material is included in the sample, the smaller the sampling units employed, the more accurate and representative will be the results. Thus, for example, in an agricultural survey, it will be more accurate to take 10 per cent. of all farms in each parish, or other small administrative unit, than to take all the farms in 10 per cent. of the parishes. This will remain true even if multi-stage sampling is adopted. It will be more accurate, for example, to take 10 per cent. of all the parishes in each county, with a second-stage sampling of the farms of each selected parish, rather than to take all the parishes in 10 per cent. of the counties, with the same degree of second-stage sampling. The reason for this is fairly obvious. The parishes in any county are likely to be more alike than are those of different counties, and if counties are used as sampling units, all the parishes in a county will be included or excluded from the sample simultaneously.

This need for small units distributed over the whole of the population often conflicts with the administrative requirements. It is clearly easier to arrange for a survey of farms in compact areas, such as parishes or counties, than to have to survey the same number of farms scattered over the whole country. The choice of a suitable balance between these two conflicting requirements is often one of the main problems in the planning of a sample survey. Furthermore, if only a small number of large units are included in the sample, whether or not there is second-stage sampling of these units, the sampling error will not be well-determined, since there will be relatively few differences between units on which to base the estimate of this.

We see, therefore, that the choice of sampling method depends not only on the relative accuracy of the different methods, but also on their practical convenience. It is important, for example, that the process of selection of the sample should not involve excessive preliminary work in the form of mapping, etc. The most suitable sampling method will therefore depend very much on the type of information that is already available on the population to be sampled ; a method which may be excellent for a country where good maps are available may be entirely useless in a country which is inadequately mapped. Again, it is important that not only should the collection of the information not involve excessive travelling, but also it should be possible to subject the field-workers to proper supervision : consequently, sampling procedures which may be excellent with postal questionnaires may be entirely unsatisfactory when the information is collected by special investigators.

CHAPTER 3

THE STRUCTURE OF VARIOUS TYPES OF SAMPLE

3.1 Definition of frame and sampling unit

In this chapter we propose to give a technical description of the structure of the various types of sample which are most commonly employed in practice, and the methods which must be followed in selecting them. The methods of obtaining estimates of the population values and of the sampling errors from the sample values will be discussed in Chapters 6 and 7.

All rigorous sampling demands a subdivision of the material to be sampled into units, termed *sampling units*, which form the basis of the actual sampling procedure. These units may be natural units of the material, such as individuals in a human population, or natural aggregates of such units, such as households, or they may be artificial units, such as rectangular areas on a map, bearing no relation to the natural subdivisions of the material.

It is not always necessary to make an actual subdivision of the whole of the material before selection of the sample, provided the selected units can be clearly and unambiguously defined. Thus, with sampling units which are rectangular areas on a map there is no need to demarcate all these areas ; they can be defined by co-ordinates, and the selected areas demarcated after selection.

Clear and unambiguous definition demands the existence or construction of some form of *frame*. In the sampling of a human population, for instance, with households as sampling units, there must be available a list of all households, and this list must be such that any household selected from it can be unambiguously located. In area sampling from maps, the maps must be such that the selected areas can be unambiguously defined on the ground.

The specification of the frame implicitly defines the geographical scope of the survey and the categories of material covered. A survey of a human population based on a list of households, for instance, will only cover those categories of the population which constitute the households included in the list. If other categories require inclusion, or if the frame is defective, special steps will have to be taken to supplement and emend it.

In statistical terminology any aggregate of values is termed a *population*, and consequently the whole aggregate of sampling units into which the material is divided is known as the *population of sampling units*. If the sampling units are aggregates of the natural units of the material, these natural units will form a further population which must be distinguished from the population of sampling units. A random sample of the sampling units will not then give a random sample of the natural units. There may also be a hierarchy of natural units, e.g. households and individuals within a household. In multi-stage sampling there is also a hierarchy of sampling units, first-stage, second-stage, etc., corresponding to the different sampling stages, and each set of units will form its own population of units.

In America the term *cluster sampling* is used to denote sampling in which the sampling units are aggregates or "clusters" of the natural units. It is also applied to two-stage sampling, in which the clusters form the first-stage sampling units, and the natural units in the selected clusters are sampled at the second stage.

Sampling units may be of the same or differing size. They may contain the same, or approximately the same, number of natural units, or they may contain widely differing numbers. The whole procedure of sampling, including the estimation of the population values and the sampling errors, is simplest when the sampling units are of approximately the same size and contain approximately the same number of natural units. Often, however, the material is such that this condition cannot be conveniently fulfilled. In particular, if the natural units are themselves of widely differing size, variation in size of the sampling units or in the number of natural units they contain is inevitable.

There is nothing in the sampling process which demands that the sampling units should be of any particular size, but, as has been explained in Section 2.9, the smaller the sampling units employed the more accurate will be the results obtained when a given proportion of the material is included in the sample.

3.2 Random sample

A random sample is the simplest type of rigorously selected sample, and is the basis of most of the more complicated sampling methods. In a random sample, after subdivision of the material into sampling units, the requisite number of units are selected at random from the whole population of units.

As has been emphasized in Section 2.3, random selection implies a strict process of selection equivalent to that of drawing lots. In practice it may be carried out either by some such process, or preferably, since adequate shuffling of cards, etc., is difficult, by the use of a table of random numbers. A small table of random numbers is given at the end of the book. The examples of this section illustrate the use of such a table.

The process of random selection may proceed in two stages. Suppose that the population is divided into groups of units containing $x_1, x_2, x_3, \ldots x_n$ units. The successive sub-totals

$$x_1 = X_1, \; x_1 + x_2 = X_2, \; x_1 + x_2 + x_3 = X_3, \; \ldots \; x_1 + x_2 + \ldots + x_n = X_n$$

are first calculated, which is easily done on a printing adding machine. The requisite number of numbers are then selected at random between 1 and X_n, numbers that occur more than once being rejected. A selected number that is greater than X_{s-1}, but less than or equal to X_s, indicates that a unit of the sth group is to be taken. Selection of a unit at random from this group, which can if convenient be made on the basis of the number already selected, will then give the equivalent of completely random selection.

This two-stage process is of value when the full numbering or demarcation of the units in all groups before sampling is laborious, since only the total number of units in each group need be known. It is of particular value when the units are artificially demarcated areas, and the total areas of natural subdivisions of the material are known. By using the process only the units in the selected groups have to be numbered or demarcated.

Example 3.2.a

Select a sample of 20 from a population of 2879 units.

Using the four-figure numbers given by the first four columns of digits in Table A1, and rejecting all numbers greater than 2879, we obtain the sample 347, 1676, 1256, 1622, 1818, 2662, 2342, 1608, 2742, 39, 1690, 1127, 1490, 2046, 526, 797, 2699, 1465, 2467, 1753.

The above procedure results in the rejection, in this example, of nearly three-quarters of the random numbers given by the table. Various devices may be used to avoid this. In the present example the simplest is to take the numbers 3001–6000 and 6001–9000 as equivalent to 0001–3000, rejecting the numbers 9001–9999 and 0000. Using the second column of four-figure numbers gives the sample 1373, 2467, 227, 2599, 2635, 1794, 1753, 378, 1234, 2632, 792, 897, 1064, 2819, 1712, 1837, 2722, 1504, 13, 2565.

If with either of the above procedures the same unit is selected a second time, the number leading to this selection is rejected, and an additional number taken.

It will be noted that neither of these samples is evenly distributed over the whole range of units. The distribution between the different thirds of the range is in fact:

Numbers		1st sample	2nd sample
1–960	. . . . .	4	5
961–1920	. . . .	10	8
1921–2879	. . . .	6	7
		20	20

Random selection will give samples that deviate somewhat from an even distribution, the actual deviations being themselves governed by statistical laws. Exact statistical tests show that about three out of four samples will have smaller aggregate deviations than the first sample, but only three out of ten will have smaller aggregate deviations than the second sample.*

Example 3.2.b

Select unit areas $\frac{1}{10}$ mile $\times$ $\frac{1}{10}$ mile at random from a rectangular area 5 miles $\times$ 4 miles.

---

* The appropriate test is that known as the X^2 test. A description of this test will be found in statistical textbooks.

There are 2000 unit areas, which can best be defined by co-ordinates 1–50 along the longer side of the rectangle, and 1–40 along the shorter side, the co-ordinates selected defining the corner of the unit area furthest from the corner of the rectangle (0, 0). The selection of a number at random between 1 and 50, and a second between 1 and 40, will therefore select a unit at random. Taking the third column of four-figure numbers (beginning 8636) and following the second of the procedures of Example 3.2.a gives the pairs of co-ordinates 36, 36 ; 12, 02 ; 16, 16 ; 14, 38 ; etc.

If points instead of unit areas are to be selected each co-ordinate range should theoretically be infinitely subdivided. The actual degree of subdivision need not usually be very fine.

The procedure of this example may be used for the selection of unit areas or points from an irregularly shaped area, provided the extreme range of each co-ordinate is included, points falling outside the area being rejected. More elaborate processes, involving less rejection, can of course be devised, but care must be taken that the probability of selection of all areas or points is equal. Thus in a triangular area, the selection of lines parallel to the base at random distances from the base, followed by the selection at random of a point within the triangle on each of the selected lines, will give a greater density of points near the apex of the triangle. The selection of points within a circle by the selection of random distances and bearings from the centre will give a greater density of points near the centre. In irregularly shaped areas, also, fractional unit areas requiring special treatment will occur at the boundaries.

Example 3.2.c

14 streets in a ward contain 25, 17, 5, 59, 64, 22, 38, 16, 21, 12, 14, 38, 17, 23 houses respectively. Make a random selection of 6 houses from all 371 houses.

The successive sub-totals are 25, 42, 47, 106, 170, 192, 230, 246, 267, 279, 293, 331, 348, 371. A table of random numbers gives the numbers 72, 128, 96, 326, 199, 202. The units 72 and 96 therefore fall in the 4th street, the unit 128 in the 5th street, the units 199 and 202 in the 7th street, and the unit 326 in the 12th street. Since $72 - 47 = 25$, and $96 - 47 = 49$, the 25th and 49th houses in the 4th street are selected, etc. The numbering of four streets, involving 199 houses, is required.

3.3 Stratification with uniform sampling fraction

In a stratified sample the population of sampling units is subdivided into groups or " strata " before selection of the sample. These strata may all contain the same number of units, or differing numbers of units. If a uniform sampling fraction is used, the same fraction of the units of each stratum is included in the sample, the units selected being chosen at random from all the units within each stratum. A stratified sample is thus equivalent to a set of random samples on a number of sub-populations, each equivalent to one stratum.

The main purpose of stratification is to increase the accuracy of the overall population estimates, and the corresponding estimates for subdivisions which are of interest : these latter may be termed *domains of study*. If there are large differences between the units in the different strata the accuracy of the overall estimates will be substantially increased, as the strata will be represented in their correct proportions, whereas in a random sample these proportions will be subject to sampling errors. It is important to recognize, however, that there may be little gain in accuracy over a random sample for some of the domains of study. The actual gain depends on the relation of the domain to the stratification structure : for a domain which consists of a single stratum the accuracy of the mean of a quantitative variate will be about the same as in a random sample, but there will be a gain for the total ; for a domain which contains several strata there will be gains both for the mean and the total ; for domains which cut across strata and which constitute only a small proportion of these strata there will be little gain either for the mean or for the total. Choice of strata, therefore, depends partly on the nature of the variability, partly on convenience, and partly on what domains of study are of particular interest.

Stratification affects the estimation of the sampling error. Since in a stratified sample only variation within strata gives rise to sampling error, it is this component of variation that requires estimation, and this can in general only be done from differences between units in the same stratum. It is therefore necessary, if an estimate of sampling error is required, that the strata be of such size that the sample contains two or more units from at least the majority of strata. In certain cases, in which the use of strata containing only a single selected unit appears advisable on account of the gain in accuracy thereby obtained, special methods, often of an approximate nature, of estimating the sampling error have to be adopted. The matter is discussed further in Section 8.15.

If the sampling units are already classified in the required strata, the selection of a stratified sample can be made in the same way as a random sample, the requisite number of units being selected at random from each stratum. If, however, the population is not so classified, selection by this method would necessitate prior classification. In this case, if the numbers of units in the different strata are known, an alternative procedure is available. This consists of selecting a sample at random ; keeping a tally, as the selection proceeds, of the numbers falling in each stratum ; and rejecting any further members of a particular stratum as soon as the requisite number for that stratum has been obtained. On the other hand, if the numbers of units in the different strata are not known, a count covering the whole population will in any case have to be made, in which case a classification which will serve as a basis for the subsequent selection of the sample may well be carried out simultaneously.

Unless all the strata contain the same number of units it will usually happen that the chosen sampling fraction will not give an exact whole number of units in each stratum. In this case the nearest whole number of units has to be taken. We may thus differentiate between the *working sampling fraction*, which with stratification with a uniform sampling fraction is the same for all strata, and the *exact sampling fractions*, which will differ slightly from the working sampling fraction. The use of the working sampling fraction in the analysis of the results leads to minor inaccuracies, but these will seldom give rise to errors of any practical importance.

It may be noted that if the numbers of units from the whole population falling in different strata are known, and a random sample is taken which is sufficiently large to ensure that adequate numbers of units are obtained from all strata, adjustment of the results so that the different strata are represented in their correct proportions will lead to practically the same accuracy as would be obtained with a stratified sample. Which of these two alternative courses is adopted in any particular case is a question of convenience. If the selection of either type of sample is equally simple it is best to use a stratified sample, as the computations are thereby simplified. In certain cases, however, the classification of units into strata may only be possible by means of information obtained in the course of the survey, in which case a random sample, with subsequent adjustment, is required. Thus, for example, in a survey of a human population, the age distribution of the whole population may be known, but prior selection of individuals of particular ages may be impossible owing to lack of information on these ages.

3.4   Multiple stratification

A population may be stratified for two or more different characteristics. If selection is made from sub-strata formed of the various combinations of the main classifications the procedure is exactly equivalent to ordinary stratification, the sub-strata being equivalent to strata. Thus we may stratify farms according to size and according to geographical regions. If the farms in each region are classified into size-groups before taking the sample then the region–size-group combinations form the individual sub-strata.

Occasionally the number of units of the population falling in each set of main strata may be known, e.g. from prior census data, but not the numbers in the various sub-strata. Thus, in the above example there may be information on the numbers of farms in the different size-groups, and also on the numbers in the different geographical regions, but not on the numbers of each size-group in each region. In such cases we may attempt the selection of a sample which will have the right proportions for each set of main strata. Such stratification may be termed *multiple stratification without control of sub-strata*. The selection of such a sample, however, presents both theoretical and practical difficulties, and the calculation of the sampling error is also troublesome.

3.5 Stratification with a variable sampling fraction

In certain types of material very considerable gains in accuracy will result if different sampling fractions are used for the different strata. The greatest accuracy for a given number of units will be attained if the sampling fractions are proportional to the within-strata standard deviations* of the units. If the sampling fractions are denoted by $f_1, f_2, \ldots$ and the standard deviations by $\sigma_1, \sigma_2, \ldots$ we have

$$\frac{f_1}{\sigma_1} = \frac{f_2}{\sigma_2} = \ldots$$

In some cases this formula may give sampling fractions greater than unity for some of the strata. If this occurs the whole of these strata are included in the sample.†

A particularly important application of the variable sampling fraction is to material stratified into size-groups. In such material the various quantitative characteristics of the units under investigation often have within-strata standard deviations which are roughly proportional to the mean sizes of the units in the different size-groups. In this case the sampling fractions should be taken about proportional to these mean sizes. If quantitative characteristics very highly correlated with size of unit are under investigation, the ranges of the size-groups may give good estimates of the relative within-strata standard deviations. The sampling fractions may then be taken proportional to these ranges. Changes with time, however, are usually by no means so highly correlated with size, and when the changes are of interest, sampling fractions proportional to the mean sizes of the size-groups will usually be best.

The above rules will determine sampling fractions which give the maximum accuracy for estimates of the population values. In cases in which the values for the individual strata are of interest, i.e. cases in which the strata themselves form domains of study, it is also important to see that all the strata are adequately represented in the sample, and for this reason the rule of strict proportionality to the standard deviations, or to the mean sizes of the size-groups, often requires some modification.‡

When several quantities are under investigation, it will usually be found that their within-strata standard deviations are not in quite the same proportions. This, however, is not a very serious problem in practice, since any sampling fractions which are somewhere near the optimal will give results which are nearly as accurate as those given by the optimal fractions. Consequently there is usually no great difficulty in choosing suitable sampling fractions which will reconcile the various conflicting requirements.§

* The meaning of this term and the method of estimation will be explained in Chapter 7.

† When variations in cost have to be taken into account the formula 8.17.b is appropriate.

‡ The situation when domains of study cut across strata is discussed in Sections 7.6.1 and 7.6.2.

§ An exact solution of this problem is given in Section 8.19.

Since, for these reasons, sampling fractions are often used which are not optimal, we have preferred not to adopt the term " optimal allocation," which has sometimes been used to denote stratified sampling with a variable sampling fraction.

The within-strata standard deviations can only be estimated from data relating to the material to be sampled, or from data derived from similar material, but general knowledge of the behaviour of material of a particular type, e.g. material stratified into size-groups, will often enable suitable sampling fractions to be chosen with all necessary accuracy. It is sometimes suggested that a preliminary survey should be undertaken merely to determine the optimal sampling fractions, but this is rarely worth while, though if a preliminary survey is being undertaken for other purposes it will of course also serve to improve the sampling fractions.

3.6 Systematic samples from lists

Although the importance of the principle of random selection in sampling has been stressed, much practical sampling is in fact not fully random in character. Thus a frequent method of selecting a sample, when a list of the units of the population to be sampled is available, is to take every qth entry on this list. This may be termed a systematic sample from the list. Other more complicated systematic procedures may occasionally be adopted for special purposes.

It is customary, and salutary, to determine the first entry by selecting a number at random between 1 and q, but this element of randomness does not convert the sample into a random one.* A systematic sample would be equivalent to a fully random sample if the list were arranged wholly at random. No lists, however, are arranged at random. The nearest approach to random order is probably provided by alphabetical lists, though even these have certain non-random characteristics : in this country, for instance, a large proportion of the Scotsmen will be found under the letter M. If every qth entry is taken, a kind of partial stratification will therefore be obtained, and the sample will be somewhat more precise than a fully random sample. Thus in a systematic sample of farms taken from a list of farms arranged by parishes, the proportion of farms drawn from each parish will be more or less constant, provided the sampling interval is small compared with the number of farms in a parish.

Owing to the lack of definition of the strata it is impossible to make a fully valid estimate of the sampling error, but provided there are no periodic features in the list the sample will not be biased. An estimate of the sampling error which is good enough for practical purposes can usually be made by regarding the sample as a sample stratified in the major subdivisions of the list,

---

* Except in the trivial sense that the sample is a random sample of 1 unit out of q units, each unit being composed of the aggregate of a set of all entries at spacing q.

ignoring any minor and ill-defined groupings. If the sampling error is estimated as if the sample were fully random an overestimate will be obtained, the inaccuracy being greater the more marked the similarity of the neighbouring entries in the list.

In general, systematic sampling from lists will be found to be quite satisfactory provided care is taken to see that there are no periodic features in the list which are associated with the sampling interval. The method is often much more convenient than random or stratified random sampling, since the labour of making a proper random selection, which in an extensive sampling scheme is often very considerable, is avoided. It must be clearly recognized, however, that the responsibility for the judgment that the material is such that systematic sampling will give satisfactory results rests with the investigator.

Selection of a systematic sample from a population which is listed in a set of short lists, each of which is numbered consecutively, requires additional precautions. Even if the order within the separate lists can be regarded as effectively random, the simple procedure of adopting the same starting point for each list can be unsatisfactory if an estimate of the number in the population is required from the sample.

Consider, for example, a set of lists (n in number) which all contain more than 250 but less than 300 entries, with a mean length of 270. If, for a 1 per cent. sample, the members numbered 50, 150 and 250 of each list are selected, the number estimated from the sample will be $100 \times 3n$, which is 11 per cent. greater than the population total of $270n$.

More importantly, with this method of selection the entries at the start of each list are always excluded ; the ends of the lists may also be under- or over-represented, depending on the distribution of list lengths—in the above example there will be under-representation. In many types of short list the initial and final entries may well differ from the remainder ; in a population census in which dwellings in each of a set of small demarcated areas are numbered consecutively from 1, for example, the entries at the beginning and end of the lists will probably be near the boundaries of the areas.

Both the above defects can be avoided by taking a random starting point between 1 and 100 separately for each list, or alternatively listing the numbers 1 to 100 in random order, and taking each in turn as a starting point. The latter method will give a more accurate estimate of the number of units in the population, and is administratively more convenient, as the same list of numbers can be used repeatedly. A further refinement is to list the numbers 1 to 100 in random order with the restriction that each pair has a total of 101.

Sampling in which the selection is wholly systematic should be clearly distinguished from sampling in which there is proper random selection of sampling units which are themselves systematic aggregates of smaller sub-divisions of the material. Thus a common method of sampling rows of potatoes has been to use sampling units consisting of every 20th plant, two such sampling units being selected at random from each row by selecting two numbers at random between 1 and 20. Such a method of sampling fulfils all the conditions required for fully valid random sampling.

3.7 An example of alternative ways of sampling highly variable material

In order to illustrate the various methods of sampling which have been discussed in the preceding sections, we will consider their application to the problem of determining the area under wheat in an English county. For this purpose the wheat acreages of Hertfordshire farms for 1939 were utilized.

The choice of problem and of basic material were dictated partly by convenience, and partly by the need for a complete set of data covering highly variable material, since an investigation of different sampling methods can be carried out most easily when data are available for the whole of the material.

It should be emphasized that this example is to be regarded as illustrative only. Sampling methods are very unlikely to be required for the determination of crop acreages in a country such as England, since all farmers make returns of their acreages each year ; the only possible use of sampling would be at the compilation stage, where it might be used to avoid the necessity of totalling the whole of the returns. The present investigation is not fully relevant to this problem, however, since the acreage of only one of the most extensively grown crops is considered, and that for only one county. Considerably greater errors, proportionately, may be expected in the less common crops.

Records were available for 2496 farms, and the acreage of wheat, and also the total acreage of crops and grass, which is virtually the total acreage of farmed land and will be termed the *size* of the farm, were abstracted for each farm. The original records were arranged by districts, by parishes alphabetically within districts, and by farmer's name alphabetically within parishes. This order was preserved in the abstract. The return for any farm, or " holding," does not necessarily relate solely to land in a given parish, but may include land in other parishes farmed by the same farmer ; farmers with two or more distinct farms may make separate returns for these farms or may include them all in a single return. The total area of wheat in the county, from the abstracted returns, was 44,676 acres, and the total area of crops and grass was 273,074 acres.*

If farms are taken as sampling units the dominant source of variation in wheat acreage will be variation in size of farm, since farms range from 1 acre to over 1000 acres, and no farm can have more than a fraction of its area under wheat. Stratification by size of farm is therefore indicated. The use of a variable sampling fraction will also be advantageous, since the wheat acreages

---

* It may be noted that these values disagree with the values shown in *Agricultural Statistics*, viz. 46,281 acres and 278,380 acres respectively. The reasons for this discrepancy need not concern us here, but it provides an illustration of the fact that disagreement between sample and complete returns must not be assumed to be necessarily solely due to sampling error.

of the large farms will be much more variable than those of the smaller farms. Further stratification by districts is possible, but is not likely to give much increase in precision unless the incidence of wheat growing in the different districts is very markedly different.   In any case a systematic method of selection from the list, which in view of the alphabetical method of arrangement will be quite satisfactory, will give the effect of stratification by districts.

For comparative purposes the following samples were taken :

(1) a random sample of 1 in 20 farms, 125 farms in all ;

(2) a stratified random sample with a uniform sampling fraction of 1 in 20 ;

(3) a stratified systematic sample with a variable sampling fraction, the fraction being approximately proportional to mean size of farm within each size-group, and chosen so as to give about the same number of farms, actually 135, as samples 1 and 2 ;   the systematic method of selection within size-groups results in approximate stratification by districts also.

The size-groups chosen, the number of farms in each group, and the sampling fractions and numbers of farms for sample 3 are shown in Table 3.7.a. For sample 2 the last two size-groups were combined.

TABLE   3.7.a—HERTFORDSHIRE   FARMS,   1939 :   SIZE-GROUPS, NUMBERS OF FARMS, AND CHOSEN VALUES OF THE VARIABLE SAMPLING FRACTION

Size-group (acres crops and grass)	No. of farms in county	Sampling fraction	No. of farms in sample
1–5	435	Nil	0
6–20	519	1/200	3
21–50	357	1/60	6
51–150	519	1/20	26
151–300	400	1/10	40
301–500	215	1/5	43
501–	51	1/3	17
	2,496		135

As pointed out in Section 3.3, the random sample can be stratified after selection so as to eliminate the effect of variation between size-groups, provided the number in each size-group of the population is known.   Variation due to size may also be eliminated by using size as supplementary information, provided information on size of farm is available for each farm in the sample, and the total area of all farms in the county is known.   Either the ratio or the regression method may be used, as explained in Chapter 6.

The estimates of the wheat acreage and of the number of farms growing wheat obtained from these samples, their estimated sampling standard errors, and in the case of the acreage the actual errors of the estimates, are shown in Table 3.7.b. The sampling standard errors, as will be explained in detail in Chapter 7, give measures of the average magnitudes of the sampling errors that may be expected with given methods of sampling and of estimation. The computations leading to these estimates are set out in Chapters 6 and 7, where tables giving the actual values of the sample units for the three samples (Tables 6.6.a, 6.5.a and 6.7.a respectively) will also be found.

TABLE 3.7.b—HERTFORDSHIRE FARMS : COMPARISON OF VARIOUS TYPES OF SAMPLES AND METHODS OF ESTIMATION

No.	Method of selection and estimation	Wheat acreage				No. of farms growing wheat		
		Estimate	Standard error	Actual error	Relative variance per farm	Estimate	Standard error	Relative variance per farm
1a	Random, direct	46,020	± 7,950	+ 1,340	100	900	± 104·6	100
1b	Random, stratified after selection	41,100	± 4,320	− 3,580	30	860	± 75·2	52
1c	Random, by ratio	41,570	± 3,940	− 3,110	25	Not calculated		
1d	Random, by regression	40,400	± 4,130	− 4,280	27	Not calculated		
2	Stratified, direct	40,220	± 4,110	− 4,460	27	1,080	± 71·6	47
3	Variable sampling fraction, direct	42,765	± 2,550	− 1,915	10	911	± 88·9	72

It is apparent from the values of the sampling standard errors for wheat acreage that, as is to be expected, both stratification and the use of a variable sampling fraction have resulted in large gains in accuracy.* The use of

* The word *accuracy* is used in this book to denote the expected accuracy of an estimate, as indicated by its sampling standard error. It is occasionally also used to denote the actual accuracy, as indicated by the actual error (usually unknown), but this should not cause any confusion.

supplementary information on size, whether by stratification after sampling, by ratio or by regression, serves to make the random sample about as accurate as the stratified sample with a fixed sampling fraction. This is also to be expected.

The numbers of units, i.e. farms, required to attain the same accuracy with different methods of sampling and estimation may be taken as roughly proportional to the squares of the standard errors, allowance being made for the greater number of units included in sample 3. These are shown in the column " relative variance," the random sample with direct estimation being set at 100. Thus stratification by size, or elimination of variation due to size in the estimation process, reduces the number of farms required by a factor of about 4, and the variable sampling fraction results in a further reduction by a factor of about $2\frac{1}{2}$.

The situation with respect to number of farms growing wheat is somewhat different. Stratification has again resulted in considerable increase in accuracy, though the gain is not so great as with acreage. The sample with a variable sampling fraction, on the other hand, is not so accurate as the ordinary stratified sample. The sampling fractions which are optimal for the determination of wheat acreage are by no means optimal for the determination of the number of farms growing wheat.

The actual errors of the estimates bear little relation to the sampling standard errors, except that they are in no case markedly larger than these standard errors. The random sample without adjustment gives the most accurate estimate of acreage, though this estimate has the largest sampling standard error, and adjustment of the random sample makes the actual error larger, though it reduces the sampling standard error. This is an illustration of the fact that an inaccurate method of sampling will sometimes by chance give an accurate estimate. The accuracy of a sampling procedure must never be judged by the magnitude of a single discrepancy ; a large discrepancy provides some evidence that a method is inaccurate, but a single small discrepancy provides practically no evidence that it is accurate.

3.8   Multi-stage sampling

In multi-stage sampling the material is regarded as made up of a number of first-stage sampling units, each of which is made up of a number of second-stage units, etc. The sampling process is carried out in stages. At the first stage the first-stage units are sampled by some suitable method, such as random or stratified sampling. At the second stage a sample of second-stage units is selected from each of the selected first-stage units, again by some suitable method, which may be the same as or different from the method employed for the first-stage units. Further stages may be added as required.

By suitable choice of sampling fractions it is often possible to keep the over-all sampling fraction (i.e. the product of the sampling fractions at the different stages) constant for different parts of the population. This leads to considerable simplification of the computations (see Section 6.18).

Multi-stage sampling introduces a flexibility into sampling which is lacking in the simpler methods. It enables existing natural divisions and subdivisions of the material to be utilized as units at the various stages, and, as pointed out in Section 2.9, it permits the concentration of the field work of censuses and surveys covering large areas. On the other hand, for the reasons there given, a multi-stage sample is in general less accurate than is a sample containing the same number of final-stage units which have been selected by some suitable single-stage process.

Multi-stage sampling also has the important advantage that subdivision into second-stage units, i.e. the construction of the second-stage frame, need only be carried out for those first-stage units which are actually included in the sample. It is therefore particularly valuable in surveys of undeveloped areas where no frame exists which is sufficiently detailed and accurate for subdivision of the material into reasonably small sampling units.

Since there are many variants of multi-stage sampling which are possible for any given type of material, careful investigation is often required before a decision as to the procedure which is best for any particular purpose can be reached. This matter will be discussed in detail in Chapter 8, after the methods of evaluating sampling errors have been described.

3.9 Sampling with probabilities proportional to size of unit

If we have areas demarcated on a map, such as fields, and a point is located at random on the map, the probabilities of the point falling within the boundaries of the different fields are clearly proportional to the areas of the fields. Consequently areas can be selected at random with probabilities proportional to their size by the simple procedure of taking random points on the map. It will be noted that such a process of selection may result in the same area being included twice or more in the sample. In this case it must be counted twice or more. We cannot, without distorting the probabilities, make a further selection in the manner followed with equal probabilities.

The principle has applications in agricultural surveys designed to determine the acreage and yield of different agricultural crops, total cultivated area, etc. All that is required for acreage is to determine the proportion of points which fall in areas of the given type. The method is therefore particularly attractive when carrying out surveys of the areas of crops, etc., by aerial survey, provided the different crops can be recognized on the photographs, since it avoids all the measurements of area which would be required if an ordinary random sample of areas were taken. The sampling of the fields with probabilities proportional to size is in this case equivalent to the sampling of small unit areas of equal size whose locations are determined by the random points. When only areas require to be determined the sizes of the fields in which the random points fall are in fact immaterial.

33

The analogy with the case of a stratified sample with a variable sampling fraction indicates that under certain circumstances greater precision may be expected from areas selected with probabilities proportional to size than will be obtained if they are selected with equal probabilities.

In the case of yield determinations, when the total acreage is known, the determinations of the yield from a sample of fields selected with probability proportional to size may always be expected to give a more accurate estimate of the mean yield per acre and total yield than will similar yield determinations on a random sample of fields irrespective of size. If the total acreage is not known then the situation is more complicated, but here again sampling with probabilities proportional to size is often advantageous.

Sampling with probabilities proportional to size of unit, or to some other known quantitative character of the units, may be carried out on other types of material by forming a cumulative or running total of the sizes of the units, and selecting numbers at random from the total of all the units in the manner of Example 3.2.c. Stratification by size and the use of a variable sampling fraction will usually be preferable in such cases, however, on the grounds both of accuracy and convenience, except in the special circumstances to be described in the next section.

3.10   Sampling from within strata with probabilities proportional to size of unit

Apart from area sampling, sampling with probabilities proportional to size of unit (pps) is mainly of use when the units are stratified according to some other characteristic: if the number of units to be selected from each stratum, or from some of them, is small, stratification by size within the main strata and the use of a variable sampling fraction are often impractical. The procedure is often used in conjunction with two-stage sampling with large first-stage units of variable size, particularly in surveys of social and economic characteristics of human populations. Here the populations of the first-stage units serve as a measure of " size ".

If the first-stage units are selected from within strata with probabilities proportional to size and the second-stage units are selected with probabilities inversely proportional to the size of the first-stage units, the overall sampling fraction will be uniform, and the sample will be *self-weighting*. Under certain conditions such self-weighting allocation of second-stage units will minimize costs (Section 8.17). With punched card machines, data from self-weighting samples could also be processed much more easily, but this advantage is of less importance with computers as differential weighting can now be readily effected.

As in area sampling, when more than one unit is required from a stratum, selection with probability strictly proportional to size can only be simply effected if a unit which happens to be selected twice is counted twice. Generally, however, when each stratum contains only a few large units, duplication of units is not desirable; instead, further units are selected with probability proportional to the sizes of those remaining. If probabilities proportional to size are assumed, therefore, some slight bias will be introduced into the results. This will rarely be of importance, but can be reduced if the " effective " probabilities of selection are ascertained. The formula for effective probabilities for samples of two units per stratum—the most important case—is relatively simple (Yates and Grundy, 1953, A''). Now that computers are available, the effective probabilities for two and possibly three units per stratum might well be furnished by programs primarily designed for the selection of samples from a computer file of data on unit sizes. Such programs avoid the tedium of having to form running totals of the unit sizes stratum by stratum.

The best way of assessing whether biases of any importance are likely to occur is by taking a series of samples from data covering all the first-stage units of a few typical strata. Such data can often be conveniently collected in the course of a pilot survey. Alternatively a simple hypothetical model of one or more typical strata can be constructed from previous knowledge.

Table 3.10.a illustrates such a hypothetical model of a single stratum. The sparsely populated units are assigned a much lower per-capita income

TABLE 3.10.a—HYPOTHETICAL PER-CAPITA INCOME DATA FOR A STRATUM OF 20 UNITS WITH VARYING POPULATION DENSITIES

Type of unit	No. of units	Population per unit (x)	Per-capita income (r)	Total income per unit (y)
A	10	1,000	200	200,000
B	6	3,000	300	900,000
C	3	6,000	400	2,400,000
D	1	10,000	500	5,000,000
Whole stratum	Units 20	Total population 56,000	Mean income 350	Total income 19,600,000

than the less numerous more densely populated units. The average per-capita income for the stratum, 350, is given by a weighted mean (with population weights) of the per-capita incomes over all 20 units, i.e. the total income for the stratum, 19,600,000, divided by the total population, 56,000. This, of course, differs widely from the unweighted mean, 275, of the per-capita incomes over all 20 units.

In this model there are only nine different types of samples of two units (since there is only one D unit), and the true probabilities of selection of each type can easily be calculated. If probabilities exactly proportional to size are assumed, the estimate of the mean r from a sample is given by the mean of the r's for the two members (Section 6.16), e.g. 250 for an AB sample. The mean of these estimates over all types of sample, weighted by the true probabilities of selection, is 346·8, giving a bias of $-3·2$ or 0·9 per cent. The variance, 4243 ($=65·1^2$) of the estimates can be calculated in a similar manner.

The bias is therefore much less than the standard error of the individual estimates, and even with 100 strata the bias only increases the mean square error of the mean over all strata by a quarter.

The bias can be reduced by the use of the effective probabilities. The effective probabilities p'_i for pps samples of two units are shown in the third column of Table 3.10.b. If instead of weights of $1/x$, weights proportional

TABLE 3.10.b—COMPARISONS OF SIZE RATIOS AND EFFECTIVE PROBABILITIES FOR SAMPLES OF TWO UNITS

| Type of unit | 20 units in stratum | | | | | 4 units in stratum | | |
	No. of units	Size ratios $x/S(x)$	Effective probabilities pps	pps/pe	Random	No. of units	Size ratios $x/S(x)$	Effective probabilities pps
A	10	·0179	·0186	·0348	·05	1	·05	·065
B	6	·0536	·0547	·0517	·05	1	·15	·186
C	3	·1071	·1060	·0771	·05	1	·3	·334
D	1	·1786	·1680	·1109	·05	1	·5	·414

to the reciprocals of the effective probabilities are introduced into the ordinary ratio estimate formula Sy/Sx (Section 6.9) this becomes $S(y/p')/S(x/p')$. We then obtain a mean estimate of 348·6, giving a bias of $-1·4$ or 0·4 per cent. Thus the bias is about halved by the use of effective probabilities.

With smaller numbers of units in a stratum the biases will be larger, as the distortion of the probabilities is greater. With only four units in the stratum, one each of A, B, C and D, for example, the bias (on a stratum mean of 425) is $-15·2$ or 3·6 per cent. This is reduced to $-3·5$ or 0·8 per cent by the use of the effective probabilities (also shown in Table 3.10.b).

The situation is entirely different if the sampling within strata is random. If, as here, there are large differences in size and marked association of r with x, the bias with only two units per stratum can be serious. The mean ratio estimate in this example, using sample estimates Sy/Sx, is 304·1, giving a bias of $-45·9$ or 13·1 per cent. The sampling variance is 6668. The remedy is to use a pooled ratio estimate from the data for several strata. With pooling over n similar strata the bias will be reduced by a factor of $1/n$; the sampling variance per stratum may be somewhat increased by variations in the mean r from stratum to stratum.

There is a further alternative method of selecting the sample which completely eliminates the bias from ratio estimates. This is to select the first unit with probability proportional to size and the remaining units with equal probability (pps/pe). The ordinary ratio estimate formula, Sy/Sx, will then give unbiased estimates. The sampling variance in this example is 6520.

Effective probabilities can be calculated for pps/pe samples from the formula

$$np'_i = p_i + (n-1)(1-p_i)/(N-1)$$

where N is the number of units in the stratum, n the number in the sample, and the p_i are the size ratios x_i/Sx.   The values of p_i' are given in Table 3.10.b.

It should be noted that with pps or pps/pe samples straight (unweighted) means of a y variate should be used with caution.   Units with large x will be selected more frequently than those with small x, and consequently, if there is any correlation between x and y the unweighted mean will be biased.   Weighting the individual y's by the reciprocals of the effective probabilities will reduce the bias but will not entirely eliminate it.

On the other hand, for quantities which show little relation to the size measure the ratio method can be grossly inaccurate and should not be used, whatever the method of selection.

In population surveys of the above type it may well happen that the census data which provide the x values are somewhat out of date, and that for the surveyed units up-to-date values are obtained in the course of the survey. In such cases a consistent set of size measures must be used as the basis of the ratio estimates.   Usually these will be the out-of-date values, but if some scheme for up-dating *all* the unit population values in the light of the survey results is devised such up-dated values could be used.

Apart from area sampling by the procedure of the last section, sampling with probability proportional to size requires knowledge of the sizes of all the units.   If only the total size of each stratum, or that of the whole population, is known, an ordinary stratified random sample has to be taken.   If, however, the sizes of the sampled units are determined in the course of the survey on the same basis as the known totals, these totals serve as supplementary information and can be used to adjust the estimate by the use of ratios, on the lines indicated in Section 6.8.

It should be noted that the example that has been taken is an extreme one, both in range of x and in the systematic variation of r with x.   In less extreme situations such as the Hertfordshire farm data discussed further in the next section, biases in ratio estimates are not likely to be of practical importance.

3.11 An example of sampling by administrative areas

Reverting to the illustrative example of Hertfordshire farms considered in Section 3.7, we may now investigate the effect of taking parishes as sampling units, or as first-stage units in two-stage sampling with farms as second-stage units.

The use of parishes as sampling units may be expected to result in a sample which is somewhat less accurate than a sample containing the same number of farms distributed over all parishes.   Nevertheless, if actual visits have to be made to the farms it may pay to use parishes as sampling units, or as first-stage units in two-stage sampling.   In analogous situations in undeveloped countries, where definition of farm boundaries may present difficulties, the complete survey of small administrative or other areas may also be better than any attempt to sample individual farms.

Inspection of the Hertfordshire data showed that the total farm area (crops and grass) included in the returns of farmers of a single parish was very variable, partly owing to variations in size of the parishes, and partly because some of the parishes were mainly urban in character. It is therefore best to use individual parishes as sampling units only if they contain a certain minimum acreage of crops and grass. The minimum chosen was 2000 acres, the remaining parishes being grouped roughly in the order in which they appeared in the alphabetical list, to form " combined " parishes containing over the minimum acreage. The effect of this combination is shown in Table 3.11.a.

TABLE 3.11.a—NUMBERS OF FARMS, PARISHES, AND " COMBINED " PARISHES
IN THE DISTRICTS OF HERTFORDSHIRE

District No.	District	No. of farms	No. of parishes	No. of parishes after combination
1	Barnet . . .	230	17	7
2	Bishop's Stortford .	316	23	16
3	East Herts. . .	564	31	20
4	Hitchin . .	553	36	25
5	St. Albans . .	218	9	7
6	Tring . . .	424	16	11
7	Watford . .	191	8	5
		2,496	140	91

Districts were used as strata in this sampling, 1 in 5 " combined " parishes being taken per district, i.e. 17 parishes in all.

Two samples were taken. In sample A the parishes were selected in the ordinary manner, with equal probability of selection for each parish. In sample B selection with probabilities proportional to size was employed. The parishes of sample B were also sub-sampled in two ways, samples B_1 and B_2. In sample B_1 a uniform sampling fraction of $\frac{1}{4}$ was taken for sampling at the second stage, with stratification by size, using the size-groups of Table 3.7.a with the last two size-groups combined. In sample B_2 a variable sampling fraction was used with values $\frac{1}{10}$ for size-group 1–50 acres, $\frac{1}{4}$ for 51–150 acres, $\frac{1}{2}$ for 151–300 acres, and 1 for over 300 acres. Sample B is given in detail in Example 6.17.

The relative efficiency of the various methods is discussed in Section 8.9. The results are summarized in Table 3.11.b. This table is similar to Table 3.7.b, except that estimated average values of the standard errors are given, and not those calculated from the actual selected samples. These latter are not sufficiently accurate for comparison owing to the small number of parishes involved.

It will be seen that a sample of 1 in 5 parishes provides results which are decidedly more accurate than a stratified random sample of 1 in 20 farms with a uniform sampling fraction, but somewhat less accurate than a similar sample with a variable sampling fraction. The stratified random sample of 1 in 20 farms is 1·29 times as accurate as sample B_1, allowing for differing numbers of farms. The similar sample with the variable sampling fraction is 1·83 times as accurate as sample B_2. Sample B is somewhat more accurate than sample A, particularly if the unbiased estimate given by the overall ratio is used for sample A. The difference is not marked, however, since the combination of parishes has created units which do not differ excessively in size.

TABLE 3.11.b—HERTFORDSHIRE FARMS : SAMPLES FOR WHEAT ACREAGE WITH "COMBINED" PARISHES AS SAMPLING UNITS OR FIRST-STAGE UNITS IN TWO-STAGE SAMPLING

| Sample | No. of stages | Method of Sampling | | Method of estimation | Estimate | Expected standard error | Actual error | Relative variance |
		1st stage	2nd stage					
A	1	Stratified by district	—	Overall ratio	41,730	± 3,080	− 2,950	100
				District ratios	41,010	± 3,010	− 3,670	95
B	1	Stratified by district, probability proportional to size	—	District ratios	46,660	± 2,870	+ 1,980	87
B_1	2	,,	Stratified random	District ratios	48,930	± 4,950	+ 4,250	259
B_2	2	,,	Variable sampling fraction	District ratios	45,600	± 3,460	+ 920	127

3.12 Multi-phase sampling

It is sometimes convenient and economical to collect certain items of information from the whole of the units of a sample and other items of information from some only of these units, these latter units being so chosen as to constitute a sub-sample of the units of the original sample. This may be termed *two-phase sampling*. Further phases may be added if required.

Multi-phase sampling is of value in several ways. Its simplest application is to the case in which the number of units needed to give the required accuracy on different items is widely different, either owing to the fact that the variability of the associated variates is different, or because the accuracy required is different. If no use is made of the relations between the different variates, such multi-phase sampling is equivalent to taking samples of different sizes for the different items.

First-phase information may also be used as supplementary information in order to improve the accuracy of second-phase information, by the same methods, ratio and regression, that are applicable where supplementary information on the whole population is available. Thus, in a crop estimation survey based on farms as sampling units, a relatively large sample of farms may be taken for the determination of the acreage of the crop, and the yields may be determined on a sub-sample only of these farms.

If the first-phase information is collected prior to the second-phase information the first-phase information may be used as a basis for the sub-sampling process, e.g. by stratification of the first-phase units for the selection of the second-phase sample, with or without the use of a variable sampling fraction at the second phase.

It will be noted that in both these latter applications of two-phase sampling the methods followed are the same as those adopted in ordinary single-phase sampling, the population being replaced by the first-phase sample ; but since the first-phase information is not known for the whole population it is itself subject to sampling error, and this must be taken into account when estimating the sampling errors of the estimates of the second-phase variates.

Multi-phase sampling differs structurally from multi-stage sampling in that in the former the same sampling units are used throughout, whereas in the latter a hierarchy of sampling units is used. Multi-phase sampling may be combined with multi-stage sampling. In a scheme for the estimation of the acreages and yields of agricultural crops, for example, a two-stage sample of farms and parishes may be taken for the estimation of acreages, and a sub-sample of these farms may be taken for the estimation of yields.

3.13   Balanced samples

If the average value of some quantitative character of the units, such as size, is known for the whole population, it is possible, provided the sizes of the individual sampling units are known, to select a sample in such a manner that the average size of the selected units is equal to the average size of all the units of the population. Such a sample will only be satisfactory if it is otherwise equivalent to a random sample, in which case it may be termed a *balanced sample*.

Balance may be employed in conjunction with stratification for some other character. In this case balance may be effected either for the whole population, or for each of the strata separately. The latter course should only be adopted if the number of units selected from each stratum is moderately large : otherwise undue restrictions will be placed on the sample which will result in the selection of a sample which is not otherwise equivalent to a random sample. On the other hand, when the strata are balanced separately more accurate estimates of the separate strata means and totals will be obtained, and the accuracy of the estimates of the overall population means and totals may also be somewhat improved.

Balancing for a known quantitative character provides an alternative to stratification by size-groups in this character. Balancing, however, will only be effective if the differences in the quantity or quantities under investigation are approximately proportional to differences in the known character, whereas stratification by size-groups will take account of any type of relationship. As will be seen in Chapter 7, the estimation of sampling errors is simpler in the case of a stratified sample.

The increased accuracy resulting from balancing can equally well be obtained, at the expense of some additional computational labour, by adjusting the results of an unbalanced sample by the use of regression in the manner explained in Chapter 6. Since the additional labour of adjustment is nearly proportional to the number of variates under investigation, the advantages of balancing as opposed to regression increase as the number of variates increases.

Balancing can also be carried out for a character which is inherently qualitative, but which for the sampling units actually employed acts as a quantitative variate because the sampling units are themselves aggregates of smaller natural units. Thus if a sample of a human population composed of two different races is being taken, the sampling units being administrative areas containing numbers of individuals, the sample can be balanced for the percentages of individuals in the two races. If the individuals were the sampling units then balance would be equivalent to stratification by races.

A balanced sample is best selected by using a process of replacement. In the first place a random or stratified random sample of the required size is selected, record being kept of the order of selection. The average value of the known quantitative character is then calculated for the sample. This will, in general, not be equal to the average for the population, indicating lack of balance. A further sampling unit is then drawn, and compared with the first unit of the original sample. If balance is improved by substitution of the new unit, this is done, otherwise the original unit is retained. The process is then repeated for the second and following units of the original sample until an adequate degree of balance is attained.

The selection of a sample balanced for more than one character presents more difficult problems, and will not be discussed here.

Balance, in the cruder form known as *purposive selection*, was at one time extensively used in sample censuses and surveys. No rigorous rules of selection were followed, however, with the result that many purposively selected samples were by no means equivalent to balanced random samples. Thus it frequently happened that the selection was confined to sampling units having values of the known quantitative character near the average. Clearly in such samples the variability of the known quantitative character, and of any other characters closely correlated with it, will be considerably less than the true variability in the population.   The sample may also be unrepresentative in other ways.

Purposive selection was often used in an attempt to avoid the necessity, which was otherwise apparent, of employing reasonably small sampling units. Thus Gini and Galvani (1929, A) selected a sample from the Italian Census Data of 1921 which consisted of all the returns of 29 out of the 214 *circondari* into which the country is divided, using seven control characters.   Agreement of the average values of other characters in the sample and population was poor, and that of the frequency-distributions of such characters was even worse. The real weakness here is the use of excessively large units, though even with smaller units the use of purposive selection without rigorous rules of selection is always liable to give unsatisfactory results.   There is, moreover, no means of judging its reliability.

For these reasons purposive selection has ceased to be extensively used, and in modern sampling work it has largely been replaced by more thorough application of the principles of stratification, etc.   Provided proper attention is paid to the process of selection, however, there is no fundamental objection to balanced samples.   These have a certain limited usefulness in some types of census and survey work, though it must be recognized that the need for the subdivision of the population into an adequate number of sampling units is in no way obviated by balancing for one or more quantitative characters.

3.14 Systematic samples from areas

A common method of sampling material continuously distributed either in space or time is to take sampling units distributed at equal intervals over the material.   The chief application in census and survey work is in the sampling of land areas.   When maps are available the sampling units can be located by superimposing a grid of points, frequently of square, or nearly square, pattern.   Such a sample may be termed a *systematic area sample*.

A systematic area sample differs from a systematic sample from lists mainly in the spatial distribution of the sampling units over the material.   Most lists do not correspond at all exactly, except for major groupings, to any physical distribution, and a systematic sample from a list therefore usually approximates much more closely to a random sample than does a systematic area sample. Different methods of estimating the sampling error are therefore appropriate in the two cases.

In general, provided there are no periodic features, a systematic area sample will be rather more accurate than a stratified random sample (with one unit per stratum) from strata consisting of rectangular blocks (or *cells*) whose centres are situated at the systematic sampling points.   In material in which the variation is of a continuous nature it is impossible to make any accurate estimate of the sampling error without taking supplementary sampling points, though if there are no periodic features an upper limit can be obtained.

If the regions near the boundary are likely to differ from the remainder of the area, as may be the case if the boundary is a natural one, such as a sea coast or a mountain range, it will be best, after locating the sampling grid at random, to demarcate the bounding lines of the cells, and sample at random the area which is not covered by complete cells, dividing it into equal or approximately equal areas and locating one sampling point at random within each of these areas. It will be convenient, if possible, to make these cell areas equal in area to those of the sampling grid, since equal weight will then be given to all sampling points. The same method of dealing with boundaries can be used if the sampling is random within rectangular cells.

Systematic sampling is entirely unsuited to material which has periodic features, but apart from this will generally provide a satisfactory method of area sampling. It has the advantage over stratified random sampling from blocks that the location of the sampling units is simpler and the results obtained provide rather better material for the construction of maps, etc. As in systematic sampling from lists, however, the responsibility for the judgment that the material is such that systematic sampling will give satisfactory results rests with the investigator.

3.15 Line sampling

In the sampling of areas certain types of information can be ascertained almost as easily for all the points on a line as they can for a set of isolated points or areas. In such cases sets of parallel lines or strips may be taken as the sampling units. In stratified random line sampling, the area is divided into rectangular blocks of convenient length and of such a width that two selected sampling lines are included in each block, their location within the block being random. If an exact estimate of the sampling error is not required, only one line need be included in each block, with correspondingly smaller blocks. In systematic line sampling the sample is made up of lines at equal intervals.

Line sampling provides an alternative method to point sampling (Section 3.9) for determining the proportions of a given area which are of different types. These proportions are given by the ratios of the aggregate of the intercepts of the different types. In area determinations of this type, whether by line or point sampling, systematic sampling can usually be adopted. The method is useful both for area determinations on the ground in undeveloped country and for the determination of areas from maps and aerial photographs. The method has been much used for forest surveys, where it is known as " cruising."

If, instead of obtaining information for all points on each of the lines, a sample of such points is taken, the sampling becomes two-stage. If the lines and the points on them are both evenly spaced the sampling is equivalent to systematic point sampling.

Line sampling of a somewhat different type is also used in order to determine the acreage of agricultural crops in areas which are well provided with roads. A route is chosen which covers the whole area as adequately as possible, and the lengths bordered by the different crops are measured. A car fitted with a special milometer can be used for this purpose. Estimates of the yield near harvest time can also be obtained in a similar manner, by stopping the car at every xth mile of a given crop, and cutting and harvesting a small area of the crop, the area being selected in some systematic manner, such as entering the crop a given number of paces at right angles to the road.

Line sampling of this type does not provide an unbiased sample, since roads are by no means randomly located with regard to agricultural crops. The results from surveys following the same route in successive years may well be comparable, however, and with calibration by more exact methods from time to time, road cruising may provide a satisfactory method of making rapid and inexpensive surveys. Similar methods based on tracks are possible in areas with only a sparse road network.

3.16 The principle of the moving observer

If counts are required of a collection of individuals who are moving about, the ordinary methods of sampling can only be applied with difficulty. Thus, to determine the number of people in a crowded street by ordinary methods would require the demarcation of a number of small areas in the street and the counting of the number of people on each of these areas. The counts need not necessarily be simultaneous, but for any one area the number of people present at a given moment has to be counted. Unless photographic methods are available, or the areas are very small, such counts are extremely difficult, since individuals are continually moving into and out of the areas and are also moving about within them.

Equally it is no use stationing observers at fixed points with instructions to count passers-by. The number of people in a street will depend not only on the numbers passing fixed points but also on the velocity of movement up and down the street. If all exits and entrances to the street are covered, and there are no people in the street at the start of the counts, the number present at any subsequent time can be determined from counts that are continuous and without error. In practice, however, errors in counting usually result in cumulative errors which invalidate the results. Thus it was found impracticable to determine the numbers in a department store by posting people at the doors to make counts.

These difficulties can be overcome by using moving instead of stationary observers. To obtain an estimate of the number of people in a street, the observer traverses the street in one direction, counting all the people he passes, in whichever direction they are moving, and deducting all the people who overtake him. He then re-traverses the street in the opposite direction, moving at the same speed and counting as before. If this is done the average of the two counts gives an estimate of the average number of people in the street during the time of the counts. If people are mostly moving in one direction the count in this direction will be reduced, but the count in the opposite direction will be correspondingly increased.   In practice the deductions required for those overtaking the observer can be kept small by moving at a speed greater than that of the majority of the crowd.

This method was used to estimate the numbers of people in streets, shops, etc., at different times of the day, in order that the adequacy of the provisions for public air raid shelters might be tested.   It was found that very dense crowds in streets and shops could be estimated with surprising ease.   Crowded streets were dealt with by teams of two or more observers, moving down the street in a transverse line, with each observer counting the people between him and the next observer.   In the large stores the floor was divided into areas which were assigned to the different observers.

The method is of general application.   It can be used, for example, to assess populations of insects or animals in a state of movement, provided all individuals can be readily seen, and provided the passage of the observer does not itself influence the movement of the individuals.

3.17   Interpenetrating samples

It is often advantageous to take two or more independent samples of a given population, using the same sampling procedure for each sample.   Such samples are called *interpenetrating samples*.

Interpenetrating samples are of value if the survey or census has to be carried out by successive stages.   This is frequently necessary when preliminary results are required quickly.   Thus in the 1942 Census of Woodlands of England and Wales, described in Section 4.25, it was necessary to obtain a preliminary estimate of the timber content with very limited field staff within six months of the initiation of the survey.   The work was therefore planned in two interpenetrating samples.   Before the field work was commenced, each unit of the first sample was subdivided into two, since it became apparent that the whole of the first sample could not be completed in the allotted time. This further subdivision itself created two interpenetrating samples of a special type.   By means of this procedure it was possible to provide a preliminary estimate of the total timber content of the whole country by the time it was required.

An incidental advantage of interpenetrating samples is that separate and independent estimates of the characteristics of the population are furnished. The agreement of such estimates is often more convincing to the layman than any statement of the sampling error.

Interpenetrating samples have a further use in that the different samples can be assigned to different investigators.   Comparison of the results provides a check of the investigators against one another.

To perform the functions outlined above, each of the interpenetrating samples must itself provide an adequate sample of the material and must be comparable with the other samples—in other words the samples must be really interpenetrating.   If this is not the case the comparisons between the different samples will be subject to relatively large errors. If, for instance, they are used to test differences between different investigators, the information obtained will be of insufficient accuracy to be of any real use. Equally, if one sample is used to provide a preliminary estimate, this estimate may well not attain the required degree of precision. Thus in an agricultural survey stratified by areas such as counties, the division of the counties into two groups, with each of the samples confined to one group only, would not be likely to give a satisfactory pair of interpenetrating samples. The separate samples would be subject to variation between counties, and would therefore be considerably less accurate than the combination of the two, from which variation between counties is entirely eliminated. The proper use of interpenetrating samples therefore necessitates increased expenditure on travelling.

3.18 Lattice sampling

If we have a square area of side p, divided into p^2 unit squares, we can select a sample of p unit squares in such a manner that every row and every column of the large square contains one of the selected unit squares. Such sampling is a special type of double stratification without control of sub-strata (Section 3.4). The rows and columns of the square can represent any two-way classification of the material in which there are equal numbers of classes in the two classifications and one unit in each sub-class.

Similar schemes are possible for three- or more way classifications. With a three-way classification, for example, with p^3 units, a sample of p units can be selected such that one unit is taken from each class of each classification. Alternatively a sample of p^2 units can be selected in such a manner that there are p units in every class of each classification, the p units belonging to any one class of any classification being so selected that one unit falls in each of the p classes of the other two classifications. A sampling scheme of this last type will be defined by a Latin square of side p^2, that is, a square pattern of p letters in which each letter occurs once and once only in each row and each column. Table 3.18 shows a 6×6 square. The rows, columns and letters of the square represent the three classifications.

Some element of randomization must, of course, be introduced in the

TABLE 3.18—EXAMPLE OF A 6×6 LATIN SQUARE

C	B	E	A	F	D
F	D	A	E	C	B
D	F	C	B	A	E
E	A	D	C	B	F
B	C	F	D	E	A
A	E	B	F	D	C

selection of actual samples.   If an estimate of error is required the type of randomization is governed by the form of the estimate (see Section 7.18.1). If no estimate of error is required a sample of p units from a square can be selected by selecting a unit from the first row at random, then selecting a unit from the $p - 1$ unoccupied columns of the second row at random, and so on. The selection of a sample of p units from a cube is similar.   For the first two classifications a sample of p units is taken from a square, as above, and p letters are then allocated to these points to indicate the third classification.   In the case of a sample of p^2 units from a cube, the rows, columns and letters of any available $p \times p$ square can be randomized among themselves.   The details of this procedure are explained below.   Examples of squares up to 12×12 are given in *Statistical Tables*, but if these are not available a randomization of one of the diagonal squares (i.e. a square with each letter on lines parallel to a diagonal) will suffice.   This may be obtained directly by allocating the letters in the first column at random, and then allocating those in the first row (except the first) at random.   The further rows may then be filled in by writing the letters in the same order as in the first row, beginning with the determined letter. No further randomization is required.

Sampling of this kind, which may be termed *lattice sampling*, is of particular use when the material to be sampled is of a type that lends itself to multiple subdivision on a square or cubic pattern.   One such type arises in sampling schemes which extend over both space and time.   An example is provided by a sampling scheme for estimating the catches of fish landed at various ports along the coast of India.   Catches are there landed at all hours of the day and night, the times of landing depending on the tide, weather, etc. On any one part of the coast the times of landing at the different ports are highly correlated.   It was therefore proposed that a sampling scheme be adopted in which every port would be sampled every day, the times of day being so chosen that for a group of p neighbouring ports a different part of the day was covered at each port.   Moreover the times of day were to be so rotated that over a period of p days all times of the day were covered at each port.   This requires a $p \times p$ Latin square in which the rows, columns and letters represent ports, days and times of day.   A more complicated example involving two-stage sampling is provided by proposals for road-traffic censuses outlined in the next section.

Various two-dimensional schemes of the lattice type were suggested by Tepping *et al.* (1943, D) under the name of *deep stratification*.   Their performance was tested out on housing data in an American city.   The sampling unit was a city block, rent and size (number of housing units) being taken as the two types of subdivision.   Each rent-size classification contained a number of blocks, one block being selected at random from each chosen cell.   If the number of blocks is to be the same in all cells the rent classification must be varied within the different size subdivisions, or vice versa.   If this is not done, the simplicity of the scheme is sacrificed.   This lessens the effectiveness of schemes of this type for sociological or economic material, particularly if the two classifications are highly correlated.

More complicated schemes, in which different probabilities of selection are assigned to different patterns, have, been discussed by Goodman and Kish (1950, A'). A scheme of the same general type appears to have been used in the Canadian Labour Force Survey (Keyfitz and Robinson, 1949, F').

3.19   Sampling on successive occasions

The types of sampling so far discussed are appropriate to a census or survey carried out on a single occasion, with the object of determining the characteristics of the surveyed population at or about a given point in time. If the population is subject to change, a survey carried out on a single occasion, however accurate, cannot of itself give any information on the nature or rate of such change. In certain types of population extraneous sources of information, such as registrations of births and deaths, may be relied on to provide information on the changes which the population is undergoing. Even in such cases the census must be repeated at intervals, both because of inaccuracies in the extraneous information, which may lead to a gradual accumulation of errors, and also because the information is rarely of such a nature that all aspects of the original census or survey can be kept up-to-date. Registration of births and deaths, for example, coupled with figures for immigration and emigration, will furnish data for the revision of the total of the population but will not enable changes in the population of separate towns and districts to be determined.

In many cases no such extraneous information on the changes that are taking place is available, and in such cases provision must be made for periodical re-survey if up-to-date information is required. A number of alternatives then present themselves :—

(1) A complete census or survey may be repeated in its original form at intervals.

(2) A sample census or survey may be repeated at intervals, a new sample being selected on each occasion without regard to previous samples.

(3) A sample census or survey may be repeated on the same sample.

(4) Part of the sample may be replaced on each occasion, the remainder being retained. If there are a number of occasions a definite scheme of replacement may be followed, e.g. one-third of the sample may be replaced, each selected unit being retained (except for the first two occasions) for three occasions.

(5) A re-survey of a sub-sample of the original sample may be made. In the case of a complete census this is equivalent to a re-survey of a sample of the whole population.

The following terms are suggested for the last four alternatives : (2) independent samples ; (3) fixed sample ; (4) partial replacement ; (5) sub-sample. It will be noted that independent samples are formally equivalent to interpenetrating samples, a fixed sample is formally equivalent to the observation of different characters (variates) on the same sample, and a sub-sample is formally equivalent to a two-phase sample. Only partial replacement has no formal equivalent in the types of sampling already described. These equivalences are of importance in that the methods of estimation will be the same for formally equivalent sampling processes.

The relative advantages of the various types of procedure depend on the relation between the variability of the units and the variability of changes in these units as well as on the relative importance of information on the population means and on the changes in these means. If, for instance, the units are very variable but the changes of all units are similar, accurate information on change can most easily be obtained by re-survey of a fixed sample of units ; provided always that proper provision is made for new entrants to the population, and for the elimination of the disturbance which results from the extinction of selected units. If, on the other hand, information on the population means is of paramount importance, partial replacement or a sub-sample will usually be preferable. A more detailed discussion, in terms of the errors to which the various estimates are subject, is given in Section 8.8.

There are two further points which must be borne in mind in connection with sampling on successive occasions. Firstly, repeated re-survey of the same units may be inexpedient, since resistance to the provision of the necessary information may be engendered, and secondly, repeated re-survey may result in modification of these units relative to the rest of the population. This can arise in many ways. In a survey of agricultural practice, for instance, visits to farms may result in the farmers concerned improving their practice through advice from the investigators : advice which, if asked for, can scarcely be refused. A more subtle example is provided by the 1942 Census of Woodlands. In this census it was considered that if the subsequent fellings and replantings were recorded on the sample areas then surveyed an adequate measure of the changes in woodland throughout the country might be obtained over some considerable period of time. It has since been suggested that the amount of felling on the sample areas may have been affected by the fact that survey information was available on these areas and not on others, with the result that these areas have been more intensively exploited.

3.20 Composite sampling schemes

Simplicity and uniformity of sampling procedure is obviously in general desirable, but there are occasions on which different methods of sampling are required for different parts of the population. In sampling a human population, for instance, some form of area sampling may be most suitable for the rural parts of the country, whereas some form of stratified random or systematic sampling based on lists of houses may be best in the towns. There is, of course, no objection to the use of such composite schemes, provided each part fulfils the requirements of good sampling procedure already laid down.

3.21   Combination of complete census and sample survey

It sometimes happens that a complete enumeration of the population can easily be made, but that detailed information on the individual units of the population can only be obtained by sampling methods. In such cases a complete enumeration will often be of value as a frame for the sample census. Thus in a census of a human population, a complete enumeration consisting of lists of households and of numbers in each household can be made. A sample of houses can then be visited by investigators so as to obtain details regarding the age, sex, etc., of the occupants. Such a sample census will not only serve to provide the required detailed information, but will also provide a partial check on the accuracy of the complete enumeration. It will not, however, provide a check on omissions from the lists of households. To carry out such a check it will be necessary to take a further sample of properly defined areas, checking that all the households in the sample areas have been included in the full census returns.

A complete census, even if it is very inaccurate, is also of the greatest use in planning a more accurate sample census. In a sample census of a human population, for instance, some knowledge of the relative sizes of different towns and villages, and of the density of population in rural areas, is essential for the proper allocation of resources. Similarly in a census of agriculture, knowledge of the amount of cultivated land in different parts of the country is necessary if excessive survey of largely uncultivated areas is to be avoided.

The information provided by an inaccurate complete census can also be used to improve the accuracy of a subsequent sample census, by the methods applicable to supplementary information which will be given in Chapter 6. Here, however, we must proceed with caution. If, for instance, a complete census of a human population consistently underestimates the population of villages of all sizes by about 10 per cent., the sample census will determine the amount of the underestimation and a common adjustment can be made. If, however, small villages are underestimated by 20 per cent. and large villages by 5 per cent. the application of a common correction will result in the underestimation of the population of small villages and the overestimation of the population of large villages. This distortion will be avoided if separate corrections are calculated for small and large villages. Unfortunately it is not always possible to be certain that all potential disturbances are taken into account. These differential inaccuracies are particularly troublesome in that they tend to be associated with the administrative areas for which separate results are required. The overall results, however, will not be materially affected by differential inaccuracies of this kind if the methods of estimation given in Chapter 6 are followed.

CHAPTER 4

PRACTICAL PROBLEMS ARISING IN THE PLANNING OF A SURVEY

4.1 Questions requiring consideration

The practical problems encountered in the planning of a sampling investigation vary greatly with the type of material and the nature of the information that is required. We shall here only concern ourselves with problems arising in the conduct of censuses and surveys, such as are required in the study of human populations, economic institutions, and agriculture. The sampling of batches of material, industrial products, etc., which is necessary in manufacturing processes of all kinds, and is broadly categorized by the term *quality control*, presents rather different problems which are not discussed in this book. The sampling problems encountered in biological research are also omitted from the discussion.

The questions that require consideration at the planning stage of censuses and surveys may be broadly classified as follows :—

(1) Specification of the purposes of the survey.
(2) Definition of the population, types of institution, or categories of material to be covered by the survey.
(3) Decision on the nature of the information to be collected.
(4) Decisions on the method of collecting the data, whether by interviewers, investigators, mail, etc., and methods of dealing with non-response.
(5) Choice of frame, or construction of a frame if none is available.
(6) Choice of sampling unit and type of sample, whether stratified, multi-stage, etc., determination of size of sample required, and method of selection.
(7) Decision on whether the survey is to be an isolated one, undertaken without intention of repetition, or is to be planned with a view to repetition at intervals.

These questions cannot be considered in isolation one from another. To a greater or less extent any decision taken on one question will influence the decisions that should be taken on the others. They should therefore be resolved jointly, or if independent decisions are made these should at least be regarded as tentative and subject to modification until the plan as a whole has been finalized.

Nor can the correct decisions be arrived at without considerable knowledge of the nature of the material to be covered, particularly its variability, both as a whole and within and between strata of various types. Knowledge is also required of the ways in which it is practicable to collect the required information with the necessary accuracy. If prior knowledge in these matters is not available a *pilot* or *exploratory survey* will be necessary. Even if there is adequate knowledge of the statistical properties of the material, pilot surveys are frequently advisable in large-scale surveys in order to test and improve field procedure and schedules, and to train field workers.

Questions arising under heads (1), (2), (3), (4) and (7) of the above list are common both to complete censuses and surveys and to sample censuses and surveys. Even here, however, the problems encountered differ considerably in the two cases, owing to the greater scope for the collection of detailed information and the execution of complicated observations by the sampling method.

The determination of the items on which information is to be collected, the degree of detail to be attempted, and the ways in which the information can best be obtained, often constitute the most difficult and crucial part of the planning of a survey. No amount of care in the planning of the sampling or skill in the analysis will compensate for failure in this respect. A survey in which the information collected does not adequately cover the field to be investigated at the best provides a partial and incomplete picture, and at the worst may be irrelevant or actively misleading.

Careful consideration must therefore be given at the outset to the purposes for which the survey is to be undertaken, the type of information it is proposed to collect, and the uses to which the information obtained will be put. In the case of large-scale surveys, which are likely to provide information that will be of value to a number of different organizations or government departments, a detailed statement on these points should be prepared. In this way those who are likely to want to make use of the results of the survey will be fully apprised of its nature, and can if necessary make suggestions for modifications before the survey is begun.

The statistician who will ultimately be responsible for the analysis and presentation of the results should, if possible, be selected and appointed at the planning stage. Similarly if the advice of a statistical expert is to be sought, this should be done, in the first instance, at the planning stage. This rule applies even in the simplest types of census. It frequently happens that such censuses are undertaken without any prior consultation with a statistical expert, whose advice is only sought when the results have been collected and the stage of analysis is reached.

4.2 Definition of the population

The categories, or types of material, which require to be included in a survey, and its geographical scope, are conditioned in broad outline by the purposes of the survey, and by administrative and research requirements related to these purposes. Within this broad outline, however, there is often a certain amount of latitude, and careful consideration should therefore be given to the inclusion or omission of marginal categories, particularly those on which the collection of information is likely to be specially difficult, or for which an adequate frame is lacking. By excluding unimportant marginal categories the task of collecting the information may often be very materially simplified, without seriously reducing the value of the results.

A census of the human population residing in a given territory, for example, should ideally include all individuals present in that territory at a particular moment, and in simple censuses an attempt is usually made to attain this end. It is often, however, difficult to obtain information on certain minor categories, such as nomads. These difficulties occur even in a complete census, but are often more marked in the case of a sample census. The question of whether such categories may be omitted entirely without serious loss should therefore be considered.

The matter becomes of even greater importance when a human population census requiring the collection of detailed and complicated information is undertaken, using skilled investigators making visits to individual members of the population. In such cases visits to members of the population with a permanent residence, even if they are absent from their residence at certain times, are relatively simple, but it is far more difficult to cover the floating elements of the population. The conduct of such a census becomes very much simpler, therefore, if these latter elements can be omitted.

In a similar manner, in the case of an agricultural census, the determination of the areas of the various crops might ideally require that all areas of the crops grown within the boundaries of the territory should be included. It may, however, be possible to exclude small areas, such as those found in gardens and holdings of very small size, without seriously reducing the value of the information. The agricultural censuses of England and Wales, for example, which are based on returns from farmers, exclude all agricultural holdings of less than one acre, and do not attempt to take account of crops grown in private gardens or allotments.

The question of whether or not minor categories should be included depends mainly on the purposes for which the information is required. A case is sometimes made for the inclusion of certain categories on which the information is intrinsically of little interest in order to ensure comparability with the results of previous censuses or surveys, or with the results of parallel surveys in other countries. Comparability within and between statistical series is obviously desirable, and lack of it can seriously reduce the value of the results, and also increase the labour of statistical analyses and the danger that those unfamiliar with the details of the various sources of information may draw wrong conclusions. Nevertheless when introducing a radically new method of collecting information, such as replacement of a complete census by the sampling method, excessive weight should not be given to past practice. It should not be forgotten that so-called complete censuses are often in themselves subject to errors of various kinds, including lack of completeness, and that such errors are often a greater source of disturbance to comparability than the omission or inclusion of a few minor categories. If there is any serious doubt whether a given category should or should not be included this may be regarded as *prima facie* evidence that the category in question differs in essentials from the other more important categories. Consequently, if it is decided that the category should be included, the results should be kept separate so that they can be summarized separately and eliminated fro.n the final estimates if required, or given special treatment in these estimates. If this is done for the first one or two surveys of the new type, comparability with previous results will be ensured, without preventing the omission of the category in subsequent surveys if this ultimately appears desirable.

The arguments in favour of the adoption of identical definitions in different countries in which conditions are radically different are even less strong. Categories which are of very minor importance in one country may be of great importance in another. Decisions as to their inclusion or omission should be taken primarily on the grounds of their importance in the country which is being surveyed, without undue regard to definitions designed to ensure formal uniformity of world statistics.

In many cases in which complete omission of unimportant categories would not be justified, they can be very conveniently dealt with by some special sampling procedure, which may be multi-phase, or may be of an entirely different type with different frame and sampling units. Thus in a human census, certain of the simpler items of information, which can be reliably furnished by neighbours or other members of the household, may be collected for absentees abroad, or a sub-sample of these absentees may be taken for a follow-up enquiry by more intensive methods. Nomads may be dealt with by instituting a supplementary sample census to deal only with this category of the population.

In a sample survey the frame adopted contains its own implicit definitions of the categories of material to be covered. If a category is not included in the frame it will either have to be omitted entirely or special steps will have to be taken to supplement the frame. Definitions of the population should therefore be considered in conjunction with the choice of frame.

4.3 Determination of the details of the information to be collected

The detailed problems which arise in deciding what information is necessary and how it can best be obtained vary widely in surveys covering different fields of enquiry and according to whether the results are required primarily for administrative or for research purposes. Full discussion of any particular case necessarily requires extensive knowledge of the subject as a whole and of the particular questions at issue, and would be out of place here, but there are certain general points which may be mentioned.

The basic problem is essentially that of the selection of the most relevant items of information or types of observation from all those which it is practicable to collect and which might conceivably have a bearing on the matters under investigation. This selection must be such that a coherent whole is obtained which covers the required field adequately, or if this is not possible at least provides information on some relevant part of it.

This basic problem is essentially the same in complete censuses and sample censuses and surveys, but the problem is more complex in the case of sample censuses and surveys, since the items of information that can be collected and the observations that can be made are themselves more complex and varied.

The best way of arriving at a satisfactory solution of this basic problem is usually as follows. In the first instance, the details of the information required to deal with the problems originally propounded are determined. The question is then considered whether there are any related problems of importance on which this information, possibly supplemented to some extent, would throw light. If this is the case the supplementary items of information required for the full elucidation of these additional problems should be determined. With the whole field mapped out in this way, the practicability of obtaining the necessary items of information covering any given set of problems can be considered, and final decisions taken in the light of the relative importance of these problems and the total load which it is considered expedient to place on the investigators and respondents in a single survey.

The details of this process vary greatly in different types of survey, but the general principle to follow in all types of survey is to see that the items of information collected form a rounded whole covering a definite subject or coherent group of subjects.

This principle is of particular importance in surveys of the questionnaire type on human populations, whether the questionnaires are filled in by the respondents themselves, or the information is elicited by field investigators. Accurate information can only be obtained in such surveys if full and willing co-operation of those providing the information is obtained. The survey must therefore have a clear purpose which can be explained to the respondents, and the questions asked must be relevant to this purpose. If additional questions dealing with unrelated subjects are included, or if the questions relating to the main enquiry seem trivial, and do not cover aspects which appear of importance to those providing the information, the survey will cease to appear as a serious enquiry into a particular subject, and will meet with unfavourable reactions, summed up in such terms as " snooping."

The matter is of importance even in enquiries which require the collection of factual information by observation and measurement by the investigators themselves, without any co-operation from respondents. If the field investigators are not imbued with a sense of the importance of their enquiry, and are overloaded with the collection of miscellaneous data, they will not give of their best. Occasionally information may be sought on points unconnected with the main survey if it is urgently needed, and considerable expense is thereby saved, e.g. in travelling, but this should be avoided as far as possible.

Occasionally, in cases in which a questionnaire would otherwise be unduly long it may be possible to split it into parts, obtaining information on one group of items from one set of respondents, and on another group of items from a second set. Certain basic items of information will be required from all respondents, and the two sets will form a pair of interpenetrating samples. The sampling has also a two-phase structure, the basic information acting as first-phase information for both sets. If this procedure is followed, however, the relationships between items of information in the two groups can only be studied for strata or other suitable domains of study, and not for the individual respondents.

Certain items of information are often required in order to ensure the proper interpretation of other items. Thus, for example, if housewives are being asked whether they prefer coal, gas or electric cooking, and the reasons for their preferences, it is essential to ascertain in some detail what experience they have had of methods of cooking other than the one they are now using, including the type of apparatus used. If this is not done the answers may be more an indication of the effectiveness of an advertising campaign in favour of one of the methods, or a condemnation of antiquated pieces of apparatus, rather than any reflection of the true relative merits of the different methods.

Information is also often required on items which, though not of primary interest, will act as supplementary information and thereby enable the precision of the results to be increased by the appropriate methods of estimation.

In reaching the decisions on the type of information required, both in broad outline and in detail, it is absolutely essential to work in collaboration with experts on the subjects which it is proposed to cover. If research or administrative experience in the subjects to be covered is lacking, it is fatally easy when designing a survey to omit some vital items of information. A simple instance of such omission is provided by the 1921 and 1931 Population Censuses of the United Kingdom. In these censuses information on age of mother at marriage, and total number of children born, which had been obtained in the 1911 Census, was not asked, with the result that the value of the information provided by these censuses for studies on changes of fertility of the population has been very seriously reduced. As a result of this lack of information it was felt to be necessary to institute a special Family Census in 1946 (Section 4.10). In this instance it can scarcely be that the need for this information was wholly overlooked, but insufficient weight must clearly have been given to this aspect of census information.

In addition to direct collaboration with experts in the various subjects, the plans for the survey should be circulated at all stages of development to the various organizations and individuals who are likely to be interested in the results. This will usually result in requests for the collection of supplementary items of information, some of which may not be necessary for the purpose for which the survey was originally planned but which will enable the results to be used for other purposes. In this way the usefulness of the survey may often be considerably increased. On the other hand, the danger of overloading the survey with the collection of miscellaneous items of information must be guarded against, and all requests should therefore be very carefully reviewed.

4.4 Inter-relations of groups of natural units

If the physical inter-relations between the members of groups of natural units of the material under survey are of interest, or if information is required for groups of natural units as a whole, then information must be collected for such groups as a whole, or at least for pairs of units from such groups. Thus if the inter-relations between the different members of a household require to be studied, it is essential to have information for pairs of individuals belonging to the same household, and it is usually best that the information should cover the whole of a household. This can be ensured by using households or dwellings as sampling units.

Another type of natural aggregate for which it is often important to obtain results as a whole is that provided by towns, villages, etc., and, in agricultural surveys, homogeneous geographical areas. This often calls for the adoption of multi-stage sampling, the natural aggregates forming the sampling units at the first stage, even in cases where the use of single-stage sampling is otherwise preferable. Thus in a survey of a human population it may be of considerable interest to contrast the results for individual towns of differing types, and to study the inter-relations existing within a single town, even when there is no need for all the towns of the country to be covered.

Similarly, if inter-relations between the behaviour of the same individuals or other natural units at different times are of interest, the survey must be designed so as to provide information covering an adequate period of time. Thus in an investigation into hours of sleep of children, it is of little value to determine the amount of sleep of a sample of children each for a single day. Such data will throw no light on the question of whether children who have a short period of sleep on a particular day tend also to go short of sleep on other days or are able to make up for this short period by longer periods on preceding or following days. In the same way, studies of nutrition in which the intake of food is determined for each individual for a single day only, although they will show whether a group as a whole is under-nourished, are incapable of revealing the degree of variation in under-nourishment between individual and individual, since individuals going short of food on a particular day may make up for such deficiencies, in whole or in part, on succeeding days.

We have stressed this point at some length because there has been a tendency, in surveys on human populations of the questionnaire type, to take the relatively easy course of asking those interviewed about occurrences which are still fresh in their mind, e.g. what happened on the previous day. This course is followed for various reasons. It may be considered that information provided about earlier occurrences will be inaccurate, or that there is a danger of overburdening respondents if an attempt is made to cover too long a period ; or the object may be to save interviewers the trouble of repeated visits which might be required to cover a period of time accurately. Actually it has been found that the use of a very short period does not necessarily lead to accurate average results : in certain circumstances there may be a tendency on the part of respondents, either consciously or unconsciously, to telescope events, and report them as happening in the given period when in fact they happened earlier. Thus a survey of crockery breakages made by asking what breakages occurred over the past week led to an entirely excessive estimate of the amount of breakage, whereas a similar survey asking for breakages over the past year gave results which checked well with production figures and the domestic stocks (Box and Thomas, 1944, D, discussion).

4.5 Practicability of obtaining the required information

So far we have been considering the problem of determining what items of information are required in order that the purposes of the survey may be fulfilled. Each item must, however, be considered in the light of the practicability of obtaining it. If the information is to be furnished in response to questions, the points for consideration are whether the respondents are sufficiently informed to be capable of giving accurate answers ; whether, if the provision of accurate answers involves them in a good deal of work, such as consulting previous records, they will be prepared to undertake this work ; whether they have motives for concealing the truth, and if so whether they will merely refuse to answer, or will give incorrect replies. If the information is to be obtained by observation or physical measurement, the points for consideration are whether the observations are such that they are within the competence of the investigators or other individuals who will be required to undertake them ; whether they will make excessive demands on the time of the investigators or others, or require excessively expensive apparatus ; and whether the owners of the surveyed material will permit the observations to be made.

Considerations of this kind will inevitably lead to modifications of what would otherwise be considered an ideal scheme. Nor can general answers be given, even within the limits of a particular field of enquiry. In countries such as the United Kingdom, for example, there is no reason to suppose that any large amount of inaccuracy is introduced into the returns of the population censuses by deliberate mis-statements. In countries not accustomed to population censuses fear that the information will be used for such purposes as taxation or conscription may lead to considerable inaccuracies. Similarly in crop-sampling work the use of small sample areas may be quite satisfactory with certain classes of field worker, but, as is shown by Table 2.5, is entirely unsatisfactory in other cases.

When the ideal requirements cannot be fully met it is sometimes possible to include other items of information, observations, or physical measurements, which, owing to their high correlation with the quantities which it is desired to determine, will serve as more or less adequate substitutes for these quantities. These substitute measures may be used for purposes of stratification or classification of the data in the final analysis, as for example when the rateable value of a dwelling is used as a substitute for the income of the household occupying it ; or they may be substitutes for measures of quantities which themselves require assessment, as for example the use of eye estimates in place of direct measurements of the yields of a standing crop. The efficiency of such substitute measures can only be properly judged by a proper statistical investigation of the relations between them and the quantities for which they are substitutes. In the case of substitutes for measures of quantities that are to be assessed, some method of calibration is essential if objective estimates of the original quantities are to be obtained. (The calibration of eye estimates is discussed in Sections 6.15 and 7.14.)

It will inevitably happen in certain cases that information which is of considerable importance will prove to be unobtainable, or unobtainable with sufficient accuracy. When such a situation arises it must be squarely faced. There is at times a tendency to attempt to collect information which, because of its nature, cannot be obtained with the necessary accuracy, and then to condemn the survey method in general because the results are of little value.

This, however, does not mean that the collection of difficult items of information should not be attempted. The sample survey procedure, because it makes possible the use of skilled investigators working on a relatively small sample, is frequently capable of eliciting reliable information on points which it would be quite impossible to include in a general enquiry. The fact that the enquiry *is* on a small sample, if known to the respondents, frequently makes them willing to give information which they would certainly not be prepared to give if the enquiry were general. In such cases it is important that the investigators should themselves be recognized as impartial and disinterested ; in particular they should not be officials of an organization which itself might make use of the information obtained to the detriment of the respondents.

Nevertheless there are subjects on which it is impossible to collect accurate information from a random sample of the population. In certain of these cases information can be collected from a selected group of individuals, e.g. individuals with whom social welfare workers are in contact. Information of this type is not necessarily valueless, but it must be clearly recognized that it is not the equivalent of information obtained from a random sample of the whole population, and any attempted generalization of the results will be of limited validity.

Attempts are sometimes made to obtain a sample from such a group of individuals which conforms more closely in certain respects to the population, e.g. in classification by age or social class, than does the group as a whole. While this may improve the sample somewhat, it still does not provide the equivalent of a random sample. On the other hand, if the whole of the group is not required, it is usually advisable to apply some rigorous form of selection rather than to permit the workers themselves to select individuals for investigation, as the latter procedure will merely introduce further unnecessary elements of bias.

In cases in which some of the items of information are difficult to collect, multi-phase sampling may be of value. It may, for instance, enable specially skilled investigators to be used for the more difficult items. Thus in a health survey medically qualified investigators may be used on a small sub-sample of a much larger sample on which more general items of information relating to health have been collected. Equally it may be used to reduce the work required to manageable proportions. Thus, in the Survey of Fertilizer Practice soil samples for chemical analysis were taken from one old-arable field, one new-arable field, and one field of permanent grass on each farm, these fields being a sub-sample of all the fields on which information on the use of fertilizers was obtained (Section 4.23).

A procedure which may reassure respondents that their anonymity is safeguarded when faced with an incriminating or highly personal question is to submit two alternative questions, the second of which is unrelated to the subject at issue.   Which of the two is to be answered is determined by some method of random choice, operated by the respondent in such a manner that the outcome is concealed from the investigator.   This is known as the *random-response* method.   It was first introduced (in slightly different form) by Warner (1965).   Provided the probability of a Yes answer to the unrelated question is known, or can be determined, the probability **p** of a Yes answer to the key question can be estimated from the overall proportion of Yes answers.

The method is obviously subject to much larger sampling errors than would result if the question were put directly.   Cochran gives an excellent summary of the different variants of the method in the third edition of *Sampling Techniques*.   Formulae for the estimation of **p** and its sampling variance are included.

The method does not necessarily completely safeguard the anonymity of all respondents.   If for example the probability of a positive answer to the unrelated question is negligible for certain respondents, a positive answer will indicate that the key question was selected.   Nevertheless the method does appear to be effective in removing inhibitions to a truthful answer, in that several cases have been reported in which substantially higher estimates of **p** have been obtained than when the same question has been put directly to a comparable sample.

4.6   Methods of collecting the information

The methods of collecting the information are to a large extent conditioned by the material under survey and the type of information required. Where the alternative possibilities exist, it may be stated as a general rule that observations are preferable to questions, and questions on facts and on past actions are preferable to questions on generalities and on hypothetical future conduct. Thus it is better to inspect a house to see if it shows signs of damp, than to ask the occupant if the house is damp ; and it is better to find out what considerations, from among the various alternatives (if any) that presented themselves, governed the selection of the house in which the occupant is living, rather than to ask what type of dwelling—house, flat, bungalow, etc.—is " preferred."

On the other hand, it is scarcely possible to state any general rule with regard to physical measurements and qualitative observations made by the investigator. Physical measurements are more objective, but qualitative observations are often more capable of summing up the salient features of a complex situation. Thus a qualitative grading by the investigator of the degree of dampness of a house is likely to be more effective than any physical measurements designed to determine the degree of dampness. Moreover, by proper standardization and calibration among investigators qualitative observations can themselves be made objective.

When the information is collected by means of a census form or questionnaire the questions which are to be asked should be considered at the planning stage, since the information obtained will depend on the exact form of these questions. Equally the exact form of any observations and physical measurements which are required should be determined.

Census forms and questionnaires may be designed either for completion by the respondent with little or no assistance from investigators, or for completion by the investigator by the aid of questions put to the respondents. In questionnaires of the latter type the investigators may be instructed to ask questions with a given form of wording, or they may be instructed to elicit information which will provide an answer to the questions of the questionnaire by enquiry and discussion without adherence to any exact form of words. Both means of eliciting information may be required in the same survey for different items of information.

Census forms and questionnaires designed for completion by the respondent may be delivered and returned by post, delivered by post and collected by an enumerator or investigator, or vice versa, or delivered and collected by an investigator. Use of the post is clearly most economical, and is the method generally followed in censuses and surveys of industrial and commercial organizations, such as censuses of production. In such cases the use of investigators will not normally have any great advantage over the post, either in ensuring more complete response or obtaining more accurate information, though occasionally in local surveys investigators may be used to explain the purposes of the survey and persuade the respondents to co-operate. In population censuses, however, investigators are normally used both in order to ensure the maximum response, and to give assistance where necessary in filling up the forms. Censuses and surveys of small-scale industrial and commercial organizations, and of farms, occupy an intermediate position, and the method used will depend to a large extent on local circumstances.

Attention must be paid to the detailed wording of all questions, even if these are only intended as guides to the investigator. If the question itself creates a wrong impression in the mind of the investigator this will undoubtedly lead to errors, even if additional explanatory notes indicate that something else is really required.

Careful thought must also be given to the order of the questions. If questions are arranged in an orderly sequence the investigator's task is much easier, and the respondent's reaction is likely to be more favourable. This applies to all forms of questionnaire, but is most important in the verbal questionnaire.

In many types of survey it is profitable to give the investigator or respondent an opportunity of making general remarks on special points. This can be done very simply by including a space for observations. Some guidance should be given on the type of observations required. Although such observations do not easily lend themselves to exact analysis they are frequently of considerable value in drawing attention to relevant facts not covered by the questionnaire itself.

The type of investigator to be employed must also be considered. Investigators should have a background knowledge of the subject under investigation, particularly in investigations of the research type. In a technical investigation into housing conditions, for instance, the investigators should have some knowledge of housing construction and of standards normally adopted in good practice. This requirement of technical knowledge in the investigators limits the scope of unspecialized teams of investigators. Such teams are suitable for carrying out *ad hoc* and routine investigations which require only relatively simple questionnaires, but they are no substitute for the more specialized teams required for investigations of a research nature involving technical questionnaires and observations or physical measurements by the investigators themselves.

In surveys requiring any high degree of technical knowledge it is usually best either to use members of existing organizations, or to appoint a small specialized team of technically qualified research investigators. The various surveys into the technical and economic aspects of agricultural practice in England and Wales, for example, are carried out by the staffs of the National Agricultural Advisory Service and the Provincial Advisory Economists. By this means teams of investigators are obtained who are technically qualified and capable of discussing the problems involved with the farmers ; at the same time the investigators themselves gain a wider knowledge of the farms of their district which is of value to them in their other work.

4.7 Methods of dealing with non-response

Unless non-response is confined to a small proportion of the whole sample the results cannot claim any general validity. Every effort must therefore be made to reduce non-response to negligible proportions.

Non-response is usually most serious in postal questionnaires. Delays in response can also sometimes be very troublesome, particularly when the results are required quickly. A rigorous system of dealing with failure to respond and delay in response must therefore be instituted at the outset. The first step is to send a follow-up letter, but if this does not produce the required effect, the possibility of using more intensive methods such as telephone calls and personal visits must be considered. These will require a special regional organization.

In censuses of industrial and commercial undertakings in which data on production, sales, labour force, etc. are required for the purposes of economic planning it is usually possible to make the returns compulsory. This is often a help in dealing with a small minority of recalcitrant institutions, particularly if pressure can be brought to bear in other ways, but it is no substitute for full and willing co-operation by the majority. Complete population censuses are usually also made compulsory, and there appears to be no logical reason why sample censuses of the same type should not also be compulsory. While this is little help in dealing with obstinate refusals, since the census authorities are not likely to wish to bring the offenders before the courts, it is an indication that the government regard the census as of importance, and to this extent is likely to act as a persuasive force with the waverers.

In censuses which are to be repeated at intervals it is particularly important to deal vigorously with non-response and delay in response at the outset, as otherwise they tend to increase progressively. If any large volume of non-response persists, or if there is any serious delay in making the returns, it is an indication that something is wrong with the census, which should either be reorganized or abandoned.

In sociological surveys using the interview method the amount of deliberate non-response is usually small. If it is not, the questionnaire and the type of investigator used should be reviewed. Revisits by special investigators can be tried, but are not likely to be very effective. In technical surveys of agriculture involving interviews with the farmers the amount of deliberate non-response is also usually small, unless the amount of information required is such that it puts too heavy a burden on the respondents.

In sociological surveys, however, initial non-response due to failure to contact the respondent can be very troublesome. There is no proper way of dealing with this except by persistent call-backs. The number of call-backs can often be reduced by enquiring of neighbours when the respondent is likely to be at home, or where he can be found so that an interview can be arranged. Call-backs are also required because the respondent, though willing to give the information, is otherwise engaged at the time of the first call.

The amount of work involved in follow-ups and call-backs can be reduced, if this appears desirable, by taking a sub-sample of those not contacted at the first (or subsequent) call, and weighting up the sub-sample in the final results. In repeated censuses, however, complete follow-ups are advantageous in encouraging better response to later censuses.

4.8 The frame

The whole structure of a sampling survey is to a considerable extent determined by the frame. The methods of survey which are suitable for a given type of material may be radically different in different territories because different types of frame have to be used. Consequently, until particulars of the nature and accuracy of the available frames have been obtained, no detailed planning of the survey can be undertaken. If no frame exists, the construction of a frame suitable for the purposes of the survey may well constitute a major part of the work of the survey.

Frames are subject to various types of defect, which may be broadly classified as follows. A frame may be:

(1) Inaccurate.
(2) Incomplete.
(3) Subject to duplication.
(4) Inadequate.
(5) Out of date.

A frame may be termed *inaccurate* if information about the units listed in it or defined by it is inaccurate. The term may also be used to cover the listing of units which do not in fact exist. Thus a ration-card list in which certain women were incorrectly described as married when they were in fact single, or in which certain individuals were included who had died, would be inaccurate in these respects.

A frame may be said to be *incomplete* when certain units of the material are omitted entirely, and be *subject to duplication* when certain units of the material are included more than once. Thus a ration-card list in which certain individuals were not included, and others were included twice, would be both incomplete and subject to duplication.

A frame may be termed *inadequate* when it does not cover all the categories of the material which it is desired to include in the survey. Thus a ration-card list which did not include the temporary residents in a district would be inadequate for a survey of the population of that district in which it was necessary to include such temporary residents.

A frame, though accurate, complete, and free from duplication at the time it was constructed, may no longer be so at the time it is required for use. Such frames may be said to be *out of date*. Errors of all the first three of the above types may be introduced through the frame being out of date.

These different types of defect have very different consequences in the defects they introduce into the sampling process. Inaccuracy in the frame, in so far as it relates to the selected sampling units, will automatically be discovered and corrected as the survey progresses, and consequently will not invalidate the results. If the information contained in the frame has been used as a basis of stratification, etc. or as supplementary information, inaccuracy in this information will result in somewhat lower accuracy in the results, but the actual accuracy attained will be assessable from the results themselves.

Incompleteness in the frame will not be discovered in the course of the survey itself, and to the extent to which a frame is incomplete the population or material will fail to be covered. Incompleteness is likely to be more serious than it appears to be at first sight, since it is often confined to units possessing some special characteristics, which may in consequence be seriously under-represented in the sample. Duplication has a similar effect, since the dupli-cated units will have a double chance of being included in the sample. There is the difference, however, that incompleteness cannot be determined or set right by an examination of the frame itself, whereas duplication may under certain circumstances be detected and corrected by such examination, though this will almost always be a tedious operation. If the sampling fraction is large and the degree of duplication is also large, the duplication may come to light in the course of the survey. Thus, with 5 per cent. duplication and a sampling fraction of 1 in 10, two out of every 210 units in the sample will on the average constitute a duplicate pair. With a sampling fraction of 1 in 100, however, only two out of every 2100 units in the sample will constitute a duplicate pair.

A frame which is inaccurate for certain purposes may be incomplete for others. Thus a ration-card list in which some of the single women were described as married would be complete, though inaccurate, if used as a frame for a survey of all women, but would be incomplete if used as a frame for the survey of single women only. Such incompleteness could be remedied by taking a sample covering all women, and rejecting those members of the sample who were found on investigation to be married.

Inadequacy of the frame will usually be known before the survey is under-taken from the specification of the frame itself. Inadequacy can in general only be dealt with by the construction of a subsidiary frame for the omitted categories.

In actual practice, frames are likely to suffer to a greater or less extent from all of the above defects. It is therefore essential at the outset of the survey to carry out a careful investigation of any frame it is proposed to use, since many defects are not at all apparent until a detailed investigation has been made. Such an investigation will naturally commence with a study of the administrative machinery by which the frame has been constructed and by which it is kept up-to-date, but may also have to include a certain amount of field work.

4.9 Frames suitable for censuses and surveys of human populations

Human populations have a tendency to aggregate in towns and villages, often with very high local densities, which makes any form of area sampling based on maps and plans subject to high variability, unless a very elaborate sampling procedure is adopted. This is most serious if the total numbers are not known, and require to be estimated from the sample, but even the proportions of the population falling in different categories will be subject to substantial errors, since different classes of the population tend to be concentrated in different areas.

Three very different types of survey of human populations may be distinguished. These are:

(1) Surveys of the census type, requiring the collection of relatively simple facts, but covering the whole population, and capable of giving separate results for small administrative areas.

(2) Surveys covering the whole population of a country, and capable of giving reasonably accurate estimates for the whole population, and possibly for certain broad subdivisions, but not for small administrative areas. Such surveys often involve the collection of more detailed and elaborate information than do those of type (1).

(3) Local surveys covering a particular town or rural area, or a few contrasted towns or rural areas, in which no attempt is made to obtain a sample which is fully representative of the country as a whole. Such surveys almost always involve the collection of detailed information by field investigators. They are usually investigations of a research nature, and may be precursors of simplified surveys on the same problems covering the whole country.

Surveys of the first type present relatively simple sampling problems, and relatively complicated administrative problems. The sampling, since it has to cover small subdivisions of the population, must generally be single-stage, usually with stratification and a uniform sampling fraction. Surveys of the third type are also relatively simple; since only limited areas have to be covered, a one- or two-stage sampling process usually suffices.

Surveys of the second type, however, present much more difficult sampling problems, and also give much greater scope for increase in efficiency by the use of the more elaborate sampling methods. Since results are not required for small areas, administrative or other areas can form the first stage of a multi-stage process, thus enabling the sampling to be concentrated in relatively few areas instead of being spread over the whole country. This condition is absolutely necessary when field investigators are used.

In fully developed areas a good deal of prior information on administrative areas is usually available. This often enables the accuracy of sampling at the first stage, which in general is the stage which contributes most to sampling error, to be substantially improved by the judicious use of stratification, supplementary information, etc. The sampling problems of this second type of survey are discussed in more detail in Section 4.18.

Frames suitable for the sampling of human populations may be broadly classified as follows:

(a) Lists of individuals in the population, or parts of it, provided for administrative purposes.

(b) Aggregates of census returns resulting from a complete census.

(c) Lists of households or dwellings in given areas.

(d) Town plans.

(e) Maps of the rural areas.

(f) Lists of towns, villages, and administrative areas, often with supplementary information of various types.

In the following sections we will give a brief description of the outstanding features, from a sampling point of view, of these various types of frame.

4.10 Frames from lists of individuals

Lists and card indexes of individuals are provided by various administrative activities, such as registration of the population, rationing, or even a recent census. Such lists are only likely to be complete, accurate, and up-to-date if the administrative machinery is very efficient, and there is some definite administrative need for the lists to be kept under constant revision. Most lists of this type cover the whole of a country on a more or less uniform basis, but they are necessarily maintained by local offices, and their accuracy may consequently vary in different parts of the country. Even when a list is sufficiently accurate to fulfil adequately the administrative purposes for which it was designed, it will often be found to have unsuspected defects which make it unsuitable as a sampling frame.

Lists of individuals are not suitable as a frame for sampling households, unless the individuals are grouped by households. If addresses are selected from a list not so grouped, the probability of selection will be proportional to the size of the household, which is rarely what is required. If such a method of selection is for any reason used, the results must be weighted in inverse ratio to the number of people in the sampled households included in the list (Section 6.16).

Examples of administrative lists are provided by the National Registration and ration-book lists maintained in the United Kingdom. The National Register, which was instituted in 1939, has probably always constituted a reasonably accurate register of the whole population, but in its early stages, at any rate, it was very defective as a local register, owing to the failure of individuals to register changes of address. This defect was later rectified by establishment of joint offices with the food offices, and insistence that any applicant for a new ration book should first have his identity card amended. This, however, did not ensure immediate registration of local changes of address, since new counterfoils were only required if the removal necessitated change of shops. Consequently local changes were often only registered at the time of the regular yearly issue of ration books.

The card index of the ration-book issues necessarily suffered from similar defects. Consequently neither of these registers formed a suitable frame for the sampling of small administrative areas, such as a single food-office district, particularly during the war when movements of population were frequent and considerable owing to air raids. On the other hand, they were and are capable of serving as a reasonably adequate frame for a sample census of the whole population.

The food-office card index was used as the frame for the 1946 Family Census of the United Kingdom.   This census was carried out by the Royal Commission on Population, with the object of providing, for married women, information on age, age at marriage, number and dates of birth of all children, and husband's occupation, information which had never previously been collected in full.   A sample of 1 in 10 of all the married women (including those widowed or divorced) was taken, by examining every tenth card, and recording the name and address if the card was for a female adult with the prefix Mrs. or with no prefix.   Unmarried women selected by this process were requested to mark the questionnaire " unmarried."   Questionnaires were dispatched by post, and collected by subsequent visit.

Since there is necessarily a time lag in cancellation of the old food-office card on removal—this is effected by notification from the food office issuing the new counterfoils—special steps had to be taken to deal with removals. This was effected by fixing a " zero " date at an interval prior to the date when the sample was taken.   The interval was chosen so as to be somewhat longer than the time taken for notification of change of address to be received at the old office.   Thus virtually all duplicate cards corresponding to changes of address prior to the zero date would have been removed.   Registrations effected after the zero date were excluded from the sample, and all cancellations bearing a date of re-registration subsequent to the zero date were sampled, the new address being recorded.   It will be seen that by this procedure all individuals entered into the sampling frame once and once only, and the only individuals for whom incorrect addresses were recorded were those who had for some reason delayed their re-registration, or for whom notification of change of address had not yet been received by the old office.   This procedure avoided all duplication except in the rare event of excessive delay in notification between offices, while obviating the necessity of any attempt to construct a fully up-to-date non-duplicated index.

4.11   Frames from complete population censuses

A complete census, in so far as it really is complete, will automatically provide an aggregate of forms which includes all the individuals in the area covered by the census.   Nevertheless complete censuses, although they would appear at first sight to provide very satisfactory frames, have a number of defects.   A complete census by its very nature can only be carried out at infrequent intervals, e.g. every ten years, and consequently the frame provided by such a census is for the greater part of its existence badly out of date.   The way in which the census information is customarily collected and analysed also tends to reduce its utility as a frame, since the information is not readily accessible, at least in the early stages during which it is being transferred to punched cards.   One of the great advantages of the food office register used in the Family Census was that the cards could be consulted and sampled in the local offices without serious disturbance of the office routine.

Many of these disadvantages can be overcome if at the time a complete census is undertaken arrangements are made to construct a proper *master sample* from which further samples can be drawn as required. The sampling unit for such a master sample should be the dwelling, and not the individual or household occupying that dwelling at the time of the census. If dwellings are adopted as sampling units the master sample will have a much greater degree of permanence than would be the case if individuals were used as sampling units. Furthermore, for most purposes a sample of households, and not of individuals scattered over all households, is required.

A complete census will provide a very suitable frame for a simultaneous sampling census in which more detailed information is collected on a sample of the population. This procedure was used in the 1940 census of population in the U.S.A. (Stephan *et al.*, 1940, C). In this census supplementary questions were asked of 1 in 20 of the individuals included in the complete census, at the same time as the main census information was collected. The procedure was thus analogous to two-phase sampling, with the exception that the first phase constituted the whole population.

Since the selection of 1 in 20 individuals for the collection of the supplementary information was done on the spot by the field investigators, certain very rigorous rules had to be instituted in order to avoid bias. The actual procedure adopted was as follows. The census forms each contained lines for 80 individuals, 40 lines on each side. Two of the lines on each side were specially marked. Five different types of form were used, with the marked lines distributed in the manner shown in Table 4.11.

TABLE 4.11—SAMPLING LINE NUMBERS IN THE 1940 U.S. POPULATION CENSUS, AND THEIR PROPORTIONS

Style	Proportion	Line numbers			
V	16	14	29	55	68
W	1	1	5	41	75
X	1	2	6	42	77
Y	1	3	39	44	79
Z	1	4	40	46	80

The investigators were instructed to enter the names of each family in a defined order, and to complete all lines of the form before commencing a new form. Actually these instructions were not always adhered to, $3\frac{1}{2}$ per cent. of the last lines (Nos. 40 and 80) being found to be blank. If the blanks extend over the earlier lines, which are not marked on the W-Z forms, but not as far as the lines marked on the V form, this will lead to a slight deficiency in the proportion in the sample of entries in line 1 and the other lines marked on the W-Z forms. This disturbance, however, is only very small, but any tendency of the investigators to alter the order in which the names were entered so as to secure a suitable person for supplementary questioning could easily give rise to more serious biases.

The danger of this type of bias is always present in this method of sampling, and can only be overcome by the most rigorous training of observers, and the imposition of rules which determine uniquely the order in which names are entered on the list.

4.12 Frames from lists of households or dwellings

Lists of households or dwellings are frequently available from such sources as rating offices, electoral registers, etc. Frames based on such lists are in many ways preferable to frames based on lists of individuals. As already mentioned, in most surveys in which the information is collected by personal visit it is advantageous, and often essential, to collect information from all members of a household, in other words to use households as sampling units.

Frames consisting of lists of dwellings also have a much greater degree of permanence, being unaffected by movements of the population. Such frames, if complete at the time of their construction, will only become incomplete to the extent that there is new building, or changes in the use of existing buildings. New building is necessarily a slow process, and the listing of new buildings usually presents no serious difficulties.

Lists of households can generally be utilized to give a frame of dwellings by taking as the sampling units the dwellings occupied by the households at the time the frame was constructed. Certain special precautions are required to ensure the inclusion of dwellings which were unoccupied at the time the list was prepared. In a town in which the dwellings are arranged in streets and in which the list is also arranged by streets this presents no particular difficulty. What has been called the *half-open interval* can be used. The procedure is as follows. When drawing the sample, the dwelling unit appearing next in the list to the selected unit is recorded, and the field investigator is instructed to see if there is any other unit on the ground between these two units, and if so to include that unit in the sample. Thus the field investigator might receive the instruction to survey No. 9 in a certain street, with No. 13 as the next recorded unit (odds only). If on visiting No. 9 he finds that No. 9A and No.11 also exist, these are also surveyed. The even numbers between 9 and 13 are not included, since the instruction " odds only " indicates that they lie on the opposite side of the street. This procedure is clearly only possible if the list is arranged in an order which corresponds to some geographical pattern on the ground. If the list is not so arranged, incompleteness of the frame cannot be corrected by the use of the half-open interval or analogous procedure. In such cases the frame will have to be amended by other means, and complete rearrangement of the list in some geographical order may be necessary.

An example of the use of this type of frame is provided by some surveys carried out during the war in certain towns in the United Kingdom by the Ministry of Home Security, to investigate disturbances to the population on account of air raids. In the English towns electoral registers were used as frames, and in the Scottish towns rating lists were so used. The electoral registers consist of printed lists of voters arranged by streets in order of dwellings, all voters in one dwelling appearing together. Each dwelling therefore has as many entries as there are voters. Consequently, selection of entries in the list with equal probability will not give an equal probability of selection in the different dwellings. This could have been overcome by subdividing the list into dwellings, and basing the sampling on these dwellings.

As the surveys had to be conducted at considerable speed, delay in selection of the sample was avoided by the device of examining every xth entry, and including the dwelling in the sample if the entry referred to the first listed member in the dwelling. This introduces a certain additional discrepancy between the working sampling fraction, $1/x$, and the actual fraction of dwellings included in the sample, which introduces errors that are appreciable relative to the sampling errors for estimates of such quantities as numbers in the population obtained by multiplying the sample total by x. Such estimates were not the primary concern of these surveys, and consequently no adjustments were required. If necessary, errors arising from this cause could have been eliminated subsequently by ascertaining the ratio of the number of dwellings included in the sample to the number in the whole register, and treating this ratio as the true sampling fraction.

In this survey the method of the half-open interval was used to deal with dwellings not included in the list, and was found to be quite satisfactory. Had there been new housing estates not covered by the register these would have had to be dealt with separately.

In certain towns the separate flats of blocks of flats were not listed, and therefore presented a difficulty, since the blocks constituted very large units whose chance inclusion or exclusion would have materially increased the sampling error. The existence of blocks of flats was, however, always apparent from the large number of voters appearing under the same address. The blocks were therefore listed, together with other large institutions, by preliminary inspection of the register, and every xth flat was selected by visit to all the blocks in turn.

4.13 Frames provided by town plans

Town maps and plans provide a useful frame for the sampling of dwellings in built-up areas. In some cases there may be detailed maps showing the location of all dwellings, but in many cases only street plans, not showing any great amount of detail, will be available.

Any town plan which gives an accurate representation of the streets will enable the town to be divided up into " blocks," i.e. areas bounded by streets. Such a plan will, therefore, provide a frame for area sampling in which the units are blocks. A sample of dwellings can then be obtained by including all the dwellings in the selected blocks. In general, however, the variability between block and block is likely to be large even after careful stratification, since there is often considerable local segregation of different classes of the population. Consequently, two-stage sampling is in general advantageous, blocks being taken as the first-stage units and dwellings as the second-stage units.

To obtain a two-stage sample in cases in which the map does not show the location of dwellings, it will be necessary to construct the second-stage frame for the blocks selected at the first stage by ground survey. This, however, is a much lighter task than the construction by ground survey of a frame for all dwellings in the city, and can frequently be done in the course of the survey itself.

In towns in which the natural blocks are of very unequal area, groupings of the smaller blocks or subdivision of the larger blocks should be performed, so as to reduce the within-strata inequalities in area. If little is known about the town, it may be advantageous to make a ground survey in order to demarcate and stratify the block units. It may even be advantageous to carry out a rough preliminary count, or make some other rough estimate of the number of dwellings in each block, as this will enable the subsequent selection of the first-stage units to be made from within strata with probabilities proportional to the estimated numbers of dwellings. This was done in parts of the Greek population census described in Section 4.16. If such preliminary estimates are not available the best that can be done is to make a selection with probabilities proportional to area, but block areas are unlikely to be very closely correlated with the number of dwellings, even within strata. In either case the second-stage sampling fractions may be taken inversely proportional to the first-stage fractions, so as to give a constant overall sampling fraction.

Whether an elaborate procedure of this kind is needed depends not only on the accuracy required but also on whether estimates of total numbers are required from the survey. Since total numbers will be highly correlated with numbers of dwellings, prior supplementary information on these numbers for the different blocks, even if only rough, will be particularly effective in reducing the sampling variability of estimates of total numbers. They are not likely to have such large effects on estimates of the proportions of the population falling in various categories.

Sampling by streets is sometimes used instead of sampling by blocks. This is usually not so satisfactory as sampling by blocks, since each block represents a clearly defined area, whereas if a street is taken, there is often doubt as to exactly what is to be included and what excluded : alleyways and court-yards having entrances from more than one street, and not shown on the street map, for example, present considerable difficulty if the sampling is by streets.

Sampling by blocks is particularly valuable for surveys of towns in which all types of building have to be covered. Second-stage sampling of any or all of the different types of building can be adopted if required, by enumerating the different types for the selected blocks after the first-stage sample has been drawn.

4.14 Frames provided by maps of rural areas

The use of maps as a frame for the sampling of rural areas presents somewhat different problems from those encountered in the sampling of towns by the aid of town plans.

If accurate and detailed maps showing all or virtually all buildings are available, rectangular areas may be used as sampling units, the buildings falling in the selected areas being examined on the ground to see whether they are dwellings, with a further examination for unmapped dwellings.

Sampling with probability proportional to the apparent number of dwellings indicated by the map is possible, but would involve counting the dwellings in all the rectangular areas. Consequently it is better, if preliminary work on the maps of this magnitude appears to be worth while, to divide the map into areas containing approximately equal numbers of dwellings, using natural boundaries as far as possible and paying particular attention to stratification.

It may be noted here that the selection of a point at random on the map and selection of the dwelling unit nearest to this point for inclusion in the sample—a method which is sometimes used—is inadmissible, since a unit which is widely separated from other units will have a much greater chance of selection than one which is close to other units.

With less detailed maps rectangular areas marked on the map will not be capable of being demarcated exactly on the ground. Natural features occurring on the maps must therefore be used as boundaries of the sampling units. This will necessarily give units of differing size. In particular, occasions will arise when it is impossible to subdivide a somewhat large area. In such cases, the area in question may be taken to represent two or more units. If any of these units are selected, a subdivision into two or more parts as alike as possible is made on the ground at the time of the survey, and the requisite number of parts selected by random choice.

In most countries, even in rural areas, there will be a number of villages of varying sizes, which are best dealt with separately by some form of stratification and two-stage sampling, since if these are included in the area sample a high degree of variability will be introduced. The use of a variable sampling fraction at the first stage, a larger proportion of the larger villages being selected, will be advantageous. A compensating reduction in the second-stage sampling fraction can be made if desired. The boundaries of all villages will require careful demarcation, as otherwise there will be ambiguity as to what should be included in the area sample.

4.15 Frames from lists of villages

In undeveloped areas the available maps are not likely to be of sufficient accuracy for area sampling. Where the population is concentrated in villages, these usually form the best first-stage sampling units. A list of villages will then serve as a suitable frame.

Even if the majority of the population is concentrated in villages there may be a residue located in the intervening countryside. If this residue owes allegiance to definite villages, the problem is relatively simple, since all that is required is the identification of the individuals belonging to the selected villages. This can normally be done by the head-men of these villages.

If no such association exists, some form of area sample of the intervening areas may be necessary. If rough maps are available, suitable areas may be demarcated by tracks, rivers, etc. If no maps are available, some form of line sampling may be possible in open areas.

If the country is not sufficiently open to be easily traversed, the construction of a rough map of the tracks may be necessary before any adequate sampling of the intervening areas can be carried out. It may be possible, however, to use these tracks without full mapping. Thus all tracks leading from villages selected for the sample may be traversed, and dwellings to which they give access included up to half-way to the next listed, but not necessarily selected, village. Such a method will only be effective with a relatively simple track system such as is met with in forest areas : intermediate junction points, for example, present special problems.

4.16 The 1946 population sample for Greece

The 1946 population sample for Greece provides a good example of the way in which sampling methods of the types discussed in the previous sections can be used to obtain speedy census data from a sample covering the whole population of a country. (Jessen *et al.*, 1947, C). *

The sample was taken by the Allied Mission for Observing the Greek Election, as part of the investigation of the accuracy of the electoral lists. A population sample was required in order to test for omissions from the electoral lists, and the opportunity was therefore taken of securing more general census data. The corresponding test for duplications and redundancies in the lists was made by examining the lists themselves and investigating a sample of names drawn from them.

The frame for the first stage of the population sample was that given by the 1940 Population Census. This census gave returns for *koinotetes*, which are small communities or groups of villages, and *demoi*, which are towns and cities, usually with more than 10,000 population. Maps were available which showed the areas included in these *koinotetes* and *demoi*, and the names and location of all the populated centres. The *koinotetes* and *demoi* were used as

---

* See also U.S. Dept. of State publ. 2522 (1946, D′) and Jessen *et al.* (1949, D′).

sampling units at the first stage of the sampling. The units were stratified according to their population in 1940, and a variable sampling fraction was used. The actual scheme is shown in Table 4.16. Selection from within strata was systematic.

The sampling of the selected first-stage units was based on lists of households within the area. These lists were either based on existing lists checked and brought up to date, or were specially prepared to show the location of the households on a map.

For the sampling of towns an additional stage was used, a sample of blocks demarcated on an existing or a constructed street plan being first taken, with a further sample of houses from within the selected blocks. Sampling was sometimes with probability proportional to estimated numbers of households, these estimates being obtained by a rapid cruise of the whole area, and sometimes with equal probability.

The sampling fraction at the final stage was in all cases adjusted so as to give a constant overall sampling fraction. When blocks were sampled with probability proportional to estimated numbers of households, this required that the sampling fraction within the selected blocks should be taken as inversely proportional to the estimated number of households in the block. Thus the parish of Agios Panteleemous, which is given as an example, was initially subdivided into 98 blocks. Before sampling, some of the smaller blocks were combined so as to give 65 combined blocks. * The total of the estimated number of households was 966. It was decided to sample three blocks, which were selected systematically by taking a sampling interval of 322 ($= 966/3$) with a random starting point of 288, using sub-totals in the manner of Example 3.2.c. This gave combined blocks containing 23, 18, and 13 estimated households respectively. Since a sampling fraction of 1/100 within the parish was required, the *estimated* number of households in the sample was $966/100 = 10$. This number was divided approximately equally between the three selected blocks (4, 3, 3), and the sampling intervals were calculated by dividing the estimated numbers in the blocks by these numbers, i.e. the intervals were taken as $23/4 = 6$, etc. This procedure gives the required constancy in the overall probability of selection. The actual number of houses in the sample will of course differ from 10 if the estimated numbers are in error: it is the overall probabilities of selection, not the numbers of houses, that must be fixed.

The ratio method of estimation was adopted, using the 1940 population data as supplementary information. The actual method used was that appropriate to sampling without stratification with probability proportional to size of unit (Section 6.16), as this was considered to be the most accurate. There does, however, appear to be some danger of the introduction of bias by this method, and the unbiased method appropriate to a stratified sample with variable sampling fraction (Section 6.11) might have been preferable.

* This procedure is the same as that used in the sampling of Hertfordshire farms by parishes, Section 3.11.

TABLE 4.16—GREEK POPULATION CENSUS : SUMMARY OF THE SAMPLE DESIGN

Size-class code	Population in 1940	Assumed average population in 1940	Sampling ratios		Number of places	
			For selection of sample places	For selection of names and households within a sample place	In size-class	In sample
			For koinotetes			
1	0– 499	350	1/100	1/5	2,147	20
2	500– 999	750	1/50	1/10	2,049	40
3	1,000–4,999	2,500	1/20	1/25	1,366	70
4	5,000 and over	7,000	1/5	1/100	54	10
				TOTALS	5,616	140
			For demoi			
5	Under 25,000	17,000	1/2	1/250	52	26
6	25,000 and over	—	1/1	1/500	22	22
				TOTALS	74	48

The survey was very successful and achieved high accuracy, the standard error of the estimate of the total population being estimated to be $\pm 2 \cdot 1$ per cent. The field work occupied 65 observer teams, each consisting of an observer, an interpreter and a driver, with a jeep, for three weeks. The entire sample and the computations were completed in 7 weeks.

4.17 Master samples

When a number of surveys covering the same population or aggregate of material are likely to be required, it is sometimes advantageous to construct a *master sample*, from which smaller samples can be drawn as required by means of a sub-sampling scheme.

The use of a master sample has a number of advantages. It enables a more accurate, complete and adequate frame to be constructed than could be justified if the frame were only required for a single survey. It simplifies the selection of samples, since in the sub-sampling only the material contained in the master sample has to be subjected to the selection process. It enables supplementary information to be obtained which is of value in improving the accuracy of the various surveys. And it enables surveys on the same material to be so planned that the same units are not selected an excessive number of times for different surveys—a matter of some importance when the information is obtained by response to questionnaires.

An example of an extensive and elaborate master sample is provided by one constructed for agriculture of the United States of America. The construction of this sample was undertaken by the Statistical Laboratory of Iowa State College, in co-operation with the Bureau of Agricultural Economics and the Bureau of the Census (King and Jessen, 1945, G).

The sampling units of this master sample consist of small areas covering the whole of the United States. The units have a mean area of about 2·5 square miles, but vary according to location and other circumstances, the mean area per state ranging from 0·71 square miles to 108 square miles. They were formed so as to contain on the average 4, 5 or 6 farms, depending on the part of the country. One-eighteenth of all the areas were selected for the master sample.

The whole of the land area of the United States was divided into three categories, called in the master sample " primary strata." These primary strata are (1) the incorporated stratum, (2) the unincorporated stratum, (3) the open-country stratum. The incorporated stratum consists of incorporated cities and towns and unincorporated places regarded as " urban " by the Bureau of the Census. The unincorporated stratum consists of all named places outside the incorporated areas which have an estimated population of 100 or more, and all other areas which appear on the map and have a population density of 100 or more persons per square mile.

The incorporated areas were defined by the corporate boundaries, of which the location could be obtained. The unincorporated areas were demarcated on the maps so as to give areas as compact as possible, while including everything that did not appear to be open country. Subject to this, the boundaries were chosen so as to be easily identified on the ground. Aerial photographs were used in some cases for this work.

The general highway and transportation maps showed with varying degrees of accuracy the location of farms and other dwellings in the open-country areas and to some extent in the smaller unincorporated places, and these were therefore used to demarcate the actual sampling units of the open-country stratum. The procedure was as follows. The numbers of farms and non-farm units were first counted in what are termed " count units " from the map. A count unit consists of a unit defined by minor civil boundaries or natural boundaries, and in general included from 6 to 30 farms. These count units were numbered, and the number of farms and the total number of dwellings including farms were marked on the map. The number of sampling units into which each count unit was to be subdivided was also decided and noted on the map ; in making this decision, consideration was given to the prevalence of natural boundaries, etc. The data for each count unit were then recorded on punched cards, and cumulative totals of the farm count, the dwelling count, and the number of sampling units, were tabulated. These cumulative totals were used to determine the count units which contained selected units, a random number between 1 and 18 being chosen as a starting-point, and the count unit containing every 18th sampling unit being selected thereafter. The count units containing selected sampling units were next subdivided on the map into the specified number of sampling units, the subdivisions being so chosen that they could be located on the ground. The units so demarcated were then numbered or counted systematically and the appropriate sampling units selected. Existing aerial photographs were used extensively for the demarcation of boundaries. In cases in which there were no suitable natural boundaries on the maps or photographs two or more units were amalgamated, subdivision and random selection being subsequently made on the ground if either unit was selected.

Somewhat different procedures, which need not be detailed here, were followed for the unincorporated and incorporated strata. For the incorporated stratum information was obtained from the Bureau of the Census on numbers of farms, etc.

In its final form such a master sample provides an adequate master sample of both farms and population, and also of the land area of the whole of the United States. Because the sampling units consist of areas, the frame will remain complete and adequate whatever changes occur in the course of time. The supplementary information provided by the number of farms and number of dwelling units will naturally become progressively more inaccurate, but major changes are likely to take place only in limited areas, and the master sample will in the course of its use reveal the extent of these inaccuracies. There will, therefore, be no difficulty in revising the sample when it appears necessary for those areas of the country where extensive changes have occurred, and this will in no way invalidate the existing sample for the rest of the country.

It will be seen that the construction of a master sample of this type is a major undertaking, and it should not be assumed that a master sample of the same type is necessarily expedient in other countries in which the conditions are different. Thus the United Kingdom large-scale Ordnance Survey maps provide an excellent frame for land area surveys, and the register of farms which is maintained for the collection of agricultural statistics provides a very complete frame of farms. If a master sample for agriculture is ever considered necessary, its construction could be based on this register and on the associated returns of farmers. The task of construction would therefore be very much simpler than would be the case if no such register existed. On the other hand, there is a need in the United Kingdom for an adequate master sample for localized population surveys. This problem is discussed in the next section.

4.18 Localized population surveys

As has already been indicated in Section 4.9, surveys are often required which will give reasonably accurate estimates for the country as a whole, but not for the separate small administrative districts. Such surveys have to be concentrated in a few localities, particularly if they are to be carried out by field investigators, since the amount of travelling would otherwise be excessive and supervision difficult. They may therefore be termed *localized* surveys. A multi-stage process must be used, the units at the first stage being administrative districts or similar areas of such size that each selected unit is capable of being covered by a single investigator or a small team of investigators.

The crux of the problem, therefore, consists of so planning the primary stage of the sampling process that the sampling error at this stage is not excessive. A secondary consideration, which must not be ignored, is that the within-strata comparisons should be sufficiently numerous to furnish a reasonable estimate of the sampling error at the first stage.

The use of stratification is obviously indicated. This stratification must in the first instance serve to differentiate between urban and rural areas. Consequently the country should be divided into large cities, into smaller urban areas, and into rural areas, in a manner somewhat similar to that employed in the master sample of the United States. The number of classes required will depend on the character of the towns and rural areas. A variable sampling fraction will be required in association with this stratification ; for most surveys it will probably be necessary to take all of the very large towns, but a proportion only of the intermediate towns, a smaller proportion of the smaller towns, and a still smaller proportion of the rural areas. Regional stratification of the smaller towns and rural areas may also be adopted as far as possible in parallel with this stratification.

These two types of stratification by themselves, however, are not likely to be entirely adequate for the urban areas, and some further form of stratification may be sought which will ensure (a) reasonably correct proportions of areas of different industrial types, and (b) reasonably correct proportions of the different social classes.

The methods by which it may be possible to ensure this will vary greatly according to the nature of the country, the type of primary unit that is adopted, and the amount of information that is available on these primary units. Administrative areas are usually most suitable from the point of view of the amount of readily available information, but they are not always ideal from the sampling point of view. As far as the United Kingdom is concerned, administrative areas appear to be the only possible type of area which can be used without a great deal of preliminary work. They will probably prove reasonably satisfactory if the boroughs and urban districts associated with the large towns are treated as parts of these towns, and sampled fairly intensively. Thus, for example, the sampling of the various parts of London and of its satellite suburban towns should be considered as a special problem separate from that of the sampling of the smaller towns in other parts of the country.

The second-stage sampling of the selected first-stage units is not likely to present any very serious problems. In the very large towns such as London, and in dispersed rural areas, two or more stages are likely to be required to avoid excessive travelling. Adjustment of the sampling fraction at the final stage to give equal overall sampling fractions is often advisable, since estimates can then be rapidly and simply obtained. Provision at the final stage for a proper rota of households to be included in the different samples, so as to avoid using the same household too frequently, is also of importance.

Much further research work remains to be done before it can be said with certainty whether a sample of this nature covering the United Kingdom is likely to be satisfactory for all purposes, or whether samples having a different structure will be required for different purposes. The importance of investigating the possibility of obtaining such a sample is clear. Without it, localized sociological and economic surveys of the general population cannot be carried out with any high, and at the same time ascertainable, degree of accuracy.

4.19   The U.S. series of employment estimates

An early example of a localized sample is that set up in the United States in 1939 to provide regular and speedy statistics on unemployment, employment and the labour force (Frankel and Stock, 1942, F). The sample was modified and improved in 1943 (Eckler, 1945a, F).

In the original sample, counties were used as the first-stage sampling units. All the 3097 counties of the United States were classified and sampled as follows :—

	Total No. of counties	Percentage of population	No. of counties in sample
Cities	9	14	9
Urban	447	50	28
Rural	2641	36	27
	3097	100	64

The 9 city counties relate to the 5 largest cities ; all these 9 counties were included in the sample. The urban counties are those with 1930 populations of 45,000 and over. In the urban and rural classes a triple stratification, each of three strata, was adopted, the bases of the three stratifications being population, administrative areas, and percentage unemployed. Divisions between each of the main strata were so chosen that approximately equal numbers of counties fell in each main stratum. There were thus 27 sub-strata for both the urban and rural classes.* One county was selected from each of these sub-strata at random, with one exception where two counties were selected.

A further two-stage process was used to sample the urban and rural areas within the selected counties. The numbers of households to be selected from the various urban and rural areas were allocated on the basis of the census population figures for these areas. This led to the gradual development of a differential bias between urban and rural areas, owing to a drift of population away from the rural areas.

The results from within a single county were aggregated without any weighting. The aggregates were then weighted according to the population of the stratum from which they were obtained.

The within-county sample was changed every 4-6 months. This was a compromise between having a constant group of households, which would give most accurate estimates of monthly changes, and having new households on each occasion so as to avoid repeated visits to the same household. It introduced a certain discontinuity into the results, which has been avoided in the modified sample by using a proper system of partial replacement of the type described in Section 3.19.

In the modified sample, which included 68 first-stage units, allocation of households on the basis of population figures was abandoned. Instead, small areas were used as units at the second stage. This eliminates bias resulting from population drift.

Several other features were also introduced in order to improve the accuracy of the results. The stratification was more detailed, selection of the primary units was with probability proportional to their populations, and the ratios between the numbers of households having certain contrasting characteristics, e.g. farm and non-farm, were adjusted in each selected first-stage unit to agree with the corresponding ratios in the stratum to which the unit belonged. This last procedure is not entirely free from danger of bias. In a unit with a relatively small proportion of farm households, for example, those that do occur may be expected to be somewhat abnormal owing to proximity to non-farm areas, and such abnormal households will consequently receive excessive weight in the final results.

In both these samples only a single unit was selected from each of the first-stage strata. Although this unquestionably increases accuracy by permitting the use of smaller strata, it has the consequence that no fully valid estimate of error is available. The best estimate is that obtained by combining the strata in pairs, and this is likely to be somewhat of an overestimate.

---

* It will be noted that the numbers of the counties in the different sub-strata were not by any means equal.

4.20   Frames suitable for special classes of a human population

Surveys of special classes taken from the whole of a human population are often required.   If a general frame covering the whole population is available, it can be used for a survey of a special class by selecting a sample from the whole population, and rejecting those members which do not fall in the required class.   If the frame itself does not contain the necessary information, this will necessitate surveying all units of the sample in order to find out which individuals are to be retained and which rejected.   If the required class is only a small fraction of the whole population there will be a large proportion of rejects, and a disproportionate amount of work is therefore required in such cases.

Consequently, if a frame covering only the required class or classes is available, this should be used in preference to a general frame.   In surveys of the labour force, for example, it is often possible to utilize unemployment insurance registers and similar records.   Such frames are often to a certain extent inadequate—all types of labour may not be included in an unemployment insurance scheme, for example—but their greater convenience frequently outweighs their defects.   Occasionally it may be considered advisable to cover the excluded classes with lower accuracy by means of a general frame.

When no partial frame exists a survey undertaken for another purpose can sometimes be used to provide one.   Thus in a recent survey of the aged carried out by the Nuffield Trust in certain towns of the United Kingdom, the records of an earlier survey by the Social Survey of the Central Office of Information covering all households were used to locate those households which contained aged people.

4.21   Frames suitable for the survey of economic institutions

Surveys of economic institutions may be divided into two general classes : those covering the whole or a large part of the commercial and industrial undertakings of a given town or country, and those covering a single type of undertaking or industry.

In surveys of the former class the use of frames constructed from maps or plans is often feasible : thus, in a general survey of factories of a given town a town plan may conveniently be used, with area sampling from this plan. Since, however, most commercial and industrial undertakings vary greatly in size, a variable sampling fraction is often required, a larger proportion of the large undertakings being selected.   A map does not provide a suitable frame for this purpose.   Even for general surveys, therefore, it is often advisable to use a special frame for the large undertakings, excluding these from the area sample, which is used to cover only the smaller undertakings.   In a survey of the factories of a town, for example, there is usually no great difficulty in drawing up a list of the larger factories.   If necessary a preliminary ground survey, with or without the aid of detailed maps, can be made.

If a particular type of undertaking or industry requires to be surveyed, the use of any form of area sampling will generally be unsatisfactory unless the units are small and widely dispersed, as, for example, occurs with retail shops. Even here shopping centres will require differentiation from other areas. In other cases a list of all the units of the given type of undertaking or industry will form a very much more suitable frame. In order that a variable sample fraction may be used it is important that such a list should contain some indication of the size of the units. If no satisfactory frame of this type exists it will often be worth while carrying out a complete census, simply for the purpose of constructing a frame and collecting a few basic facts about the given type of undertaking or industry. Such a census is usually more effective if it is on a compulsory basis.

When a frame provided by a complete census is required for repeated surveys, the problem of keeping it up-to-date must be considered. This is usually best effected by keeping a register of the undertakings concerned, and making a regulation that requires all changes to be reported. For the purposes of sampling, the most important type of change which requires to be recorded is that of new entries. Failure to report other changes will merely result in inaccuracies in the frame.

4.22 Market research and opinion surveys

Market research includes not only investigation into consumer reactions to goods and services and to advertising campaigns, but also investigations into consumer needs. In the case of consumer reactions information is mainly required on opinions, while in investigations of consumer needs factual information will also be required. Market research surveys can therefore be carried out in the same manner as sociological surveys of the questionnaire type, using the methods which have already been described.

This is also true of other surveys of public opinion, but in many opinion surveys and also in certain types of investigation into consumer reactions, the requirements are somewhat different from those for sociological investigations. Speed is often essential, and changes in the percentage of individuals holding a given opinion are frequently of more interest than the absolute value of the percentage holding that opinion at any one time.

To meet these requirements, and to reduce costs to a minimum, what is known as the *quota method* of sampling has been developed. This method is a variant of purposive selection. Interviewers are given definite quotas of people in different social classes, of different age-groups, etc., and are instructed to obtain the requisite number of interviews in each quota. Additional instructions, which are designed to prevent excessively unrepresentative selection within the allotted quotas, may also be given on mode of contact, etc. The interviews themselves are sometimes carried out by house-to-house visits, sometimes by interviewing people in the streets and other public places, and occasionally even by telephone.

It is clear that, however accurately the quotas are fulfilled, such samples cannot be regarded as the equivalent of random samples. Consequently the danger of bias is always present, and the quota method must therefore be ruled out as a suitable method of investigation for precise enquiries in which unbiased results are required. Moreover, if there is a change in conditions, a quota sample which has previously adequately reproduced the characteristics of the population may cease to do so. Consequently, the fact that a quota system has consistently given reliable results over a period of years is no guarantee that it will also do so in the future.

The striking failure of public opinion polls to predict the results of the 1948 American Presidential Election—in contrast with previous successes— has been explained as being due to a shift of opinion during the closing weeks of the campaign, but it is possible that this failure was in part attributable to defects in the sampling procedure. It is generally agreed that the votes of organized labour influenced the outcome of the election to a much greater extent than usual, and it may well be that, in spite of the quota system, the samples were very deficient in factory workers and other trade union labour. The mere fact that a quota system is designed to give the correct proportion of workers, or even of different classes of workers, does not necessarily ensure that those included are representative, as regards the way they vote, of the workers as a whole. Consequently the results may be biased in elections in which the different types of worker vote very differently.

On the other hand, if used with skill the quota method may give sufficiently accurate results in simple enquiries where only general indications of the opinions held are required. If the samples are taken in the same manner on different occasions, and circumstances remain broadly the same, it may also provide a not-too-inaccurate measure of changes of opinion.

Apart from the problem of obtaining a representative sample, there is the inherent difficulty in opinion surveys that an individual's opinion on a given subject is frequently both ill-defined and liable to change. Moreover, on certain subjects the respondents may be unwilling to voice their true opinions. Opinions are also held with very different degrees of intensity, which there is no easy way of measuring. Much of the information provided by public opinion polls is therefore of doubtful significance.

4.23 Frames for agricultural censuses and surveys

Agricultural censuses and surveys can be carried out in collaboration with the farmers, or in certain circumstances by direct observation without contacting the farmers. The latter method is in general only applicable to surveys of agricultural crops, and then only if all particulars required are ascertainable by inspection. For censuses and surveys of livestock the collaboration of the farmer is usually necessary, the essential difference being that livestock is mobile whereas crops are immobile. Collaboration is also obviously required if information relating to the farm as a whole is needed. In many countries contact with the farmer is advisable even for crop surveys, because exception may well be taken to the examination of a crop without the farmer's permission.

If a census or survey is to be conducted by contacting the farmer, the farm will usually form the sampling unit at some stage of the sampling process. A frame covering farms will therefore be required. Such frames are provided either by lists of farms, or by some form of area sampling which serves to locate the farmhouses. Frames based on maps, etc., which are suitable for the sampling of human populations in rural areas (Section 4.14) are equally suitable for the sampling of farms.

If contact with the farmer is not necessary maps can be used directly as a frame for crop surveys. Their use for this purpose is discussed in the next section. Even in this case, however, farms may well provide the best available frame.

In crop surveys the natural unit for many purposes is the field and not the farm. In cases where it appears advisable to obtain information for some only of the fields of a farm under a given crop, a further stage will have to be introduced into the sampling process. This inevitably results in a somewhat complicated sampling structure with different sampling fractions for the different parts of the sample, which in turn introduces complications into the analysis of the results, at least if unbiased estimates are required.

An example of this type of survey is provided by the Survey of Fertilizer Practice, carried out in various counties of England and Wales from 1942 onwards (Yates *et al.*, 1944, G). The objects of this survey are to determine the way in which farmers manure the different crops, and the relation of this manurial practice to the fertilizer requirements of the soil, in so far as these can be determined by the current methods of chemical soil analysis.

The method of selecting the samples is as follows. For each county a systematic sample of farms is selected from the Ministry of Agriculture's addressograph list, maintained for the purpose of collecting the agricultural statistics on crop acreages and livestock. This list is arranged alphabetically by farmer's name and parish, and shows the total acreages (crops and grass) of each farm. A variable sampling fraction is used, with three size-groups, about 100 farms being selected from an average county. Larger samples are taken from counties which can be subdivided into districts containing different types of farming.

Each selected farm is visited by a field investigator, who is a member of the Provincial Advisory Staff. All the fields of the farm are listed in consultation with the farmer according to their crops, and also according to whether they have been recently ploughed out from grass (*new* and *old* arable). In the earlier surveys one field of each crop was selected at random from all the old-arable fields, and similarly for all the new-arable fields. One permanent grass field was also selected. In the later surveys one field in three of each crop has been selected from each of these categories. From each group of selected fields one old-arable, one new-arable and one permanent grass field is selected at random, and soil samples taken for chemical analysis.

For the selected fields information is obtained from the farmer on the cropping over the previous four years, and the amounts and chemical composition of the fertilizers, farmyard manure and lime applied in each year of this period. In some of the later surveys only a single year has been covered. When necessary the fertilizer merchants are consulted in order to obtain information on the chemical composition of the fertilizers.*

The methods of analysis adopted in this survey are illustrated in Example 6.19.

4.24   Use of maps as frames in agricultural surveys

If accurate large-scale maps showing the field boundaries are available, the point method of sampling is very suitable for crop surveys in which contact with the farmer is not necessary. The fields will then act as sampling units, and selection will be with probability proportional to size. Provided the whole of a selected field is under a single crop, all that is necessary for acreage estimates is to ascertain the crop, no determination of area being required (Section 3.9). If more than one crop is being grown on a selected field, the proportions of the area under the different crops must be determined, but eye estimates will usually be adequate for this purpose.

In this type of work two-stage sampling will often be advisable in order to save travelling, and also to avoid having to handle an excessive number of maps. Thus in the United Kingdom the 6-inch Ordnance Survey quarter-sheets (3 miles × 2 miles) might provide suitable first-stage units, a fairly dense grid of points being taken over the selected sheets.

If selection with equal probability of irregularly-shaped areas such as fields is required, these areas must each be defined by a single point, such as the most northerly point of the area. The map is then divided into sampling units consisting of rectangular areas, a number of which are selected with equal probability, the fields with defining points in the selected areas being included ʒin the sample. Only the selected rectangular areas need be demarcated. If more convenient, circular areas whose centres are located at random (or systematically) may be used in preference to rectangular areas. Rectangular areas have the formal advantage that the whole of the area is included once and once only in the aggregate of sampling units, but this is not of great practical importance.

It should also be noted that with this method of selection the sampling units consist of groups of fields whose defining points are included in a single rectangular area, and not the fields themselves. Rectangular areas which contain no defining points must be counted as units of zero area. It can be shown that when the rectangular areas are small and mostly contain one or no defining point, the sampling errors of estimates of crop acreages are greater

---

* The survey still continues, and now covers the whole of England and Wales on an annual basis. Minor changes in the sampling design have been made from time to time. For recent accounts see Church and Webber (1971) and Church and Lewis (1977).

with this method of sampling than with the point method. The point method is therefore preferable under these circumstances.

On the other hand, if the rectangular areas are large relative to the sampled fields—as will be the case, for example, if whole sheets of a map are surveyed—the use of defining points in this manner saves splitting fields which are cut by the map boundaries. Some slight additional variance will be introduced unless the total area of all fields is determined and used as supplementary information.

Maps have been extensively used as frames for the estimation of the acreages of crops in surveys conducted by the Calcutta Institute of Statistics (Mahalanobis, 1944, A ; 1946, A ; 1940, H ; 1945, H ; 1946, H). The method followed is to demarcate square areas located at random on the maps, and to survey all fields covered in whole or in part by these areas. The areas of the whole and part fields are determined from the maps by measurement. This measurement of areas and their subsequent summation might be avoided by the use of point sampling : if each square area were replaced by a square pattern of 9 or 16 points, for example, it would appear that the loss of accuracy would be small (see Example 8.10.c).

4.25 The 1942 Census of Woodlands

The 1942 Census of Woodlands covering England and Wales provides an example of the use of maps as a frame. The object of the survey was primarily to determine the volumes of standing timber of various types in the country, and their broad regional location, in order to estimate the amount of available home-grown timber and to plan its utilization.

The sample was initially planned to be taken in two parts, each consisting of 5 per cent. of the total land area. The sampling units were 6-inch Ordnance Survey quarter-sheets (3×2 miles), systematically located on two inter-penetrating 12×10-mile rectangular grids, one for each part. All areas of woodland of over 5 acres on the selected quarter-sheets, and 1 in 5 of the areas under 5 acres, were surveyed. Areas of woodland cut by the boundary of the map were surveyed if their southernmost point was included in the selected map, areas subdivided by rides marked on the map being treated as separate areas for this purpose. The land areas covered by the selected maps were also inspected in the course of the survey to determine any new plantings since the map was last revised, thus correcting for any incompleteness in the frame.

The woodland areas were divided by inspection on the ground into " stands," each of which represented a homogeneous area of woodland. The boundaries of these stands were demarcated on the maps so that their areas could be determined, and a representative plot was chosen from each stand on which all or a sample of the trees were measured. Representative plots were used instead of random plots because reasonably accurate volume figures for the individual stands were required. Control of the bias introduced by the use of representative plots was effected by determining the quantities of converted timber actually obtained from surveyed stands felled in the course of ordinary forestry operations.   This procedure served also as a check against any errors in the assumed wastages on conversion.   A further control by the measurement of randomly selected plots on a sub-sample of stands was also planned, but was not in fact carried out.

The total area of woodland was determined independently from the areas coloured green on the 1-inch Ordnance Survey sheets (see Example 7.18). Errors in the 1-inch sheets were allowed for by comparing the selected 6-inch sheets with the 1-inch sheets after survey.   The final estimates of volume were calculated from the volumes per acre determined from the surveyed areas and the total areas determined as above.

A first estimate of volumes was required within six months of the decision to undertake the survey, and it was thought that with the teams available the first part of the survey could be completed and the estimates prepared within this time.   Before field work commenced, however, it became apparent that the original programme could not be adhered to.   Each selected quarter-sheet was therefore roughly divided into two halves as similar as possible, and one half of each sheet was selected at random for the first part of the survey, giving a $2\frac{1}{2}$ per cent. sample of all woodlands in the country.   Subsequent calculations of the sampling errors showed that this $2\frac{1}{2}$ per cent. sample was quite adequate for the determination of general policy, which was the first objective of the survey.

The survey was then completed in two further parts, first the remaining halves of the first set of quarter-sheets, and secondly the other set of quarter-sheets.   The three parts therefore consisted of $2\frac{1}{2}$ per cent., $2\frac{1}{2}$ per cent. and 5 per cent. respectively of all the woodlands of the country.   In addition certain heavily wooded areas were completely surveyed.

The history of this survey demonstrates the extreme flexibility of sampling surveys, and the way in which they can be made to yield preliminary results of ascertainable reliability.   By the procedure of first surveying a properly selected quarter of the whole sample, it was possible to obtain the preliminary estimates in the required time in spite of unexpected delays in the commencement of the survey.

4.26   Frames for undeveloped areas

If no accurate maps are available, exact location of previously demarcated small sample areas on the ground will be impossible.   Alternative methods must therefore be employed.

For completely undeveloped areas such as natural forests the line method of sampling is very suitable, provided the terrain and vegetation is such that the lines can be followed on given compass bearings without an undue amount of deviation.   Distances along the lines can be determined by some simple measuring device such as a rope, or even by pacing.   Where volume measurements are required small areas can be demarcated at given distances along each line.

Some frame for the location of the lines is necessary. This can often be provided by existing mapped roads or other tracks, but it is by no means impossible to construct a secondary frame as the survey proceeds by the use of cross traverses, using any available tie-in points. Except where maps are to be constructed, no great accuracy in the location of the lines is required, since it is only necessary that they be located in an unbiased manner with a density which is the same for the different parts of the area, or, if not the same, is determinable.

In areas in which a line on a fixed bearing cannot be followed, any attempt at complete and unbiased coverage must necessarily be very expensive. Often, however, a sufficiently unbiased sample of natural vegetation will be obtained by traversing existing tracks and taking sample areas at suitable intervals by offsets at right angles to the tracks. If a map of these tracks is not previously available it may be worth constructing one by rope and sound or similar rough surveying technique.

Crop surveys in partially developed areas without adequate maps present somewhat different problems. If the cultivated areas are located in the neighbourhood of villages, a two-stage sampling process will probably be required, a sample of villages being taken at the first stage. Since the total area of cultivated land associated with a village is likely to be closely correlated with the population figures, these (if known) should be treated as supplementary information. If not known the feasibility of making a simultaneous population census should be considered, since information on cultivated areas will be of more value if it can be related to population figures. In this case the sampling may well be two-phase, a larger sample being taken for the determination of population.

The survey of the cultivated areas associated with the selected villages will require the construction of second-stage frames. If the line or point method of sampling is practicable this is likely to be the simplest method of dealing with compact areas of cultivation. Outlying fields will in this case have to be enumerated and sampled separately.

In many cases enumeration of all fields will be the only practicable method. The preparation of a sketch map will then be advisable. A certain percentage of the enumerated fields can be measured for area, and the crops determined if this has not been possible at the mapping stage. If the cropping is known, stratification by crop should be made before the selection of the sample for area measurements. A frame of this kind may remain serviceable, with some revision, over a number of years. It will also serve to locate the samples required in a crop estimation scheme.

4.27 Use of aerial survey photographs

When no maps are available the possibility of using aerial survey photographs as a frame for agricultural and land utilization surveys should be borne in mind. Although it is unlikely to be practicable to make an aerial survey simply for the purpose of providing a frame for sample surveys, it is often possible to utilize a survey that has been undertaken or is contemplated for other purposes.

Any aerial photographs covering the area are likely to provide an adequate frame, though the use of aerial survey photographs even for a frame is not as simple as it appears at first sight. The mere handling of the photographs covering any large area is a somewhat difficult task which demands an adequate and properly trained office staff. Moreover, aerial photographs are subject to variations of scale (and also distortion) due to tilt and changes of altitude of the aircraft. The stated scale is therefore not always correct, and the scale sometimes exhibits disconcerting variations even over different parts of the same mosaic. The precaution should therefore be taken of checking the scale by means of measurements on the ground in a sufficient number of instances to make certain that no important source of error is introduced.

Various methods of sampling can be used in conjunction with aerial photographs. If crop acreages have to be determined, point sampling is suitable. After the points have been marked on the photographs the fields in which these points fall must be identified on the ground and the crops growing on them recorded. In order to avoid excessive travelling it will almost certainly be worth using a two-stage process, the units at the first stage being rectangular areas which can be demarcated on the photographs, with a number of points taken within each of the selected areas.

If line sampling is required, the lines can first be demarcated on the photographs, and subsequently surveyed on the ground. In certain circumstances it may be possible to make the intercept measurements on the photographs, using the ground survey merely to determine the characteristics of the various intercepts.

If areas such as fields, the boundaries of which are recognizable on the photographs, are to constitute the sampling units, they may be selected with probabilities proportional to their sizes by point sampling. If there are likely to be ambiguities in the definition of the boundaries the units should be demarcated before selection.

If natural units such as farmhouses which depend on point locations are to be selected, small rectangular or circular areas may be used as sampling units in the same manner as in selection from a map.

In certain cases aerial photographs may provide the necessary information without any ground survey work. It is usually possible, for example, to recognize cultivated areas on the photographs, and the total cultivated area may consequently be determined directly from the photographs. In certain cases it may even be possible to differentiate between the different crops. In these cases the total cultivated area and the proportions of the area under the different crops can be determined by sampling of the photographs, point or line sampling being used as convenient. If desired, adjustments for variations in scale can be made by varying the spacing of the points or lines.

In some cases the differentiation between the different crops on the photographs may be only partial, or subject to error. In such cases a sub-sample of the points classified on the photographs can be re-classified by ground survey. The information provided by the photographic classification will then serve as supplementary information. By this procedure the amount of ground survey necessary may be very considerably reduced. The examination of stereo-pairs may be a considerable aid to the classification of certain types of area, particularly forest areas.

If an aerial survey is specially undertaken for the purpose of a sample census or survey, it is possible to reduce the amount of photography by taking parallel strips of photographs separated by unphotographed areas, but aerial photographs taken in this manner will not be of much use for mapping purposes. If no map frame is available, a few cross-strips will have to be taken to provide links between the separate strips. Too much reliance must not be placed on the accuracy of the location of the strips unless special navigational aids are installed.

4.28 Crop estimation

The total yield of a crop can be regarded as the product of its acreage and the mean yield per acre. These two quantities may therefore be estimated separately. Estimates and forecasts of the mean yield per acre must of course be related to the conventions adopted in the estimation of acreage, particularly with regard to areas on which the crop has failed or been abandoned.

The estimation of acreage has already been discussed in the preceding sections, and in this section we shall therefore mainly be concerned with the problem of the estimation of the mean yield per acre.

There are a number of ways in which estimates of the mean yield per acre of a crop, or the total yield, may be obtained. These may be broadly classified as follows :—

(1) Reports from crop-reporters, who, at or subsequent to harvest, make returns to a central authority of their estimates of the average yields of the crop in their own districts, these estimates being based in the main on general impressions, discussions with farmers, etc.

(2) The harvesting of small sample areas of the crop immediately prior to the main harvest.

(3) Eye estimates of the yields of a sample of fields, with subsequent calibration of these eye estimates by comparison with the actual yields of some at least of the sample fields.

(4) Co-operation with the farmers at harvest time so that accurate yield figures may be obtained from a sample of fields as they are harvested.

(5) Returns by farmers of the yields of their crops.

(6) Market returns, export statistics, etc.

If necessary, Methods (2) and (3) may be combined in a two-phase sampling scheme, eye estimates being taken from a comparatively large sample of fields, with crop-cutting samples from a smaller sub-sample of these fields.

These various methods all have their advantages and disadvantages. Method (1), that of crop-reporters, is the one commonly adopted by countries with long-established and stable systems of agriculture. Its success depends on the ability of the individual crop-reporters to make accurate and unbiased estimates of the average yields of their districts. The method is not objective, and no assessment of its accuracy can be made unless independent estimates, provided by some other method of known or ascertainable accuracy, are available for comparison. Doubt is often cast on estimates provided by the method because of disagreement with market returns, etc., and their lack of objectivity makes it impossible to say which set of estimates is at fault.

Even if crop-reporters are reasonably accurate on the average over a run of years, estimates for particular years or particular districts may be subject to considerable errors. There seems to be a general tendency, for instance, to underestimate yields in good years and overestimate them in bad years. The accuracy attained may also be very different for the different crops. Moreover, spurious long-term trends may be introduced through gradual changes in the standards of the reporters, and this considerably reduces the value of the estimates as a measure of the improvement or deterioration of the agriculture of a country. Any sudden change in an agricultural system, such as the introduction of new varieties, or the bringing into cultivation of new land, may introduce disturbances into previously satisfactory estimates.

Method (2), the harvesting of small sample areas, is theoretically capable of providing a completely objective estimate of the mean yield per acre of the standing crop at harvest time ; it will not, of itself, provide any estimate of the losses at or subsequent to harvest. In practice, however, serious bias may arise in a number of ways if proper precautions are not taken. These sources of bias, and the practical details of the method, are discussed further in the next section.

Method (3), that of eye estimates, has the advantage that on certain types of crop such estimates can be relatively rapidly made, and consequently a larger sample of fields can be visited in a given time. The difficulty of having to transport and thresh a large number of samples, which often arises with Method (2) is also avoided. Some results of a trial of this method on wheat are given in Example 6.15. The method is not suitable for root crops such as sugar beet and potatoes, since it is difficult to judge the yields from inspection of the tops. In such crops, however, there are no transport and threshing problems, since the samples can be weighed in the field.

If calibration of the eye estimates from the farmers' yields is not practicable, or if the calibration is found to vary substantially from year to year, a few specially-trained field workers can be used to take crop-cutting samples in order to calibrate the eye estimates of each investigator at the time of harvest.

Methods (4) and (5) require the co-operation of the farmer. Method (4) differs from Method (5) in that in Method (4) the harvesting is done in the presence of an investigator, and if necessary with assistance, such as the provision of a threshing machine, whereas in Method (5) reliance is placed entirely on the farmer to provide accurate yield figures. Owing to delays of threshing, etc., Method (5) is not likely to provide estimates till some time after harvest.

Estimates from market returns, export statistics, etc. (Method 6) provide a useful basis for comparison with estimates by other methods, but such returns will only exceptionally give an accurate estimate of the actual yields, since the amount of the crop passing through the market is likely to vary very considerably in different circumstances.

In Methods (2) and (3), which require field investigators, the question must be considered whether the survey should cover the whole of the country or whether it should be confined to certain districts only, using a two-stage sampling process. If only an estimate of the yield of the whole country or of large districts is required, comparatively few fields will need to be sampled, and a single-stage process for the selection of fields will result in a very dispersed sample. A two-stage process will avoid this difficulty at the cost of introducing a between-districts component of variation into the sampling error.

Finally, it should be emphasized that crop estimation, though theoretically simple, presents many practical difficulties. The introduction of a satisfactory scheme where none exists, or the provision of objective estimates to check existing subjective estimates, requires continuous work over a number of years by a properly established team of workers. Except for preliminary investigations, crop-estimation projects should therefore not be undertaken unless continuity can be maintained. Nor should an existing method of estimation be abandoned or disturbed until a better alternative has been evolved and kept in operation for some time. If the old and new methods are run in parallel for a number of years it will be possible to assess the reliability of the old method and its degree of bias, a task which will be quite impossible if it is discontinued before adequate comparative data have been obtained.

4.29 Estimation of yield by the harvesting of sample areas

As mentioned in the previous section, the estimation of the mean yield per acre of an agricultural crop by the harvesting of small sample areas presents many practical difficulties, and the results may be biased in a number of ways if the proper precautions are not taken.

Errors can occur through faulty selection of the fields, through faulty sampling of the selected fields, through failure to take samples from the fields at dates sufficiently near harvest, or through failure to sample some of the selected fields owing to their having been harvested before they were visited.

If rigorous means of selection are employed there is no reason why the selection of the fields should be faulty. If, however, the cruise method is used, fields being taken at equal distances along a given route in the manner described in Section 3.15, the estimate will almost certainly be appreciably biased, though this bias may be reasonably constant from year to year if the same route is followed each year. On the other hand, the use of the cruise method overcomes the difficulty of ensuring that the visits to the fields are made sufficiently near harvest, and also eliminates the risk of missing fields through their having already been harvested. All that is necessary is to traverse the route at sufficiently close intervals of time, stopping the car at each sample point and examining the crop to see if it has reached a sufficiently mature stage for a sample to be taken.

An alternative procedure which is sometimes used is to follow the prescribed route and take a sample from all or a given fraction of the fields that are actually being harvested. This, however, may introduce an additional component of bias, since, unless special precautions are taken, limitations of time will result in the inclusion of a greater proportion of the fields which are harvested very early or very late.

With crops that do not have to be fully mature at harvest, e.g. potatoes, samples will normally have to be taken somewhat before maturity, unless information is available from the farmers as to when they intend to lift. With such crops, however, it is usually possible to estimate the amount of growth between the time of taking the sample and date of harvest ; this latter date can then be determined by a subsequent visit. In the potato crop, for example, investigation has shown that the weight of tops provides a fair indication of the amount of further growth that may be expected.

The cruise method of sampling, therefore, provides a method of crop estimation which, though theoretically more liable to bias than a proper random selection of fields, may in practice give more satisfactory results, particularly in the estimation of yields per acre. It is also likely to be considerably more economical in travel. Which method is most suitable will depend largely on local conditions, and must be the subject of local investigation.

Bias in the estimation of the yields of the actual fields can arise from improper location of the samples and from cutting a larger area of the crop than the true unit area. An example of such bias has already been given in Section 2.5.

Edge effects are also liable to give rise to bias, since in an irregularly shaped field it is impossible without a great deal of labour to locate samples properly at random over the whole of the area. The method described in Example 3.2.b is clearly impracticable, and no simple method of traversing the field has been devised which will give equal probability of selection over the whole field. In practice, however, a systematic method of selecting the sample is quite adequate. The important thing is to see that the location of the sample units is as objective as possible.

The determination of the bias arising from headlands, lower yields at the

94

edges of the field, and errors in estimation of the area of the field—the U.K. Ordnance Survey areas, for example, include farm roads, hedges and ditches— can be made if required by more rigorous supplementary observations on a small percentage of the fields. Often, however, the separate determination of these components of bias is of no great practical interest, since the losses at and after harvest will also affect the total amount of the crop that is finally available for consumption, and the total bias is best determined by comparison with the farmers' reported yields on a sub-sample of the fields, or by determining these yields in co-operation with the farmer.

Two methods of locating the sample units have been found convenient in practice in this country. The first is to traverse the field diagonally from corner to corner, using one or both diagonals, and locating samples at equal intervals along these diagonal lines. The interval required can be calculated by pacing the diagonal or making an eye estimate of the number of paces. Errors in the eye estimates are of little consequence, since the exact number of sampling units is immaterial. Alternatively, if the crop is in rows the field can be traversed along the rows. The length of one end is paced from corner A to corner B, the field being entered at a distance of one-quarter of this length from corner B. It is then traversed along this row to the other end of the field, and a return row is selected by the same procedure. A suitable number of sampling units is taken at each traverse in the same manner as in the case of a diagonal traverse. In this method of sampling it is advisable to step laterally across a given small number of rows after each sampling unit has been taken, since a given row may fall wholly on a particularly good or bad strip of the field, e.g. on a ploughman's " land."

Whatever the exact method of traversing the field, it is of the utmost importance that the location of the rows and of the sampling units should be made without inspection of the crop in the neighbourhood. This can be done quite effectively by counting paces and digging in the heel when the requisite number of paces have been taken, but the field workers must be thoroughly trained in this procedure.

A good deal of work has been done on the most suitable size and shape of the sampling units. In this country experimental tests have shown that 4–6 contiguous quarter-metre row lengths are suitable for cereal crops, with 6–10 units per field ; for potatoes 4 units each of 6 ft. of row have been tested on a large scale. Mahalanobis (1946, A), working in India, used three or four concentric circles of 2–8 ft. radius, each annular ring being harvested separately so as to provide a control of field workers—with ordinary workers a bias is regularly found with the smallest circle—and this gives a check that the samples have really been taken in the prescribed manner.

The best size and shape of the sampling units depends very much on the nature of the crop and local conditions, such as type of field worker, whether the crop is sown or planted in rows or broadcast, variability within fields, available equipment for threshing and transport, etc. Local investigation should therefore always be undertaken if any extensive work is contemplated. On the other hand, it should be recognized that the sampling error of individual fields is usually small relative to the variation from field to field, and consequently the introduction of a crop-estimation scheme need not await the results of such investigation ; any reasonably efficient method will give satisfactory results provided bias is avoided. (See Section 8.12.)

4.30 Crop forecasting

The term *forecast* is here used to denote an estimate of the yield of a crop furnished at some date well before harvest. The term is sometimes used to indicate estimates made by crop reporters at or even shortly after harvest, since such estimates are usually subject to later revision in the light of information received from farmers. Such estimates, however, are better termed *preliminary* estimates, in contrast to the revised or *final* estimates.

There is some confusion also between forecasts and estimates of acreages, forecasts of mean yields per acre, and forecasts of total yields. Once the crop is sown the determination of the acreage, apart from crop failures, is a matter of estimation and not of forecasting, and any forecast of the total yield is usually best presented in the form of an estimate of the total acreage and a forecast of the mean yield per acre.

There are three main methods of crop forecasting. Forecasts can be provided by crop reporters, they can be based on meteorological data such as rainfall obtained prior to the date of the forecast, or they can be based on observations and physical measurements of the growing crop, alone or in conjunction with meteorological data.

Meteorological data do not directly provide forecasts of the yields. If they are to be used as a basis of such forecasts, reliable data on both the yields and the meteorological events must be collected over a number of years, and the crop-weather relations evaluated. The same is true if observations and measurements on the growing crop are to be used. The evaluation of these relations requires the application of the method of statistical analysis known as " regression analysis." We shall not describe this method here, but it may be well to emphasize that its application is not entirely simple, and the advice of a mathematical statistician experienced in this type of work should therefore be sought.

It must not be assumed that it will be possible to evolve a prediction formula which will give satisfactory results, even if accurate and extensive data are available. In the first place, the yield of a given crop is influenced by meteorological and other events up to and sometimes after harvest, and this may introduce too great a degree of uncertainty into yields predicted some months before harvest to make the prediction of any value. In the second place, although meteorological factors undoubtedly account for a good deal of the variation in crop yields, they are not by any means the only factors. Changes in variety, insect pests, plant diseases, exhaustion of the fertility of the soil, changes in the type of land under crop, changes in the amount of fertilizers, and many other factors may also exert a major influence. Thirdly, meteorological effects are often somewhat complex, and it may therefore be impossible to determine them from a set of data extending over a limited number of years ; owing to the similarity of weather conditions over large areas, data from a number of districts in any one year are only a partial substitute for data extending over a number of years.

One advantage of using measurements of crop growth instead of relying wholly on meteorological observations is that the crop is thereby used as its own integrator of meteorological and other effects up to the time of the measurements. Frost and flood damage, for instance, are clearly better assessed, once they have occurred, by survey of the crop than by examination of meteorological records. The selection of the particular types of observations and measurements which are likely to give an adequate basis for forecasting is a problem on which further scientific research is required, particularly in the case of grain and other seed crops. In the case of root crops investigation has already shown that the amount of growth made by the tubers or roots, coupled with some measure of the extent to which the plant is still growing, e.g. weight of tops, are likely to give satisfactory results.

Since the evolution of a satisfactory method of crop forecasting demands a knowledge of the actual yields over a period of years, an investigation of suitable methods can be combined with an objective crop-estimation scheme. Once the observations and physical measurements which are likely to give useful information have been decided, all that is necessary is to take these measurements on a sub-sample of the fields which will subsequently be selected for sampling at harvest. In the initial stages it may be better to carry out the observations on a special sample of fields, rather than on the more scattered sample which will be suitable for crop estimation proper. More intensive investigations can also be carried out on experimental plots on which different varieties are sown, and which are subject to different cultural treatments and sowing dates. Experimental plots by themselves, however, are not likely to provide all the information required for the evolution of a suitable forecasting scheme, since the variation from field to field in a given district is often quite large, and the inclusion of a number of fields in the district usually gives a much more adequate representation of the average meteorological effects in that district than will a single field.

If observations and measurements are to be made on the growing crop, a sampling scheme will have to be devised in order that single plants or small areas may be selected for measurement. A method of selecting wheat shoots for height measurements, for example, is described in Section 2.4. The principles to be followed in the location of the sampling units in the fields or experimental plots are similar to those which operate in the selection of samples for yield estimates.

4.31 Censuses of road traffic

Statistics for road traffic, such as total vehicle-miles, passenger-miles and ton-miles, can be obtained either from returns made by vehicle operators or from counts and other observations of vehicles passing selected points of the road network. Statistics of tons loaded and length of journey can only be simply obtained from returns by vehicle operators. The sampling problems of the former type of census are straightforward, and need not be further discussed here, but those arising in road traffic counts present a number of special features which are of general interest.

In addition to their use in estimating the volume of road traffic, traffic counts are of value in road planning. For this purpose counting points can best be located at strategic points in the road network, so as to estimate the volume of traffic on particular stretches of road. The counts themselves may be confined to a particular week chosen as typical, or possibly to two or three such weeks at different times of the year. Nor is there any need to cover periods of the day for which the traffic is known to be light. Classification into types of vehicle will frequently be required, but not information on loads carried. Consequently, unless information on the origin and destination is required it will not be necessary to stop vehicles. If automatic counting devices are installed the whole period covered may be sampled for purposes of classification of vehicles.

For the purpose of estimating the total volume of road traffic a different procedure is necessary. Unbiased estimates of quantities such as passenger-miles, vehicle-miles and ton-miles, cannot be obtained if the counting points are located at strategic points in the network. Instead some method of random or systematic location is necessary.*

If points are located at random on the roads of a network with a density of 1 per k miles and from counts made at these points it is found that $n_1, n_2 \ldots$ vehicles pass the points in a given period, then an unbiased estimate of the total vehicle-miles in the period is given by

$$\text{vehicle-miles} = k \, (n_1 + n_2 + \ldots) = k \, S \, (n)$$

Similarly an unbiased estimate of the ton-miles is $k \, S \, (W)$, where W is the total of the loads of all the vehicles passing a given point in the given period. It will be necessary to stop at least a sample of the vehicles passing the chosen points to ascertain the loads carried.

In order to locate points at random on the network a running total of the lengths of all the roads and sections of road comprising the network may be made in such a manner that each piece of road is included once and once only. If the total length is L miles, any number l less than L then defines a unique point on the road network. If j numbers are selected at random between 1 and L these numbers will define points on the road network which will be located at random, each mile point having an equal chance of being selected. To give a density of one point per k miles, we take $j = L/k$.

Instead of locating the points at random they may be located systematically,

---

* The proposals which follow are based on a note prepared for a Working Party of the Inland Transport Committee of the U.N. Economic Commission for Europe.

i.e. at equal intervals with regard to the running total, using a random starting point h between 1 and k and selecting the points $h, h + k, h + 2k$ . . . .

In practice it will be advisable to use a variable sampling fraction, with a considerably higher density of points on the more important roads. For this purpose the roads must be stratified according to their importance. When this stratification has been made, separate running totals can be constructed for the different strata. If any large area is to be covered it will also obviously be advisable to divide the area into regions and treat each region separately.

If desired, certain types of road such as roads within city boundaries, and minor roads in built-up areas, can be excluded entirely. This will automatically exclude the traffic on these roads from the estimates.

Sampling may be used in various other ways to increase the accuracy of the results with a given expenditure of effort. Lattice sampling is particularly useful. Instead of carrying out continuous counts, for example, counts may be made at different hours of the day at different points, a rotation being arranged so that all periods of the day are equally covered, and that different periods are covered at the same point on different days. Thus the counts on a group of 12 points can be so arranged that on 12 consecutive days each 2-hour period is counted on one and one only of the 12 days, and that on each day every 2-hour period is covered at some point of the group. Since there is likely to be a considerable amount of variation between points, a rotation of this type may be expected to increase the accuracy considerably, since a much larger number of points can be included for the same amount of counting.

There will of course be little advantage in using this type of sampling for counts if automatic counting devices are available, but the procedure will still be of value for sampling to ascertain loads, etc. The number of vehicles that have to be stopped may also be reduced by examining only a fraction of the vehicles passing during the time a point is under observation. Care must be taken to see that bias is not introduced by this procedure. If every third vehicle is examined, for example, there must be no element of choice, i.e. the count must be based on the order of arrival. If the fraction is varied from time to time owing to variation in traffic density each such variation must be noted so that the correct raising factor can be used for each part of the data.

It is essential for objective estimates that the whole of the day and night is covered. If the volume of traffic is substantially reduced during the night, however, it may be advantageous to use a variable sampling fraction here also, covering the night period less intensively than the day period. Similarly if an objective estimate of the total annual volume of traffic is required the different parts of the year must be properly sampled.

4.32 Determination of the size of sample when the sample is fully random

As has been indicated in Chapter 3, the size of sample required to achieve a given accuracy depends on the variability of the material and the extent to which it is possible to eliminate the different components of this variability from the sampling error. In this and the following section we will describe the procedure which is appropriate for determining the size of a random sample, and indicate the general relationship between the errors of a random sample and other types of sample. Detailed consideration of the more involved types of sampling must be deferred till Chapter 8, where the comparative accuracy of the various types of sampling is discussed.

In the discussion of sample size we shall require the concept of *standard error*. As already explained in Section 3.7, the sampling standard error of an estimate is a measure of the average magnitude of the random sampling error to be expected in that estimate. It also provides an indication of the frequency with which errors of various magnitudes may be expected to occur (Section 7.3). In rough general terms, one-third of the actual sampling errors will be greater than the standard error, and one-twentieth will be greater than twice the standard error.

In the case of a fully random sample from a large population the formula for the standard error of the estimate of the proportion of *units* of a given type, i.e. having a given *attribute*, is very simple.* If $\mathbf{p}$ is the proportion of units of the given type in the whole population, and $\mathbf{q} = 1 - \mathbf{p}$ is the proportion not of the given type, the standard error of the proportion of units of the given type in a random sample of n units (which provides an estimate p of $\mathbf{p}$) is given by

$$\text{standard error of } \mathbf{p} = \sqrt{\frac{\mathbf{p\,q}}{n}}$$

The formula holds unchanged if the proportions are replaced by percentages. Thus

$$\text{standard error of } (\mathbf{p}\%) = \sqrt{\frac{(\mathbf{p}\%)\,(\mathbf{q}\%)}{n}}$$

The full line of Fig. 7.4 provides a graphical representation of the standard errors given by this formula. If 20 per cent. of the units of the population are of the given type, for example, the standard error of the percentage of units in a sample of 100 is

$$\sqrt{\frac{20 \times 80}{100}} = 4.0$$

which is the value given by the full line. Thus estimates in the range 20 ± 4 per cent., *i.e.* between 16 and 24 per cent., will be obtained in two-thirds of all samples of 100 units, and estimates in the range $20 \pm (2 \times 4)$ per cent., *i.e.* between 12 and 28 per cent., will be obtained in nineteen-twentieths of all samples. If a sample of 1000 is taken, the standard error will be 1·26, and

* The adjustment for finite sampling, required when the population is not large relative to the sample, is given in Section 8.1.

estimates between 18·7 and 21·3 per cent. will be obtained in two-thirds of the samples (Section 7.4).

When dealing with estimates of quantities such as means and totals it is best for our present purpose to work in terms of the percentage standard errors. The *percentage standard error* of the estimate of a quantity is the standard error of the estimate expressed as a percentage of the true value of the quantity.

The percentage standard error of the estimate of the total number of units having the given attribute in the population is the same as the *percentage* standard error of p. From the above formula we see that this percentage standard error is given by

$$\text{percentage standard error} = 100\sqrt{\frac{q}{n\,p}} = 100\sqrt{\frac{(q\%)}{n\,(p\%)}}$$

The percentage standard errors given by this formula are shown by the dotted line in Fig. 7.4.

In a population in which 20 per cent. of the units possess the given attribute, for instance, the percentage standard error with a sample of 100 is*

$$100\sqrt{\frac{80}{100 \times 20}} = 20 \text{ per cent.}$$

If there are 10,000 units in the whole population, 2000 will possess the given attribute. The standard error of an estimate of this number from a sample of 100 will therefore be $2000 \times 20/100 = 400$, *i.e.* in two-thirds of the samples estimates between 1600 and 2400 will be obtained.

The formula may be re-written so as to give the number n required for the sample when the required standard error of p or the required percentage standard error is known. We have

$$n = \frac{p\,q}{(\text{required standard error of p})^2}$$

$$= \frac{10,000\,q}{p\,(\text{required percentage standard error})^2}$$

The proportions p, q and p may be replaced by the corresponding percentages without change.

If, for example, we are sampling a population in which it is believed that about 20 per cent. of the units are of a given type, and it is required to determine this percentage with a standard error of 1 per cent (i.e. a percentage standard

* This result can also be obtained directly from the actual standard error of the percentage, which is $100 \times 4/20$, i.e. 20 per cent.

error of 5 per cent.), we shall require to take a sample with a number of units given by

$$n = \frac{20 \times 80}{1^2} = \frac{10{,}000 \times 80}{20 \times 5^2} = 1600$$

These formulæ hold only for a random sample in which the sampling units are the units for which the proportion having a given attribute requires to be estimated. In sampling a human population, for example, the sampling unit may be the household and not the individual. In this case the standard errors of proportions of individuals determined from the sample will be larger—often substantially larger—than those given by putting n equal to the number of individuals in the above formulae (Section 7.7 and Example 7.8.b).

Note also that the simple rule that the standard error of the difference of two quantities is the square root of the sum of the squares of the separate standard errors does not hold for the differences between the proportions, $\mathbf{p}_1$, $\mathbf{p}_2$, of the population possessing different attributes (Section 7.4). If $\mathbf{p}_1$ and $\mathbf{p}_2$ together comprise most of the population the actual standard error will be considerably larger.

The corresponding formulæ for a quantitative character are equally simple. We then have for the estimate of the mean or total from a random sample of n from a large population

$$\text{percentage standard error} = \frac{\text{percentage standard deviation of a unit}}{\sqrt{n}}$$

$$n = \frac{(\text{percentage standard deviation of a unit})^2}{(\text{required percentage standard error})^2}$$

In order to use these formulæ we require an estimate s of the *standard deviation* (or standard error) of a unit in percentage terms. The standard deviation of a unit is the measure of the variability of the units amongst themselves, and can be determined from a frequency distribution of the unit values. Table 7.1 provides an example of such a frequency distribution, and Example 7.1.b gives the method of calculating the standard deviation from this distribution. For these data (family incomes) $s^2 = 276{,}290$ and therefore $s = 526$. The value of the mean is 1629·1, and the percentage standard deviation is therefore $100 \times 526/1629 \cdot 1 = 32 \cdot 3$. The number required in the sample to give a standard error of 5 per cent. in the estimate of the mean income per family is therefore

$$n = \frac{32 \cdot 3^2}{5^2} = 42$$

For a standard error of 1 per cent. the number required would be 1050.

A similar calculation for the wheat acreages of the random sample of Hertfordshire farms (Table 7.2), already described in Section 3.7, is given in Example 7.2.b. In this case $s = \sqrt{1351 \cdot 4} = 36 \cdot 8$. The mean wheat acreage per farm is 18·6, and the percentage standard deviation is therefore $100 \times 36·8/18·6 = 198$. The percentage standard deviation is very large because there are a large number of farms growing little or no wheat. In order to determine the total wheat acreage of an area with a 5 per cent. standard error from a random sample of farms, therefore, we shall require $198^2/5^2 = 1570$ farms.

From the above formulæ we see that the standard errors of estimates derived from random samples of different sizes taken from the same population are inversely proportional to the square roots of the numbers in the samples. Conversely, to reduce the standard errors of the results in a given ratio we require to increase the size of the sample by the square of the ratio. Thus, in order to halve the standard errors of the results we must multiply the size of the sample by 4.

There is one further point to note when considering the errors affecting the mean or total of a quantitative variate. If we are concerned with means relating to subdivisions of the population, the relevant standard deviations per unit will be those within these subdivisions, not the standard deviation for the whole population. The standard deviations for subdivisions may be substantially less than that for the population. Thus in a survey of the physical characteristics of children of different ages, the variation of heights and weights within age-groups will be much less than the variation over all ages. Within limits, the finer the grouping by age the greater the reduction, but of course with a given number of children in the sample the finer the grouping the smaller will be the number in each group. When considering what groups to adopt, a balance therefore has to be struck.

The estimates of the standard errors of subdivision totals must also be based on the standard deviations within subdivisions, but here there is an additional contribution to error due to errors in the number of units in the subdivisions if these are estimated from the sample (Section 7.7).

4.33   Some general rules on size of sample

From the above discussion it will be seen that the calculation of the size of sample required to attain a given accuracy is a relatively simple matter when a random sample is taken. With the more involved types of sampling the calculations are more complicated, and more must be known of the material that is being sampled. Further complications arise when, as is usually the case, estimates are required for subdivisions of the population, and for contrasts between these subdivisions, i.e. for domains of study.

Calculation of the accuracy which would be attained by a random sample is, however, often a useful preliminary guide to the size of sample likely to be required in the more involved types of sampling. If only one type of sampling unit is under consideration, the reduction in numbers of units required with the more complicated types of sampling is determined by the fraction of the total variability which is removed by the imposition of restrictions such as stratification or by the use of supplementary information. It is frequently possible to form a rough idea of the likely reduction from a general knowledge of the characteristics of the material. Thus in a survey designed to determine crop acreages, using farms as sampling units, it is to be expected that stratification by size of farm and the use of a variable sampling fraction in conjunction with such stratification will each give considerable increase in accuracy over a random sample. This is confirmed by the results already given in Section 3.7.

When sampling units of different types or of alternative sizes are under consideration, the situation is more complicated, as is shown by the results already presented in Section 3.11. This is true also of multi-stage sampling.

The following general rules may be of value. Rules 1 to 5 are applicable to the case in which only one type of sampling unit is under consideration, rules 6 and 7 to the case in which more than one type of sampling unit is being considered.

(1) The use of stratification, a variable sampling fraction or supplementary information may in general be expected to increase the accuracy. Consequently the calculation of the number of units required in the case of a random sample gives an upper limit to the number of units required in any reasonable form of sampling using the same sampling units.

(2) Stratification will only increase the accuracy substantially if there are marked differences between the different strata. The increases are usually larger for quantitative characters than for qualitative characters, i.e. attributes (Table 3.7.b). There will be little reduction in error from stratification for domains of study cutting across strata (Section 7.6.2).

(3) A variable sampling fraction can greatly increase the accuracy when the units vary greatly in size, or more generally in variability from stratum to stratum. Fractions which increase the accuracy for quantitative characters may reduce it for qualitative characters (Table 3.7.b).

(4) The use of supplementary information can greatly increase the accuracy in appropriate cases, and often serves as an alternative to stratification (Table 3.7.b).

(5) Since there must be at least one unit per stratum, more detailed stratification is possible with larger samples. In such circumstances the increase in accuracy with increasing size of sample will be more rapid than is indicated by the square-root law. Conversely, for samples of a given accuracy the advantage of stratification may be reduced by the fact that reduction in the size of the sample necessitates an increase in the size of the strata (Section 8.15).

(6) If sampling units of type A consist of aggregates of sampling units of type B (e.g. households and individuals), the use of sampling units of type A in place of units of type B will usually result in lower accuracy for a given amount of material in the sample (Table 3.11.b and Section 7.11).

(7) If multi-stage sampling is used, more final-stage units will be required than will be the case with single-stage sampling of the final-stage units (Tables 3.7.b and 3.11.b).

All the above rules are indicative only. The quantitative gains in accuracy or reduction in number of units required in any particular case must be evaluated by the methods described in Chapter 8. The final decision as to the type of sampling to be adopted necessarily depends on the relative accuracy of the various methods and their relative costs.

It is advisable at the planning stage to consider as far as possible the form in which the results require to be presented. In more complicated surveys, particularly of the exploratory and research type, the results themselves will in part suggest the form in which they require to be presented, but in simple types of survey the form of presentation can often be laid down in considerable detail. This is a help in verifying that the sample is sufficiently large to cover the required domains of study adequately.

4.34   Pilot and exploratory surveys

From what has been said in the previous sections of this chapter it will be apparent that there are many points on which decisions can only properly be reached after preliminary investigations in the form of a *pilot survey* have been carried out. On material of which nothing is initially known, *e.g.* in surveys of undeveloped territory, a preliminary *exploratory survey* may be required before any proper pilot survey can be undertaken. In addition to providing general information, such an exploratory survey may be used to construct a first-stage frame.

A pilot survey has two main objects : firstly, the provision of information on the various components of variability to which the material is subject, and secondly the development of field procedure, the testing of questionnaires and the training of investigators. A pilot survey may also provide data for the estimation of the various components of cost of the different operations involved in the survey, e.g. interview time, time of travel, etc. Knowledge of such costs is required not only as a basis for general estimates of cost, but also in order to determine what type and intensity of sampling will be most efficient.

A further function of pilot surveys is to determine the most effective type and size of sampling unit. In a crop-cutting survey involving the harvesting of small areas, for instance, we may require to determine the best size and shape for these areas. In order to investigate the variability of different types and sizes of unit it is necessary to be able to form aggregates which represent the largest units which are of interest. Thus, if areas ranging from, say, $\frac{1}{2}$ ft. $\times$ 1 ft. to 3 ft. $\times$ 3 ft. are under consideration, it is necessary, or at least preferable, to harvest randomly distributed areas of 3 ft. $\times$ 3 ft. in sections of $\frac{1}{2}$ ft. $\times$ 1 ft. (Section 8.14).

Pilot surveys will not normally be required for material on which there is considerable previous survey experience, since every survey provides information on the variability of the material surveyed, and this can often be used in the planning of further surveys. Thus the 1942 Census of Woodlands (Section 4.25) was planned on the basis of experience gained in the 1938–9 Census of Woodlands. Even in such cases, however, new questionnaires and new methods of observation and measurement should be tested on a more or less random sample of the material before being put into operation.

The testing of the field procedure by means of a pilot survey is discussed in Chapter 5, and its planning needs no special comment. The planning of a pilot survey to provide relevant information on the various components of variation is rather more difficult. The finer points will be fully apparent after the estimation of efficiency has been discussed (Chapter 8), but a few fundamental points may be made here.

At first sight it might be thought that a fully random sample would be a satisfactory form of sample for a pilot survey. This, however, is not necessarily the case. As a simple example we may consider the survey of material in which the use of a stratified sample is likely to be appropriate. In this case we shall require to determine the components of variation within strata. This can only be done from a fully random sample if the sample is sufficiently large relative to the number of strata for the majority of strata to contain at least two units.

This difficulty can be overcome by adopting some form of multi-stage sampling, so that the whole of the pilot sample is concentrated in a few of the strata. The primary stage of this multi-stage process need not necessarily be very rigorous. Thus in a survey covering a human population concentrated in villages, it may be considerably more convenient to use certain villages for the pilot survey rather than others. Provided that sufficient is known of these villages to indicate that they are fairly typical there is no serious objection to their use. Similarly in area sampling it may be sufficient to confine the pilot survey to districts conveniently situated with regard to the main and regional headquarters, provided there is some assurance that the different types of district are properly represented.

Within the towns or areas selected for the pilot survey a fairly intensive survey can be made. If necessary a further stage may be introduced into the sampling process. Thus the survey may be confined to a selection of blocks in a city, instead of the whole city being sparsely covered. By this means, it is possible to obtain data which cover selected areas with a density which is of the same order as that which will be adopted in the final survey. If this is done the various possible types and sizes of strata can be effectively investigated.

The concentration of the pilot sample into selected areas should not be pushed to extremes. It is better to have adequate cover of a representative sample of the data than highly detailed cover of small and possibly non-representative sections of it. Detailed cover will in any case not be required if the material is such that very small strata are known to be impracticable or of no value.

When the possibilities of multi-stage sampling have to be investigated, the problem of designing a pilot survey is more difficult. The component of variability that governs the sampling error for the whole population is the variability of first-stage units (Section 7.17). Only if an adequate number of first-stage units are represented in the pilot sample will it be possible to make any reliable estimate of their variability. Moreover the first-stage units must be selected at random from within the first-stage strata that will be finally adopted.

For this reason a much more extensive pilot survey is required for a multistage survey if any reliable preliminary estimate of the sampling error is to be obtained. In surveys in which comparatively few first-stage units are used, such as localized surveys on human populations (Section 4.18), the prior determination of the expected accuracy from pilot survey data is likely to prove to be impracticable. Reliance will then have to be placed on previous experience or on estimates of error from previously available data. This, however, is not quite so serious as it appears at first sight, since multi-stage samples with relatively few first-stage units are in general used most frequently in surveys of the research and investigational type, or in surveys which are repeated at intervals. In such cases the first few surveys can be used to provide data on the various components of variation, and the design can be modified if necessary in the light of this information.

Even if the first-stage sampling error cannot be determined by means of a pilot survey, such a survey can be made to furnish reliable information on the errors to be expected at the second and subsequent stages. This will enable the survey to be planned so that the necessary accuracy is obtained on comparisons between first-stage units, which frequently form important domains of study in surveys of this type.

Elaborate pilot surveys are not likely to be worth while in small-scale surveys. It is usually better to proceed with the actual survey work, even if the design adopted is not as efficient as would be possible if a full-scale pilot survey were first undertaken. If a series of small-scale surveys of similar type have to be undertaken the earlier surveys will themselves act as pilot surveys for the later surveys, the design of which can be modified in the light of the experience gained.

CHAPTER 5

PROBLEMS ARISING IN THE EXECUTION AND ANALYSIS OF A SURVEY

5.1 Types of problem

The problems arising in the execution of sample censuses and surveys are for the most part similar to those encountered in complete censuses. We shall therefore not discuss these problems in detail, but merely draw attention to some of the points which are of particular importance when sampling is used.

The various phases of the work subsequent to the planning stage may be broadly classified as follows :—

    (1) Setting up of the general administrative organization.
    (2) Design of forms.
    (3) Selection, training and supervision of the field investigators.
    (4) Control of the accuracy of the field work.
    (5) Arrangements for follow-up in the case of non-response.
    (6) Abstraction and coding of the information.
    (7) Statistical analysis.
    (8) Reporting.

In certain types of survey no action may be required under some of these heads. In a survey conducted by postal questionnaire, for example, there will be no field investigators unless they are required to deal with cases of non-response.

5.2 Administrative organization

The administrative organization required will depend very much on the nature and scale of the census or survey, and on the area to be covered.

The main field task for which an extensive administrative organization is required is the supervision of the investigators, or the carrying out of follow-up enquiries in cases in which there is no staff of investigators to undertake this work. The main administrative task at headquarters is the supervision of the computing and clerical staff engaged in the abstraction and analysis of the completed forms.

Every opportunity should be taken to utilize existing administrative and office organizations. When the survey covers a large area, supervision from a central office is likely to be difficult and in such cases it is best to establish regional offices. Very frequently some existing organization can be used for this purpose.

It is not necessary for the computing and clerical staff at headquarters to be administered by the same organization as the field staff. In many cases it is convenient to use some existing statistical organization to carry out the analysis, and to utilize some administrative organization with regional offices to supervise the field work. Often an organization can be employed which already has contact with the respondents, or which has on its staff individuals who are suitably qualified to act as field investigators.

5.3 Design of forms

Most careful attention should be given to the detailed design of the various forms that will be used in the course of the census or survey, especially the forms on which the observations and answers to questions are recorded. This applies also to the instructions and explanatory notes which accompany the forms.

The content of the forms for the recording of the information is determined by the information that is required, and has already been discussed. They may be forms designed for completion by the recipients with little or no assistance, questionnaires which form the basis of interviews, or forms on which observations and measurements taken in the field are recorded by the field investigators.

Each type of form presents its own difficulties of design. The simplest is that on which observations and physical measurements are recorded by the field investigators themselves. In this case, the chief points to observe are that the form is convenient to use, and that the results are set out in such a manner that they are convenient to abstract. Figures which have to be summed by the field investigators, for example, should be arranged vertically and not horizontally, as the investigators will not be using calculating machines.

In surveys which involve observations and physical measurements it will almost always be necessary to supply field investigators with a separate set of instructions. Consequently there is no need for the form to carry its own full explanation, though it should of course be made as self-explanatory as possible. Experience has shown that instructions to field investigators should be very detailed, and should cover all possible points of uncertainty or ambiguity. Provision should also be made for revision and amendment as need arises, since it is extremely difficult to draw up a set of instructions which are completely unambiguous and deal with all possible contingencies.

In forms of the census type, designed for completion by the recipients without assistance, very careful attention must be paid to the exact wording both of the questions and explanatory notes, so that there is no doubt in the mind of the recipient as to what is required. Detailed and lengthy explanations should be avoided as far as possible. Such explanations as have to be given should if possible appear in conjunction with the question to which they refer. The common practice of giving detailed explanatory notes on the back of a form is not very satisfactory, since it frequently results in the respondent filling in the whole or portions of the form without consulting these notes. Forms of this type should, if possible, carry a brief explanation of the reasons for the census. Even if this has been given in the press and elsewhere it is unlikely that all recipients will in fact have seen it.

In forms of the questionnaire type designed for completion by field investigators the investigators must be instructed whether the questions are to be put in the exact form given, or whether they can be asked in a general form. As already stated, in most cases the general form is more suitable, but in questions on opinions, where different forms of wording may be expected to affect the answer, it may be necessary to adhere to an exact form.

With the general form of question explanatory notes are often required in order to make clear to the investigators exactly what information is required. Such explanatory notes can either appear on the questionnaire itself or be given in a separate set of instructions.   The latter course results in a much more compact form of questionnaire and is suitable when full-time investigators are used. The former course is more likely to ensure that all investigators are in fact aware of what is really required and is best when the investigators are carrying out the survey in the course of other duties. In a lengthy questionnaire this will necessitate the questionnaire being in the form of a booklet. Such questionnaires are more bulky and costly, and frequently entail more work in the coding of the results, but are nevertheless frequently preferable in these circumstances.

The other important consideration in the design of forms is to provide for easy conversion of the data into machine readable form. If the data are to be transcribed by key-punching, e.g. onto punched cards, the information that has to be punched should be arranged so as to be readily accessible to the punchers, with minimal preliminary coding or other clerical work. This aspect of form design is discussed in more detail in Section 5.8.

For certain types of survey, key punching can be eliminated entirely by the use of forms from which marks can be read optically or magnetically. This is particularly suitable for surveys in which simple qualitative information has to be recorded.   Instruments are also available for automatically recording quantitative measurements such as weights, temperatures, etc.   Information from these and other sources can then be assembled on a single file, unit by unit, on the computer.

5.4   Tests of questionnaires and investigators

Questionnaires should if possible be given a preliminary trial. This often reveals unexpected faults and difficulties.   If the information is to be obtained by personal interview this test is best arranged in two parts :

(a) a trial by investigators who are fully experienced in questionnaire work, and who are conversant with the problems under investigation ;

(b) a trial by investigators of the type that are to be employed in the survey.

The first trial will serve to determine whether the questionnaire is in the form most suitable for eliciting the required information from the respondents, and the second trial will provide information on whether the questions and associated instructions are understood by and within the capability of the investigators.

In certain cases it may be worth making rigorous tests of different forms of the same question to see whether there are any material differences in the answers received. Since the question cannot be put in both forms to the same respondent, this must be done by the use of interpenetrating samples, using the same investigator or investigators for both forms. In order to eliminate the effect of any progressive change in the investigators or the respondents the tests of the two forms should proceed simultaneously. In the same way the difference between two or more investigators using the same form of question can be tested.

More elaborate and precise tests of differences resulting from different forms of the same question and different investigators can be carried out by using the methods developed in the design of experiments. Thus if two forms P and Q of a question and three investigators A, B and C require to be tested, groups or *blocks* of six respondents may be used. The blocks should be chosen in such a manner that the respondents within each block are as alike as possible, using any available prior information. The six question–investigator combinations PA, QA, PB, QB, PC, QC are then assigned at random to the respondents of each block. This design is technically known as a *2 × 3 factorial design in randomized blocks*. By this device differences between forms of question and between investigators are simultaneously tested. Information is also obtained on what are known as the *interactions* between forms of question and investigators, i.e. on whether the differences between the forms of question are different for the different investigators, and vice versa. The grouping of respondents into blocks ensures that errors due to differences between respondents are eliminated as far as possible ; the randomization enables the standard errors of the comparisons to be calculated by the methods appropriate to the analysis of replicated experiments (see for example Fisher's *Design of Experiments* or Snedecor's *Statistical Methods*).*

Investigations of this kind can be carried out in the course of an actual survey, but they are normally better undertaken as a special investigation or as part of the pilot survey, since information on different forms of question will be required at the planning stage, and it is usually inadvisable to complicate the field procedure of a large survey. Routine tests of differences between different investigators may, however, be incorporated without undue complication in the actual survey by means of interpenetrating samples.

5.5 Selection, training and supervision of field investigators

Field investigators may be specially appointed, they may be members of existing staffs appointed for other work but over whom authority can be

---

* An interesting investigation of the differences between three groups of interviewers (two belonging to professional organizations, one of university students) has been carried out by the Department of Research Techniques, London School of Economics (Durbin and Stuart, 1951, D', Booker and David, 1952, D'). A factorial design was employed.

exercised, or they may be individuals asked to undertake the work on a voluntary basis or for a small honorarium.

The problem of selection arises primarily in the case of investigators appointed specially for the work. In order to secure a suitable type of person preliminary tests should if possible be made of all applicants, and the early work of newly appointed investigators should be carefully watched and supervised. In large-scale censuses and surveys, proper training courses should be arranged. If a pilot survey is undertaken this provides a valuable opportunity for training, and every attempt should be made to build up the team of investigators at this stage rather than later, even if this involves a certain amount of additional expense.

It is of the greatest importance that investigators, once they have been trained and are found suitable, remain in the job. Every effort must therefore be made to see that the pay is adequate, and that the work is made as attractive as possible. In the case of the interview type of survey, investigators are sometimes paid on piece rates at so much a completed questionnaire. This is in general unsatisfactory, since it tends to lead to skimped work and to irregularities such as substitution of one respondent for another.

It should not be forgotten that field work of the interview type is very arduous and is found by almost all investigators to involve considerable mental strain. Hours of work are also likely to be irregular, since if excessive non-response is to be avoided some evening interviews are almost inevitable. Investigators should therefore not be expected to work excessively long hours, and should if possible be given a rest on other work from time to time. It is often advantageous to bring full-time investigators to headquarters at intervals and use them for office work such as abstraction and analysis of the results. This not only serves to provide a break from field work, but also enables them to gain a much better insight into the purposes of their work.

Whatever the conditions of work and form of payment, there must be adequate field supervision. The supervisors should themselves undertake field work from time to time, so that they are in a position to appreciate the difficulties of the work, and should also contact the workers while they are actually in the field. Provision should be made for personal contacts not only between supervisors and the field investigators, but also between supervisors and the headquarters staff. In long-term surveys it is also often advantageous to arrange conferences of the investigators from time to time at which difficulties can be discussed and the whole progress of the survey reviewed.

5.6 Control of the accuracy of the field work

The best assurance that the field work shall be accurate is that the investigators are thoroughly trained in their work, and are capable, conscientious, and keen. Nevertheless it is important even with the best investigators to keep a close watch on the progress of the work.

In certain cases, particularly in surveys involving observations and physical measurements, it is possible to arrange a system of field checks by the supervisors. These should preferably be carried out on a random sub-sample of units, and should in any case be conducted in such a manner that the investigators cannot know which parts of their work will be checked. Checks of this type will not usually be possible in the interview type of survey, as it is clearly impracticable to ask for the same information twice from the same individual.

A preliminary examination of the completed forms must be made as soon as possible after they are completed. In this way defective work, in so far as it reveals itself in the forms themselves, is brought to light immediately, and remedial action can be taken. If the census or survey is such that a large volume of work is turned in by each field investigator, and it is not considered necessary to give individual scrutiny to all the returns, a proper sample of the work of each investigator should be scrutinized as a routine matter.

The investigators should themselves be instructed to carry out any simple numerical calculations that are required on the forms, and also to look through the forms before sending them in to see if they are in satisfactory order. On the other hand extensive revision of the forms should not be permitted. The preparation of fair copies by the investigators is in general undesirable, since it leads to copying errors and also makes any judgment on the quality of the work more difficult. If fair copies are permitted the originals should be returned together with the fair copies, and a certain percentage at least should be checked for copying errors and other changes.

If the questionnaire is such that the investigator has to furnish or amplify the answers to some of the questions from notes taken at the interview this should be done immediately after the interview rather than at the end of the day, even if this course is somewhat inconvenient. In general, however, it is best for the information to be written down in its final form at the interview, any supplementary observations by the investigator being given under a separate heading.

If comparisons between the different investigators by means of interpenetrating samples have been arranged, the comparative results must be made available as quickly as possible, in order that effective action may be taken if discrepancies are discovered. On the other hand the use of interpenetrating samples should not be made an excuse for the relaxation of other forms of control. Interpenetrating samples are not likely to reveal minor defects in an individual investigator, and they will certainly not reveal faults which are common to all investigators. They should therefore be regarded as a check against major defects in individual investigators rather than as a complete control of all investigators.

5.7 Arrangements for follow-up in the case of non-response

The follow-up arrangements will naturally vary very greatly according to the type of census or survey.

In the case of a postal questionnaire they will often involve an entirely different organization from that which is employed to carry out the survey itself. Since postal follow-ups from headquarters are of limited utility, some form of local organization which can deal with non-respondents by telephone and personal visit is required.

In surveys using field investigators careful instructions must be issued in order to be sure the follow-up arrangements are properly carried out. Specific warnings should be given against such practices as substitution of neighbouring households when there is no response. If the follow-up is to be made on a sub-sample only of the non-respondents exact instructions for taking this sub-sample must be given, so that it can be obtained as soon as the non-respondents are known. In general some very simple sampling method, such as taking of every qth non-respondent, is adequate. Such a procedure has the advantage that a list for follow-up can be prepared at the time of the original non-responses, if necessary by the field investigator concerned.

5.8 Conversion of data to machine readable form

Before the introduction of computers, punched card machinery (primarily sorters, sorter-counters and tabulators) had to be used for the preparation of tables for all but the smallest surveys. The limitations of such equipment necessitated a good deal of preliminary coding so that the information could be arranged on the cards in compact numerical form.

Punched cards are still the most commonly used medium for the transcription of data from small and medium-sized surveys to machine readable form. If the data are to be processed by computer, however, there is no need to arrange the data in any special order or in compact form on the cards. Convenience of recording and simplicity of punching should be the objectives. Nor indeed is there any need for cards. The development of microprocessors enables modern key-punching equipment to incorporate facilities for error correction, such as the equivalent of back-spacing when a character is mispunched, and the imposition of simple checks before the data for a unit are accepted. Magnetic tape or disc can then replace cards as convenient intermediate storage between key punch and computer. A further alternative is to transmit the data directly to the computer from a computer terminal, using the standard text-editing facilities that are available on most computers.

Computers have eliminated much of the earlier need for preliminary clerical work, as the computer can itself readily recode literal (alpha-numeric) data, condense numerical and non-numerical data by grouping, and perform numerical calculations on the data for individual units to produce new (*derived*) variates.

For qualitative items for which there are only a few admissible alternatives " pre-coding " is often convenient ; for this, lists of the alternatives, together with the relevant codes, are provided on the survey form, with instructions to circle the code which is applicable. If the alternatives are too numerous to list completely, or if provision is required for unforeseen alternatives, a partial list can be provided, with the addition of " Other (specify) " which can be coded in the office before punching.

An alternative to precoding items with numerous alternatives is to record the answers in literal form and use abbreviated literal codes that are sufficiently simple to be memorized by the punch operators ; or the recorders can themselves insert the appropriate codes. Thus agricultural crops can be coded by their first three or four letters. A list of admissible codes is then supplied to the computer to enable it to translate the codes into numerical form. Literal codes should not, however, be used unless there is a real advantage in them, as purely numerical data require somewhat less punching skill ; certain types of punching error in mixed numerical and literal data can also give trouble when read by commonly used computer subroutines. There would, for example, be no point in coding sex as M,F instead of precoding it as 1,2.

In some cases quantitative information is only required as a basis for classification. Precoding into the required classes, e.g. income classes, can then be used. In general, however, if reasonably precise quantitative information can easily be obtained it is better to record and punch this, as new (derived) classification variates can readily be formed by the computer by grouping the original data in any required manner. The data can then be regrouped if this is required for any reason, e.g. for comparison with the results of some other similar survey. Often also the most suitable grouping can only be determined when the results of the survey are available.

Occasionally information on a quantitative item is most easily provided in alternative forms by different respondents. Thus income may be reported as the weekly wage, or monthly or annual salary. To avoid danger of arithmetical errors and for ease of recording it is best to record these in the forms given, distinguishing them by some auxiliary code. They can then be converted to some common form by the computer. Similarly, if measurements have to be made using non-decimal scales, e.g. heights in feet and inches, they should be so recorded and punched. Their conversion to some form of decimal units can be left to the computer.

5.9 Multiple non-exclusive responses

Questions admitting multiple non-exclusive alternative responses, such as " Which newspaper(s) do you read ? " present special problems. They are, of course, equivalent to a set of questions each with two exclusive alternatives, e.g. " Do you read *The Times* : Yes/No ? ", etc., and can be so treated. The answers, for ten alternatives, say, can thus be recorded by punching a string of ten 0's and 1's.

Alternatively, multiple punching in a single column can be adopted, positive answers to the alternatives being precoded 0, 1, . . . 9 respectively. If more than ten alternatives are required, two or more columns can be used. The top two holes of a column can also be used, but as the punch codes for these are symbols, which differ on different computer installations, it may be judged better to omit them.

Multiple punching gives more compact representation on the cards, and simpler punching when there are likely to be only one or two positive answers per respondent. It does, however, complicate the computer processing of the data. It is obviously not suitable for multiple response questions in which more than simple Yes/No answers are required, e.g. Regularly/Occasionally/Never in the above newspaper question. The various alternatives must then be treated as a set of questions.

5.10   Missing information

Codes have to be provided for missing information. Entries on the forms are simplified if a standard code is adopted for all missing items, e.g. the letter X, or X and Y if refusal to answer and lack of knowledge, for example, require to be differentiated. Such coding serves equally well for qualitative and quantitative items, but has the disadvantage that numerical items cannot be read by the standard computer subroutines for reading numerical data.

An alternative, if special subroutines for handling such data are not available, is to use numerical codes. As almost all numerical survey data are non-negative, -1 and -2 provide suitable unknown value codes. These, however, require two columns, and for items requiring one column only it may be preferable to use single digit codes for unknown values. There is some advantage for qualitative data in making such codes consecutive, e.g. if there are six alternatives coded 1 to 6, unknown values can be coded 7 ; an " unknown " class can then be readily included in tables if required.

In questions with non-exclusive alternatives a missing answer implies that all the items are unknown. If the items are coded as separate variates these could all be coded as unknown, but it is better to assign an additional variate in which failure to answer is recorded. The remaining variates can then all be skipped or punched as 0. If multiple punching is used, an additional multipunched code can be assigned to " not answered ".

Blanks should not in general be used to indicate unknown values, as they cannot be distinguished from zero on most machines if the standard Fortran subroutine is used for reading the cards. They can, however, be used for questions which are not applicable when non-applicability is indicated by an answer to some other question. This permits strings of non-applicable questions to be skipped when punching, which is particularly convenient if there are tabulating (skip) facilities on the punching equipment. Blanks can also be used to separate groups of variates so as to improve legibility if the data are listed.

5.11   Card layouts

If cards are used it is usually best to allocate specific columns on the card to the various items of information. This is known as *fixed format*. Alternatively *free format* can be used. In free format the items are punched in order, and separated by spaces, or distinguished in some other recognizable manner.

Free format has advantages for questionnaires in which a negative answer to a (key) question indicates that several subsequent questions are irrelevant. If a skip indicator is provided, e.g. $20 to indicate that 20 items are omitted, punching this will enable the machine to proceed directly with the next relevant item. It is important, also, to have a symbol to indicate the end of each record, otherwise the omission or duplication of an item in a record will cause utter confusion in the reading of subsequent records.

One advantage of fixed format is that if the cards are listed the values for each separate item will appear neatly in a column, thus making visual comparison between cards much easier, particularly if only one card per unit is required.

If, with fixed format, more than one card is required for a unit, the different cards must be distinguished by card-type numbers punched in some chosen column, and all cards appertaining to a unit must carry the same unit reference (numerical or alpha-numeric). Similarly if the data are hierarchical, e.g. data on households and on individuals within each household, with the data for the different levels of the hierarchy on different cards, the cards for the different levels must be assigned separate card-type numbers, with appropriate unit references.

When deciding on the most suitable card layouts and types of coding for a survey, it is advisable to bear in mind the types that can be readily interpreted by the survey programs which are available. For small surveys, in particular, the advantages to be expected from more elaborate coding may be outweighed by the extra programming work required to enable the computer to interpret them.

On the other hand, if, through lack of liaison, the survey analyst is confronted with a set of cards that contain coding which is not directly interpretable by the available programs, the easy course of asking that the data should be repunched should be eschewed. There is no real difficulty in writing an *ad hoc* program which will produce a computer file of the data in more acceptable form ; non-experts in computer programming will, however, be well advised to invoke assistance from someone skilled in the art.

5.12 Analysis and presentation of the numerical results

The types of analysis that are required and the way in which the numerical results can best be presented depend on the nature and purposes of the survey. At one extreme there are large-scale surveys of the census type, such as population censuses, which have as their primary object the presentation of factual information, often in considerable detail, for administrative purposes, or as an immediate guide to policy decisions. At the other extreme there are investigational surveys, usually on a much smaller scale, which require much more penetrating statistical analysis of the results.

The treatment of the results of sample censuses and surveys is similar in most respects to that of complete censuses and surveys. If, however, the sampling fractions are different for the different units, the appropriate weights have to be applied at some stage of the calculations.   The utilization of supplementary information will also necessitate adjustment of the basic totals and means.   The appropriate formulae for these operations, which differ according to the type of sampling adopted, are given in Chapter 6.

In sample censuses and surveys we may well also require estimates of the sampling errors.   In addition, investigations of the relative efficiency of different sampling methods may be undertaken in order to improve the efficiency of future surveys on the same or similar material.   These matters are discussed in Chapters 7 and 8.

Many computer survey programs will produce standard errors or standard deviations relating to tabular values in tables of means or totals on demand. These must, however, unless the program includes provisions for the specification of the sample design, be based on the assumption that the sample is fully random, and will be inappropriate to a greater or lesser degree for more complex designs and also even for simple random samples if the sampling units are aggregates of the units on which the tables are based.   As will be seen from Chapter 7, the computation of sampling errors in such designs is fraught with complications.   If standard errors are included in the results, therefore, it is advisable to indicate how they were derived.   This at least will enable experts in the subject to judge whether they are relevant to their requirements.   If special programming is needed for their calculation this should be tested on small samples of data and checked on a desk calculator.

Tabulation of the results is often all that is required in many types of survey, and is an essential preliminary step in almost all surveys.   Modern electronic computers have greatly increased the speed and accuracy with which well-arranged and well-annotated tables can be produced directly from the basic information, once this is available to the machine in the form of a data file.   It is important to realize, however, that time is needed for the detailed specification of the required tables and the auxiliary computations, and for subsequent testing to see that these are correct.   In any but the simplest types of survey, therefore, this should not await the receipt of the full data.

If a pilot survey is being conducted, provision should be made for a trial analysis of the pilot results.   This often suggests improvements in the content of the tables and in their form of presentation, and indicates how the final data file should best be constructed.

The degree to which an analysis can be planned at the outset depends very much on the type of survey.   In a simple census type of survey the categories of information required may be determined at the outset by administrative requirements, and in this case the whole of the analysis can often be planned in advance.   In surveys of the investigational type, however, a step-by-step approach is often necessary, particularly when causal relations are being explored.

Even in surveys of the census type the information collected often serves as a suitable basis for more detailed and critical statistical investigations.

The possibility of such investigations should be borne in mind when planning the content of the data file, so that any item of information will be available in accessible form if required.   Items should not in general be omitted merely on the ground that they are not required for the primary analysis ; the general aim should be to summarize the whole of the relevant information in coded form, so that should new needs arise or should the primary analysis indicate that further analysis is likely to be of value, the work can be undertaken without reference to the original records.

5.13   Control of accuracy

The attainment of a high standard of accuracy in the processing of survey material demands careful organization and scrupulous attention to detail at all stages.   An effective and efficient system can only be devised by careful consideration of the types of error that can occur, their effects if undetected, and the ways in which they can be controlled.

Errors can arise at all the various stages in the collection and tabulation of survey information :

(1) Collecting the information and recording it on the original forms.
(2) Coding of items that are not precoded or in suitable form for direct transcription.
(3) Transcription to machine readable form, e.g. by the use of punched cards.
(4) Calculation of functions of the original variates required for tabulation.
(5) Formation of tables, and their presentation in reports.

5.14   Checks on the basic data

Electronic computers greatly facilitate checks on the basic data, many of which were difficult or impracticable on punched-card equipment.   Qualitative characteristics can be examined for logical inconsistencies, such as that a male has borne children.   Quantitative items can be checked in various ways : values of the basic variates, or of functions of them such as ratios, can be tested against defined limits ; scatter diagrams of variates believed to be highly correlated can be prepared to see if there are any outliers ; if the sum of several variates should be approximately or exactly equal to another variate this can be checked.   Illegitimate codes, and various types of punching error such as non-digit characters in numerical fields and displacement of fields through the introduction or omission of a space or other character in the punching, can be detected.

If cards are used and there is more than one card type, either because the data are hierarchical or because there is more than one card per unit, a preliminary check can be made that all the required types are present and in the correct order, and that cards relating to the same unit are grouped together. Such a preliminary sequence and reference check is essential for any large body of data, as any error in card type sequence due to a missing or misplaced card, or a mispunched card type number, may cause serious trouble in subsequent work.

The imposition of checks on the basic data of the above types (*validation*) provides an overall indication of the quality of the recording, coding, and transcription of the data (stages 1 to 3 above).   Validation cannot, of course, provide effective checks on all items of information, and is no substitute for adequate detailed checks on the preliminary operations of coding and punching. Since, however, the basic information is in general subject to errors of various kinds, it may be sufficient in large-scale work to impose sample checks on these operations.   Much depends on the nature of the information, and the uses to which it is to be put.

5.15   Correction of errors in the basic data

The action required when errors or dubious values are detected depends on the nature of the data, and the frequency and types of error.   A single misclassification in a table of counts, for example, will often produce an entirely trivial error in the results, but a misclassification in a table of means may produce a serious error when there are large differences between the means of the different classes.   Similarly mispunching a quantitative item, e.g. 610 for 110, can produce a serious error in the resultant mean or total of the class into which the unit falls.

Most survey tabulation programs incorporate various checks that can be used to exclude automatically certain types of anomalous values from the tables. If few errors are revealed, therefore, it may well be that little or no correction of the data file is worth while.   Unnecessary correction of trivial or unimportant errors should be avoided, as this is tedious and time-consuming and carries the risk of introducing further and more serious errors.   If a reference disagreement is indicated in a sequence and reference check, for example, it is usually immediately apparent whether this is due to a misplaced card or a mispunched reference ; if the latter there will be no point in correcting it if the references are not further referred to.

If correction or exclusion of erroneous or dubious values is necessary the following options are available :

(a) the units in which they occur can be excluded from the subsequent analysis ;
(b) the values can be recoded so that they will be recognized as " missing ";
(c) estimated (*imputed*) values can be substituted.

Any of these options can be adopted either without reference to the original records, or after correction of any punching and other clerical errors revealed by such reference.   In general, reference should be made for at least a sample of the errors, as only in this way can the accuracy of the recording, coding, and punching be separately assessed.

If correction or exclusion of erroneous records is judged to be necessary the records in which the faults occur must be located. This presents no serious problems for tests based on values of a single unit only; location of the units which give rise to outliers on a scatter diagram, however, cannot be done from the diagram itself. These points are discussed in more detail in Chapter 10, as also are methods of imputation, and the precautions to be taken when rejecting or imputing values.

5.16 Checks on the subsequent computations

Apart from errors in the program used for analysis, modern computers can be relied on to do their arithmetic correctly. Consequently the types of error that have to be guarded against in items 4 and 5 of Section 5.13 are the incorrect specification of what is required. Such errors, of course, may be very serious and may not be immediately obvious from a casual inspection of the final tables. Thus derived variates (functions, groupings, etc., of other variates) may be incorrectly specified, raising factors may be incorrect, or there may be raising by the wrong variate, class-names may be specified in the wrong order, etc. Such specifications should therefore be very thoroughly checked. In addition, for important work a trial run with a small sample of test data is advisable ; the resultant tables can then be checked on a desk calculator, preferably independently by a second person, to see that they are what is expected. If the derived variate computations are at all complicated it is well to give these a preliminary check on some test data.

Large tables that are required for presentation in reports should, if possible, be reproduced in a form that is suitable for direct photographic reproduction. This saves much time and labour in the final stages, and avoids the danger of copying errors.

Some of the derived variates that are required for table formation may well have been required also for validation. It may therefore be advantageous to include them, and others such as group variates, in an edited data file, together with other particulars, such as variate and class-names, before starting the main tabulations. This is particularly advantageous in surveys involving large and complicated questionnaires with elaborate coding, and when many different sets of tables and other analyses have to be prepared from the file at different times, possibly by different workers. Subsequent operations are thereby considerably simplified, and errors and misunderstandings avoided. This aspect of data file preparation is further discussed in Chapter 10.

The final tables should always be scanned for anomalous values. This provides a further check on the quality of the original information and the preparation of the data file.

5.17 The use of sampling in the statistical analysis

In certain cases it is possible to attain the necessary accuracy and at the same time to reduce the volume of numerical and machine work by analysing a sample of the available data. Such sampling may be applied to the data from a complete or sample census or survey.

At first sight the use of sampling in this manner appears illogical, since it might be argued that if the collection of the information on the whole of the population or on a large sample was justified, its inclusion in the analysis is also justified. This, however, is not always the case, since a complete census or a large sample may have been taken in order to furnish information on individual units or on small groups of units, while the further analysis may be required to elicit information which does not require to be broken down in detail. Moreover it does sometimes happen that excessively large samples are taken which can well be reduced before analysis.

As already mentioned, the analysis of a sample of the returns is also of use in providing preliminary results for a complete or sample census, even though the whole of the material will ultimately require analysis.

Furthermore, when a large sample has been taken for administrative purposes, supplementary analyses of the investigational type can often best be undertaken on a sub-sample of the original sample. The reduction in the total volume of material to be handled is of particular value in such analyses, since they often require the application of relatively complicated statistical processes. Special points which emerge and on which a higher accuracy is desired can be re-tabulated subsequently by using the whole or a larger sub-sample of the material.

The actual technique of obtaining a sample suitable for analysis is usually relatively simple. For many purposes a systematic sample of every qth return is all that is required. In some cases, however, the use of a variable sampling fraction is advisable. This is particularly the case in the analysis of census returns referring to economic institutions, factories, farms, etc., since these are usually of very variable size.

An example of an analysis of this type is provided by the National Farm Survey of England and Wales (Ministry of Agriculture, 1944; G), which covered all holdings in England and Wales of over 5 acres. Sampling was not used in the survey because records were required for each individual farm, both for administrative purposes and for detailed studies of small areas. A map of the boundaries of each farm, for example, was one item of information which was collected.

For the purpose of obtaining a general summary of the results by counties, types of farming, etc., the analysis of the whole of the material was unnecessary. The holdings were therefore divided into size-groups and a systematic sample stratified for counties and size-groups was taken, using a variable sampling fraction for size-groups. The sampling fractions, and numbers of holdings in the population and in the sample, are shown in Table 5.17.a.

Had a uniform sampling fraction been used in place of a variable sampling fraction, a sample over twice as large would have been required to give results of the same accuracy on such items as the percentage of land under different systems of tenure. By the use of a variable sampling fraction results of ample

TABLE 5.17.a—ANALYSIS OF THE NATIONAL FARM SURVEY :
CONSTITUTION OF SAMPLE

Size-group (acres)	Average size (acres)	No. of holdings	Sampling fraction (per cent.)	No. of holdings in sample
5– 25	12	101,450	5	5,072
25–100	55	111,360	10	11,136
100–300	165	65,210	25	16,302
300–700	413	11,150	50	5,575
Over 700	1,035	1,430	100	1,430
		290,600	(13·6)	39,515

accuracy were obtained from an analysis covering only one-seventh of all the holdings. This not only considerably reduced the amount of coding and machine work, but also enabled work to proceed as soon as the information for the sample farms had been assembled and abstracted. In consequence it was possible to make the results of the analysis available a year or two sooner than would have been the case had the whole of the material had to be abstracted before analysis.

A further example is provided by the 1 per cent. sample of the returns for the 1951 Population Census of Great Britain (General Register Office, 1952 ,C'). The object of this sample was to provide data for a preliminary tabulation of certain aspects of the census information for the 1,225 census areas in England and Wales and the 1,026 areas in Scotland. The example is of interest in that it illustrates the problems arising in systematic sampling from short lists, which were briefly discussed in Section 3.6.

The general procedure of this census was as follows. Each census area was divided into enumeration districts (O.E.D.s), numbered consecutively from 1 within each area. In all there were 49,318 O.E.D.s in England and Wales with an average content of 268 households, and 9,730 in Scotland with an average content of 149 households. Each habitation in an enumeration district was listed before the actual census by complete traverse by the enumerator concerned, and the census schedules were subsequently numbered in list order. Each schedule covered one household. Institutions, etc., likely to contain more than 100 persons were listed separately.

To obtain the 1 per cent. sample the enumerators were instructed to provide a copy of each household schedule bearing a number ending in 25 if their O.E.D. number was odd, and 76 if even. Individuals in institutions, etc., were sampled similarly by taking lines on the schedules with numbers ending in 25 or 76.

A comparison of the sample and the full census is shown in Table 5.17.b.

TABLE 5.17.b—1951 CENSUS OF GREAT BRITAIN : COMPARISON OF THE 1 PER
CENT. SAMPLE AND THE PRELIMINARY COUNT

	Sample	1/100th of full census (preliminary count)	Excess or defect of sample	
			Amount	Per cent.
Great Britain—				
Total population . . .	488,395	488,411	− 16	− ·00
Households in O.E.D.s .	146,628	146,539	+ 89	+ ·06
England and Wales—				
Total population . . .	437,158	437,450	− 292	− ·07
Households in O.E.D.s .	131,973	131,998	− 25	− ·02
Scotland—				
Total population . . .	51,237	50,961	+ 276	+ ·54
Households in O.E.D.s .	14,655	14,541	+ 114	+ ·78

For England and Wales there is good agreement for the total number of households, but for Scotland there is an excess of 0·78 per cent. in the sample. This discrepancy was in fact of no consequence as the sample was adjusted so as to bring not only the estimate of the number of households but also that of the population in each census area into agreement with the preliminary count. This was done by the addition, removal or substitution of complete households, using a mechanical selection procedure. The required additions and removals were each less than 1 per cent. of the households in the sample.

As was pointed out in the report, the discrepancy for Scotland can be attributed to the fact that in census districts containing an odd number of O.E.D.s the selection numbers 25 and 76 will not be balanced, as that for the last O.E.D. will always be 25. If the terminal digits of the lengths of the O.E.D. lists are uniformly distributed over the range 00 to 99 this will lead to an average excess of approximately $\frac{1}{4}$ of a household for half the census districts. The expected excess for Scotland is therefore $\frac{1}{8} \times 1026$, i.e. $+ 128$. The discrepancy in the number of households for Scotland is therefore fully accounted for by this bias, but a similar adjustment for England and Wales would increase that discrepancy from −25 to −178.

Actually the terminal digits cannot be assumed to be uniformly distributed, which introduces further complications. The effect of this is discussed in Section 7.18, which gives a method of evaluating the expected error due to bias and the random sampling error if the distribution is known.

5.18   Adjustment of the results to compensate for defects in the sample

When the sampling procedure is defective in one respect or another, attempts are sometimes made to adjust the results in order to compensate for the defects. Thus it may happen that owing to defects in the selection of the sample or in the collection of the information, different classes of the population are found to be represented in incorrect proportions in the final sample. In such cases it is possible to adjust the results by weighting the different classes in such a manner as to compensate for the errors in the proportions.

This procedure must be clearly distinguished from the procedure of stratification after selection mentioned in Section 3.3. The validity of the latter procedure depends on the fact that the sample as a whole is random and therefore the selection from within strata is also random. If the proportions in the different classes are different because of defects in the sampling procedure, however, it is most unlikely that the selection from within these classifications will be fully random. Any adjustment of the type envisaged, therefore, although it may somewhat improve matters, must not be expected to eliminate by any means the whole of the defects.

Stratification after selection is a special case of the use of supplementary information of all kinds. Such adjustments, whether planned at the outset or decided on subsequently after examination of the data, are quite justified. The essential difference between these adjustments and between adjustments of the same type made in order to compensate for defects in the sampling procedure is that in the former the selection is random, except for permissible restrictions, whereas in the latter it may be biased in various ways.

In general, if the sampling procedure is defective it is best to report the results obtained without adjustment. At the same time data should be given indicating, so far as is possible, the deviations of the sample from the expected distributions. Thus if the proportions in the different classes of a classification are known for the population these may be presented alongside the parallel classification of the sample. Similarly the sample means of quantities for which the population means are known may be presented for comparison. Occasionally an adjustment of some of the more important values derived from the sample may be considered worth while, but in such cases the unadjusted results should also be presented.

The above remarks apply primarily to samples for which the sampling procedure is markedly defective. In cases in which there are slight defects, such as a minor degree of non-response, the application of some small adjustment, if this appears necessary, is more justified. If such adjustments are made, however, the fact should be clearly stated and their magnitude should be indicated.

The simplest way of dealing with non-response is to regard the non-respondents as similar to the remainder of the sample, i.e. to treat the sample as if it were a sample on a smaller number of units. With a stratified sample, the non-respondents in each stratum can be treated as the equivalent of respondents in that stratum. Alternatively some other appropriate classification can be used, as indicated above.

If follow-up methods have been used and there has been a good response to the follow-up, initial non-respondents who subsequently respond can be treated as a sub-sample of all initial non-respondents and weighted accordingly.

It is clear that if there is any difference between respondents and non-respondents the final non-respondents may be expected to be more like the initial non-respondents than the general population. This procedure was first, so far as I know, suggested by Professor D. V. Glass, and was used by him in the analysis of the Family Census (Section 4.10).

In this survey those who failed to provide the enumerator with the required information were sent a letter further explaining the purposes of the survey and requesting that the form be sent direct to the Royal Commission. Of the 230,000 initial non-respondents (i.e. 17 per cent. of the whole sample), 50,000 responded to this appeal. This 50,000 therefore constituted a sample (though a non-random one) of the 230,000, and the first 12,000 of the 50,000 replies were combined with the remainder of the sample with a weight of 230/12. This procedure was found to give overall birth-rates which corresponded very closely to those already known from other sources, whereas the original sample gave birth-rates which were substantially too high, owing to the fact that the majority of the initial non-respondents were women with few or no children.

5.19   Inferences from tabulations of survey data

Tables of the results of a complete census or survey will, apart from errors of recording, etc., give factual information on the counts, means and totals relating to the various subdivisions of the population shown in the tables. In sample censuses and surveys the tabulated values are subject to sampling errors, but apart from this are similarly factual in nature.

Factual statements relating to the population, and to contrasts between the various subdivisions, can therefore be made without qualifications other than those due to errors of ascertainment and inaccuracies due to sampling errors. When, however, we attempt inferences on cause and effect we are on much more uncertain ground. Over-simplified tables, though factually correct, can very easily lead to conclusions which differ markedly from those which emerge when more detailed cross-tabulations are examined. The two examples below illustrate this. They are here examined by approximate methods, suitable for a desk calculator. These provide an introduction to the more advanced treatment described in Chapter 9.

It must be clearly recognized, moreover, that even when all the relevant information from a survey is taken into account, inferences on causal relations must always be tentative. Thus the apparently higher yields of some of the potato varieties in the second example may be due in part or in whole to other undetermined and possibly unascertainable factors, as for instance that the farmers growing them are farming better land and are carrying out their farming operations with greater skill.

5.20   Illustrative examples : (a) seasonal incidence of milk fever

The first example is based on the results of a large survey on calf-rearing (Leech *et al.*, 1960). In the course of this survey information was obtained on

126

the incidence of milk fever in the parent cows. The older cows are much more subject to this disease—the incidence was found to be only 0·41 per cent. in the first two lactations, but increased to 9·38 per cent. in the sixth and subsequent lactations. There were also marked seasonal differences in incidence.

The first column of Table 5.20.a shows the percentage incidence in the different seasons aggregated over all lactations, i.e. the numbers affected

TABLE 5.20.a—MILK FEVER : ESTIMATES OF SEASONAL DIFFERENCES OF PERCENTAGE INCIDENCE

	Overall means	Means of sub-class means Unweighted	Means of sub-class means Weighted	Likelihood estimates
Jan.–April	2·67	3·33 (2·08)	2·22 (2·10)	2·03
May–July	4·17	6·20 (3·87)	3·99 (3·77)	3·78
Aug.–Sept.	4·25	7·97 (4·97)	5·38 (5·09)	5·26
Oct.–Dec.	3·50	6·23 (3·88)	4·06 (3·84)	3·82
Mean (wtd)	3·46	5·55 (3·46)	3·66 (3·46)	3·46

TABLE 5.20.b—PERCENTAGES OF COWS AFFECTED BY MILK FEVER AND NUMBERS OF CALVINGS

Lactation

	1–2	3	4	5	6+	Overall
			(a) *Percentage affected*			
Jan.–April	0·42	1·45	3·15	5·57	6·07	2·67
May–July	0·17	3·08	7·37	9·43	10·94	4·17
Aug.–Sept.	0·66	4·89	9·40	10·93	13·98	4·25
Oct.–Dec.	0·35	2·93	6·07	9·61	12·19	3·50
Overall	0·41	2·75	5·58	8·06	9·38	3·46
			(b) *Calvings*			
Jan.–April	3806	2137	1744	1184	2010	10881
May–July	2352	1006	706	488	841	5393
Aug.–Sept.	3169	1003	617	366	522	5677
Oct.–Dec.	5117	1740	1252	791	1042	9942
Overall	14444	5886	4319	2829	4415	31893

in the different seasons divided by the total numbers of calvings in those seasons. From these figures it might be concluded that the percentage incidence was about the same in May–July and August–September. This is indeed factually correct and is the information which, taken in conjunction with the number of calvings in the different seasons, is required to estimate the likely demands on the veterinary services. If, however, we examine the percentage incidence for the separate lactations a different picture emerges. These percentages are shown in Table 5.20.b, but their relations can be seen much more readily if the table is displayed in graphical form. This is done in Fig. 5.20. From this diagram it is immediately apparent that at each lactation the greatest incidence occurs in August–September, and that the incidence in May–July is about equal to that in October–December.

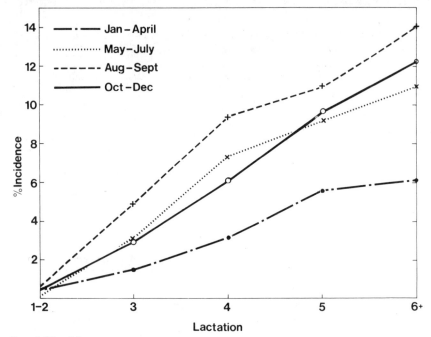

FIG. 5.20 —MILK FEVER: RELATION OF PERCENTAGE INCIDENCE TO LACTATION NUMBER
AND SEASON OF THE YEAR

This discrepancy between the overall mean incidence and the incidence in the separate lactations is due to seasonal variation in the distribution of calvings over the different lactations, i.e. to *disproportionate frequencies* in the table of counts. This can be seen directly from the table of numbers of calvings (Table 5.20.b), but is more readily seen if the numbers are expressed as percentages of the total numbers for each season. These are shown in Table 5.20.c. Bearing in mind the steady increase in percentage incidence with increasing lactation number, it is easy to see that this seasonal difference in the distribution of calvings explains the discrepancy.

Two incidental points may be made here: first, that graphical presentation often displays the information contained in a table much more clearly than do the numerical values; second, that tables of counts are usually best shown as percentages of some appropriately chosen margin.

TABLE 5.20.c—MILK FEVER : DISTRIBUTION OF CALVINGS OVER
THE DIFFERENT LACTATIONS (per cent.)

	Lactation					Calvings	
	1–2	3	4	5	6+	No.	Per cent. of total
Jan.–April	35	20	16	11	18	10881	34·1
May–July	44	19	13	9	16	5393	16·9
Aug.–Sept.	56	18	11	6	9	5677	17·8
Oct.–Dec.	51	18	13	8	10	9942	31·2
Overall	45	18	14	9	14	31893	100·0

Although Fig. 5.20 gives a good visual picture of the seasonal effects, it does not directly provide an estimate of their magnitude in quantitative terms. If the values in the body of the percentage incidence table (Table 5.20.b) are of sufficient accuracy, we can take the straight means of these, or alternatively we can take weighted means with weights proportional to the numbers in the different lactation classes, e.g. the percentages shown in the last line of Table 5.20.c. The use of weighted means requires a little more computation, but has two advantages over straight means. It is more accurate, since greater weight is on the average given to cells containing the greater numbers of units. It also gives estimates that refer to a hypothetical distribution of units which conforms, as far as the marginal distributions are concerned, with that actually existing, whereas straight means represent a uniform distribution over the whole table.

Columns 2 and 3 of Table 5.20.a give the results of these calculations. The value of the weighted mean for January–April, for example, is given by

$$(0·42 \times 45·3 + 1·45 \times 18·5 + \ldots)/100 = 2·22$$

The unweighted values are all substantially larger than the overall means, as is to be expected. Fig. 5.20 shows that the seasonal differences in the percentage incidence are roughly proportional to their magnitude : for comparative purposes, therefore, they may be adjusted proportionally to bring their weighted mean to equality with the population mean, 3·46. The adjusted figures are shown in brackets.

Neither of these sets of estimates makes full use of the available information, as the relative accuracy of the individual percentages is not fully taken into account. This can be done by fitting a suitable model by maximum likelihood ; in this case a suitable model is one which represents each sub-class percentage as the product of two factors, the value of one being dependent on the lactation, the other on the season. To fit such a model a computer program is necessary ; the basis of the method is described in Chapter 9.

The likelihood estimates are given in the last column of Table 5.20.a. Both the unweighted and weighted means, after adjustment, differ little from the likelihood estimates ; the average of the discrepancies of the weighted estimates is about half that for the unweighted estimates.

The substantial agreement of the above three sets of estimates does not, of course, give any measure of the accuracy with which the seasonal effects have been determined. All the estimates are based on the same data, and are therefore subject to the same sampling errors. The discrepancies merely provide rough indications of the loss of accuracy due to the use of estimation processes which do not make the best use of the data, i.e. in statistical terminology are not fully efficient.

5.21 Illustrative examples : (b) yields of potato varieties

The data for the second example are from results obtained in the course of an investigation into the blackening of potatoes on cooking, the data on yields being collected from the farmers when the samples were taken, together with a considerable amount of information on fertilizers, cultural practices, etc. Approximately 180 fields were selected in each region. The selection within regions was not strictly random, but can be regarded as substantially so for the purpose of the present discussion. The sample was confined to the five named varieties, but was not stratified by varieties.

Table 5.21.a gives the mean yields, classified by variety and region, together with the numbers of fields on which they were based. The marginal means of this table indicate that there are substantial differences between varieties, and also that yields are greater in the more northerly regions. The distribution of varieties over the different regions is very irregular, however, and consequently the apparent regional differences may be distorted by varietal differences, and vice versa.

This is confirmed by inspection of the sub-class means. Scotland, for example, gives higher yields than every other region for all varieties except Arran Banner, of which there are only 8 fields in Scotland. The Northern region, on the other hand, does not show any consistent differences from the other English regions. The high value of the marginal mean for the Northern region is due, in part at least, to the fact that very little King Edward, which yields about 2 tons per acre less than the other varieties, was grown in this region.

In this example quantitative estimates cannot be obtained directly from means of the sub-class means as there are four empty cells in the table. We can, however, circumvent this difficulty by inserting estimated values, calculated on some plausible basis. As there are no marked differences between the English regions, it is reasonable to take means over these regions, weighted by the numbers of fields. These are shown in brackets in Table 5.21.a. The value for Kerr's Pink, for example, is given by

$$(7{\cdot}61 \times 36 + 6{\cdot}61 \times 13)/49 = 7{\cdot}34$$

Unweighted and weighted means can now be calculated as in the milk fever example. The results are shown in Table 5.21.b, together with the least-squares estimates, here equivalent to the maximum likelihood estimates, since an additive instead of a product model is appropriate (Section 9.6).

130

TABLE 5.21.a—POTATO SURVEY : TWO-WAY CLASSIFICATION OF THE DATA BY
VARIETIES AND REGIONS

Means of yields per acre (tons)

	Scot.	North	E. Mid.	South	West	All
Majestic . .	9·46	8·19	8·42	8·15	8·27	8·38
King Edward .	7·65	5·71	6·34	5·87	5·49	6·25
Great Scot .	9·22	7·57	(7·76)	8·17	7·78	8·23
Arran Banner .	9·12	9·24	(9·12)	7·22	9·54	9·12
Kerr's Pink .	8·29	7·61	(7·34)	(7·34)	6·61	7·90
All . .	8·52	8·05	7·49	7·27	7·64	7·78

Numbers of fields

	Scot.	North	E. Mid.	South	West	Total
Majestic . .	37	75	104	101	76	393
King Edward .	42	14	85	66	43	250
Great Scot .	18	14	—	6	18	56
Arran Banner .	8	38	—	9	29	84
Kerr's Pink .	69	36	—	—	13	118
Total .	174	177	189	182	179	901

There is little difference in this example between the weighted and unweighted means for regions. Those for varieties are almost identical as the
weights are approximately equal. The regional means agree reasonably well
with the least-squares estimates, and correctly indicate that the Northern
region is similar to the other English regions. The varietal means have
corrected the overestimate of the Kerr's Pink variety, but have seriously
underestimated the yield of Arran Banner. This is mainly a consequence of
the excessive weight given to the very low yield of 7·44, based on only 9 fields,
in the Southern region.

A further dodge which is sometimes useful is to pool classes which inspection indicates are similar before taking the means of the sub-class means. In
this example the English regions can reasonably be pooled, thus avoiding the
complication of the empty cells and reducing the inaccuracies due to sparsely

131

occupied cells. The results are shown in the last column of Table 5.21.b. This gives a better estimate for Arran Banner.

An alternative procedure, when a comparison is required between two classes only of a factor which is cross-classified by one or more additional factors, is to take a weighted mean of the differences of the means of each pair of sub-classes. Maximum accuracy will be attained when the weights are inversely proportional to the squares of the standard errors of these differences.

TABLE 5.21.b.—POTATO SURVEY : ESTIMATES OF VARIETAL AND REGIONAL DIFFERENCES

| | Overall means | Means of sub-class means | | Least-squares estimates | From pooled English Regions |
		Unweighted	Weighted		
Majestic	8·38	8·50	8·49	8·50	8·49
King Edward	6·25	6·21	6·21	6·27	6·29
Great Scot	8·23	8·10	8·09	8·11	8·04
Arran Banner	9·12	8·85	8·84	9·27	9·12
Kerr's Pink	7·90	7·44	7·43	7·41	7·53
Scotland	8·52	8·75	8·76	8·77	8·76
North	8·05	7·66	7·49	7·59	
E. Midlands	7·49	7·80	7·73	7·77	7·55
South	7·27	7·35	7·33	7·33	
West	7·64	7·54	7·37	7·45	
Mean (wtd. or unwtd.)	7·78	7·82	7·73	7·78	7·78

With independent samples in each sub-class the square of the standard error of a difference is equal to the sum of the squares of the standard errors of the two means (Section 7.5). Under certain circumstances, which will be apparent from a study of Chapter 7, and in particular when the selection from within sub-classes is effectively random and the standard deviation per unit is constant, the standard errors are inversely proportional to the square roots of the numbers of units in the sub-classes (Section 7.1). In this case the reciprocals of the weights must be taken proportional to the sums of the reciprocals of the pairs of sub-class numbers, i.e. if n_1 and n_2 are a pair of sub-class numbers the weight can be taken equal to w, where

$$\frac{1}{w} = \frac{1}{n_1} + \frac{1}{n_2}$$

The procedure may be illustrated by using it to estimate the difference between the Scottish and Northern regions. The calculations are shown in Table 5.21.c. The weight for Majestic, for example, is given by $1/w = 1/37 + 1/35$. The weighted mean, $+1\cdot08$, is obtained by dividing the total of wz by the total of w. With a modern desk calculator calculations of this kind can be performed very rapidly, without writing down any intermediate values.

TABLE 5.21.c—POTATO SURVEY : ESTIMATE OF DIFFERENCE OF SCOTTISH AND NORTHERN REGIONS FROM WEIGHTED MEAN OF VARIETAL DIFFERENCES

	Difference z	Weight w	wz
Majestic	$+1\cdot27$	25	$+31\cdot75$
King Edward	$+1\cdot94$	10	$+19\cdot40$
Great Scot	$+1\cdot65$	8	$+13\cdot20$
Arran Banner	$-0\cdot12$	7	$-0\cdot84$
Kerr's Pink	$+0\cdot68$	24	$+16\cdot32$
Total	—	74	$+79\cdot83$
Weighted Mean	$+1\cdot08$		

A comparison of the various estimates of the above difference is given in Table 5.21.d. Provided that the true difference between the two regions is the same for all five varieties, the estimates (2), (3) and (4) are all estimates

TABLE 5.21.d—POTATO SURVEY : ESTIMATES OF THE DIFFERENCE BETWEEN THE SCOTTISH AND NORTHERN REGIONS

	Estimate	Efficiency per cent.
Varietal effects not eliminated		
(1) Difference between overall means	0·47	—
Varietal effects eliminated		
(2) Difference between unweighted means of sub-class means	1·09	68
(3) Difference between weighted means of sub-class means	1·27	70
(4) Weighted mean of sub-class differences	1·08	92
(5) Least squares	1·18	100

of this difference. The same is true of the least-squares estimate, provided the above condition holds for all five regions. The efficiencies of the various estimates, i.e. the reciprocals of their relative variances, are then those given in the table (see Example 7.5.a and Section 8.1).

If the above condition does not hold, the estimates are affected differentially, the first three in a straightforward manner by the differences in the weights, the last in a more complex manner. As is shown however in Example 7.5.a and in Section 9.7, the variation in the apparent regional differences for the different varieties in this example is only slightly greater than would be expected from sampling errors.

To conclude, it is clear from the above results that although examination of sub-class means on a desk calculator often gives useful indications of distortions in the marginal means of tables due to cross effects, weighted or unweighted means of them cannot be relied on to give estimates of the effects of the factors which approximate closely to the best estimates if the numbers in some of the sub-classes are small. It is important, therefore, to have available a suitable computer program for the derivation of least-squares and maximum likelihood estimates, particularly when, as is commonly the case, the effects of more than two factors have to be disentangled. The uses of such a program are illustrated in Chapter 9.

CHAPTER 6

ESTIMATION OF THE POPULATION VALUES

6.1 Possibility of alternative estimates

In this chapter we shall deal with the derivation of estimates of the population values from the numerical results obtained in the sampling. A simple example of such an estimate is provided by the arithmetic mean of the sample values of a random sample. It is well known that this mean provides an estimate of the mean of the population from which the sample was drawn, though it will not, owing to sampling errors, be exactly equal to the mean of the population.

The arithmetic mean of the sample values is not the only possible estimate of the population mean. We might, for instance, take the median, *i.e.* the central value, or the geometric mean, *i.e.* the antilogarithm of the mean of the logarithms of the sample values, or even the mean of the highest and lowest values in the sample.

In addition to estimates such as the mean and the median, which can be derived from a given set of values independently of any supplementary information associated with these values, there are further alternative estimates which can be derived by taking account of such supplementary information as is available, either qualitative or quantitative. Thus, as has been mentioned in Section 3.3, if the numbers of units from the whole population falling in the different strata of some stratification are known, a random sample can be adjusted so that the different strata are represented in their correct proportions. Similarly, supplementary information on a quantitative character can be used in various ways to provide estimates which will in general be more accurate than the simpler estimates which do not utilize this information.

In deciding which is the best estimate for any given type of sampling three different criteria have to be considered. These are, absence of bias, accuracy (or, as it is technically known, efficiency), and computational convenience. In the case of a random sample, if the population values are normally distributed—the meaning of this term is explained in Chapter 7—the arithmetic mean will provide an estimate which is both free from bias and, apart from supplementary information, of maximum accuracy. It is also sufficiently simple computationally for practical use. More important, the mean will remain an unbiased estimate of the population mean whatever the form of the distribution of the population values, though it will not necessarily be the most accurate estimate that could be devised. The mean also has the incidental advantage that the sampling errors to which it is subject can be relatively easily assessed, and are not greatly dependent on the form of the distribution of the population values.

In this book we do not propose to do more than list what appear to be the most useful estimates for any given type of sampling, and give examples of the computations involved. Any discussion of questions of bias and relative efficiency requires advanced mathematical statistical theory. In general the recommended estimates are free from any important source of bias ; where this is not the case the circumstances in which bias can arise are indicated.

Estimates are required not only for the whole population, but frequently also for the different parts of it which constitute domains of study. The formulæ of estimation which are applicable to the whole population are in general applicable also to the separate domains, and need not be discussed separately. In certain cases adjustments which can be applied to the population estimates cannot be applied to the estimates for the different domains. Thus if the population mean of a supplementary variate is known, but not the means for the different domains of study, adjustment by means of supplementary information can be applied only to the estimates of the whole population. In like manner the gain in accuracy due to stratification is sometimes different for the population and domain estimates. If the domains cut across the strata, for example, the errors of the domain estimates may only be slightly reduced by the stratification.

6.2   Notation

It is important, both in discussion of the problem of estimation and in the mathematical formulæ, to make a clear distinction between estimates of the population values and the population values themselves. The present convention in mathematical statistics is to denote the population parameters by Greek letters and the corresponding estimates by the corresponding Latin letters. This convention, however, is difficult to apply consistently, and is in any case more appropriate for infinite hypothetical populations than for the finite populations met with in sampling. In the present manual we have for the most part adopted the convention of denoting the population values by bold type, the corresponding estimates of these values by Gill Sans type, and values appertaining to the selected sampling units by ordinary italic type. Thus, with a quantitative character or *variate* y, the values for the selected sampling units will be denoted by y (with or without suffices as necessary), the mean of these values will be denoted by $\bar{y}$ (following the ordinary convention), the estimate of the mean of the population by $\mathsf{\bar{y}}$, and the true mean of the population by $\mathbf{\bar{y}}$. With a random sample we shall have $\mathsf{\bar{y}} = \bar{y}$, but $\mathsf{\bar{y}}$ differs from $\mathbf{\bar{y}}$ by the sampling error. Totals for the population are indicated by capitals, summation over the sample values by S, and summation over the different strata by Σ.

In certain types of estimation we shall be concerned with the use of supplementary information, such as size of unit, which is known not only for the selected sampling units but also for the whole of the population, or in the case of two-phase sampling, for the larger number of units selected at the first phase. A variate representing quantitative supplementary information will be denoted by x.

Even when information on a variate is not available for the whole population, it may be necessary to make an estimate of the population values for some standardized value of this variate. The letter x will also be used to denote a variate of this type.

The following is a list of the principal symbols employed in this chapter :—

b,	estimated regression coefficient.
f,	working sampling fraction.
$\mathbf{f}$,	exact sampling fraction.
g,	working raising factor $(= 1/f)$.
g,	exact raising factor $(= 1/\mathbf{f})$.
i (suffix),	denotes values belonging to a particular stratum i.
n,	number of units in the sample.
N, N,	number of units in the population, and its estimate.
p, p,	proportion of units in the population possessing a given attribute, and its estimate.
r,	ratio y/x.
$\bar{r}$, $\bar{r}$,	ratio $\mathbf{Y}/\mathbf{X}$, and its estimate.
S,	summation over the units of the sample.
S_i,	summation within stratum i.
Σ,	summation over the strata.
u,	number of units in the sample possessing a given attribute.
U, U,	number of units in the population possessing a given attribute, and its estimate.
x,	supplementary quantitative variate, such as size of unit.
X, X,	total of x for the population, and its estimate.
y,	quantitative variate under investigation.
Y, Y,	total of y for the population, and its estimate.
$\bar{r}$, $\bar{x}$, $\bar{y}$,	means of r, x, y, for the sample.
$\bar{x}$, $\bar{y}$, $\bar{x}$, $\bar{y}$,	means of x and y for the population, and their estimates.

6.3 General rules

There are certain fundamental rules of estimation which apply to all types of sampling. These are :—

Rule 1—The population total of a quantitative variate

To estimate totals for the population multiply all sample values by their raising factors (equal to the reciprocals of the sampling fractions) and sum the raised results.

Rule 2—Number of units in the population

To estimate the number of sampling units in the population follow Rule 1, scoring each selected sampling unit as 1.

Rule 3—The population mean of a quantitative variate

Divide the estimated total of the variate for the population by the estimated number of units in the population.

Rule 4—Proportion (or percentage) of units possessing a given qualitative character

Proceed as for a quantitative variate, scoring all units possessing the given character as 1 and all others as 0. Divide the estimated total score by the estimated number of units in the population.

Rule 5—Ratio of two quantitative variates

Estimate the totals of the two quantitative variates for the population by Rule 1 and take the ratio of these totals. (Rules 3 and 4 are special cases of Rule 5.)

In cases in which the probability of selection of all units is the same (uniform sampling fraction or, in the case of multi-stage sampling, uniform overall sampling fraction), the first four rules can be condensed into the simple general rule that means and proportions in the population are estimated by the corresponding means and proportions in the sample, and totals and numbers in the population are estimated by multiplying the corresponding totals and numbers by the common raising factor.

The above rules cover most of the methods of estimation discussed in this manual except those involving regression, which cannot easily be summarised in simple rules. They give rise to the formulæ of estimation set out in the following sections of this chapter.

6.4 Random sample

Number :

$$N = gn$$

N will be equal to N except for minor discrepancies due to the use of a working sampling fraction which does not give an integral number of sampling units. If N is known then the true sampling fraction f equals n/N and the true raising factor g equals N/n.

Mean of a quantitative variate :

$$\bar{y} = \bar{y} = \frac{1}{n} S(y) \qquad (6.4.a)$$

Total of a quantitative variate :

$$Y = g\, S(y) = N\bar{y}$$

or more accurately, if N is known, and differs from N,

$$Y' = g\, S(y) = N\bar{y} \qquad (6.4.b)$$

Proportion possessing a given attribute :

$$P = \frac{u}{n}$$

138

Number possessing a given attribute :

$$U = gu = Np$$

or, more accurately,

$$U' = gu = Np \qquad (6.4.c)$$

The same formulæ of estimation will hold for systematic samples from lists, etc.

Example 6.4.a

In a housing survey of a town a systematic sample from a list of all houses was taken with a sampling fraction of 1/50. 627 houses out of a total of 8491 in the sample were classified as defective. What is the estimated number and percentage of defective houses in the town ?

$$\text{Percentage defective} = 100\,p = 100 \times \frac{627}{8491} = 7\cdot38 \text{ per cent.}$$

$$\text{Total number defective} = U = 50 \times 627 = 31{,}350$$

Example 6.4.b

If the values in Table 6.4 are taken to represent measurements on a random sample of 20 objects, selected from a batch of such objects with a sampling

TABLE 6.4—SAMPLE OF 20 MEASUREMENTS

6·2	8·0	8·2	11·0
13·8	12·0	8·7	10·3
8·0	10·7	8·5	14·6
7·6	9·1	10·1	8·0
10·3	10·4	9·3	9·0

fraction of 1/25, estimate the mean measurement of the batch, and the total of all the measurements of the batch.

$$N = 25 \times 20 = 500$$

$$S(y) = 193\cdot8$$

$$\bar{y} = \frac{1}{20} \times 193\cdot8 = 9\cdot69$$

$$Y = 25 \times 193\cdot8 = 4845$$

If the number in the batch is known to be 507, a slightly more accurate estimate of the total is

$$Y' = 507 \times 9\cdot69 = 4913$$

6.5 Stratified sample with uniform sampling fraction

The formulæ for a random sample hold, except that if the numbers in the different strata N_i are known, and differ from N_i, the formula 6.4.b is replaced by

$$Y' = \Sigma (N_i \, \bar{y}_i) \qquad\qquad (6.5.a)$$

and the formula 6.4.c by

$$U' = \Sigma (N_i \, p_i) \qquad\qquad (6.5.b)$$

with corresponding slight increases in accuracy in $\bar{y}$ and p, if they are derived from these estimates by division by N.

Example 6.5

Table 6.5.a shows the wheat acreages of the stratified random sample of 1 in 20 Hertfordshire farms described in Section 3.7. Estimate the total wheat acreage of the county and the mean acreage of wheat per farm (*a*) from the data of the sample alone, (*b*) given the total number of farms in each size-group. Estimate also the number of farms growing wheat.

TABLE 6.5.a—HERTFORDSHIRE FARMS, 1939: ACREAGES OF WHEAT IN A STRATIFIED RANDOM SAMPLE OF 1 IN 20 FARMS (STRATIFIED BY ACREAGES OF CROPS AND GRASS)

Size-group	3		4		5		6
Acres	21–50		51–150		151–300		301–
No. of farms	18		26		20		13
	8	0	49	19	20	56	72
	0	5	10	14	24	18	92
	0	0	27	4	30	17	69
	0	0	33	0	59	32	78
	0	0	4	12	17	71	51
	0	9	30	13	70	48	84
	0	0	0	0	80	70	0
	8	0	0	16	0	62	102
	5	0	13	28	36	0	13
			0	5	0	0	92
			27	23			158
			10	22			62
			24	3			0
TOTAL	35		386		710		873

Size-group 1 (1–5 acres): 22 farms, no wheat.
Size-group 2 (6–20 acres): 26 farms, 7 acres of wheat on 1 farm.

The results are summarized in Table 6.5.b. The estimate of the total area of wheat in the county from the sample is

$$Y = 20 \times 2011 = 40,220 \text{ acres}$$

The mean area of wheat per farm is

$$\bar{y} = 2011/125 = 16 \cdot 1 \text{ acres}$$

The number of farms growing wheat is

$$U = 20 \times 54 = 1080$$

TABLE 6.5.b—SUMMARY OF SAMPLE OF TABLE 6.5.a

Size-group, acres	No. of farms in sample	Farms with wheat in sample		Wheat acreage in sample		No. of farms in county	Total wheat acreage
		No.	Pro-portion	Total	Mean		
1–5	22	0	·000	0	0·0	435	0
6–20	26	1	·038	7	0·3	519	160
21–50	18	5	·278	35	1·9	357	680
51–150	26	21	·808	386	14·8	519	7,680
151–300	20	16	·800	710	35·5	400	14,200
301–	13	11	·846	873	67·2	266	17,880
ALL	125	54	·432	2,011	16·1	2,496	40,600

If the total number of farms in each size-group is known, the estimate of Y can be calculated with slightly more accuracy by using the size-group means, as shown in the last three columns. This gives an estimate Y' of the total area of wheat of 40,600 acres, and a mean area per farm $\bar{y}'$ of 40,600/2496 = 16·3 acres. The gain in accuracy is here quite trivial, since the variation within each stratum is large relative to the mean of that stratum.

The number of farms growing wheat can be estimated similarly from the proportions in the size-groups, giving

$$U' = 0 \times 435 + 0 \cdot 038 \times 519 + \ldots = 1083$$

Again the gain in accuracy is trivial.

6.6 Random sample, stratified after selection

The means of, or proportions in, the different strata must be calculated separately, and formulæ 6.5.a and 6.5.b used, with division by N for estimates of $\bar{y}$ and p.

Example 6.6

Table 6.6.a shows the data, including acreages of crops and grass, for the random sample of 1 in 20 Hertfordshire farms described in Section 3.7. Estimate the total area of wheat and the number of farms growing wheat (*a*) directly from the sample, (*b*) by stratification by size, given the total numbers of farms in the size-groups of Table 6.5.b.

TABLE 6.6.a—HERTFORDSHIRE FARMS, 1939: ACREAGES OF CROPS AND GRASS (1ST COLUMN), AND OF WHEAT (2ND COLUMN), OF A RANDOM SAMPLE OF 1 IN 20 FARMS (CLASSIFIED BY DISTRICTS AFTER SELECTION)

District 1 15 farms		District 3 40 farms			District 4 24 farms		District 5 4 farms		District 6 24 farms	
188	16	370	67	40	0	11	0	4	0	8	0
60	0	26	0	28	0	6	0	312	102	87	14
192	0	369	58	221	59	543	80	8	0	6	0
48	0	212	45	31	0	822	265	11	0	44	0
44	0	153	20	6	0	654	112	———	———	4	0
79	33	287	44	34	0	3	0	335	102	614	72
14	0	28	0	316	75	158	50			192	20
465	92	14	0	116	33	4	0			10	0
197	0	4	0	4	0	68	27	*District 7*		24	0
163	0	17	0	409	102	55	12			2	0
198	0	2	0	6	0	4	0	10 farms		9	0
78	0	3	0	115	0	2	0			3	0
6	0	7	0	19	0	192	24	128	5	2	0
35	0	6	0	274	6	4	0	4	0	120	24
168	0	335	82	3	0	491	24	46	0	58	0
———	———	4	0	144	0	224	28	181	20	20	0
1,935	141	1	0	3	0	280	75	17	0	30	0
		4	0	482	62	90	0	24	0	197	6
		180	0	156	28	3	0	10	0	14	3
District 2		120	11	302	71	3	0	36	0	32	6
				———	———	6	0	12	0	2	0
8 farms				4,851	763	4	0	89	0	285	29
						161	80	———	—	138	0
8	0					246	60	547	25	126	0
294	29					———	———			———	———
597	107					4,034	837			2,027	174
8	0										
2	0							GRAND TOTAL,			
200	65							125 farms :—		15,114	2,301
14	0										
262	58										
———	———										
1,385	259										

TABLE 6.6.b—HERTFORDSHIRE FARMS, 1939 : ESTIMATION OF WHEAT ACREAGE
FROM THE RANDOM SAMPLE OF 1 IN 20 FARMS (TABLE 6.6.a) STRATIFIED BY
SIZE-GROUPS AFTER SELECTION

Size-group acres	No. in sample	Farms with wheat		Acreage of wheat		No. of farms in county	Total for county
		No.	Pro-portion	Total	Mean		
1–5	25	0	0	0	0	435	0
6–20	26	1	·038	3	0·1	519	50
21–50	16	1	·062	6	0·4	357	140
51–150	17	8	·471	159	9·4	519	4,880
151–300	26	20	·769	762	29·3	400	11,720
301–	15	15	1·000	1,371	91·4	266	24,310
	125	45	·360	2,301	18·4	2,496	41,100

(a) Total area of wheat = 2301 × 20 = 46,020 acres.
   Number of farms growing wheat = 20 × 45 = 900.

(b) Classifying the data by size-groups (crops and grass) the numbers and
   totals shown in Table 6.6.b are obtained. The mean wheat acreage
   is then calculated for each size-group, multiplied by the total number
   of farms in that size-group, and the products summed, giving an
   estimated total wheat acreage of 41,100. Similarly, using the proportion
   of farms with wheat instead of the mean acreage for each size-group,
   the estimated number of farms growing wheat is found to be
   ·038 × 519 + ·062 × 357 + . . . = 860.

6.7   Stratified sample (variable sampling fraction)

$$N = \Sigma \, (g_i \, n_i)$$
$$Y = \Sigma \, (g_i \, S_i \, (y))$$
$$\bar{y} = Y/N$$
$$U = \Sigma \, (g_i \, u_i)$$
$$p = U/N$$

If the N_i are known, the alternative formulæ 6.5.a and 6.5.b, with division
by N for $\bar{y}$ and p, are slightly more accurate.

Example 6.7

Table 6.7.a shows the Hertfordshire farm data for the stratified systematic sample with a variable sampling fraction described in Section 3.7. Estimate the total wheat acreage and the number of farms growing wheat.

TABLE 6.7.a—HERTFORDSHIRE FARMS, 1939 : STRATIFIED SYSTEMATIC SAMPLE OF WHEAT ACREAGES, WITH A VARIABLE SAMPLING FRACTION (CLASSIFIED BY DISTRICTS)

Size-group :	1–5	6–20	21–50	51–150		151–300			301–500			501–	
Sampling fraction :	Nil	1/200	1/60	1/20		1/10			1/5			1/3	
No. in sample :	0	3	6	26		40			43			17	
District													
1	—	—	0	0 6	30	17 28	18	0	172 56	0	92	114	
2	—	0	10	0 40	0	30 55	16 62	50	50 63 186 104	49 72 124	121 100 105	119 107 101 160	
3	—	—	0 0	25 10 0	5 0	0 0 0 61	0 41 24	77 42 25	67 58 78 86	22 75 94 97	5 51 126	195 120	
4	—	0	17	28 8 5	24 0	42 54 44	0 60 6	24 75	88 94 121 18	65 115 80 40	58 98 92 120	268 265 112 240 209	260 260 155 168
5	—	—	—	0 27	0	22 27	31 0	32	66	142	26	—	
6	—	0	0	0 0 19	0 0	38 56	0 17	0 29	0 0			72	
7	—	—	—	0 14	0	0	60		16			—	
TOTAL	—	0	27	214		1,163			3,292			2,925	

The calculations are shown in Table 6.7.b. They follow the same lines as before, except that the sample total for each stratum must be raised separately. Using the working sampling fractions we obtain estimates of 42,765 acres of wheat and 911 farms growing wheat.

TABLE 6.7.b—SUMMARY OF SAMPLE OF TABLE 6.7.a

Size-group acres	No. in sample	No. with wheat	Total acreage	Raising factor	Raised totals		Mean acreage per farm
					No.	Acreage	
1–5	0	—	—	—	—	—	—
6–20	3	0	0	200	0	0	0
21–50	6	2	27	60	120	1,620	4·5
51–150	26	12	214	20	240	4,280	8·2
151–300	40	30	1,163	10	300	11,630	29·1
301–500	43	40	3,292	5	200	16,460	76·6
501–	17	17	2,925	3	51	8,775	172·1
	135				911	42,765	

6.8 Use of supplementary information in estimation

As already indicated, supplementary information on a quantitative character, the values of which are known for all the units of the population, can be used as the basis of stratification, or for the adjustment of an unstratified sample by stratification after selection. Alternatively, as mentioned in Section 2.8, such information can be used directly without stratification. Two methods, the *ratio method* and the *regression method*, are available. In either case only the total or mean of the supplementary variate for the whole population need be known (in addition to the values for the selected sampling units). The ratio method is simpler computationally, but the regression method is in certain circumstances more accurate.

In the ratio method, the ratio of Y/X in the population is estimated from the sample, the estimated ratio being multiplied by the total X of x for the population to give the estimated total Y of y for the population. The method of estimation must be such that bias is avoided. As already explained in Section 2.6, the appropriate estimate of the ratio for a random sample is $S(y)/S(x)$ or $\bar{y}/\bar{x}$. More generally, Rule 5 of Section 6.3 will give an unbiased estimate, though in certain cases separate values of the ratio may be estimated for the different strata as described in Sections 6.10 and 6.11.

In the regression method the average change of y for unit change of x (known as the *regression coefficient*) is estimated, and this coefficient is used to adjust the sample results for any discrepancy between the mean size of unit in the sample and in the population.*

The contrast between the ratio and regression methods is illustrated in Fig. 6.8. The data plotted are those of Table 6.12. The dots represent the x and y values of the sample points, the sample mean $(\bar{x}, \bar{y})$ being M. Q' represents the known population mean $\bar{\bar{x}}$ of the supplementary variate, which differs from $\bar{x}$ by QQ'. The line OMD through the origin and the mean represents the ratio $\bar{y}/\bar{x}$ given by the sample, and the ordinate $P_1 Q'$ of the point P_1 on this line, equal to $(\bar{y}/\bar{x}) \bar{\bar{x}}$, gives the adjusted estimate of the population mean by the ratio method. The regression line AMB also passes through the mean, and has a slope b equal to the regression coefficient.

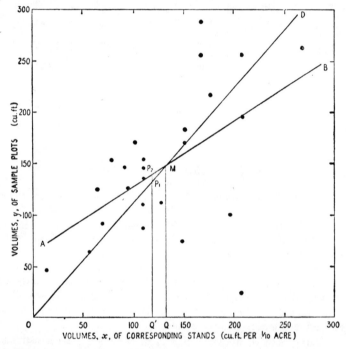

FIG. 6.8—USE OF SUPPLEMENTARY INFORMATION : RATIO AND REGRESSION METHODS
(DATA OF TABLE 6.12)

This line has the property that the sum of the squares of the vertical distances of the sample points from it is minimum. The adjusted estimate by the regression method is given by the ordinate $P_2 Q'$ of the point P_2, and equals $\bar{y} + b (\bar{\bar{x}} - \bar{x})$.

* More complicated functions, such as regression on two or more variates or the double ratio described in Section 8.6.1, can be used where appropriate.

The regression method therefore differs from the ratio method in that in the former the straight line which best fits the sample values is taken, whereas in the latter the line through the origin is taken. When the supplementary variate x represents size of unit, the true regression line generally passes through the origin, though curvature of this line may result in the best-fitting straight regression line not passing through, or even very close to, the origin. Nevertheless, in most census work in which x represents size, or some variate closely correlated with size, the greater simplicity of the ratio method outweighs any small gain in accuracy resulting from the use of regression.

It may be noted that in large samples the regression line can be plotted by grouping the data according to the x values and plotting the means of y for the different groups.

The formulæ for regression have been included in this book, not because it is expected they will be very commonly used in census work, but because the regression method represents an important part of sampling procedure, without which no account of sampling methods would be complete, and because the calculation of the sampling errors to which a balanced sample is subject can only be made by use of the regression concept.

If the population mean of x is estimated from observations at the first phase of two-phase sampling, the observations for y being obtained at the second phase, the same formulæ of estimation hold, the estimate $\bar{x}_1$ being substituted for $\bar{x}$.* If, however, the sampling is single-phase, and the estimate $\bar{x}$ from the sample is substituted for $\bar{x}$, the formula $Y = \bar{y}$ appropriate to a random sample without supplementary information is obtained. In other words, there is no gain in accuracy in the population estimate unless $\bar{x}$ is known or alternatively is estimated from a larger sample than is available for y.

In addition to their use in the adjustment of the population estimate of the mean or total of y when the mean or total of x is known for the population, or is determined from the first phase of two-phase sampling, ratios and regressions are of use for the purpose of obtaining estimates of the means of y for some standardized value of x. Hence comparisons of different parts of the population can be made, freed from the effects of variation in the average values of x. In the case of ratios this is equivalent to comparing the values of ratios themselves : thus in an agricultural survey we may consider such quantities as number of sheep per 100 acres instead of number of sheep per farm. Regression enables similar standardization to be made in cases in which the ratio method is inappropriate ; in a nutrition survey, for example, it may well be found that the amount of malnutrition varies with size of family, but the relation will not be proportionate. In large-scale surveys, however, standardization of this type can equally well be made by using the size-group means, thus avoiding the trouble of calculating regression coefficients.

The formulæ in the following sections are given for a quantitative variate. Formulæ for the proportion of units possessing a given attribute can be derived

---

* Note that the observations on x for the units for which y is also observed are part of the first-phase information, and must be included when calculating this estimate.

by scoring each unit as 1 if it has the attribute and zero otherwise. If the supplementary variate x represents size, the proportion of the attribute per unit size may be required, in which case a unit of size x will be scored x if it has the attribute, and zero otherwise.

Example 6.8

The data of Table 6.8 are extracted from the *Report of the National Farm Survey of England and Wales* (Ministry of Agriculture and Fisheries, 1946, G). They give the rents of holdings per acre of crops and grass, classified by size-groups, in Berkshire and Cornwall. Calculate rents per acre standardized for size of holding, in the proportions in which the different size-groups occur in the whole country.

TABLE 6.8—RENTS PER ACRE FOR BERKSHIRE AND CORNWALL

Size-group acres	Rent per acre (shillings)		Proportionate areas of different size-groups in whole country
	Berkshire	Cornwall	
5–25	53	55	1
25–100	32	31	6
100–300	24	22	10
300–700	20	18	4
700–	17	14	1
Overall	23	28	22

The proportionate areas for the whole country are shown in the last column. The standardized rent for Berkshire, using these areas as weights, is $(1 \times 53 + 6 \times 32 + \ldots)/22 = 26$ shillings per acre, and that for Cornwall is 25 shillings per acre.

Inspection of the table shows that although the overall rent per acre is considerably less for Berkshire than for Cornwall, there is little difference between the two counties for the different size-groups, the rents for Berkshire being in general somewhat greater than those for Cornwall. This is brought out, in a single contrast, by the standardized rents. The lower overall rent per acre for Berkshire is in a certain sense accounted for by the greater proportion of large farms in that county.

This example illustrates both the use and the danger of standardization of this type. The standardized rents eliminate the effects of differences in average

148

size on the average rent, which in so far as they are due to greater concentration of buildings on the smaller holdings, greater demand for smaller holdings, etc., do not represent differences in value of the land. It would be incorrect, however, to assume that were the size-distributions of farms in the two counties the same the overall rents per acre would be the same. Part of the difference is due to the tendency of poorer land to be farmed in larger units, and such land would not command the full increase in rent which is apparently attracted by smaller farms if it were divided into smaller units.

6.9 Ratio method : random sample

$$\bar{y} = \frac{S(y)}{S(x)} \bar{x} = \frac{\bar{y}}{\bar{x}} \bar{x} \qquad (6.9)$$

$$Y = \frac{S(y)}{S(x)} X$$

$$\bar{r} = \frac{S(y)}{S(x)}$$

The formula for $\bar{y}$ may be used for obtaining the " standardized " value y_0 of y for a standard value x_0 of x, or for estimation in two-phase sampling using an estimate $\bar{x}_1$ obtained from first-phase information.

Example 6.9.a

Estimate the total area of wheat from the data of Example 6.6 by the ratio method, given that the total area of crops and grass in the county is 273,074 acres.

Crops and grass in sample $= S(x) = 15{,}114$ acres

Wheat in sample $= S(y) = \quad 2301$ acres

Estimate of wheat acreage in county $= \dfrac{2301}{15{,}114} \times 273{,}074$

$$= 0{\cdot}15224 \times 273{,}074$$
$$= 41{,}570 \text{ acres.}$$

Example 6.9.b

Table 6.9 gives the numbers of persons belonging to 43 kraals which formed a random sample of the 325 kraals in the Mondora Reserve in Rhodesia, and also the numbers of persons absent from these kraals. (The data form part of the results of a sample census of the Hartley District, and have kindly been made available by Dr. J. R. H. Shaul.) Estimate the percentage of persons absent from the reserve, and the numbers of persons belonging to the reserve and absent from the reserve.

149

TABLE 6.9—DATA FROM A RANDOM SAMPLE OF 43 KRAALS : TOTAL NUMBER OF PERSONS (INCLUDING ABSENTEES), x, AND NUMBER OF ABSENTEES, y

x	y	x	y	x	y	x	y
95	18	89	7	75	12	159	36
79	14	57	9	69	16	54	26
30	6	132	26	63	9	69	27
45	3	47	7	83	14	61	2
28	5	43	17	124	25	164	69
142	15	116	24	31	3	132	41
125	18	65	16	96	45	82	10
81	9	103	18	42	25	33	8
43	12	52	16	85	35	86	22
53	4	67	27	91	28	51	19
148	31	64	12	73	13	3,427	799

$$\text{Percentage absent} = 100\,\bar{r} = \frac{799}{3427} \times 100 = 23\cdot3 \text{ per cent.}$$

$$\text{Number belonging to reserve} = \mathsf{X} = \frac{325}{43} \times 3427 = 25{,}902.$$

$$\text{Number absent from reserve} = \mathsf{Y} = \frac{325}{43} \times 799 = 6039.$$

$$\text{Number present in reserve} = 25{,}902 - 6039 = 19{,}863.$$

This example, though superficially similar to Example 6.4.a, is structurally different, in that the sampling units consist of kraals and not individuals. This affects the estimation of the sampling error (see Section 7.8).

6.10   Ratio method : stratified sample with uniform sampling fraction

(*a*) When the ratio is assumed to be the same for all strata :
The formulæ for a random sample hold.

(*b*) When the ratio is permitted to assume different values for the different strata :

Treat each stratum separately, using the formulæ for a random sample, and build up the population estimates by summation of the estimates for the separate strata, with division by N or N for the population means.

This gives $$\mathsf{Y} = \Sigma \left(\frac{S_i(y)}{S_i(x)} \mathsf{X}_i \right)$$ (6.10)

etc.

The choice between method (*a*) and method (*b*) depends on :

(1) Numbers in the different strata—method (*b*) can only be used if the numbers of units from the individual strata are sufficiently large to give reasonably accurate determinations of the values of ratio for the separate strata : if the numbers are small and there is correlation

between r and x, the method will be biased. (This objection does not hold if selection with probabilities proportional to x is used—see Section 3.10.)

(2) The degree of variation in the ratio between the different strata—the greater this variation the greater will be the gain in accuracy by the use of method (b) ; if the variation is small, method (a) may be the more accurate.

(3) Computational convenience—method (a) is simpler, since only one value of the ratio is involved.

Method (b) can also be used for a random sample stratified after selection, provided the population totals of x for each stratum are known.

Example 6.10

Estimate the total area of wheat from the data of Example 6.6 by the ratio method, stratifying the data by districts, and using different values of the ratio for the different districts.

TABLE 6.10—HERTFORDSHIRE FARMS, 1939 : ESTIMATION OF WHEAT ACREAGES FROM THE RANDOM SAMPLE OF 1 IN 20 FARMS BY THE RATIO METHOD AFTER STRATIFICATION INTO DISTRICTS

District No.	Sample				District crops and grass X_i	Estimated district wheat $r_i X_i$
	No.	Wheat $S_i(y)$	Crops and grass $S_i(x)$	Ratio r_i		
1	15	141	1,935	·0729	22,932	1,670
2	8	259	1,385	·1870	43,591	8,150
3	40	763	4,851	·1573	57,263	9,010
4	24	837	4,034	·2075	73,946	15,340
5 & 7	14	127	882	·1440	40,905	5,890
6	24	174	2,027	·0858	34,437	2,950
	125	2,301	15,114		273,074	43,010

The computations are shown in Table 6.10. The neighbouring districts of St. Albans (5) and Watford (7) (each of which contains rather a small number of farms) have been combined.

6.11 Ratio method : stratified sample with variable sampling fraction

(a) Ratio the same for all strata :

$$\bar{r} = \frac{\Sigma \{ g_i\, S_i\,(y)\}}{\Sigma \{ g_i\, S_i\,(x)\}}$$

$$\bar{y} = \bar{r}\,\bar{x}$$

$$Y = \bar{r}\,X$$

(b) Ratio different for different strata :

Proceed in the same manner as for a fixed sampling fraction. The sampling fraction does not enter into the calculations.

6.12 Regression method : random sample

The equation of the regression line is

$$y_l = \bar{y} + b\,(x - \bar{x})$$

where

$$b = \frac{S\,(y - \bar{y})\,(x - \bar{x})}{S\,(x - \bar{x})^2} \qquad (6.12.a)$$

Hence

$$\bar{y} = \bar{y} + b\,(\bar{x} - \bar{x}) \qquad (6.12.b)$$

$$Y = N\,\bar{y}$$

If **N** is not known exactly, it must be estimated from the sample.

Note that if b is put equal to $S\,(y)/S\,(x)$ formula 6.9 is obtained, and if put equal to 0 formula 6.4.a is obtained. All values of b will give unbiased estimates, and consequently any value b_0 which appears appropriate to the data under analysis may be used. Thus, taking $b_0 = 1$ is equivalent to the use of the differences $y - x$. The regression method furnishes the value of b which gives the most accurate estimate of $\bar{y}$, using a formula of type 6.12.b, at the cost of some additional computational labour.

Regressions may be used for standardization and in two-phase sampling in the same way as ratios.

Example 6.12.a

Obtain an estimate of the total area of wheat from the data of Example 6.6, using the regression method.

We have

$$\bar{x} = 120 \cdot 912 \qquad \bar{y} = 18 \cdot 408 \qquad N = 2496 \qquad \bar{x} = 109 \cdot 405$$

$$S\,(x^2) = 5{,}061{,}734 \qquad\qquad S\,(xy) = 902{,}958 \qquad\qquad S\,(y^2) = 207{,}261$$

$$\bar{x}\,S\,(x) = 1{,}827{,}464 \qquad \bar{y}\,S\,(x) = \bar{x}\,S\,(y) = 278{,}219 \qquad \bar{y}\,S\,(y) = \ \ 42{,}357$$

$$S\,(x - \bar{x})^2 = 3{,}234{,}270 \qquad S\,(x - \bar{x})\,(y - \bar{y}) = 624{,}739 \qquad S\,(y - \bar{y})^2 = 164{,}904$$

The method of calculation of the sums of squares and products is explained in Section 7.1. The sum of squares of y will be required in the calculation of the sampling error.

We then have

$$b = \frac{624,739}{3,234,270} = 0 \cdot 19316$$

$$\bar{y} = 18 \cdot 408 + 0 \cdot 19316\,(109 \cdot 405 - 120 \cdot 912) = 16 \cdot 185$$

$$Y = 2496 \times 16 \cdot 185 = 40,400 \text{ acres}$$

Example 6.12.b

Table 6.12 gives the measured volumes of timber on 25 systematically located plots of 1/10 acre, and eye estimates of the volumes per 1/10 acre in

TABLE 6.12—MEASURED VOLUMES, y, ON 25 SAMPLE PLOTS, AND EYE ESTIMATES, x, OF CORRESPONDING STANDS (CU. FT. PER 1/10 ACRE)

y	x	y	x	y	x	y	x
170	102	195	208⎫	153	79	169	152⎫
47	14	255	208⎭	216	177	182	152⎭
64	57	135	110⎫	125	65	74	148
91	70	146	110⎮	100	196	24	207
126	95	154	110⎮	287	167	255	167
146	92	110	110⎭	261	268	3,684	3,302
87	110	112	128			147·36	132·08

the stands in which they occurred. If more than one sample plot occurs in a stand this is indicated by a bracket, but the observations have been treated as independent in the subsequent computations. The data, which refer to conifer stands of uniform age and over 20 years of age in two counties, were obtained in the course of the 1938–9 Census of Woodlands. They are plotted in Figure 6.8. The total area of conifer stands over 20 years of age in these two counties was 5124 acres, and the total volume of timber, from eye estimates of all these stands, was 6,110,000 cu. ft., i.e. 1192 cu. ft. per acre. Obtain unbiased estimates of the total volume of this class of timber from the above data.

The mean of the measured volumes on the sample plots provides an unbiased estimate of the volume per acre, and the estimate of the total volume, based on the measurements of the sample plots only, is consequently

$$147 \cdot 36 \times 10 \times 5124 = 7,551,000 \text{ cu. ft.}$$

The ratio method gives

$$1192 \times 5124 \times 1473 \cdot 6/1320 \cdot 8 = 6,814,000 \text{ cu. ft.}$$

Elimination of possible bias in the eye estimates by taking the difference between the measured volumes and eye estimates on the sample plots (equivalent to the use of the regression method with an arbitrary coefficient $b_0 = 1$) gives

$$5124 \ (1473 \cdot 6 - 1320 \cdot 8 + 1192) = 6{,}891{,}000 \ \text{cu. ft.}$$

The regression method proper requires the calculation of the regression coefficient. We find

$$S \ (y - \bar{y})^2 = 115{,}266 \quad S \ (y - \bar{y}) \ (x - \bar{x}) = 52{,}069 \quad S \ (x - \bar{x})^2 = 82{,}296$$
$$b = 52{,}069/82{,}296 = 0 \cdot 63270$$
$$Y = 5124 \{ 1473 \cdot 6 + 0 \cdot 63270 \ (1192 - 1320 \cdot 8) \} = 7{,}133{,}000 \ \text{cu. ft.}$$

The relative accuracy of these various methods of estimation is discussed in Example 7.12.b.

The above data are of course only a small part of the full data for the survey. Examination of the whole of the data for the above two counties gave a value of b of $0 \cdot 55$. The bias in the eye estimates, which are too low, though not very large, is apparent in the above data. The average bias over the whole survey was decidedly larger, and misleading results would have been obtained by using the eye estimates without correction for bias from properly measured and randomly located sample plots.

There is of course the possibility—if the location of the sample plots has not been objectively carried out, or if the measurements have been carelessly made, e.g. by the inclusion of trees whose centres do not lie within the demarcated sample area—that the sample plots will themselves be biased. The sample plots used in this survey were somewhat small, and the use of larger plots, particularly in the case of hardwoods, possibly with second-stage sampling of trees for measurement, would have reduced the risk of bias of this nature. The surveyors were well trained, however, and thoroughly appreciated the need for objectivity, and on examination it appeared that serious bias from this cause could be ruled out. The results of a later survey of England and Wales confirmed the correctness of the earlier survey.

6.13 Regression method : stratified sample with uniform sampling fraction

(a) When the regression coefficient is assumed to be the same for all strata :

The formulæ for a random sample hold, except that formula 6.12.a is replaced by

$$b = \frac{\Sigma \ \{ S_i \ (y - \bar{y}_i) \ (x - \bar{x}_i) \}}{\Sigma \ \{ S_i \ (x - \bar{x}_i)^2 \}}$$

(b) When the regression is permitted to assume different values in the different strata :

Proceed as in the ratio method.

6.14 Regression method : stratified sample with variable sampling fraction

(a) Regression the same for all strata :

$$\bar{y} = \bar{y}_w + b\,(\bar{x} - \bar{x}_w)$$

where

$$b = \frac{\Sigma\{\lambda_i\,S_i\,(y - \bar{y}_i)\,(x - \bar{x}_i)\}}{\Sigma\{\lambda_i\,S_i\,(x - \bar{x}_i)^2\}}$$

the λ_i being numerical weighting coefficients, and

$$\bar{y}_w = \frac{\Sigma\{g_i\,S_i\,(y)\}}{\Sigma(g_i\,n_i)}$$

with a similar expression for $\bar{x}_w$. $\bar{y}_w$ and $\bar{x}_w$ are the estimates of $\bar{y}$ and $\bar{x}$ that would be obtained from the sample if there were no supplementary information on x (see Section 6.7).

If the regressions within strata are truly linear, with identical values of the regression coefficient, then the most accurate estimate of b will be obtained if the λ_i are taken inversely proportional to the residual within-strata variances of y about the regression lines. If the regression coefficients are different for the different strata, then the component of error due to the assumption of equality of regression coefficients will be minimized by taking λ_i proportional to g_i^2. Any set of λ_i will give a virtually unbiased estimate of $\bar{y}$, and detailed investigation of the theoretically best values to adopt is seldom worth while. For most work λ_i may be taken as unity if all the strata contain material of similar variability, i.e. if the variable sampling fraction arises from extraneous causes not connected with the variability of the material, and equal to g_i if the sampling fractions have been chosen so as to minimize the sampling error. Under certain conditions $\lambda_i = g_i^2$ would be best in this case, but under other conditions this would give excessive weight to the strata with small sampling fractions.

(b) Regression different for different strata :

Proceed as in the ratio method.

6.15 Use of regression to calibrate eye estimates

It sometimes happens that eye estimates or similar subjective measurements, x, can be made on a properly selected and unbiased sample of the population, but that the objective measurements y, which are required to calibrate these estimates, can only be carried out on a non-random sub-sample of the original sample. The eye estimates cannot then be used as supplementary information in the manner of Example 6.12.b, since any bias in the sub-sample used for the objective measurements would be reflected to a greater or less extent, depending on the value of b, in the population estimate derived from the regression.

155

In this case the regression of x on y, instead of y on x, must be calculated, and the equation of estimation must be replaced by

$$\bar{y} = \bar{y} + \frac{1}{b'}(\bar{x}_1 - \bar{x}),$$

when b' is the regression coefficient of x on y, $\bar{y}$ and $\bar{x}$ are the means for the sub-sample, and $\bar{x}_1$ is the mean of the eye estimates for the original sample.

This procedure is subject to certain limitations. Firstly, the sub-sample, though non-random for the whole of the population, must be effectively random for units having any given value of y. If, for example, there is a tendency to select units which, for a given value of y, have high values of x serious bias may result. Thus, in a crop-estimation scheme, if eye estimates are made on a random sample of fields, and if reliance is placed on returns by farmers of the actual yields of some of these fields, any tendency on the part of the farmers to return only the yields of fields which have turned out better than their appearance would indicate will lead to an overestimate of the yield. On the other hand, the omission of a greater proportion of the low-yielding than of the high-yielding fields from the sub-sample will not bias the results, provided this omission is conditioned only by the final yield and not by the previous appearance or the value of the eye estimate.

Secondly, for accuracy in the final estimate, the eye estimates must be reasonably accurate in the sense that variation about the regression line must be small, and the line itself must have an adequate slope. If the regression line is curved, this curvature can only be allowed for in the estimation formula if the variation about the regression line is negligible. Otherwise bias will be introduced. The use of the best fitting linear regression line, however, will avoid this source of bias.

Example 6.15

In order to test the accuracy of eye estimation as a method of estimating the yields of cereal crops shortly before harvest, a trial survey of the wheat crop of Hertfordshire was undertaken in 1940. Two observers were employed, one of whom visited 47 farms, observing 110 fields, and the other 16 farms, observing 37 fields. The whole set of farms constituted a systematic sample of 1 in 12 farms, excluding those growing less than 5 acres of wheat in 1939, a random sub-sample of fields being taken on the larger farms. The actual yields, as determined by the farmers, were subsequently obtained for as many of the observed fields as possible, and these were used to calibrate the eye estimates. The relation between the eye estimates, x, and the actual yields per acre, y, for the first observer are shown in Fig. 6.15 for the 37 fields for which yields were obtained. Obtain an estimate of the mean yield per acre for the part of the county covered by this observer.

The regression coefficient, b', of x on y, calculated from the unweighted values of x and y for the 37 fields, is 0·6926, the regression equation being

$$x_l = 30{\cdot}00 + 0{\cdot}6926 \, (y - 28{\cdot}78)$$

This is shown by the full line in the figure. The dotted line represents the line that would be obtained if there were no errors in the eye estimates. It will be seen that there is a tendency to underestimate high yields and overestimate low yields. The other observer and the farmers gave very similar results.

The mean of the eye estimates $\bar{x}_1$ for the whole of the first observer's sample, and that for the eye estimates $\bar{x}$ and the yields $\bar{y}$ of the fields for which yields were available, weighted according to the acreages of the fields, with an additional raising factor if all the fields on the farm are not sampled, are, in bushels per acre,

$$\bar{x}_1 = 30{\cdot}12 \qquad \bar{x} = 30{\cdot}13 \qquad \bar{y} = 28{\cdot}95$$

Hence, since $1/0{\cdot}6926 = 1{\cdot}444$, the final estimate of the yield per acre is

$$\bar{y} = 28{\cdot}95 + 1{\cdot}444 \, (30{\cdot}12 - 30{\cdot}13) = 28{\cdot}94$$

The adjustment is here negligible, since $\bar{x}_1$ and $\bar{x}$ are almost identical.

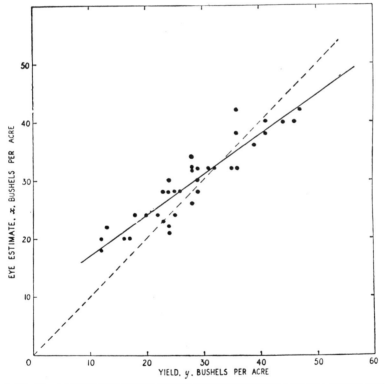

FIG. 6.15—RELATION BETWEEN EYE ESTIMATES OF THE YIELDS AND THE ACTUAL YIELDS OF 37 FIELDS OF WHEAT IN HERTFORDSHIRE

6.16 Sampling with probabilities proportional to size of unit

(*a*) Size, *x*, of all units of the population known, or $\mathbf{X}$ known :

In this case *x* acts as a supplementary variate, and the ratio method will in general be appropriate. Since the probability of selection is proportional to *x*, raising factors proportional to $1/x$ must be introduced into the formulæ already given. This leads to the formulæ

$$\bar{r} = \frac{1}{n} S \left(\frac{y}{x} \right) = \bar{r}$$

$$\mathbf{Y} = \bar{r}\,\mathbf{X}$$

$$\bar{y} = \bar{r}\,\bar{x}$$

In other words the unbiased estimate of the population value of the ratio is given by the arithmetic mean of the ratios from the selected sampling units.

(*b*) Total size $\mathbf{X}$ of the population not known :

In this case $\mathbf{X}$, as well as $\mathbf{Y}$ and $\bar{r}$, have to be estimated from the sample. Selection has to be made by some such process as randomly or systematically locating points on a map, and points not falling in the units under consideration must be taken into account. If n_0 is the total number of sampling points, and $\mathbf{A}$ is the total area covered by the sampling grid, we have

$$\bar{r} = \bar{r}$$

$$\mathbf{X} = \mathbf{A}\, n/n_0$$

$$\mathbf{Y} = \bar{r}\,\mathbf{X} = \bar{r}\,\mathbf{A}\, n/n_0$$

Alternatively, if $\mathbf{A}$ is not known exactly the density *d* of points per unit area may be used. We have

$$\mathbf{A} = n_0/d$$

$$\mathbf{X} = n/d$$

If the sampling is two-phase, with n_0' points (density d') at the first phase, of which n' fall in the units under consideration, n, n_0 and d must be replaced by n', n_0' and d' in the above formulæ.

Example 6.16.a

In a survey to estimate the area and yield of a crop, systematically located points at a density of one per 4 square miles are taken, and the yields per acre of the fields in which the points fall and which carry the crop are determined by the harvesting of small areas. 8317 points in all are obtained, of which 529 fall in fields carrying the crop. The arithmetic mean of the yields per acre of the selected fields is 15·7 cwt. per acre. Estimate the total area and yield of the crop.

A density of 1 per 4 square miles is equal to 1/2560 per acre. Hence

$$\text{area} = \mathbf{X} = 529 \times 2560 = 1{,}354{,}000 \text{ acres}$$

$$\text{yield} = \mathbf{Y} = 15{\cdot}7 \times 1{,}354{,}000 \text{ cwt.} = 1{,}063{,}000 \text{ tons}$$

Example 6.16.b

If, in addition to the yield data of Example 6.16.a, a further 24,938 points were surveyed for type of crop only, giving an overall density, with the 8317 points of the yield survey, of one point per square mile, and 1673 of the fields so located were found to carry the crop in question (in addition to the 529 fields above), obtain revised estimates of the total area and yield of the crop.

This is an example of two-phase sampling. The two sets of points together constitute the first phase, for area of crop, and the 529 points for which yield samples were taken constitute the second phase.

We therefore have

$$n_0' = 24{,}938 + 8317 = 33{,}255$$

and

$$n' = 1673 + 529 = 2202$$

The density at the first phase is $1/640$ per acre. Consequently

$$\text{area} = X = 2202 \times 640 = 1{,}409{,}000 \text{ acres}$$
$$\text{yield} = Y = 15 \cdot 7 \times 1{,}409{,}000 \text{ cwt.} = 1{,}106{,}000 \text{ tons}$$

6.17 Sampling from within strata with probabilities proportional to size of unit

In this case the sizes of all units will be known. As pointed out in Section 3.10, if more than one unit is selected from some or all of the strata, the same unit not being selected twice, the probabilities will not in fact be exactly proportional to size, and slight bias will be introduced. The ratios from the selected units are meaned separately for each stratum, giving equations of estimation

$$\bar{r}_i = \frac{1}{n_i} S_i \left(\frac{y}{x} \right) = \bar{r}_i$$
$$Y = \Sigma \{ \bar{r}_i X_i \}$$
$$\bar{r} = Y/X$$

Example 6.17

Estimate the acreage of wheat and the number of farms growing wheat in Hertfordshire from the sample of parishes described in Section 3.11.

The data from the sampled parishes are shown in Table 6.17.a, and the further computations for wheat acreage in Table 6.17.b.

The computations for number of farms growing wheat follow exactly the same lines, using the ratio of number of farms growing wheat to acreage of crops and grass for each parish. These computations are left as an exercise to the reader.

159

TABLE 6.17.a—HERTFORDSHIRE WHEAT : SAMPLE OF 17 " COMBINED "
PARISHES SELECTED FROM WITHIN DISTRICTS WITH PROBABILITY PROPORTIONAL
TO SIZE

District	1	2				3			
Wheat	164	766	701	503	311	228	249	686
Crops and grass	3,350	3,040	3,440	2,040	2,370	3,330	2,290	2,930
Ratio	·049	·252	·204	·247	·131	·068	·109	·234

District	4					5	6		7
Wheat	558	775	495	565	862	818	225	738	290
Crops and grass	2,300	4,430	2,890	2,420	4,160	3,470	2,520	3,740	3,060
Ratio	·243	·175	·171	·233	·207	·236	·089	·197	·095

TABLE 6.17.b—ESTIMATION OF WHEAT ACREAGE FROM THE DATA
OF TABLE 6.17.a

District	Mean ratio wheat/crops and grass	Acreage of crops and grass in district	Estimated acreage of wheat
1	·049	22,932	1,120
2	·234	43,591	10,200
3	·136	57,263	7,790
4	·206	73,946	15,230
5	·236	24,964	5,890
6	·143	34,437	4,920
7	·095	15,941	1,510
TOTAL		273,074	46,660

The use of the ratio of number of farms growing wheat to the total number
of farms in the district would be equally admissible if the selection of parishes
had been made with constant probability, but when the probability of selection
is taken proportional to size of unit, the ratio to size, however defined, must
be taken for all variates.

6.18 Multi-stage sampling, no supplementary information

In multi-stage sampling the process of estimation can be carried out stage
by stage, using the appropriate methods of estimation at each stage. It is often
more convenient, however, to combine all stages in a single process of
estimation.

Thus the combined or *overall* raising factor g for any sub-unit in two-stage sampling is given by the product of the first-stage raising factor g' of the main unit in which it occurs and the second-stage raising factor g'' of the particular sub-unit, i.e.

$$g = g' g''$$

Hence the general formula for Y, when there is no supplementary information, is

$$Y = S(gy)$$

where the summation is taken over all units, with similar formulæ for N, etc.

If the combined raising factors for a group of units are equal then the computations will be simplified by summing these units before multiplication. In particular, if all the combined raising factors are equal, the sample can be treated for purposes of estimation as if it were an ordinary random or stratified random sample with uniform sampling fraction.

6.19 Multi-stage sampling with supplementary information

(*a*) Ratio method, ratio the same for whole population :

$$\bar{y} = \frac{S(gy)}{S(gx)} \bar{x}$$

etc., where $g = g' g''$.

(*b*) Ratio method, ratio different for different parts of the population :

Many variants are possible. All can be resolved by proceeding stage by stage by the methods already outlined. The danger of introducing bias if the number of units on which the ratios are based is small must be recognized.

(*c*) Regression method :

Regressions will usually be employed at the first stage of the sampling, in which case the regression coefficient or coefficients will be calculated in the manner appropriate to the type of sampling adopted at this stage, using the values of the totals of x and y for each main unit estimated from the second-stage sampling.

If regression is used at the second stage the procedure for stratified samples can be used, treating the selected first-stage units as if they were strata.

(*d*) Sampling with probability proportional to size :

An important case is that in which the first-stage units are sampled from within strata with probability proportional to size, and the second-stage sampling fractions are chosen so as to give a uniform overall sampling fraction. In this case the use of the mean ratios $\bar{r}_i'$ at the first stage of the estimation process (Section 6.17) will be found to be equivalent to the direct estimation from the second-stage units by means of the overall raising factors, i.e. $Y = S(gy)$.*

---

* The estimation of error of this type of sampling is described in Section 8.13.

Example 6.19

From the data of Table 6.19.a, obtained in the course of the Survey of Fertilizer Practice, estimate the average dressing of nitrogenous fertilizer on sugar beet in Norfolk.

The two-stage sampling procedure of this survey has been described in Section 4.23. Information is lacking from a few farms, mainly owing to changes in tenancy. Since this affects the small farms to a greater extent than the large farms the adjusted sampling fractions shown in the table have been used. These are equal to the number of farms on which information is available divided by the total number of farms in the size-group. The second-stage sampling fractions are given by the reciprocals of the number of fields, since one field is selected on each farm. The combined raising factor for the sampled field of the first farm in Table 6.19.a is therefore 105, that for the sampled field of the third farm is $105 \times 3 = 315$, etc.

TABLE 6.19.a—SURVEY OF FERTILIZER PRACTICE: DATA ON THE APPLICATION OF NITROGENOUS MANURES TO SUGAR BEET ON OLD ARABLE LAND IN NORFOLK

No. of fields	Acreage		Cwt. N per acre	No. of fields	Acreage		Cwt. N per acre	No. of fields	Acreage		Cwt. N per acre
	Total	Sample			Total	Sample			Total	Sample	
	Small farms				Medium farms				Medium farms (*contd.*)		
1	2	2	·68	2	13	10	·42	1	12	12	·16
1	6	6	·63	4	40	5	·63	1	6	6	·30
3	5	2	·55	1	8	8	·15	2	14	7	·42
2	4	3	·42	3	14	4	·42	3	21	6	·14
1	5	5	·36	2	51	11	·63	1	8	8	·42
1	3	3	·21	2	16	10	·90	2	10	2	·49
1	4	4	·21	3	19	7	·21	1	4	4	·30
1	2	2	·42	3	31	10	·84	1	4	4	·30
2	6	2	·42	3	39	4	·63	3	14	7	·45
2	8	4	·90	2	19	13	·52	6	42	10	·36
1	2	2	·52	1	9	9	·52	1	6	6	·21
2	4	3	·15	1	20	20	·42	1	4	4	·30
1	6	6	·70	5	26	7	·30	2	19	8	·42
2	5	4	·36	2	8	6	·54		Large farms		
2	6	4	·56	4	20	8	·21	1	8	8	0
1	2	2	0	1	4	4	·42	1	48	48	0
1	4	4	·21	1	20	20	·72	3	19	4	0
1	5	5	·48	2	7	4	·36	3	56	24	·68
2	3	2	·10	4	32	11	·63	2	22	5	·36
1	7	7	·32	2	16	8	·82	4	30	5	·42
1	2	2	·80	2	6	3	·63	3	20	5	·90
1	4	4	·30	2	20	10	·56	6	126	29	·75
				1	7	7	·57	6	28	6	·63

Number of farms without sugar beet on old arable land: small (6–50 acres), 8; medium (51–300 acres), 11; large (301– acres), 5.

Sampling fractions (adjusted for absence of information): small, 1/105; medium, 1/59; large, 1/30.

The average dressing of nitrogen must be obtained by calculating the raised total of the amount of nitrogen applied $S(gy)$ and the raised total of the acreage sampled $S(gx)$. The amount of nitrogen applied to a field is given by the product of the acreage and the rate per acre. The three size-groups are best kept separate in the computation. For the first size-group, therefore, applying the second-stage raising factors, we have

$$S(g''y) \doteq S(g''rx) = 1 \times 2 \times 0.68 + 1 \times 6 \times 0.63 + 3 \times 2 \times 0.55 + \ldots$$
$$= 1.36 + 3.78 + 3.30 + \ldots$$
$$S(g''x) = 1 \times 2 + 1 \times 6 + 3 \times 2 + \ldots$$
$$= 2 + 6 + 6 + \ldots$$

This gives the results shown in Table 6.19.b. Applying the first-stage raising factors to the total nitrogen and total acreage, we obtain the average dressing of nitrogen per acre :

$$\bar{r} = \frac{46.13 \times 105 + \ldots}{104 \times 105 + \ldots} = \frac{28{,}512.04}{58{,}229} = .490 \text{ cwt. per acre}$$

TABLE 6.19.b—ESTIMATION OF AVERAGE DRESSING FROM RAISED RESULTS

Size-group	Total nitrogen $S(g''y)$	Total acreage $S(g''x)$	Nitrogen per acre $S(g''y)/S(g''x)$	First-stage raising factor g'
Small . .	46.13	104	.444	105
Medium . .	285.41	601	.475	59
Large . .	227.64	395	.576	30

The data presented comprise only a small part of the information collected, and the above method of estimation therefore demands a good deal of computation. For certain purposes comparative figures may be obtained from the straight averages of the dressings per acre on the sampled fields.

These averages are given in Table 6.19.c. The first-stage raising factors have not been used in this calculation, since the larger farms have more and larger fields in sugar beet, so that inequality in the omitted second-stage raising factors more than compensates for the difference in the first-stage raising factors.

It will be noted that the mean dressings are less than those previously obtained for all size-groups, indicating the possibility that farms with little sugar beet, which are overweighted in the straight averages, are using less nitrogen per acre than those with a large amount of sugar beet. The data are too variable, however, to determine with certainty from this sample alone whether this is really a bias or is due to random sampling errors. The large

TABLE 6.19.c—ESTIMATION OF AVERAGE DRESSING
FROM UNWEIGHTED MEANS

Size-group	No. of farms	Nitrogen per acre	
		Sum	Mean
Small . .	22	9·30	·423
Medium . .	36	16·32	·453
Large . .	9	3·74	·416
All . .	67	29·36	·438

difference for the large farms, for example, is due to farm 8 having a very large acreage of sugar beet. The relative accuracy of the two methods of estimation, apart from bias, is discussed in Example 7.17.

In the Survey of Fertilizer Practice the second method oɪ estimation was used in investigation of secondary points, *e.g.* comparison of different types of farms. For the more important estimates, such as mean dressings per acre, a modification of the first method was used, the total acreages of sugar beet on the farms being taken as the raising factors for the second stage. This method of estimation is slightly more accurate than the first method given above, but will be biased if there is any tendency for farmers to apply heavier (or lighter) dressings to their large fields. There is no evidence that any appreciable bias does in fact arise from this cause, but even so it is perhaps doubtful whether there is much advantage in using this method of estimation rather than the unbiased method given above. The method would have been unbiased had selection of fields within a farm been made proportional to area, but this would have demanded somewhat more elaborate methods of selection in the field.

6.20 Systematic and balanced samples

The methods of estimation described in the preceding sections are also appropriate for systematic and balanced samples. Samples of these types without other restrictions, for instance, can be treated as if they were random samples for the estimation of the population values (but *not* for the sampling errors) ; if there is stratification the procedure for stratified random samples holds. An example of a systematic stratified sample with variable sampling fraction has already been given (Example 6.7).

Certain estimation processes are naturally inappropriate to systematic and balanced samples. If the process of selection in a systematic sample is such, for example, that stratification is automatically introduced, there will be no gain from stratification after selection. Equally in a balanced sample the variate for which balance has been effected will be of no further value as supplementary information—the balance ensures that the corrections based on regression, or ratio, will be zero, whatever the value of the regression coefficient. If each stratum is balanced separately, then the corrections for the different strata will all be zero, even if the regression coefficient or ratio varies from stratum to stratum.

In systematic samples of material which varies in a continuous manner, some gain in accuracy may result from the use of what are known as *end-corrections*. These corrections are made by assigning to the boundary observations weights which depend on their distance from the boundary. In systematic one-dimensional sampling of a line AB (Fig. 6.20), for example, with

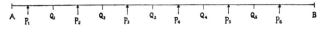

FIG. 6.20—SYSTEMATIC SAMPLE, P_1, P_2, . . . P_6, OF THE LINE AB

sampling points located at P_1, P_2, . . . P_6, if Q_1, Q_2, . . . Q_5 are the mid-points of P_1P_2, etc., we may regard the observations at P_2, P_3, P_4, P_5 as estimates covering the lengths Q_1Q_2, Q_2Q_3, etc. The observations at P_1 and P_6 can similarly be regarded as estimates covering the lengths AQ_1 and Q_5B. Consequently the weights assigned to P_1 and P_6, relative to that assigned to P_2, P_3, etc., will be AQ_1/Q_1Q_2 and Q_5B/Q_1Q_2.

The same principle, or an adaptation of it, might be applied in the case of two-dimensional systematic sampling of areas. End-corrections, however, are not likely to be of much value in the type of material usually dealt with in census and survey work, and we shall not discuss them further here, beyond mentioning that if the regions near the boundary differ from the remainder of the area, the use of end-corrections, instead of separate treatment of the boundary regions in the manner outlined in Section 3.14, will lead to biased estimates.

6.21 Sampling on two successive occasions

The most straightforward procedure for estimating the values of the population mean on two successive occasions is to treat each occasion separately, following whatever method of estimation is appropriate to the sample obtained on that occasion, regardless of the values obtained on the other occasion. Such estimates may be termed *overall* estimates.

With independent samples on the two occasions, or with the same fixed sample on each occasion, the overall estimates will contain virtually all the available information, but where a sub-sample is taken on the second occasion, or there is partial replacement of the sample on the second occasion, the situation is more complicated.

If the sample on the second occasion is confined to a sub-sample of the original sample, change will be most simply estimated from the differences of the units included in the sub-sample only. An estimate of the population mean or total on the second occasion is similarly obtained by adding the estimated change to the overall estimate on the first occasion. The most accurate estimate of the population mean on the second occasion will, however, be obtained by calculating the regression of the sample values for the second occasion on the corresponding values for the first occasion, and using the sample values for the first occasion as supplementary information. The procedure is exactly the same as has been already outlined for use of supplementary information by the regression method, the method appropriate to the type of sampling used being followed. The most accurate estimate of the change can then be obtained by taking the difference of this estimate of the mean from the overall estimate for the first occasion.

The formulæ for this procedure are as follows. Denote the sample values on the first occasion by x and those on the second occasion by y, the values belonging to units included in both samples by x', y', and those included on the first occasion only by x''. If a fraction λ of all the units included on the first occasion are taken on the second occasion, and a fraction μ equal to $1 - \lambda$ are omitted, then, for a random sample,

$$\bar{\mathrm{x}} = \lambda \bar{x}' + \mu \bar{x}''$$
$$\bar{\mathrm{y}} = \bar{y}' + b\,(\bar{\mathrm{x}} - \bar{x}')$$
$$\phantom{\bar{\mathrm{y}}} = \bar{y}' + \mu b\,(\bar{x}'' - \bar{x}')$$

where $\bar{\mathrm{x}}$ is the overall estimate for the first occasion and $\bar{\mathrm{y}}$ the adjusted estimate for the second occasion. The change is consequently estimated by

$$\bar{\mathrm{y}} - \bar{\mathrm{x}} = \bar{y}' - \bar{x}' - \mu\,(1 - b)\,(\bar{x}'' - \bar{x}')$$

The calculation of the regression coefficient is based on the values of the units which are included on both occasions.

If the changes of individual units are small compared with the differences between units, i.e. if the correlation between units on the two occasions is very close, as is likely to be the case when this type of sampling is adopted, b will be nearly equal to unity, and the estimate of change will differ little from that, $\bar{y}' - \bar{x}'$, derived from the units included on both occasions only. Equally the estimate of the population mean or total will differ little from that obtained by adding the estimate of change to the overall estimate on the first occasion.

When a sample of the same size is taken on each occasion with partial replacement, a fraction μ of the units being replaced and a fraction λ being retained, the sample units which are retained can be used to furnish an estimate $\bar{y}_1$ of the population mean on the second occasion by the regression method already given. In addition there will be a further independent estimate $\bar{y}_2$, equal to $\bar{y}''$, derivable from the sampling units which are included on the second occasion only. The most accurate estimate $\bar{y}_w$ will be provided by a weighted mean of these two estimates. The correct weights are $\lambda/(1 - \mu^2 r^2)$ and $\mu(1 - \mu r^2)/(1 - \mu^2 r^2)$, where r is the *correlation coefficient* between the unit values on the first and second occasions.

The correlation coefficient is calculated in the same manner as the regression coefficient b, using the values of the units common to both occasions, with the exception that instead of dividing by a quantity of the type $S(x - \bar{x})^2$ we divide by the corresponding quantity of the type $\sqrt{\{S(x - \bar{x})^2 S(y - \bar{y})^2\}}$. Thus, for a pair of random samples,

$$r = \frac{S(x' - \bar{x}')(y' - \bar{y}')}{\sqrt{\{S(x' - \bar{x}')^2 S(y' - \bar{y}')^2\}}}$$

In the more complicated types of sampling the sums of squares and products are modified in the same manner as in the calculation of b.

We thus have

$$\bar{y}_w = \frac{\lambda}{1 - \mu^2 r^2}\,\bar{y}_1 + \frac{\mu(1 - \mu r^2)}{1 - \mu^2 r^2}\,\bar{y}_2$$

$$= \frac{\lambda}{1 - \mu^2 r^2}\{\bar{y}' + b(\bar{x} - \bar{x}')\} + \left(1 - \frac{\lambda}{1 - \mu^2 r^2}\right)\bar{y}''$$

If the numbers in the sample on the two occasions are not the same the above formula takes the modified form

$$\bar{y}_w = \frac{n'\{\bar{y}' + b(\bar{x} - \bar{x}')\} + n''(1 - \mu r^2)\bar{y}''}{n' + n''(1 - \mu r^2)} \qquad (6.21.a)$$

where n' is the number of units re-sampled on the second occasion, n'' is the number of new units, and μ is the proportion of units sampled on the first occasion which are not re-sampled on the second occasion.

An estimate of the change can similarly be obtained by taking the weighted mean of the two estimates, $\bar{y}' - \bar{x}'$ and $\bar{y}'' - \bar{x}''$. The weights to be assigned to these estimates are $\lambda/(1 - \mu r)$ and $\mu(1 - r)/(1 - \mu r)$, so that

$$\text{Change} = \frac{\lambda}{1 - \mu r}(\bar{y}' - \bar{x}') + \frac{\mu(1 - r)}{1 - \mu r}(\bar{y}'' - \bar{x}'') \qquad (6.21.b)$$

This estimate of the change will differ from that given by the difference of $\bar{y}_w$ and the overall estimate on the first occasion. The reason for this is that once the sample for the second occasion has been taken, a more accurate estimate of the population mean on the first occasion is possible by using the information provided by the sample on the second occasion as supplementary information. If this revised estimate $\bar{x}_w$ is calculated, then the estimate of change given above will be very nearly equal to $\bar{y}_w - \bar{x}_w$. The slight discrepancy arises from the fact that unless the variances on the two occasions are equal the estimate of change given above is not quite the most accurate possible.

It will be noted that when r is equal to 1, the above estimate of change is equal to $\bar{y}' - \bar{x}'$, whereas if r equals 0 the estimate is equal to the difference of the overall estimates of the population means. Similarly, if r equals 0, $\bar{y}_w$ equals the overall mean of y, and if the values of each unit are the same on both occasions ($\bar{x}' = \bar{y}'$, $r = 1$, $b = 1$), $\bar{y}_w$ equals the mean of all the sample values included on each occasion, each value being included once only.

A further practical point arises in connection with the estimation of b. If the variability on the two occasions is the same, both the regression coefficients, y on x and x on y, will be equal to the correlation coefficient. In most material for which sampling on successive occasions of the type under consideration is likely to be used, the variability on the different occasions may be expected to be very similar, and for such material it is best to replace b by r, as the latter is less subject to errors of estimation.

Example 6.21

The percentage solids-not-fat in two successive months for all the 16 cows in a herd which were in their 2–6 months of lactation in one, at least, of these months were observed to be :

Cow	.	.	1	2	3	4	5	6	7	8
November	.		8·82	8·94	9·86	8·90	9·00	9·13	8·90	9·02
December	.		—	—	—	—	8·98	8·66	8·68	8·86

Cow	.	.	9	10	11	12	13	14	15	16
November	.		9·46	9·52	9·28	9·22	—	—	—	—
December	.		9·30	9·50	9·13	9·32	9·38	8·78	9·10	9·04

Estimate the change in percentage solids-not-fat between the two months, the mean percentage for December and the revised mean percentage for November.

The sampling is here due to natural causes, and the number of cows in their 2–6 months of lactation will therefore not remain completely constant. The above two months were selected from more extensive records.

We find :

$$S(x') = 73\cdot53 \qquad S(y') = 72\cdot43$$
$$S(x'') = 36\cdot52 \qquad S(y'') = 36\cdot30$$
$$\bar{x}' = 9\cdot1912 \qquad \bar{y}' = 9\cdot0538 \qquad \bar{y}' - \bar{x}' = -0\cdot1374$$
$$\bar{x}'' = 9\cdot13 \qquad \bar{y}'' = 9\cdot075 \qquad \bar{y}'' - \bar{x}'' = -0\cdot055$$
$$\bar{x} = 9\cdot1708 \qquad \bar{y} = 9\cdot0608 \qquad \bar{y} - \bar{x} = -0\cdot11$$
$$S(x' - \bar{x}')^2 = 0\cdot3435 \qquad S(x' - \bar{x}')(y - \bar{y}) \qquad S(y' - \bar{y}')^2 = 0\cdot6742$$
$$= 0\cdot4076$$

$$r = 0\cdot4076/\sqrt{(0\cdot3435 \times 0\cdot6742)} = 0\cdot847$$

It will be noted that the value of b is greater than unity, which illustrates the point made above that r provides a better estimate of the regression in material of this kind. The estimation of r should normally be based on more extensive data, though no very high accuracy is required—100 pairs of values will be fully adequate. In the present case the more extensive data confirm that the above value of r is about correct.

We also have $\lambda = \tfrac{2}{3}$, $\mu = \tfrac{1}{3}$, and thus

$$\frac{\lambda}{1 - \mu^2 r^2} = 0{\cdot}724 \qquad \frac{\mu\,(1 - \mu r^2)}{1 - \mu^2 r^2} = 0{\cdot}276$$

$$\frac{\lambda}{1 - \mu r} = 0{\cdot}929 \qquad \frac{\mu\,(1 - r)}{1 - \mu r} = 0{\cdot}071$$

$\bar{y}_w = 0{\cdot}724\,\{9{\cdot}0538 + 0{\cdot}847\,(9{\cdot}1708 - 9{\cdot}1912)\} + 0{\cdot}276 \times 9{\cdot}075 = 9{\cdot}0471$

Change $= 0{\cdot}929\,(-\,0{\cdot}1374) + 0{\cdot}071\,(-\,0{\cdot}055) = -\,0{\cdot}1315$

The estimate of the mean for November, revised on the basis of the December values, is

$\bar{x}_w = 0{\cdot}724\,\{9{\cdot}1912 + 0{\cdot}847\,(9{\cdot}0608 - 9{\cdot}0538)\} + 0{\cdot}276 \times 9{\cdot}13 = 9{\cdot}1786$

giving the check, $9{\cdot}0471 - 9{\cdot}1786 = -\,0{\cdot}1315$. Agreement is here exact, since the two regression coefficients, y on x and x on y, have both been taken equal to r.

6.22 Sampling on a number of successive occasions

The formulæ of estimation given in the last section cover all cases of sampling on two occasions only. When sampling is carried out with partial replacement on more than two occasions no such simple general solution is possible, but certain approximate solutions, which are very similar in form to those for sampling on two occasions, are likely to be sufficient for most practical purposes.

In a sampling scheme which is repeated at intervals it is generally desirable to provide as accurate an estimate as possible of the population mean on each occasion without any revision of the estimates for previous occasions. Suppose that $\bar{y}_h$ is the most accurate estimate which can be obtained for occasion h, taking into account the results of the sampling up to and including this occasion h, and that $\bar{y}_{h-1}$ is the similar estimate for occasion $h - 1$, taking into account the results up to and including occasion $h - 1$. Subject to certain limitations, $\bar{y}_h$ and $\bar{y}_{h-1}$ are related by a formula of the type

$$\bar{y}_h = (1 - \varphi)\,\{\bar{y}_h{}' + r\,(\bar{y}_{h-1} - [\bar{y}_{h-1}'])\} + \varphi\,\bar{y}_h{}'' \qquad (6.22.\text{a})$$

where suffices indicate the occasion, single dashes units common to occasions h and $h - 1$, the mean on the earlier occasion being distinguished by square brackets, and double dashes units occurring on occasion h only.

The limitations are that a given fraction of the units is replaced on each occasion, that the variability on the different occasions and the correlation r between successive occasions are constant, and that the correlation between occasions two apart is r^2, that between units three apart is r^3, etc. This last condition is only necessary when units are included for more than two occasions, and no great loss of accuracy will occur under normal circumstances if it does not hold exactly.

The value of φ depends on the value of r, on the fraction μ replaced on each occasion, and on the number of occasions h on which samples have already been taken. With increasing h, φ rapidly tends to a limiting value, which depends only on r and μ. This limiting value is

$$\varphi = \frac{-(1 - r^2) + \sqrt{[(1 - r^2)\{1 - r^2(1 - 4\lambda\mu)\}]}}{2\lambda r^2}$$

The values for $h = 2$ have been given in the previous section. For practical purposes the limiting value of φ may be used for all occasions after the second (the above formula for φ is due to Patterson (1950, A')).

When the value of r has been determined, the values of φ can be calculated and formula 6.22.a used.

For most practical purposes $\bar{y}_h - \bar{y}_{h-1}$ will provide an adequate estimate of the change between occasions $h - 1$ and h. If change is of particular

TABLE 6.22—PERCENTAGE SOLIDS-NOT-FAT : ADJUSTMENT OF SAMPLES TAKEN ON SUCCESSIVE OCCASIONS

	January	February	March	April	May	June
$\bar{y}_h$	5 9·400	11 9·090	9 9·111	10 9·059	8 9·211	11 9·345
$\bar{y}'_h$	—	4 9·288	8 9·122	8 9·086	3 9·060	7 9·326
$\bar{y}''_h$	—	7 8·977	1 9·020	2 8·950	5 9·302	4 9·380
$\{\bar{y}_h\}$	—	9·341	9·163	9·188	9·226	9·322
$\bar{y}_h$	9·400	9·122	9·151	9·152	9·262	9·338
$[\bar{y}'_h]$	4 9·335	8 9·072	8 9·025	3 8·947	7 9·267	—
$\bar{y}_h - [\bar{y}'_h]$	+ ·065	+ ·050	+ ·126	+ ·205	− ·005	—
φ'	—	·603	·084	·151	·472	·275
From differences	9·400	9·353	9·403	9·464	9·577	9·636

interest, however, formula 6.21.b may be used.* This latter estimate will of course not agree exactly with $\bar{y}_h - \bar{y}_{h-1}$ and will therefore lead to apparent inconsistencies in the summary of the results.

It sometimes happens that the sampling scheme, though broadly following a partial replacement procedure, gives rise to some inequality of numbers of units on the different occasions. This can be allowed for by substituting for φ the value φ' given by

$$\varphi' = \frac{n_h''}{\mu n_h} \varphi \qquad (6.22.b)$$

where n_h is the number of units on occasion h, and n_h'' is the number of units not included on the previous occasion.

Example 6.22

Similar data on percentage solids-not-fat to those of Example 6.21 are given in abstract form in Table 6.22 for the months January–June. Only the 3–5 months of lactation are included. Obtain estimates of the mean percentage in successive months.

The table shows for each month the overall mean $\bar{y}_h$, the mean of cows occurring in the previous month $\bar{y}_h'$, and the mean of new cows $\bar{y}_h''$. The numbers of cows on which these means are based are also shown. The mean for the month $h-1$ of cows occurring in months h and $h-1$ is shown in the line $[\bar{y}_h']$ in the column for month $h-1$. Thus 9·335 and 9·288 are the means for January and February of the four cows occurring in both these months.

Summation of the sums of squares and products of deviations of pairs of entries for successive months from January to December gives an overall value for r of 0·811, so that r^2 is 0·657. The similar calculation of the correlation between months two apart gives r' equal to 0·746. The assumption that r' equals r^2 therefore somewhat underweights the information obtainable from occasions two apart.

The average value of μ over a long period will be 1/3, but considerable fluctuations in numbers occur from month to month. For $\mu = 1/3$ the value of φ for occasions subsequent to the second is

$$\varphi = \frac{-(1 - 0·657) + \sqrt{(1 - 0·657)\{1 - 0·657\,(1 - 4.2/3.1/3)\}}}{2 \times 2/3 \times 0·657} = 0·252$$

Hence for March

$$\varphi' = \frac{1}{9/3}\,0·252 = 0·084$$

etc. These values are shown in Table 6.22.

---

* To obtain the most accurate possible estimate of change the information from occasions prior to $h-1$ would have to be taken into account. The procedure has been investigated by Patterson (1950, A′).

For the second occasion formula 6.21.a may be used. This is of the same form as formula 6.22.a, and gives

$$\varphi' = \frac{7\,(1 - 0\cdot657/5)}{4 + 7\,(1 - 0\cdot657/5)} = 0\cdot603$$

Had the value for $\mu = 1/3$ been calculated, and corrected by means of formula 6.22.b, we should have obtained

$$\varphi = 1 - \frac{2/3}{1 - 0\cdot657/9} = 0\cdot281$$

$$\varphi' = \frac{7}{11/3}\,0\cdot281 = 0\cdot536$$

which does not differ greatly from the correct value. Equally φ differs little from the value $0\cdot252$, obtained above for subsequent occasions.

The remainder of the calculations follow a standard pattern. The quantity $\{\bar{y}_h\}$, equal to $\bar{y}_h' + r\,(\bar{y}_{h-1} - [\bar{y}_{h-1}'])$, is calculated, and the weighted mean of $\bar{y}_h''$ and $\{\bar{y}_h\}$ taken, with weights equal to φ' and $1 - \varphi'$. Thus for February

$$\{\bar{y}_h\} = 9\cdot288 + 0\cdot811 \times (+ 0\cdot065) = 9\cdot341$$
$$\bar{y}_h = 0\cdot603 \times 8\cdot977 + 0\cdot397 \times 9\cdot341 = 9\cdot122$$

The overall estimates $\bar{y}_h$ and the estimates from differences $\bar{y}_h' - [\bar{y}_{h-1}']$ are shown for comparison. The differences show a tendency to cumulative errors, which is to be expected even with close correlation.

It will be seen that once a value for r has been determined, and provision has been made to abstract the means $\bar{y}_h'$, $\bar{y}_h''$, and $[\bar{y}_{h-1}']$, the calculations are very simple, and can easily be undertaken for large-scale surveys, even when a number of different quantities require estimation.

CHAPTER 7

ESTIMATION OF THE SAMPLING ERROR

7.1 Sampling errors of a random sample

The general principles involved in the estimation of sampling errors can best be made clear by considering the error of a random sample drawn from a large population.

Consider first a sample consisting of a single unit. Let the mean of the population be $\bar{y}$, and let the deviations of the individual values from this mean be z_1, z_2, . . . , so that $z_1 = y_1 - \bar{y}$, $z_2 = y_2 - \bar{y}$, . . . Then the actual error in the estimate of the mean from a sample of one unit will be z_r, where r is the selected unit.

The mean of all the z's is zero, and therefore the average of the errors of the estimates from a large number of samples of one unit (having regard to the signs of the errors) will approximate to zero. This is equivalent to saying there is no bias in the estimate.

In order to obtain a measure of the magnitude of the expected error we must therefore obtain some form of average of the z's which does not take account of sign. One simple measure which might be taken is the average of all the z's without regard to sign, but an alternative measure, which has a number of statistical advantages, is provided by the mean of the squares of all the z's. This is termed the *mean square deviation* of y or the *variance* of y, and is denoted by $\mathbf{V}(y)$ or σ^2, and its estimate by $\mathsf{V}(y)$ or s^2. The square root of this variance is termed the *standard deviation* σ of a single unit.

In the same way, if a sample contains a number of units we may define the sampling variance of an unbiased estimate, say $\bar{y}$, derived from such a sample as the mean of the squares of the actual errors of a large number of samples of the same size. This variance will be denoted by $\mathbf{V}(\bar{y})$, or, if estimated, by $\mathsf{V}(\bar{y})$. The square root of this variance is generally termed the *standard error* of the estimate, and will be denoted by $\mathbf{S.E.}(\bar{y})$, or, if estimated, by $\mathsf{S.E.}(\bar{y})$. The term standard error is also sometimes applied to the standard deviation of a single unit, particularly when the deviations are in the nature of errors of observation.

The standard error of the estimate of the population mean derived from a sample of one unit is therefore equal to the standard deviation of a single unit. If a sample of two units r and s is taken, the actual error of the estimate of the population mean will be

$$\bar{y} - \bar{y} = \tilde{y} - \bar{y} = \tfrac{1}{2}(z_r + z_s)$$

The standard error of the estimate will therefore be given by the square root of the average value of

$$\tfrac{1}{4}(z_r + z_s)^2 = \tfrac{1}{4}(z_r^2 + z_s^2 + 2z_r z_s)$$

If the population is large the average value of $z_r z_s$ is zero, as can be seen if

173

we consider a series of samples having the same first unit r and different second units s. The average values of z_r^2 and z_s^2 are both σ^2. Hence the average value of the above expression is $\frac{1}{2}\sigma^2$. Consequently the standard error of the estimate is $\sigma/\sqrt{2}$.

It will be noted that the above argument does not depend on the form of the distribution of the z's—there is no need, for example, for positive and negative deviations to be equally frequent. It does, however, require that each unit of the sample shall be randomly and independently selected. If, for instance, there were a tendency to select a second unit with a deviation similar to the first unit the average value of $z_r z_s$ would not be zero.

The argument can easily be extended to a sample of n units, for which the variance and standard error of the estimate will be found to be

$$\mathbf{V}(\bar{y}) = \frac{1}{n}\mathbf{V}(y), \qquad \text{S.E.}(\bar{y}) = \frac{1}{\sqrt{n}}\sigma \qquad (7.1.\text{a})$$

We thus have the important general result that *the standard error of the estimate of the mean of a large population from a random sample is inversely proportional to the square root of the number of units in the sample.*

The standard error of the estimate of the total follows immediately from the rule that if l is any multiplier the standard error of $l\bar{y}$ is equal to l times the standard error of $\bar{y}$, provided l is not subject to sampling variation. Thus

$$\text{S.E.}(Y) = \text{S.E.}(gn\bar{y}) = gn\{\text{S.E.}(\bar{y})\} = g\sigma\sqrt{n} \qquad (7.1.\text{b})$$

Although $S(y)$ is not itself an estimate it is often convenient to consider its sampling variance or standard error. From the above rule

$$\mathbf{V}\{S(y)\} = n\mathbf{V}(y), \qquad \text{S.E.}\{S(y)\} = \sigma\sqrt{n} \qquad (7.1.\text{c})$$

The standard error of an estimate can be expressed as a percentage of the population value of the estimated quantity. This form of expression is useful, as the percentage standard error is unaffected by the units in which the estimate is expressed, and the percentage standard error of the mean, of the total of the sample, and of the estimate of the population total are all equal. Similarly the standard deviation of a single unit can be expressed as a percentage of the. mean value of a single unit. This is sometimes termed the *coefficient of variation.* Denoting it by $\sigma\%$, we have, in a large population,

$$\text{S.E.}\%(\bar{y}) = \text{S.E.}\%\{S(y)\} = \text{S.E.}\%(Y) = (\sigma\%)/\sqrt{n}$$

Thus in a population with a percentage standard deviation per unit of 20 per cent., the percentage standard error of the estimate of the population mean or total from a sample of 100 will be 2 per cent., that from a sample of 400 will be 1 per cent., etc.

In order to estimate the standard error of the mean or total in numerical terms an estimate of the value of σ will be required. This can be obtained from the deviations $y - \bar{y}$ of the numerical values of the selected units from their mean. $y - \bar{y}$ will be nearly, though not exactly, equal to z, and to a first approximation an estimate of σ^2 will therefore be provided by the mean square deviation $S(y - \bar{y})^2/n$. Actually the sum of the squares of the deviations from the sample mean is always less than the sum of the squares of the deviations from the population mean, as can be seen from the identity

$$S(y - \bar{y})^2 = S(y - \bar{\bar{y}})^2 - n(\bar{y} - \bar{\bar{y}})^2$$

The average value of the first term on the right-hand side is $n\sigma^2$, and the average value of the second term is σ^2, since $\bar{y} - \bar{\bar{y}}$ is the error in the estimate of the mean. Thus $S(y - \bar{y})^2$ has an average value of $(n - 1)\sigma^2$, and consequently an estimate of σ^2 is given by

$$s^2 = \frac{1}{n-1} S(y - \bar{y})^2 \qquad (7.1.d)$$

The divisor $n - 1$ is technically known as the number of *degrees of freedom* associated with the estimate of error, and is equal to the number of independent comparisons that can be made between n values.

The calculation of the sum of the squares of the deviations $S(y - \bar{y})^2$ is best done from the sum of the squares of the values themselves. By this procedure the calculation and squaring of the individual deviations, which often involve fractional values, is avoided. One of the expressions

$$S(y - \bar{y})^2 = S(y^2) - n\bar{y}^2$$
$$= S(y^2) - \bar{y} S(y)$$
$$= S(y^2) - \frac{1}{n} \{S(y)\}^2$$

is used. The last term of each of the three expressions is usually termed "the correction for the mean." In calculating it from one of the first two expressions, $\bar{y}$ must be taken to at least as many significant figures as are required in the correction. For this reason the last expression is often the most convenient.

Sometimes it pays to use some convenient round number y_0 as a working mean, in which case we have

$$S(y - \bar{y})^2 = S(y - y_0)^2 - n(\bar{y} - y_0)^2$$

etc.

If a calculating machine is available the individual squares should not be written down—the sum of squares can be obtained directly by squaring the numbers successively without clearing the machine.

The calculation of the sum of squares from grouped data is illustrated in Examples 7.1.b and 7.2.b.

Example 7.1.a

Estimate the standard error of the estimate of the mean of the population (assumed large) of which the values of Table 6.4 are a sample.

The computations are as follows:

$$n = 20 \qquad S(y^2) = 1959 \cdot 12$$
$$S(y) = 193 \cdot 8 \qquad \bar{y}\, S(y) = 1877 \cdot 92$$
$$\bar{y} = 9 \cdot 69000 \qquad S(y - \bar{y})^2 = \overline{81 \cdot 20}$$

$$s^2 = \frac{81 \cdot 20}{19} = 4 \cdot 274 = 2 \cdot 07^2$$

$$\text{S.E.}\,(\bar{y}) = \sqrt{\frac{4 \cdot 274}{20}} = \pm\, 0 \cdot 462$$

Example 7.1.b

Table 7.1 gives the distribution of family income in a sample of 162 white families in Norfolk–Portsmouth, Virginia. Calculate the mean income of the sample and the sampling standard error of this mean.

TABLE 7.1—ANNUAL NET INCOME OF A SAMPLE OF 162 WHITE FAMILIES IN NORFOLK–PORTSMOUTH, VIRGINIA, 1934–6

Annual net income $	No. of families	Working units	Calculation		Calculation by successive summation	
			Total $(2) \times (3)$	Sum of squares $(2) \times (3)^2$ $= (3) \times (4)$	Total	Sum of squares
(1)	(2)	(3)	(4)	(5)	(6)	(7)
600–	10	− 3	− 30	90	10	10
900–	23	− 2	− 46	92	33	43
1,200–	40	− 1	− 40	40	73	116
1,500–	32	0	0	0	116	
1,800–	28	+ 1	+ 28	28	57	105
2,100–	20	+ 2	+ 40	80	29	48
2,400–	4	+ 3	+ 12	36	9	19
2,700–	2	+ 4	+ 8	32	5	10
3,000–	1	+ 5	+ 5	25	3	5
3,300–	2	+ 6	+ 12	72	2	2
	162		− 11	495	105	358

With grouped data of this type it is best to use the group interval as the working unit and the central value of one of the central groups as the working mean. The group \$1500–1799 (central value, \$1649·5, since the data were rounded off to the nearest dollar before grouping) has been chosen. The calculation of the total and sum of squares in these units is shown in columns 4 and 5 of the table. The mean of the sample in working units is therefore $-11/162$, *i.e.* $-0·06790$, and in the proper units is

$$1649·5 - 0·06790 \times 300 = 1629·1$$

The sum of squares of the deviations in the working units is

$$495 - 0·06790 \times 11 = 494·25$$

and in the proper units is therefore $494·25 \times 300^2$, *i.e.* 44,482,000. Hence, dividing by 161, $s^2 = 276,290$, and the sampling standard error of the mean income is $\sqrt{(276,290/162)} = \pm 41·3$.

If no calculating machine, or only an adding machine, is available the alternative form of calculation shown in columns 6 and 7 may be preferred. Column 6 is formed from column 2 by successive summation from the ends. Column 7 is similarly formed from column 6. Note the check $73 + 57 + 32 = 162$ for column 6, and the checks of the final values for column 7 from the totals of column 6. The total in working units is then given by the difference of the totals of the two halves of column 6, *i.e.* by $105 - 116 = -11$. The sum of squares is obtained by doubling the total of column 7 and deducting the sum of the totals of the two halves of column 6, i.e. by $2 \times 358 - 105 - 116 = 495$.

7.2 Sampling from a finite population

The above theory requires modification in two respects if the population is not large. In the first place it is best to define σ^2 as

$$\sigma^2 = \frac{1}{N - 1} S_p (y - \bar{y})^2$$

where S_p denotes summation over the whole population. This is equivalent to regarding the population as itself a random sample from an infinitely large population with variance σ^2. With this definition of σ^2 formula 7.1.d for s^2 stands without modification. In certain textbooks and scientific papers the alternative definition with divisor N is adopted. This introduces the factor $(N - 1)/N$ into the formula for s^2, and leads to other complications in the discussion of the errors of sampling from finite populations which are avoided by the use of the first definition.

In the second place, the formula for the standard error of the estimate of the mean or other estimate requires modification by the introduction of the factor $\sqrt{(1 - f)}$, or more strictly $\sqrt{(1 - \mathbf{f})}$. Thus we have

$$\textbf{S.E.} \; (\bar{y}) = \sigma \sqrt{\frac{1 - f}{n}} \qquad (7.2)$$

That the introduction of some factor of this kind is necessary is obvious, since if the whole population is included in the sample ($f = 1$) the sampling error will be zero. The actual factor can be deduced by an extension of the algebraic analysis given above.

It should be noted that the factor $\sqrt{(1-f)}$ should not be introduced when testing the difference between the means of two sampled populations to see whether, for example, they are subject to different causal agencies. In this case we are concerned to determine whether there is a real and consistent difference running through all the units of the two populations: in other words, we wish to test whether the two samples can reasonably be regarded as random samples from a single infinitely large parent population, or whether they have to be regarded as samples of two different parent populations.

Example 7.2.a

Estimate the standard errors applicable to the estimates obtained in Example 6.4.b.

From Example 7.1.a, $s^2 = 4 \cdot 274$, and consequently

$$\text{S.E.} (\bar{y}) = \sqrt{\left\{ \frac{4 \cdot 274}{20} \left(1 - \frac{1}{25}\right) \right\}} = \pm 0 \cdot 453$$

and since

$$Y = 500 \, \bar{y}$$
$$\text{S.E.} (Y) = 500 \times 0 \cdot 453 = \pm 226$$
$$\text{S.E.} (Y') = 507 \times 0 \cdot 453 = \pm 230$$

Example 7.2.b

Estimate the sampling error of the wheat acreage from the random sample of 125 farms of Table 6.6.a.

We find:

$$n = 125 \qquad S(y) = 2301 \qquad \bar{y} = 18 \cdot 4080$$
$$S(y^2) = 207{,}261 \qquad S(y - \bar{y})^2 = 164{,}904 \qquad s^2 = 164{,}904/124 = 1329 \cdot 9$$
$$\text{S.E.} (\bar{y}) = \sqrt{\{1329 \cdot 9 \, (1 - \tfrac{1}{20})/125\}} = \pm 3 \cdot 18$$
$$\text{S.E.} (Y) = 20\sqrt{\{125 \times 1329 \cdot 9 \, (1 - \tfrac{1}{20})\}} = 20 \times 125 \times 3 \cdot 18 = \pm 7950$$

The calculation of the mean and of the sum of squares of deviations may alternatively be carried out by grouping the data. The groups should be so chosen that the distribution within any group containing a substantial number of values is reasonably even. A grouping interval of 10 acres is here convenient, but because of the large number of zeros these must be included in a separate group.

The grouped data and the calculation of the total and sum of squares are shown in Table 7.2. The calculations are carried out in terms of working values in units of the grouping interval, and a working mean of 24·5 (the mean of group 4) is taken. The total is obtained from column 4 and the sum of squares from column 5. The mean in terms of the grouping interval is therefore

$$(- 209·75 + 136)/125 = - 73·75/125 = - 0·590$$

and in terms of the proper units is

$$\bar{y} = 24·5 - 0·590 \times 10 = 18·60$$

Similarly

$$s^2 = (1719·2 - 73·75 \times 0·590) \times 10^2/124 = 167,570/124 = 1351·4$$

The rest of the computations proceed as before.

TABLE 7.2—CALCULATION OF THE MEAN AND VARIANCE FROM GROUPED DATA
(WHEAT ACREAGES OF TABLE 6.6.a)

Acres	Number	Working value	Total (2) × (3)	Squares (2) × (3)^2
0	80	− 2·45	− 196·00	480·2
1–	5	− 1·95	− 9·75	19·0
10–	4	− 1	− 4	4
20–	11	0	− 209·75	
30–	2	+ 1	+ 2	2
40–	2	+ 2	+ 4	8
50–	4	+ 3	+ 12	36
60–	4	+ 4	+ 16	64
70–	4	+ 5	+ 20	100
80–	3	+ 6	+ 18	108
90–	1	+ 7	+ 7	49
100–	3	+ 8	+ 24	192
110–	1	+ 9	+ 9	81
....	....	....	....	....
260–	1	+ 24	+ 24	576
	125		+ 136	1719·2

This is the " pencil and paper " approach. With a modern desk calculator there is no point in grouping quantitative data merely to calculate the mean and standard deviation. Grouping, however, gives information on the form of the distribution which is additional to that conveyed by the mean and standard deviation, and enables exceptional features to be readily grasped.

In the present data the large number of zeros, the relatively large number of values between 20 and 30, and the single very high value of 265, are immediately apparent.

7.3   The normal law of error

The above analysis shows that it is possible, from the numerical values of the selected sampling units, to estimate the standard error of the estimate of the mean of the population. This gives us a measure of the average error to be expected. The analysis has not, however, given us any indication of the frequency with which errors of different magnitudes may be expected to occur.

It is a matter of common observation that in most material which is subject to quantitative variation large deviations tend to occur less frequently than do small deviations. In much material, also, positive and negative deviations occur with about equal frequency. The exact distribution of the deviations of individual sampling units will, of course, vary considerably in different types of material, but it is a fortunate circumstance that, over a wide range of distributions of the parent material, the errors to which estimates such as the mean, total, etc., are subject are distributed approximately according to what is known as the *normal law of error*, i.e. in a *normal distribution*. Other things being equal, the larger the sample on which the estimate is based, the more closely is the law followed. If the deviations of the original material are normally distributed, the errors of the estimate of the mean, etc., will conform exactly to a normal distribution.

In a normal distribution the frequency with which deviations within the infinitesimal range z to $z + dz$ may be expected to occur is given by the expression :

$$f(z)\, dz = \frac{1}{\sigma \sqrt{(2\pi)}}\, e^{-z^2/2\sigma^2}\, dz$$

where σ is the *standard deviation*, and e is the base of Napierian logarithms, $2 \cdot 71828$ approximately.

Fig. 7.3 shows normal distributions with standard deviations $\sigma = 1$ and $\sigma = 2$. The vertical scale represents the frequency with which deviations within a range of $0 \cdot 1$ of z occur per 1000 values. Thus the ordinate at $z = 0$ for $\sigma = 1$ is $39 \cdot 9$, which indicates that on the average $39 \cdot 9$ values per 1000 may be expected to have deviates having values between $-0 \cdot 05$ and $+0 \cdot 05$. The value $39 \cdot 9$ can be derived from the above formula by putting $\sigma = 1$, $z = 0$, $dz = 0 \cdot 1$ and multiplying by 1000.

From the figure it will be seen that positive and negative deviations of a given magnitude occur with equal frequency, and that large deviations are much less frequent than small ones. We are, however, in general not so much concerned with the frequency of a deviation of any particular magnitude, as with the frequency with which deviations greater than a given magnitude may be expected to occur. These latter frequencies, which correspond to areas in Fig. 7.3, are shown in Table A.2 at the end of the book, for various values of z/σ. The area for $z/\sigma = 1$ is shaded for both curves.

From Table A.2 it will be seen that 61·7 per cent. of all values have a deviation or error (positive or negative) greater than one-half the standard deviation or standard error, 31·7 per cent. of all values have a deviation greater than the standard deviation, but only 4·6 per cent. have a deviation greater than twice the standard deviation, and only 0·27 per cent. will have a deviation greater than three times the standard deviation. Consequently, if we know

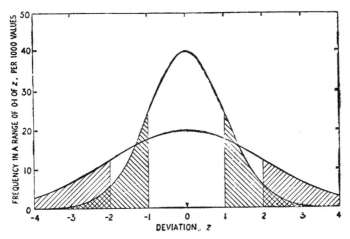

FIG. 7.3—NORMAL FREQUENCY DISTRIBUTIONS WITH STANDARD DEVIATIONS $\sigma = 1$ AND $\sigma = 2$. THE FLATTER CURVE IS THAT FOR $\sigma = 2$.

The shaded areas represent the total frequencies of the values for which the actual deviations are greater than the standard deviation.

that an estimate is subject to the normal law of error and has a given standard error, we can assign probable limits of error to this estimate. If limits of plus or minus twice the standard error are taken then in only 4·6 per cent. of the cases will the actual error lie outside these limits.

An alternative form of this statement, utilizing *fiducial probability*, which has certain logical advantages, is as follows.

If the true value of the mean is equal to the estimate minus twice the standard deviation (the lower limit of error) then a value of the estimate as high as, or higher than, that actually observed will be obtained in only 2·3 per cent. of all samples. (The values of Table A.2 are divided by 2 since deviations in only one direction are involved.) Similarly, if the true value of the mean is equal to the estimate plus twice the standard error (the upper limit of error), a value of the estimate as low as, or lower than, that actually observed will be obtained in only 2·3 per cent. of all samples. For limits of plus or minus once the standard error the corresponding percentages are 16 per cent. These closer limits are useful as indicating the region within which, or in the fairly close neighbourhood of which, the mean is likely to lie.

As pointed out above, we usually only have an estimate of the standard error, which is itself subject to error, the accuracy being dependent on the number of degrees of freedom, and statements in the above form are therefore not exact.

The effect of inaccuracy in the estimate of error on fiducial statements can be allowed for by the use of what is known as the *t distribution*, instead of the normal distribution. In general, however, inaccuracies due to paucity of data are not sufficiently great for this to be necessary in census work. More important is the fact that with a number of types of sample frequently employed, e.g. systematic samples and stratified samples with one unit from each stratum, fully valid estimates of error are not available. The estimates of error actually obtained in such cases are usually overestimates of the sampling standard errors, and any exact fiducial statement is therefore impossible.

In a random sample of n ($n-1$ degrees of freedom) from a normal distribution with standard deviation σ the standard error of s is given by

$$\text{S.E. } (s) = \frac{\sigma}{\sqrt{\{2\,(n-1)\}}} \tag{7.3}$$

If the material conforms approximately to the normal law of variation, an estimate based on 50 degrees of freedom will therefore determine the sampling error with a standard error of 10 per cent. and an estimate based on 200 degrees of freedom with a standard error of 5 per cent. If the material does not conform to the normal law the accuracy may be substantially less.

Example 7.3

Assign limits of error to the estimate of the mean of the population (assumed large) of the values of Table 6.4. (The values are actually a random sample from a normal distribution with mean 10 and standard deviation 2.) Show that they are distributed in the expected manner. Calculate also the standard error of the estimate of the standard deviation.

The results obtained in Example 7.1.a indicate that the true value of the mean is not likely to lie far outside the range $9\cdot69 \pm 0\cdot46$, *i.e.* $9\cdot23$–$10\cdot15$, and is fairly certain to lie within the range $9\cdot69 \pm 2 \times 0\cdot46$, *i.e.* $8\cdot77$–$10\cdot61$.

The distribution of the 20 values is given in Table 7.3. Integral values have been allotted $\frac{1}{2}$ and $\frac{1}{4}$ to the two appropriate classes. Each interval in this grouping is equal to $0\cdot5\,\sigma$. From Table A.2 the proportionate frequency of observations with deviations greater than $0\cdot5\,\sigma$ (positive or negative) is $0\cdot6171$, and consequently the expected frequencies of observations between 9 and 10 and between 10 and 11 are each $20 \times \frac{1}{2}(1-0\cdot6171) = 3\cdot83$. Similarly the proportionate frequency of observations greater than $\pm 1\cdot0\,\sigma$ is $0\cdot3173$, and consequently the expected frequencies between 8 and 9 and between 11 and 12 are each $20 \times \frac{1}{2}(0 \cdot 6171 - 0 \cdot 3173) = 3 \cdot 00$. In this manner all the expected frequencies shown in Table 7.3 can be calculated. The observed frequencies conform satisfactorily to the expected frequencies.

TABLE 7.3—OBSERVED AND EXPECTED FREQUENCIES IN THE SAMPLE
OF TABLE 6.4 FROM A NORMAL DISTRIBUTION

Range	< 6	6-7	7-8	8-9	9-10	10-11	11-12	12-13	13-14	14-15	> 15	Total
Observed	0	1	2·5	5	2·5	5·5	1	0·5	1	1	0	20
Expected	0·45	0·88	1·84	3·00	3·83	3·83	3·00	1·84	0·88	0·33	0·12	20·00

From formula 7.3 the standard error of the estimate s is

$$\text{S.E.}(s) = \frac{2}{\sqrt{(2 \times 19)}} = \pm 0 \cdot 324$$

The actual value of s, $2 \cdot 07$, is therefore closer to the true value than will occur on the average in samples of 20.

7.4 Qualitative variates

From the procedure developed for random samples it will be seen that the estimation of the sampling errors of estimates derived from a quantitative variate can be divided into two distinct stages, the first being the estimation of the variability of the individual sampling units (or more strictly the part of the variability which contributes to sampling error), and the second the derivation of the standard errors of the estimates in terms of the variability of the individual sampling units.

The same principles hold when the variate under consideration is qualitative. In the case of a random sample, however, the variability of an attribute of the sampling units depends only on the proportion of units possessing the attribute in the population. Hence in random samples no estimate of the variability of the individual sampling units is required.

For a random sample from a large population, if $q = 1 - p$, we have

$$V(p) = \frac{pq}{n}, \qquad \text{S.E.}(p) = \sqrt{\frac{pq}{n}}$$

If the population is finite and the sampling fraction f is appreciable the formula becomes, to all necessary accuracy,

$$\text{S.E.}(p) = \sqrt{\{pq(1 - f)/n\}}$$

The exact expression is obtained by replacing $(1 - f)$ by $(N - n)/(N - 1)$.

Similarly

$$V(u) = npq(1 - f)$$

and hence

$$\text{S.E.}(U) = g\sqrt{\{npq(1 - f)\}} = N\{\text{S.E.}(p)\}$$

183

Fig. 7.4 shows the way in which the standard error of the estimated percentage S.E. (100 p) varies with the percentage 100 p in samples of 100 and 1000 from a large population. The actual values of S.E. (100 p) are shown by the full line, while the dotted line gives the percentage standard error 100 S.E. (100 p)/100 p.

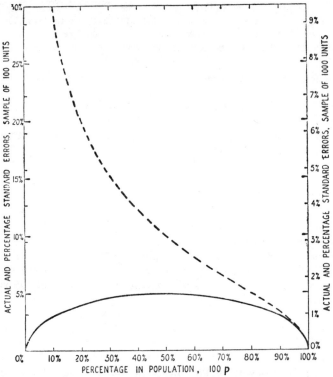

FIG. 7.4—STANDARD ERRORS OF THE ESTIMATED PERCENTAGE OF UNITS HAVING A GIVEN ATTRIBUTE, AND OF THE ESTIMATED NUMBER HAVING THE ATTRIBUTE, FOR DIFFERENT PERCENTAGES OF UNITS HAVING THE GIVEN ATTRIBUTE IN THE POPULATION

The full line shows the actual standard error of the estimated percentage, and the broken line the *percentage* standard error of the estimated number. This is equal to the percentage standard error of the estimated percentage. The scales shown are for samples of 100 and 1000. For a sample of 10,000 divide the values of the left-hand scale by 10, etc.

The standard errors obtained with larger samples for which the sample number is a power of 10 can also be read from the figure by dividing one of the scales by the appropriate power of 10. Thus for a sample of 10,000 the scale for the sample of 100 is divided by 10, since $\sqrt{(10,000/100)} = 10$.

The actual standard error has its maximum value at $p = 0.5$. At this point the standard error of the estimated percentage with a sample of 100 is 5·0, and with a sample of 1000 is 1·58, i.e. if the true percentage is 50 per cent. the value of the estimated percentage will usually lie between 40 per cent. and 60 per cent. with a sample of 100, and between 47 per cent. and 53 per cent. with a sample of 1000. Expressed in percentage terms the standard errors and limits at this point are double the above values. As the percentage in the population decreases from 50 per cent. the actual standard error of the estimated percentage also decreases, but the percentage standard error continues to increase. With $100\,\mathbf{p} = 20$ per cent. the actual standard error with a sample of 100 is 4·0 and the percentage standard error 20 per cent. ; with $100\,\mathbf{p} = 5$ per cent. they are 2·2 and 44 per cent. respectively. Thus, while quite a small sample serves to verify that the proportion in a population possessing a given attribute is small, the determination with any accuracy of the actual number possessing the attribute requires a relatively large sample when the proportion is small.

In estimating the sampling error the proportion in the population $\mathbf{p}$ can be replaced by its estimate p from the sample, i.e. by the proportion in the sample. This results in a certain amount of error in the estimate of variability, since the proportion in the sample will not in general be exactly equal to that in the population, but in large samples, such as are commonly met with in census work, this is not likely to be of much importance. Exact treatment of the problem is possible by use of Table VIII.1 of *Statistical Tables for Biological, Agricultural and Medical Research.*

If a random sample from a population is divided into several classes, A, B, . . ., the proportions p_a, p_b, . . . of the units falling in these classes give estimates of the corresponding proportions in the population, and the formulae of Section 7.4 enable standard errors to be calculated for p_a etc. However, the standard error of the difference between p_a and p_b cannot be calculated directly from the standard errors of p_a and p_b, because the sampling errors affecting p_a and p_b are not independent. If the sample contains an excess of class A the probability that it contains a deficit of class B is increased ; if there are only two classes, for example, the deficit of class B must exactly equal the excess of class A.

As explained in the next section, the lack of independence is taken into account by including a covariance term. The covariance here is:

$$\mathrm{cov}\,(\mathsf{p}_a,\,\mathsf{p}_b) = -\,\frac{\mathbf{p}_a\,\mathbf{p}_b}{n}$$

Thus, if $\mathbf{q}_a = 1 - \mathbf{p}_a$, etc.,

$$\mathbf{V}(\mathsf{p}_a - \mathsf{p}_b) = \mathbf{V}(\mathsf{p}_a) + \mathbf{V}(\mathsf{p}_b) - 2\,\mathbf{cov}(\mathsf{p}_a,\,\mathsf{p}_b)$$
$$= \frac{\mathbf{p}_a\,\mathbf{q}_b}{n} + \frac{\mathbf{p}_b\,\mathbf{q}_b}{n} + 2\,\frac{\mathbf{p}_a\,\mathbf{p}_b}{n}$$

For calculation, $\mathbf{p}_a$ and $\mathbf{p}_b$ must be replaced by their estimates p_a and p_b. n may optionally be replaced by $n-1$, both here and in the formula for $\mathbf{V}(\mathsf{p})$.

It must be clearly recognized that the above formulæ hold only when the units of which the proportion possessing a given attribute is being assessed are themselves the sampling units, and the sample is a random one from the whole population. In a stratified random sample the formulæ apply to each stratum taken separately. In other cases, e.g. multi-stage sampling, and all types of sampling with supplementary information, the variability no longer depends only on the proportions in the population or strata. Thus, for example, the formulæ are not applicable to the proportion of farms growing a given crop when two-stage sampling by administrative districts, and by farms within the selected districts, has been adopted. Equally they are not applicable to the proportion of individuals of a given race in a human population, when the sampling has been by households ; since the whole of a household is usually of the same race, the variability will clearly be greater than if a random sample of individuals had been taken.

When the above formulæ do not hold, the variability of the individual sampling units must be assessed in the same manner as with a quantitative variate, scoring the qualitative variate 1 or 0. (See Example 7.8.b.)

Example 7·4·a

Estimate the sampling errors of the estimates of Example 6.4.a.
We have $p = 0.0738$, $q = 1 - 0.0738 = 0.9262$. Hence

$$\text{S.E. } (p) = \sqrt{\left[\frac{0.0738 \times 0.9262}{8491}\left(1 - \frac{1}{50}\right)\right]} = \pm 0.00281$$

S.E. (percentage defective) $= 100 \times 0.00281 = \pm 0.281$
S.E. (total number defective) $= \text{S.E. } (U) = 50 \times 8491 \times 0.00281 = \pm 1190$

Thus the percentage defective, 7·38 per cent., has a standard error of ± 0.28, which implies, taking limits of plus or minus twice the standard error, that the true percentage defective probably lies between 6·8 per cent. and 7·9 per cent. Similarly the number defective probably lies between 29,000 and 33,700. Note that the standard error expressed as a percentage of the percentage defective or number defective, i.e. what is ordinarily called the percentage standard error, is

$$\text{S.E. } \% (p) = \text{S.E. } \% (U) = \frac{0.281}{7.38} \times 100 = \frac{1190}{31,350} \times 100 = \pm 3.8 \text{ per cent.}$$

These standard errors are likely to be slight overestimates, since the sample was in fact systematic.

Example 7·4·b

The final poll of the British Institute of Public Opinion in the 1951 election, based on a sample of 2,300 individuals, gave a forecast of the voting (excluding 3·5 per cent. who gave no indication of the way they would vote) as follows (Durant and Gregory, 1951, E′).

Conservative	.	.	.	.	49·5 per cent.
Labour	.	.	.	.	47·0 ,,
Liberal	.	.	.	.	3·0 ,,
Others	.	.	.	.	0·5 ,,

If the sample were a random one, what would be the standard error of the difference between Conservatives (A) and Labour (B) ?

The effective number in the sample is 96·5 per cent. of 2,300, i.e. 2,220. Hence

$$V(p_a) = \frac{0·495 \times 0·505}{2219} = ·000113$$

$$V(p_b) = \frac{0·47 \times 0·53}{2219} = ·000112$$

$$\text{cov}(p_a, p_b) = -\frac{0·495 \times 0·47}{2219} = -·000105$$

$$V(p_a - p_b) = ·000113 + ·000112 - 2(-·000105) = ·000435 = ·0209^2.$$

Thus the predicted Conservative percentage majority of 2·5 per cent. would have an estimated standard error of 2·09 per cent. If the covariance term had been omitted the estimate of the standard error would have been 1·50 per cent., which is substantially below the correct value.

It should be noted that the samples in polls of this kind are actually quota samples, and the random component of error will thereby be reduced by the stratification thus introduced. The amount of the reduction can be estimated if the numbers and voting intention of the different strata (quota categories) are known, using the formulae of Section 7.6.1. The reduction will only be substantial, however, if the differences in voting intention of the different strata are very substantial (see Example 8.2.a). The non-random components of error in forecasts of this kind have been discussed in Section 4.22.

7.5 General formulae for standard errors of functions of estimates

If we have a number of estimates y_1, y_2, y_3, . . . with sampling errors which are independent, the sampling variances being $V(y_1)$, $V(y_2)$, $V(y_3)$, . . . and we form a linear function of the y's :

$$L = l_1 y_1 + l_2 y_2 + l_3 y_3 + \cdots$$

where the l's are any multipliers whose values are not influenced by the sampling, the sampling variance of L is given by

$$V(L) = l_1^2 V(y_1) + l_2^2 V(y_2) + l_3^2 V(y_3) + \cdots \qquad (7.5.a)$$

The condition of independence is important. The sampling errors of two estimates will be independent if the estimates are derived from sets of values which are themselves independent. Estimates derived from samples of different populations, or from different strata of the same population, are consequently independent, as are estimates derived from different samples of a large population. Estimates derived from two different variates belonging to the same sampling units are not in general independent, since such variates are likely to be correlated, high values of the one being associated with high (or low) values of the other in the same sampling units.

A number of important simple formulæ are derivable from the above general formula.

The standard error of a multiple of an estimate is the same multiple of the standard error of the estimate:

$$V(ly_1) = l^2 V(y_1) \tag{7.5.b}$$
$$\text{S.E.} (ly_1) = l \, \text{S.E.} (y_1)$$

This formula has already been used in Section 7.1.

The standard error of the difference of two independent estimates is the square root of the sum of the squares of the standard errors of the estimates:

$$V(y_1 - y_2) = V(y_1) + V(y_2) \tag{7.5.c}$$
$$\text{S.E.} (y_1 - y_2) = \sqrt{[\{\text{S.E.} (y_1)\}^2 + \{\text{S.E.} (y_2)\}^2]}$$

The standard error of the sum of a number of independent estimates is the square root of the sum of the squares of the standard errors of the estimates:

$$V(y_1 + y_2 + y_3 + \ldots) = V(y_1) + V(y_2) + V(y_3) + \ldots \tag{7.5.d}$$
$$\text{S.E.} (y_1 + y_2 + y_3 + \ldots) = \sqrt{[\{\text{S.E.} (y_1)\}^2 + \{\text{S.E.} (y_2)\}^2 + \{\text{S.E.} (y_3)\}^2 + \ldots]}$$

which may be expressed by the rule that "variances are additive."

The standard error of the estimate of the mean of a large population can also be derived from the formula.

Weighted means are a type of linear function which occurs frequently in statistics. The general form of a weighted mean is

$$\bar{y}_w = \frac{w_1 y_1 + w_2 y_2 + \ldots}{w_1 + w_2 + \ldots}$$

where the w's are the weights. Knowing the variances of the y's, the variance of $\bar{y}_w$ can be calculated, provided the y's are independent. Two cases are of frequent occurrence.

(1) $V(y_1) = V(y_2) = \ldots = V(y)$
We then have

$$V(\bar{y}_w) = \frac{S(w^2)}{\{S(w)\}^2} V(y) \tag{7.5.e}$$

(2) $V(y_1) = \lambda/w_1$, etc., where λ is a constant.
We then have

$$V(\bar{y}_w) = \frac{S(w)}{\{S(w)\}^2} \lambda = \frac{\lambda}{S(w)} \tag{7.5.f}$$

This is the form of weighted mean which is used when we wish to obtain the most accurate combined estimate from a number of independent estimates of the same quantity whose relative variances are known. The weights are taken equal (or proportional) to the reciprocals of the variances, and the variance of the weighted mean is given by the reciprocal of the sum of the weights (or a multiple of this reciprocal).

A further type of weighted mean is that in which the weights w are in the nature of supplementary information, the quantities y and w both being determined from the individual sampling units, with the variances of the y's related in some unknown manner to the w's, and $\bar{y}_w = S(wy)/S(w)$.

In order to obtain an unbiased estimate of $V(\bar{y}_w)$, whatever the variance law, the squares of the deviations of the y from $\bar{y}_w$ must be weighted in proportion to w^2 before summation. For a random sample, if

$$Q = Sw^2(y - \bar{y}_w)^2$$

and

$$s_q^2 = Q/(n-1)$$

we have

$$V(\bar{y}_w) = \frac{(1-f)\,ns_q^2}{\{S(w)\}^2} \qquad (7.5.\text{g})$$

It may also be noted that if the variance of y for given w can be regarded as constant over the range of w, and there is also no variation in the mean value of y for given w over the range other than that ascribable to random variation in y, the *efficient* estimate of the variance of y is given by the ordinary formula

$$V(y) = S(y - \bar{y})^2/(n-1) \qquad (7.5.\text{h})$$

and formula 7.5.e may be used to estimate $V(\bar{y}_w)$, with the introduction of the factor $(1-f)$. If the variance of y for given w is inversely proportional to w then the efficient estimate of the variance of y is given by

$$\left. \begin{aligned} Q' &= Sw(y - \bar{y}_w)^2 = Swy^2 - \bar{y}_w\,Swy \\ s_q'^2 &= Q'/(n-1) \end{aligned} \right\} \qquad (7.5.\text{i})$$

$s_q'^2$ is an estimate of λ, and formula 7.5.f can be used for estimating $V(\bar{y}_w)$, with the introduction of the factor $(1-f)$. Either of these estimates will be biased if the true variance law is different from that assumed or the other condition does not hold. They should therefore not be used without careful consideration.

The mean ratio $\bar{r}$ used in the ratio method of estimation is an example of a weighted mean of the above type, since $\bar{r} = S(y)/S(x) = S(xr)/S(x)$, and we therefore substitute r for y and x for w. This case is discussed in more detail in Sections 7.8–7.11, which deal with the estimation of errors in the ratio method in both random and stratified samples. Normally formula 7.5.g will be used to estimate $V(\bar{r})$, but under certain circumstances formulæ 7.5.h and 7.5.e may be employed.

The approximate formulæ for the standard errors of the product and the ratio of two estimates whose sampling errors are independent may also be noted. These are given by

$$V(y_1 y_2) = y_2^2\, V(y_1) + y_1^2\, V(y_2) \tag{7.5.j}$$

$$V\left(\frac{y_1}{y_2}\right) = \left(\frac{y_1}{y_2}\right)^2 \left(\frac{V(y_1)}{y_1^2} + \frac{V(y_2)}{y_2^2}\right) \tag{7.5.k}$$

These formulæ are only satisfactory if $V(y_1)$ and $V(y_2)$ are small relative to y_1^2 and y_2^2 respectively.

If the estimates y_1, y_2, y_3, . . . are not independent the concept of covariance must be introduced. The covariance between two estimates is the mean product deviation, and is estimated in exactly the same manner as is the variance of each of the estimates, with the exception that the sum of squares of the deviations of a single variate is replaced by the sum of products of the deviations of the two variates. If the covariance between y_1 and y_2 is denoted by $\operatorname{cov}(y_1, y_2)$ the additional terms

$$+ 2l_1 l_2\, \operatorname{cov}(y_1, y_2) + 2l_1 l_3\, \operatorname{cov}(y_1, y_3) + 2l_2 l_3\, \operatorname{cov}(y_2, y_3) + \ldots$$

must be introduced into the formula for $V(L)$. This gives the additional term $-2\, \operatorname{cov}(y_1, y_2)$ in the formula for $V(y_1 - y_2)$. The corresponding additional term in $V(y_1 y_2)$ is $+2 y_1 y_2\, \operatorname{cov}(y_1, y_2)$, and that in $V(y_1/y_2)$ is $-2\, \operatorname{cov}(y_1, y_2)/y_1 y_2$ within the bracket.

If y_1, y_2, y_3, . . . are derived from different variates belonging to the same sampling units, e.g. measurements of different characters, the variance of any linear function L can, if desired, be estimated directly by calculating a value L for each sampling unit separately and estimating $V(L)$ from these values in the manner appropriate to a single variate. This obviates the calculation of the variances and covariances of the individual variates. The same method can be followed with products and ratios, subject to the same limitations as those given above for formulæ 7.5.e and 7.5.f. If the errors of a number of functions are required, however, it is best to calculate the variances and covariances (Example 7.8.b).

The regression and correlation coefficients can be expressed in terms of the variances and covariance. We have the relations $b = \operatorname{cov}(x, y)/V(x)$, and $r = \operatorname{cov}(x, y)/\sqrt{\{V(x).V(y)\}}$.

In the more complicated types of sampling, discussed later, the estimation of covariance is again exactly parallel to the estimation of the corresponding variances, the squares being replaced by products wherever they occur.

Example 7.5.a

Calculate standard errors for the various estimates of the regional and varietal differences between the yields of potatoes given in Tables 5.21.a, 5.21.b and 5.21.c, given that the variance of the yield per acre of any one variety in any one region is 4·22, and that the standard deviation is therefore $\pm 2\cdot05$.

The standard errors of the regional–varietal means of Table 5.21.a are obtained by dividing the above standard deviation by the square roots of the numbers of fields. Thus for Majestic in Scotland the standard error is $2 \cdot 05/\sqrt{37} = \pm 0 \cdot 34$. The standard errors are shown in Table 7.5.

TABLE 7.5—POTATO SURVEY : STANDARD ERRORS OF REGIONAL-VARIETAL MEANS

	Scotland	North	E. Midlands	South	West
Majestic . .	$\pm 0 \cdot 34$	$\pm 0 \cdot 24$	$\pm 0 \cdot 20$	$\pm 0 \cdot 20$	$\pm 0 \cdot 24$
King Edward .	$\pm 0 \cdot 32$	$\pm 0 \cdot 55$	$\pm 0 \cdot 22$	$\pm 0 \cdot 25$	$\pm 0 \cdot 31$
Great Scot .	$\pm 0 \cdot 48$	$\pm 0 \cdot 55$	—	$\pm 0 \cdot 84$	$\pm 0 \cdot 48$
Arran Banner .	$\pm 0 \cdot 72$	$\pm 0 \cdot 33$	—	$\pm 0 \cdot 68$	$\pm 0 \cdot 38$
Kerr's Pink .	$\pm 0 \cdot 25$	$\pm 0 \cdot 34$	—	—	$\pm 0 \cdot 57$

These standard errors enable the differences between the individual means to be examined more critically. The difference between Scotland and the Northern region for Arran Banner, for example, is at first sight anomalous, being $- 0 \cdot 12$. The standard error of this difference is $\sqrt{(0 \cdot 72^2 + 0 \cdot 33^2)} = \pm 0 \cdot 79$. This difference, therefore, does not conflict very seriously with the other differences.

On the other hand the difference between this difference and the largest positive difference, that for King Edward, is $+ 1 \cdot 94 - (- 0 \cdot 12) = + 2 \cdot 06$. This quantity has a standard error of $\sqrt{(0 \cdot 32^2 + 0 \cdot 55^2 + 0 \cdot 72^2 + 0 \cdot 33^2)} = \pm 1 \cdot 02$. It might therefore be judged unlikely, on this evidence alone, that the difference has arisen by chance, since Table A.2 shows that a difference of $2 \cdot 0$ times its standard error would arise by chance in less than 1 in 20 times. In statistical terminology the difference is *significant* at the 1 in 20 level of significance. This conclusion, however, is subject to the qualification that we have here picked the two extreme differences out of 10 possible pairs. A combined test* of all 5 differences shows that they are not exceptionally variable. A more comprehensive test of the differences of the whole table confirms this verdict. It may be noted, however, that Arran Banner is only $\frac{1}{3}$ as common in Scotland as in the Northern region, whereas King Edward is 3 times as common in Scotland as in the Northern region. The observed differences are therefore in the direction that would be expected if the varieties were grown in the regions to which they were most suited.

The standard errors of the means of sub-class means of Table 5.21.b, except for those that involve estimated values, may be calculated in a similar manner. The standard error of the unweighted mean for Scotland, for example, is $\frac{1}{5}\sqrt{(0 \cdot 34^2 + 0 \cdot 32^2 + \ldots)} = \pm 0 \cdot 203$. The corresponding value for the Northern region is $\pm 0 \cdot 188$. The difference between the two regions,

* The weighted sum of squares of deviations gives $\chi^2 = 5 \cdot 74$ with 4 degrees of freedom.

$8.55 - 7.\overline{46} = +1.09$, therefore has a standard error of $\sqrt{(0.203^2 + 0.188^2)} = \pm 0.277$. The weighted mean for Scotland is given by $(9.46 \times 393 + 7.65 \times 250 + \ldots)/901 = 8.76$, and its standard error is therefore, from formula 7.5.a,

$$\sqrt{[(393^2 \times 0.34^2 + 250^2 \times 0.32^2 + \ldots)/901^2]} = \pm 0.191$$

That for the Northern region is 7.49 ± 0.196, and the standard error of the difference is ± 0.274.

Table 5.21.c provides an example of a weighted mean with the weights so chosen that the most accurate combined estimate is obtained. Formula 7.5.f is therefore appropriate, and λ represents the variance of a single field, i.e. $\lambda = 4.22$. Hence the variance of the weighted mean difference $= 4.22/74 = 0.0570$, and the standard error is therefore ± 0.24.

The ratios of the squares of the above standard errors to that for the least-squares estimate give the efficiencies of the various estimates shown in Table 5.21.d.

The standard errors of the last column of Table 5.21.b cannot be evaluated exactly, as the pooling of regions is based on the assumption that there are no differences of any importance between these regions. In so far as this is not the case additional errors will be introduced, and the standard errors calculated on the assumption that there are no differences will therefore be underestimates of the true errors.

The standard errors of the means of the pooled regions can be calculated from the numbers of fields on which each mean is based. Thus that for Majestic is $\sqrt{(4.22/356)} = \pm 0.11$. The standard errors of the Scottish means have already been given in Table 7.5. The standard error of the weighted mean for Majestic is therefore given by

$$\sqrt{\left(\frac{727^2 \times 0.11^2 + 174^2 \times 0.34^2}{901^2}\right)} = \pm 0.11$$

Similarly the standard errors for the other four varieties are found to be ± 0.13, ± 0.28, ± 0.24 and ± 0.24.

The standard errors of the least-squares estimates are discussed in Section 9.8.

Example 7.5.b

Calculate standard errors for the estimates of seasonal differences of milk fever given in Table 5.20.a.

These can be calculated in a similar manner to those of Table 5.21.b. Standard errors (or variances) are first calculated for the sub-class percentage incidences from the numbers of calvings on which they are based. If the observed percentage, 0.42, is taken as an estimate of the sub-class population percentage for lactation 1–2 in January–April, the estimated variance of the percentage incidence is, from the first formula of Section 7.4,

$$(0.42 \times 99.58/3806) = 0.0110$$

The standard error for the adjusted unweighted mean (adjustment factor, $3\cdot46/5\cdot55=0\cdot623$) for January–April is therefore given by

$$0\cdot623\sqrt{[(0\cdot0110 + 0\cdot0669 + \ldots)/5^2]} = 0\cdot123$$

and that for the weighted mean by

$$0\cdot945\sqrt{[(0\cdot0110 \times 14444^2 + \ldots)/31893^2]} = 0\cdot122$$

The observed sub-class percentages are subject to proportionally large sampling errors, particularly in lactations 1–2. Better estimates of the population sub-class percentages can be obtained by taking the products of the seasonal and lactation percentages estimated from the means of sub-class means and dividing by the general mean percentage. Thus, for example, using the weighted means, we obtain the value $2\cdot10\times0\cdot38/3\cdot46=0\cdot23$ instead of $0\cdot42$ for the first cell of the table, giving an estimated variance of $0\cdot0060$ instead of $0\cdot0110$. This refinement is not likely to have any large effect in the estimation of the final standard errors—that for the weighted mean becomes $0\cdot118$ instead of $0\cdot122$.

7.5.1 Contrasts between domains : random sample

In investigations involving subdivisions of a population (domains of study) we may require, in addition to estimates of the domain values, the sampling errors of these estimates, and also the sampling errors of the differences between the different domains. This section gives the relevant formulae for the sampling variances and covariances of various commonly required estimates when the sample is random. For convenience, formulae for ratio estimates, which are discussed in more detail in Sections 7.8 and 7.9, are included here.

In a random sample the part of the sample constituting a particular domain will be equivalent to a random sample of that domain, with the exception that the number of units in the domain sample will be subject to sampling variation. The sampling standard error of the domain mean of a quantitative variate will therefore depend on the number n_a of domain units actually occurring in the sample, and on the variability of these units *within* the domain.

The sampling variation of n_a will not on the average increase the sampling error of the domain mean, but the sampling error will vary from sample to sample—samples in which the domain happens to be well represented will clearly give more accurate estimates.

The formulæ for estimates of the domain values are exactly analogous to those for the population values given in Section 6.4. If the suffix a denotes domain A, so that n_a, for example, is the number of units in the sample falling in that domain, and $\bar{y}_a$ and $S_a(y)$ are the mean and total of all the y values of the domain A units in the sample, we have—

$$\mathsf{p}_a = p_a = \frac{n_a}{n} \qquad (7.5.1.a)$$

$$\mathsf{N}_a = gn_a = \mathsf{N}\,\mathsf{p}_a \qquad (7.5.1.b)$$

$$\bar{y}_a = \bar{y}_a = \frac{1}{n_a} S_a (y) \qquad\qquad (7.5.1.\text{c})$$

$$Y_a = g\, S_a (y) = N_a\, \bar{y}_a \qquad\qquad (7.5.1.\text{d})$$

$$\bar{r}_a = \frac{S_a (y)}{S_a (x)} \qquad\qquad (7.5.1.\text{e})$$

The first two formulæ have already been given in Section 6.4 with a slightly different notation ($n_a = u$, $p_a = p$, $N_a = U$). The additional symbol p_a for the proportion of units in the domain A is introduced for convenience. As before a small, but usually trivial, gain in accuracy can be obtained by replacing N by N.

In addition to the proportion p_a of all units belonging to domain A we are often concerned with the proportion of the units of domain A that possess a given attribute, or with their total number. This proportion, which may be denoted by h_a, must be clearly distinguished from p_a. Following the notation of Section 6.4, the corresponding total number in the population may be denoted by U_a, and that in the sample by u_a. Then—

$$h_a = h_a = u_a/n_a \qquad\qquad (7.5.1.\text{f})$$

$$U_a = g u_a = h_a\, N_a \qquad\qquad (7.5.1.\text{g})$$

The estimation of the variances of $\bar{y}_a$, Y_a and $\bar{r}_a$ differs in two respects from the estimation of the variances of $\bar{y}$, Y, and $\bar{r}$. The first cause of difference is that only the component of variance of y within the domain A, and not the total variance of y in the population, contributes to the variance of the estimates $\bar{y}_a$ and Y_a. This component may be denoted by $s_a{}^2$. We have

$$s_a{}^2 = \frac{S_a (y - \bar{y}_a)^2}{n_a - 1} \qquad\qquad (7.5.1.\text{h})$$

The variances within the different domains will not only differ from the total variance, being in general less than this total variance, but may also differ amongst themselves. If, however, the population is divided into a large number of different domains, and the nature of the material is such that the variances within the different domains may be expected to be approximately equal, a pooled estimate of this variance over all domains may be adopted, using the analysis of variance technique in the manner of Section 7.7.

In a similar manner the variance of $\bar{r}_a$ depends on the variance $s_{ra}{}^2$ within the domain A about the ratio line for that domain. We have

$$s_{ra}{}^2 = \frac{S_a (y - \bar{r}_a x)^2}{n_a - 1} \qquad\qquad (7.5.1.\text{i})$$

The second cause of difference lies in the fact that the variance of the total Y_a is increased because N_a is subject to variation. Moreover the numbers of selected sample units n_a and n_b in two domains are negatively correlated, the covariance of n_a and n_b being $- n^2\, p_a\, p_b/(n - 1)$, and this gives rise to negative covariances between p_a and p_b, N_a and N_b, Y_a and Y_b, and U_a and U_b.

Furthermore, for the reason given in Section 7.2, the factor $\sqrt{(1-f)}$, which makes allowance for sampling from a finite population, should normally be omitted from the variance formulæ when comparisons between different domains are being made in investigational studies.

Putting $q_a = 1 - p_a$, the resultant formulæ for the variances and co-variances of the estimates are as follows :—

$$V(p_a) = \frac{p_a\, q_a\, (1-f)}{n-1} \tag{7.5.1.j}$$

$$\text{cov}(p_a,\, p_b) = -\frac{p_a\, p_b\, (1-f)}{n-1} \tag{7.5.1.k}$$

$$V(N_a) = \frac{g^2\, n^2\, p_a\, q_a\, (1-f)}{n-1} \tag{7.5.1.l}$$

$$\text{cov}(N_a,\, N_b) = -\frac{g^2\, n^2\, p_a\, p_b\, (1-f)}{n-1} \tag{7.5.1.m}$$

$$V(\bar{y}_a) = \frac{1-f}{n_a}\, s_a^2 \tag{7.5.1.n}$$

$$\text{cov}(\bar{y}_a,\, \bar{y}_b) = 0 \tag{7.5.1.o}$$

$$V(Y_a) = \frac{g^2\, n\, (1-f)\, \{n_a\, q_a\, \bar{y}_a^2 + (n_a-1)\, s_a^2\}}{n-1} \tag{7.5.1.p}$$

$$\text{cov}(Y_a,\, Y_b) = -\frac{g^2\, n^2\, (1-f)\, p_a\, p_b\, \bar{y}_a\, \bar{y}_b}{n-1} \tag{7.5.1.q}$$

$$V(\bar{r}_a) = \frac{1-f}{n_a\, \bar{x}_a^2}\, s_{ra}^2 \tag{7.5.1.r}$$

$$\text{cov}(\bar{r}_a,\, \bar{r}_b) = 0 \tag{7.5.1.s}$$

$$V(h_a) = \frac{h_a(1-h_a)\, (1-f)}{n-1} \tag{7.5.1.t}$$

$$\text{cov}(h_a, h_b) = 0 \tag{7.5.1.u}$$

$$V(U_a) = \frac{g^2 n^2(1-f)\, p_a h_a(1-p_a h_a)}{n-1} \tag{7.5.1.v}$$

$$\text{cov}(U_a, U_b) = -\frac{g^2 n^2(1-f)\, p_a p_b h_a h_b}{n-1} \tag{7.5.1.w}$$

If each of the domains of study with which we are concerned forms only a small fraction of the whole population, the covariances will be small relative to the corresponding variances, but if the fractions are large the covariances will be of some importance and must be taken into account when calculating the differences between estimates for the different domains.

7.6 Stratified random sample with possibly unequal variances within strata

Since in a stratified sample differences between the sampling units in the different strata are eliminated from the sampling error, in estimating this error we require not the total variance of the sampling units over the whole population, but the variances of the sampling units within the different strata.

A simple example will illustrate the difference. Suppose we have a large population of which 25 per cent. of the units have the value 8, 50 per cent. have the value 10, and 25 per cent. have the value 12. The mean of the population is 10, and 50 per cent. of the values have a deviation of 0 from the mean, the remaining 50 per cent. having a deviation of ± 2. The mean square deviation or total variance is therefore $0\cdot5 \times 0 + 0\cdot5 \times 2^2 = 2$. Suppose now the population is divided into two strata, the first containing all the units with value 8 and one-half the units with value 10, the second one-half the units with value 10 and all the units with value 12. The mean of the first stratum is 9, and all units in it have a deviation of ± 1. The mean square deviation or variance within the first stratum is therefore 1. The same holds for the second stratum.

In this example the strata are of the same size and the within-strata variances are equal. Examples can easily be constructed in which this is not the case, but even if the variances are unequal an average within-strata variance can be calculated, and in a large population this will always be less than the total variance if there are differences between the strata means.

If the sample numbers in all strata are sufficiently large for the within-strata variances per sampling unit to be separately estimated, the sampling variances of the means or totals of the individual strata can be estimated separately, and the sampling variance of the population mean or total, which is a linear function of these means or totals, can be obtained by the use of the formulæ of Section 7.5.

This method is valid even if there is inequality in the within-stratum variance per sampling unit from stratum to stratum, and is applicable to all types of stratification, including stratification with a variable sampling fraction and stratification after selection.

In general it is best to build up the variance of the population estimate under consideration by calculating the variances of the component parts, and adding these variances, or the correct multiples of them, the same steps being followed as in the calculation of the estimate itself. We will therefore not give formulæ for the variances of all the different estimates set out in Sections 6.4 and 6.5, but will illustrate the derivation of such formulæ by obtaining that for $V(\bar{y})$ in the case of a variable sampling fraction.

We have $\bar{y} = \Sigma\{g_i S_i(y)\}/N$. If σ_i^2 is the variance within the ith stratum, $V\{S_i(y)\} = n_i \sigma_i^2(1 - f_i)$, and hence by formula 7.5.a

$$V(\bar{y}) = \Sigma\{g_i^2 n_i \sigma_i^2 (1 - f_i)\}/N^2 \qquad (7.6.\text{a})$$

For $V(\bar{y})$ the σ_i^2 will be replaced by their estimates s_i^2.

If all the sampling fractions are equal we have, since $N = gn$,

$$V(\bar{y}) = (1 - f) \Sigma (n_i s_i^2)/n^2 \qquad (7.6.b)$$

The examples which follow will illustrate the details of the methods to be followed in the cases that are ordinarily met with in practice.

If we require the sampling errors of estimates applicable to domains of study which cut across the strata the situation is more complicated. The required formulae are given in the next section. At this point we may note that in the case of a stratified sample with uniform sampling fraction, an approximate estimate of the sampling error of a domain mean will be obtained by treating the sample as if it were stratified for the domain in question, and not stratified for the strata which cut across the domain (see Example 7.7.b).

Example 7.6.a

Estimate the sampling errors of the estimates of Example 6.5.

The computations for acreages are set out in Table 7.6.a. The various steps are as follows.

The sums of squares of deviations from each stratum mean, $S_i (y - \bar{y}_i)^2$, are first calculated from the sample values given in Table 6.5.a, using the method of Example 7.1.a. The estimated within-strata variances s_i^2 are then calculated by dividing by $n_i - 1$. Multiplying these by $(1 - f)/n_i$ gives

TABLE 7.6.a—ESTIMATION OF SAMPLING ERRORS OF THE WHEAT ACREAGES OF EXAMPLE 6.5

Size-group	n_i	$n_i - 1$	$S_i(y - \bar{y}_i)^2$	s_i^2	$V(\bar{y}_i)$	S.E.$(\bar{y}_i)$	$V\{S_i(y)\}$	$V(Y_i')$
1–	22	21	0					
6–	26	25	47	1·9	·069	± 0·26	47	19,000
21–	18	17	191	11·2	·593	± 0·77	192	76,000
51–	26	25	4,051	162·1	5·92	± 2·43	4,004	1,595,000
151–	20	19	13,899	731·5	34·75	± 5·89	13,898	5,560,000
301–	13	12	23,370	1947·5	142·3	± 11·93	24,052	10,069,000
	125	119					42,193	17,319,000

the variances of the strata means $V(\bar{y}_i)$, and taking the square roots gives the sampling standard errors S.E. $(\bar{y}_i)$ of these means. The means themselves are tabulated in Table 6.5.b.

To obtain the sampling errors of the population estimates we must bear in mind the method by which these estimates were arrived at. The estimate Y

of the population total was obtained by multiplying the sample total by the raising factor. We therefore require the variance of the sample total 2011. This will be equal to the sum of the variances of the strata totals. These variances $V\{S_i(y)\}$ are obtained by multiplying s_i^2 by $n_i(1-f)$. The square root of the sum of the variances is then taken and multiplied by 20. Thus

$$\text{S.E.}(Y) = 20 \times \sqrt{42{,}193} = 20 \times 205 \cdot 41 = \pm 4110$$

Similarly

$$\text{S.E.}(\bar{y}) = 205 \cdot 41/125 = \pm 1 \cdot 64$$

In the case of Y' each $V(\bar{y}_i)$ is multiplied by the square of the number of farms in the size-group, given in Table 6.5.b. Thus

$$142 \cdot 3 \times 266^2 = 10{,}069{,}000$$

Taking the square root of the sum,

$$\text{S.E.}(Y') = \sqrt{17{,}319{,}000} = \pm 4160$$

and hence

$$\text{S.E.}(\bar{y}') = 4160/2496 = \pm 1 \cdot 67$$

It will be noted that although Y' is slightly more accurate than Y the standard error given by the above calculation is slightly greater. There are two reasons for this. In the first place in calculating S.E. (Y) we have neglected the errors introduced by the use of a working sampling fraction. The average errors from this cause are equivalent to errors introduced by rounding off the numbers in the different strata of the sample to whole numbers. In the second place, use of the exact sampling fractions will result in slightly different contributions to the error variance from the different strata, which may result in raising or lowering the estimate of the standard error.

The computations for number of farms growing wheat follow the same pattern, with the exception that the variance of each size-group total $V(u_i)$ is estimated from the proportion of farms growing wheat in that size-group. The calculations for S.E. (U) are given in Table 7.6.b. For the largest size-group, for example,

$$V(u_i) = 13 \times 0 \cdot 84615 \times 0 \cdot 15385 \times 19/20 = 1 \cdot 608$$

We then have

$$\text{S.E.}(U) = 20 \times \sqrt{12 \cdot 83} = \pm 71 \cdot 64$$

If the sampling errors of the different strata means are not required, the above calculations can be simplified slightly by omitting the factor $(1-f)$ till the final stage of the computation.

Example 7.6.b

Estimate the sampling errors of the estimate of wheat acreage and number of farms growing wheat obtained by stratification after selection of the random sample of Example 6.6.

TABLE 7.6.b—ESTIMATION OF THE SAMPLING ERROR OF THE NUMBER OF FARMS
GROWING WHEAT FROM EXAMPLE 6.5

Size-group (acres)	n_i	u_i	p_i	$V(u_i)$
1–5	22	0	0	0
6–20	26	1	·03846	·913
21–50	18	5	·27778	3·431
51–150	26	21	·80769	3·837
151–300	20	16	·80000	3·040
301–	13	11	·84615	1·608
				12·829

The computations follow the same lines as those for S.E. (Y′) in Example
7.6.a. The values obtained are :

$$\text{S.E. } (Y') = \pm 4320 \text{ acres}$$
$$\text{S.E. } (U') = \pm 75\cdot2 \text{ farms}$$

The value for S.E. (Y′) is slightly greater than that for the similar estimate
of Example 7.6.a. The difference, however, is not a precise measure of the
relative accuracy of stratification before and after selection. In the first place,
since different samples are involved, there are differences in the estimates of
the within-strata variances. In particular, the estimate of the variance for
size-group 301– is much greater in the random sample because of the one
high value 265. These differences are merely errors of estimation and are
not a reflection of the relative accuracy of the two samples. On the other hand,
the two largest size-groups, which have the highest variances, happen by
chance to have more than the proportionate number of farms in the random
sample, and this particular random sample will therefore tend to give a more
accurate estimate with stratification than will a stratified sample. On the
average, however, random samples stratified after selection will give slightly
less accurate values than stratified samples.

7.6.1 Contrasts between domains : stratified sample with uniform or variable sampling fraction

Three cases arise. The domains of study may consist of strata or groups
of strata, they may consist of parts of a single stratum, or they may cut across
strata. The first and second cases present no new problems. In the first
case, since the part of the sample belonging to any domain constitutes a random
or stratified random sample of that domain, the methods developed in the
previous chapters apply. Moreover, since the sampling of the different strata is
independent there will be no covariance between the estimates for the different
domains. In the second case, the relevant part of the sample constitutes a

random sample of the stratum concerned, and the formulae of Section 7.5.1 are therefore appropriate.

In the third case, in which the domains cut across strata, the variances will be affected in much the same manner as in a random sample, except that in this case $V(\bar{y}_a)$ also loses its simple form, and there is negative covariance between $\bar{y}_a$ and $\bar{y}_b$, $\bar{r}_a$ and $\bar{r}_b$, and h_a and h_b. Putting $\lambda_i = g_i^2(1-f_i)\,n_i/(n_i-1)$ the formulae (corresponding to those of Section 7.5.1) may be written

$$V(N_a) = N_a^2 V(p_a) = \Sigma \lambda_i\, n_i\, p_{ia}\, q_{ia} \tag{7.6.1.a}$$

$$\mathrm{cov}(N_a, N_b) = N_a N_b\, \mathrm{cov}(p_a, p_b) = -\Sigma \lambda_i\, n_i\, p_{ia}\, p_{ib} \tag{7.6.1.b}$$

$$N_a^2\, V(\bar{y}_a) = \Sigma \lambda_i\{n_{ia}\, q_{ia}(\bar{y}_{ia}-\bar{y}_a)^2 + (n_{ia}-1)\, s_{ia}^2\} \tag{7.6.1.c}$$

$$N_a N_b\, \mathrm{cov}(\bar{y}_a, \bar{y}_b) = -\Sigma \lambda_i\, n_i\, p_{ia}\, p_{ib}(\bar{y}_{ia}-\bar{y}_a)(\bar{y}_{ib}-\bar{y}_b) \tag{7.6.1.d}$$

$$V(Y_a) = \Sigma \lambda_i\{n_{ia}\, q_{ia}\, \bar{y}_{ia}^2 + (n_{ia}-1)\, s_{ia}^2\} \tag{7.6.1.e}$$

$$\mathrm{cov}(Y_a, Y_b) = -\Sigma \lambda_i\, n_i\, p_{ia}\, p_{ib}\, \bar{y}_{ia}\, \bar{y}_{ib} \tag{7.6.1.f}$$

$$X_a^2\, V(\bar{r}_a) = \Sigma \lambda_i\{n_{ia}\, q_{ia}(\bar{y}_{ia}-\bar{r}_a\, \bar{x}_{ia})^2 + (n_{ia}-1)\, s_{ia}^2\} \tag{7.6.1.g}$$

$$X_a X_b\, \mathrm{cov}(\bar{r}_a, \bar{r}_b) = -\Sigma \lambda_i\, n_i\, p_{ia}\, p_{ib}(\bar{y}_{ia}-\bar{r}_a\, \bar{x}_{ia})\, \bar{y}_{ib}-r_b\, \bar{x}_{ib}) \tag{7.6.1.h}$$

$$N_a^2\, V(h_a) = \Sigma \lambda_i\{n_{ia}\, h_{ia}(1-h_{ia})+q_{ia}(h_{ia}-h_a)^2\} \tag{7.6.1.i}$$

$$N_a N_b\, \mathrm{cov}(h_a, h_b) = -\Sigma \lambda_i\, n_i\, p_{ia}\, p_{ib}(h_{ia}-h_a)(h_{ib}-h_b) \tag{7.6.1.j}$$

$$V(U_a) = \Sigma \lambda_i\, n_i\, p_{ia}\, h_{ia}(1-p_{ia}\, h_{ia}) \tag{7.6.1.k}$$

$$\mathrm{cov}(U_a, U_b) = -\Sigma \lambda_i\, n_i\, p_{ia}\, p_{ib}\, h_{ia}\, h_{ib} \tag{7.6.1.l}$$

The formulæ with a factor (N_a^2 etc.) on the left-hand side are derived from the general formula for a ratio estimate (formula 7.5.k, including a covariance term) and are therefore not exact.

As before, the covariances are of relatively little importance if each domain covers only a small part of each stratum. In this case the q_{ia} will be nearly unity. As mentioned in Section 7.6 and illustrated in Example 7.7.b, an approximate estimate of the sampling error of the domain means (and ratio estimates) will then be obtained by treating the sample *as if it were stratified for the domains but not for the strata*, provided the sampling fraction is uniform. A similar estimate for the errors of the domain totals will be obtained by treating the sample in the same manner, but omitting the corrections for the means, i.e. by replacing $S_a(y-\bar{y}_a)^2$ by $S_a(y^2)$.

This simplification does not hold when the sampling fraction is variable. In this case the full formulæ must be used.

Example 7.6.1.a

A large population is made up of an equal number of units of two types. The value of a variate y is $c-d$ for all units of type 1, and $c+d$ for all units of type 2. Half the units of each type belong to domain A and the other half to domain B. Compare the values of $V(\bar{y}_a)$ and $V(\bar{y}_a-\bar{y}_b)$ for a random sample of n units with those for a stratified random sample with uniform sampling fraction, with the types constituting the strata.

The mean of each domain is c, and the within-domain variances, s_a^2 and s_b, are both d^2. n_a and n_b will both have an average value of $\frac{1}{2}n$, and if n is reasonably large can be taken as equal to $\frac{1}{2}n$. Hence, for the random sample, from formulæ 7.5.1.n and 7.5.1.o we have

$$V(\bar{y}_a) = 2d^2/n$$
$$cov(\bar{y}_a, \bar{y}_b) = 0$$

and hence

$$V(\bar{y}_a - \bar{y}_b) = 4d^2/n$$

In the stratified sample the within-strata variances for the two domains, s_{1a}^2 etc., are all zero ; $\bar{y}_{1a} - \bar{y}_a = -d$, $\bar{y}_{2a} - \bar{y}_a = +d$, etc. ; $g_1 = g_2 = g$ say, and $N_a/g = n_a = \frac{1}{2}n$; $n_{1a} = \frac{1}{4}n$ and $p_{1a} = q_{1a} = \frac{1}{2}$ approximately. Hence, substituting in formulae 7.6.1.c and 7.6.1.d and taking $n_i/(n_i-1)$ as 1, we obtain

$$V(\bar{y}_a) = d^2/n$$
$$cov(\bar{y}_a, \bar{y}_b) = -d^2/n$$

and hence

$$V(\bar{y}_a - \bar{y}_b) = 4d^2/n$$

The variance of $\bar{y}_a$ is therefore halved by stratification, but the variance of $\bar{y}_a - \bar{y}_b$ is unchanged.

Example 7.6.1.b

In the National Farm Survey (Section 5.17) for the county of Hereford the farms of the sample (excluding size-group 1) were classified into the following domains:—

				Percentage of arable land
A	Mainly grass	....	....	0–29·9
B	Intermediate	....	....	30–49·9
C	Mainly arable	....	....	50–100

Estimate the variances and covariances of the numbers of farms and the total and mean acreages of crops and grass in the various domains.

The basic data are given in Table 7.6.1.a (there were no farms in size-group 5). As an example we may give an outline of the computation of the variances and covariances of the mean acreages. The first step is to prepare tables of p_{ia}, $\bar{y}_{ia}$ and s_{ia}^2. These are given in Table 7.6.1.b. Tables of q_{ia} and $(\bar{y}_{ia} - \bar{y}_a)$ (not reproduced here) will also be required. For each size-group the term of any particular variance or covariance can then be computed. Finally the relevant terms can be added and divided by N_a^2, etc., to give the variances and covariances.

The results are shown in Table 7.6.1.c. It will be seen that the variances of Y_a and $\bar{y}_a$ are both substantially greater for the separate domains than would be expected from the corresponding variances for the whole county.

The variances and covariances for the separate domains can be used to calculate the variance of the corresponding estimate for the whole county,

TABLE 7.6.1.a—NUMBERS, TOTAL ACREAGES AND SUMS OF SQUARES

Size-group	Domain A	B	C	Total
	Number of farms, n_{ia}, etc.			
2	72	62	39	173
3	79	155	108	342
4	13	61	25	99
Raised total	1,062	1,362	872	3,296
	Total acreage, $S_{ia}\,(y)$, etc.			
2	3,708	3,553	2,054	9,315
3	11,570	27,410	18,800	57,780
4	4,860	22,460	9,410	36,730
Raised total	93,080	190,090	114,560	397,730
	Sum of squares, $S_{ia}\,(y^2)$, etc.			
2	225,980	243,987	124,252	594,219
3	1,838,900	5,409,900	3,582,000	10,830,800
4	1,850,400	8,537,800	3,649,900	14,038,100

TABLE 7.6.1.b—VALUES OF p_{ia}, $\bar{y}_{ia}$ AND $s_{ia}{}^2$

Size-group	g_i	p_{ia}	p_{ib}	p_{ic}
2	10	·4162	·3584	·2254
3	4	·2310	·4532	·3158
4	2	·1313	·6162	·2525

Size-group	$\bar{y}_{ia}$	$\bar{y}_{ib}$	$\bar{y}_{ic}$	$\bar{y}_i$
2	51·50	57·31	52·67	53·84
3	146·46	176·84	174·07	168·95
4	373·85	368·20	376·40	371·01
$\bar{y}_a$, etc.	87·65	139·57	131·38	120·67

Size-group	$s_{ia}{}^2$	$s_{ib}{}^2$	$s_{ic}{}^2$	$s_i{}^2$
2	493·2	661·9	423·0	538·7
3	1851·4	3654·2	2891·7	3135·0
4	2792·3	4468·4	4499·0	4192·8

TABLE 7.6.1.c—VARIANCES AND COVARIANCES

	Variances				Covariances		
	A	B	C	All	A, B	A, C	B, C
N_a, etc.	4559	4668	3662	0	− 2783	− 1776	− 1885
$Y_a/100$, etc.	3394	6111	4528	2208	− 2028	− 1257	− 2627
$\bar{y}_a$, etc.	12·72	21·14	34·49	2·03	− 6·19	− 5·83	− 9·15

which thus provides a check. For number and total acreage the agreement should be exact. For mean acreage the agreement is only approximate, since the relevant formulae are not exact. We find, in fact, $(1062^2 \times 12·72 + \ldots - 2 \times 1062 \times 1362 \times 6·19 - \ldots)/3296^2 = 2·70$, compared with the correct value of 2·03.

Note that the covariances increase the standard errors of $\bar{y}_a - \bar{y}_b$, $\bar{y}_a - \bar{y}_c$, $\bar{y}_b - \bar{y}_c$ appreciably. The values neglecting and including the covariance terms are 5·8, 6·9, 7·5 and 6·8, 7·7, 8·6 respectively.

Note also that omission of the correction factors $1 - f_i$ for finite sampling, as may be appropriate in investigational work, will further increase the variances for the mean acreages.

7.6.2 Errors in the estimation of the proportions of the population total attributable to different domains

In many cases in which a quantitative variate is being studied we are interested in the proportions or percentages which are attributable to different domains, rather than the actual totals for the domains. Thus in the National Farm Survey, described in Section 5.17, interest attached to the proportion of the total farm land that was tenant-occupied, the proportion that was farmed by full-time farmers, etc.

Estimates of such proportions are given very simply for all types of sampling by dividing the estimate Y_a of the total for the domain A by the estimate Y of the total for the whole population. Thus, denoting the estimate of the proportion by P_a, we have

$$P_a = \frac{Y_a}{Y}$$

If the population total is known from other sources we may use P_a to provide an alternative estimate Y_a' of the domain total which will in general be more accurate than Y_a. The formula is

$$Y_a' = P_a Y$$

Estimation, therefore, presents no new problems, but the estimates of error will be affected by the covariance between Y_a and Y.

Case (a)   Domains not cutting across strata

When a domain A comprises one or more complete strata there will be no covariance between Y_a and Y_{a-}, where Y_{a-} is the estimate of the total of the

203

remainder of the population. We also have $Y_a + Y_{a-} = Y$. Consequently $V(Y) = V(Y_a) + V(Y_{a-})$, and $\mathbf{cov}(Y_a, Y) = V(Y_a)$. The ordinary formula for the variance of a ratio then gives

$$Y^2 \, V(P_a) = (1 - 2\,P_a)\, V(Y_a) + P_a^2 \, V(Y)$$
$$= Q_a^2 \, V(Y_a) + P_a^2 \, V(Y_{a-})$$

where $Q_a = 1 - P_a$. If there are more than two domains the first form is most suitable for computation.

The covariance between the proportions for two mutuálly exclusive domains A and B can be similarly deduced from the formula for the covariance of two ratios, which in the notation of Section 7.5 is

$$\text{cov}\left(\frac{Y_1}{Y_2}, \frac{Y_3}{Y_4}\right) = \frac{Y_1 \, Y_3}{Y_2 \, Y_4}\left[\frac{\text{cov}(Y_1, Y_3)}{Y_1 \, Y_3} - \frac{\text{cov}(Y_1, Y_4)}{Y_1 \, Y_4} - \frac{\text{cov}(Y_2, Y_3)}{Y_2 \, Y_3} + \frac{\text{cov}(Y_2, Y_4)}{Y_2 \, Y_4}\right]$$

We thus find

$$Y^2 \, \text{cov}(P_a, P_b) = -\{P_b \, V(Y_a) + P_a \, V(Y_b) - P_a \, P_b \, V(Y)\}$$

It is easily verified that when A and B together make up the whole population $-\,\text{cov}(P_a, P_b) = V(P_a) = V(P_b)$ as it should. More generally, if the population is divided into a number of domains the sum of all the variances plus twice the sum of all the covariances should equal zero. This provides a useful check when all variances and covariances are required.

Case (b)   Random sample and stratified sample with domains of study cutting across strata

In this case there will be covariance between Y_a, Y_b, Y_c, . . . It is best to calculate first the variances and covariances of Y_a, Y_b, Y_c, . . . from the formulae given in Sections 7.5.1. and 7.6.1. The variances and covariances of the proportions may then be obtained from the formulae—

$$Y^2 \, V(P_a) = V(Y_a) - 2\,P_a \, \text{cov}(Y_a, Y) + P_a^2 \, V(Y)$$
$$Y^2 \, \text{cov}(P_a, P_b) = \text{cov}(Y_a, Y_b) - P_b \, \text{cov}(Y_a, Y) - P_a \, \text{cov}(Y_b, Y) + P_a \, P_b \, V(Y)$$

with

$$\text{cov}(Y_a, Y) = V(Y_a) + \text{cov}(Y_a, Y_b) + \text{cov}(Y_a, Y_c) + \ldots \text{etc.}$$

Example 7.6.2

Estimate the variances and covariances of the percentage of land in Hereford attributable to the three types of farm of Example 7.6.1.

The percentages are

$$100\, P_a = 23\cdot40$$
$$100\, P_b = 47\cdot79$$
$$100\, P_c = 28\cdot80$$

From Table 7.6.1.c the variance and covariance matrix of Y_a, Y_b, Y_c is

$$10^4 \times \begin{matrix} 3394 & -2028 & -1257 \\ -2028 & 6111 & -2627 \\ -1257 & -2627 & 4528 \end{matrix}$$

$$\begin{matrix} 109 & 1456 & 644 & 2209 \end{matrix}$$

The column totals give cov (Y_a, Y) etc., and the grand total gives $V(Y)$.
We thus have

$$100^2 \, V(P_a) = \frac{(3394 - 2 \times 0.2340 \times 109 + 0.2340^2 \times 2209) \times 10^8}{397,730^2}$$

$$= 2.190$$

100^2 cov (P_a, P_b)

$$= \frac{(-2028 - 0.4779 \times 109 - 0.2340 \times 1456 + 0.2340 \times 0.4779 \times 2209) \times 10^8}{397,730^2}$$

$$= -1.374$$

The full variance and covariance matrix is

$$\begin{matrix} 2.190 & -1.374 & -0.816 \\ -1.374 & 3.302 & -1.928 \\ -0.816 & -1.928 & 2.744 \end{matrix}$$

with the check that each column total is zero.

7.7 Pooled estimate of error : the analysis of variance

In a stratified sample with equal sampling fractions all the f_i are equal
If in addition the within-strata variances σ_i^2 are equal, by putting $\Sigma (n_i) = n$
we obtain the simplified formula

$$\text{S.E. } (\bar{y}) = \sigma_1 \sqrt{\frac{1-f}{n}}$$

This is the same as the formula for a random sample, with the exception that
σ is replaced by σ_1, the common within-strata standard deviation.
The most accurate estimate of σ_1 will be obtained if the estimates of σ_i^2
from the various strata are weighted according to the numbers of degrees of
freedom on which they are based. This is equivalent to adding the sums of
squares of deviations from the strata means and dividing by the sum of the
associated degrees of freedom, i.e. by $n-t$.
It is worth noting here an identity which relates the above sum of squares
to the sum of squares of deviations from the general mean. This is

$$\Sigma \, S_i (y - \bar{y}_i)^2 + \Sigma \, n_i (\bar{y}_i - \bar{y})^2 = S (y - \bar{y})^2$$

This is easily verified if we recognize that

$$\Sigma \, n_i (\bar{y}_i - \bar{y})^2 = \Sigma \{ \bar{y}_i S_i (y) \} - \bar{y} S (y)$$

The first term on the right-hand side is the sum of the products of the means and totals of the separate strata, i.e. the " corrections for the means " for the separate strata, and the second term is the " correction for the general mean."

The arithmetical computations can be conveniently arranged in the form of what is known as the *analysis of variance*. This is shown schematically in Table 7.7.a.

TABLE 7.7.a—ANALYSIS OF VARIANCE BETWEEN AND WITHIN STRATA

	Degrees of freedom	Sum of squares	Mean square
Between strata .	. $t-1$	$\Sigma \bar{y}_i S_i (y) - \bar{y} S (y)$	A
Within strata .	. $n-t$	$\Sigma S_i (y - \bar{y}_i)^2$	$B = s_1^2$
Whole sample .	. $n-1$	$S (y^2) - \bar{y} S (y)$	$C = s^2$

The most convenient form of computation, at least when the numbers in the different strata are small, is to calculate the sums of squares for the whole sample and between strata, and obtain the sum of squares within strata by subtraction. The mean squares are then obtained by division by the degrees of freedom. Only the mean square within strata, s_1^2, is required for the present purpose. The mean square for the whole sample approximates closely to an estimate s^2 of the variance per unit that would result from random sampling of the whole population. The interpretation of the mean square between strata, A, is discussed in Section 8.10.1.

A further simplification is possible when each of the strata contains only two sampling units. In this case the sum of squares within strata can be calculated directly from the differences of the y's of the pairs of units within each stratum. If these differences are denoted by d, the sum of squares will be $\frac{1}{2} S (d^2)$, and consequently, since there are t differences each contributing one degree of freedom,

$$s_1^2 = \tfrac{1}{2} S (d^2)/t$$

The analysis of variance has many applications, not only in sampling but in other fields of statistics. As its name implies, it provides a way of determining the different components of variance to which a given type of material is subject. As such it is of particular value in investigations of the efficiency of different types of sampling. Its uses in this connection will be explained in Chapter 8.

The above discussion has been based on the assumption that the within-strata variances are equal. In practice this is not likely to be exactly true, though the data at our disposal may be insufficient to determine the true variance law with any accuracy. It is therefore important to ascertain what is the position if a pooled estimate of variance is used when the variances are in fact unequal.

If all the sampling fractions are equal we have, from equation 7.6.b,

$$V(\bar{y}) = \frac{1-f}{n} \; \frac{\Sigma (n_i \; s_i^2)}{n}$$

The second factor is a weighted mean of the s_i^2, with weights equal to n_i. In the pooled estimate of error described above s_1^2 is a weighted mean of the s_i^2 with weights proportional to $n_i - 1$. Unless the numbers in the different strata are very small, and associated in magnitude with σ_i^2, there will be little difference between the two estimates. Use of the pooled estimate of error in the estimation of the error of the population mean and total will not, therefore, introduce any serious disturbance when the sampling fractions are equal, even when the within-strata variances are very unequal. On the other hand, the use of a pooled estimate to determine the errors applicable to the mean or total of part of the population, e.g. a single stratum mean, may be very misleading, and it is therefore best, when there are marked differences in the within-strata variances, and the numbers in the different strata are not too small, to keep the error estimates separate, as has been done in Example 7.6.a.

There is, of course, nothing sacrosanct in weighting by the degrees of freedom ; these weights merely give the most accurate estimate when the variances are equal, and enable the analysis of variance technique to be used. If the n_i are too small for separate estimates of the within-strata variances to be of value, we can still use a pooled estimate, weighting by n_i if this appears advisable.

The situation is completely different with a variable sampling fraction. In this case equation 7.6.a shows that weights proportional to $g_i^2 \, n_i \, (1 - f_i)$ are required. The pooled estimate of variance may therefore be decidedly misleading, even with quite small differences in the within-strata variances. Consequently in this case separate estimates with proper weighting should always be used.

Example 7.7.a

Estimate the sampling errors of the estimate of the total wheat acreage and number of farms growing wheat from the stratified systematic sample with a variable sampling fraction of Example 6.7.

Since the systematic selection was from a list arranged by districts the sample will substantially be stratified by districts as well as size-groups. The numbers in the 6–20 and 21–50 size-groups are too small for the district stratification to be effective, but for the larger size-groups the within-districts variance is required, instead of the overall variance within the size-group.

The available number of degrees of freedom in each district and size-group is small, and inspection of the values of Table 6.7.a shows that there are no marked differences in variability in the different districts. We may therefore appropriately use a pooled estimate of variance for each size-group.

The district totals and means for the largest size-group are shown in Table 7.7.b.

TABLE 7.7.b—DISTRICT TOTALS AND MEANS FOR SIZE-GROUP 501–

District :	1	2	3	4	5	6	7	All
No. of farms	1	4	2	9	0	1	0	17
Total . .	114	487	315	1937	—	72	—	2925
Mean . .	114	121·75	157·5	215·2222	—	72	—	172·0588

The analysis of variance is given in Table 7.7.c. The sum of squares between districts is $114 \times 114 + 487 \times 121 \cdot 75 + \ldots - 2925 \times 172 \cdot 0588 = 40,698$. The total sum of squares is $114^2 + 119^2 + 107^2 + \ldots - 2925 \times 172 \cdot 0588 = 72,067$. Subtraction gives the within-district sum of squares, and division by the degrees of freedom the mean squares.

TABLE 7.7.c—ANALYSIS OF VARIANCE BETWEEN AND WITHIN DISTRICTS OF THE
WHEAT ACREAGES OF SIZE-GROUP 501– (DATA OF TABLE 6.7.a)

	Degrees of freedom	Sum of squares	Mean square
Between districts . .	4	40,698	10,174
Within districts . .	12	31,369	2,614
Whole size-group . .	16	72,067	4,504

The within-district mean square is substantially less than the overall mean square, indicating the greater similarity of farms within a district and consequent gain in accuracy by stratification.

The size-groups 51–, 151–, and 301– can be analysed in the same manner. There is some difference in mean squares for size-group 301–, but little difference for the other two size-groups. For size-groups 6– and 21– the overall variability within size-groups can be taken.

The remainder of the computations are set out in Table 7.7.d. They follow the same lines as Example 7.6.a, with the exception that the factors $(1 - f_i)$ are different, and that the variance of each group total must be multiplied by the square of the raising factor for that group, and the resultant variances added, in accordance with formula 7.5.a. The estimated standard error of the total acreage is thus $\sqrt{6,486,000} = \pm 2550$.

It will be noted that the acreage of wheat in the smallest size-group has been assumed to be zero, and that the estimated zero error variance of the second size-group is based on only two degrees of freedom, and is therefore very inaccurately determined. It is clear, however, from the nature of the material and the trend of the variances in the larger size-groups that this variance must be small.

TABLE 7.7.d—ESTIMATION OF SAMPLING ERRORS OF THE ESTIMATES
OF EXAMPLE 6.7

Size-group (acres)	n_i	s_i^2	$V\{S_i(y)\}$	g_i^2	$V(Y_i)$
1–5	0	....	....	....	----
6–20	3	0	0	40,000	0
21–50	6	53·5	320	3,600	1,152,000
51–150	26	159·2	3,930	400	1,572,000
151–300	40	564·2	20,310	100	2,031,000
301–500	43	1,703	58,580	25	1,464,000
501–	17	2,614	29,630	9	267,000
	135				6,486,000

In the computation of the standard error of the number of farms growing wheat, allowance should also strictly be made for the stratification by districts. If the number of farms in each size-group district sub-class were large this could be done by calculating the variance of each size-group total of farms growing wheat by the method of Example 7.6.a. The numbers in many of the sub-classes are so small, however, that the approximation resulting from using the estimated proportions p to calculate the variances will be unsatisfactory. In this case it will be sufficient to ignore the district stratification, calculating the variance of each size-group total of farms growing wheat from the proportion in that size-group, and then proceeding in the same manner as for wheat acreage. The resultant standard error will be found to be $\pm 88\cdot9$.

Example 7.7.b

Estimate the sampling standard errors of the regional and varietal marginal means of potatoes shown in Table 5.21.a, and also that of the overall mean, given that the pooled estimate of variance within sub-classes is 4·22.

As mentioned in Section 5.21, the sample can be regarded as stratified by regions but not by varieties. To obtain the standard error of a particular regional mean, therefore, we require the standard deviation or variance of the values contributing to this mean, including the varietal component. This can be obtained by constructing an analysis of variance within and between varieties, including only the data for the particular region. That for the Scottish region is shown in Table 7.7.e. The mean square within varieties is 4·22. The sum of squares between varieties can be obtained from Table 5.21.a by the mean × total rule, or its equivalent mean² × no. of observations ; the need to calculate the regional mean 8·52 to greater accuracy can be avoided by working directly

209

with deviations.   Thus we find $(9 \cdot 46 - 8 \cdot 52)^2 \times 37 + \ldots = 79 \cdot 83$.*   The table can then be completed, giving an estimate of 4·58 for the unit variance, and therefore a sampling standard error of $\sqrt{(4 \cdot 58/174)} = \pm 0 \cdot 16$ for the Scottish regional mean.   The unit variance for the other four regional means can be calculated similarly.   The five values are 4·6, 4·9, 5·3, 5·3, 6·1.

The varietal domains cut across strata, and therefore formula 7.6.1.c should strictly be used for calculating their unit variances.   The major contributions to the variances, however, here come from within sub-classes, and consequently no great error will result from proceeding as if the sample were stratified for varieties but not for regions.   This gives values of 4·3, 4·6, 4·5, 4·5, 4·4.

An analysis of variance for the whole of the data into within- and between-regions, ignoring varieties, can be very simply constructed from the regional means if the total sum of squares (or the equivalent unit variance or standard deviation) is known.   This analysis is given in the first section of Table 7.7.f. The first line is obtained from the regional means.   The sum of squares within regions can then be obtained by subtraction.   This is equal to the sum of the

TABLE 7.7.e—POTATO SURVEY : ANALYSIS OF VARIANCE OF YIELDS, SCOTTISH REGION ONLY

	Degrees of freedom	Sums of squares	Mean square
Between varieties	4	79·83	
Within varieties	169	713·18	4·22
Total	173	793·01	4·58

total sum of squares of Table 7.7.e and those of the four similar tables for the other regions.   Thus the mean square within regions, 5·25, provides a pooled estimate of the separate unit variances for the regional means given above. As the sample is stratified by regions, 5·25 is the relevant unit variance for estimating the sampling standard error of the overall mean, which is thus $\sqrt{(5 \cdot 25/901)} = \pm 0 \cdot 076$.   Had the sample been fully random the estimated standard error would have been $\sqrt{(5 \cdot 42/901)} = \pm 0 \cdot 078$.   The gain in accuracy is here small.   If it were possible to stratify by varieties instead of regions, the gain would be substantially greater, as is shown by the second section of Table 7.7.f, and if by both still greater, as the relevant unit variance is then 4·22.

The relation between the overall unit variance and that within classes is shown in tabular form in the third section of Table 7.7.f.   This can be used to deduce either of the variances given the other.

Any two independent mean squares in an analysis of variance can be tested

* If δ is the rounding error of the mean of a set of values, the error due to using the rounded mean to calculate the " correction for the mean " is $2n\delta\bar{y}$.   In the extreme case here, with $\delta = 0 \cdot 005$, the error in the sum of squares would be 14·9.   Using deviations from the rounded mean gives an error of $n\delta^2$ only.

TABLE 7.7.f—POTATO SURVEY : ANALYSIS OF VARIANCE OF YIELDS, ALL REGIONS

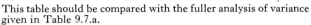

	Degrees of freedom	Sum of squares	Mean square
Between regions	4	173·7	43·42
Within regions	896	4701·3	5·25
Total	900	4875·0	5·42
Between varieties	4	887·3	221·82
Within varieties	896	3987·7	4·45
Total	900	4875·0	5·42
Between sub-classes	20	1161·4	58·07
Within sub-classes	880	3713·6	4·22
Total	900	4875·0	5·42

This table should be compared with the fuller analysis of variance given in Table 9.7.a.

to see if they differ significantly by means of the F-test.*   The separate tests of regions and varieties provided by Sections 1 and 2 of Table 7.7.f are, however, not fully relevant, because the regions and varieties are nor *orthogonal,* i.e. the regions contain differing proportions of the different varieties, and consequently the regional means are differentially affected by varietal differences, and vice versa.   A fuller analysis of variance, which provides separate tests for regions and varieties, is given in Section 9.7.

The standard errors obtained above may be compared with those for the unweighted and weighted means of the sub-class means obtained in Example 7.5.a.   Those for the Scottish region are $\pm 0·20$ and $\pm 0·19$.   Elimination of the cross-effects has here increased the standard errors.   This will not always happen.   If the within sub-class variances are sufficiently small the standard errors will be decreased.

Display of the results in analysis of variance form enables the contributions of the various sources of variation to the sampling errors to be readily assessed. It should not be regarded as a routine procedure for the calculation of sampling errors.   Most computer programs for survey analysis are capable of producing standard deviations or standard errors for the marginal means or individual cells of a table, at least for random samples.   Uninhibited use of this facility, however, often results in a plethora of unwanted standard errors.   If standard errors have not been obtained, or have not been reported, the above example shows how, provided the underlying variability is similar in different parts of the population, approximations to them can be deduced knowing only the overall variance of the sample.

* F is the ratio of the two mean squares.   Tables of its significance levels, tabulated against the associated degrees of freedom, are widely available.

7.8 Ratio method : random sample

In order to calculate the variance of $\bar{r}$ the correlation between the values of x and y for the same sampling unit must be taken into account. From formula 7.5.k, with the additional covariance term and allowance for finite population, we obtain for a random sample

$$V\,(\bar{r}) = \frac{1-f}{n}\,\bar{r}^2\left(\frac{V\,(y)}{\bar{y}^2} - \frac{2\,\mathbf{cov}\,(xy)}{\bar{x}\bar{y}} + \frac{V\,(x)}{\bar{x}^2}\right)$$

This is an approximate formula, but is accurate enough for practical purposes in the cases met with in sampling.

The formula can be put in an alternative form, which somewhat simplifies the approach to more complicated cases such as stratified samples. If we denote by Q the sum of the squares of the deviations of the y's from the values given by the ratio line (*OMD* of Fig. 6.8), we have

$$\begin{aligned}
Q &= S\,(y - \bar{r}x)^2 \\
&= S\,\{(y - \bar{y}) - \bar{r}\,(x - \bar{x})\}^2 \\
&= S\,(y^2) - 2\bar{r}\,S\,(xy) + \bar{r}^2\,S\,(x^2) \\
&= S\,(y - \bar{y})^2 - 2\bar{r}\,S\,(x - \bar{x})\,(y - \bar{y}) + \bar{r}^2\,S\,(x - \bar{x})^2
\end{aligned}$$

the last two expressions being those which are suitable for computation.

If we now take s_q^2 to represent the estimated mean square deviation from the true ratio line, we have

$$s_q^2 = Q/(n - 1)$$

We then find

$$V\,(\bar{r}) = \frac{1-f}{n\bar{x}^2}\,s_q^2 = \frac{1-f}{\{S\,(x)\}^2}\,ns_q^2$$

$$V\,(Y) = \frac{X^2}{\{S\,(x)\}^2}\,(1 - f)\,ns_q^2$$

$$V\,(\bar{y}) = \frac{\bar{X}^2}{\bar{x}^2}\cdot\frac{1-f}{n}\,s_q^2$$

The first of the above formulæ is equivalent to formula 7.5.g.

The analogy with the case of a random sample without supplementary information can now be seen. Apart from the factor $\bar{X}^2/\bar{x}^2$, which will be approximately unity, the only difference is that the sum of the squares of the deviations of y from the mean $\bar{y}$ of the sample is replaced by the sum of the squares of the deviations of y from the ratio line of the sample.

The variance of a standardized estimate y_0 will be obtained by replacing $\mathbf{x}$ by x_0 in the above formula.

In the case of two-phase sampling the first-phase estimate $\bar{x}_1$ of $\bar{x}$ will be subject to sampling errors. If the sampling is random for both phases the variance of $\bar{y}$ will be

$$V\,(\bar{y}) = \frac{1}{n_2}\left(1 - \frac{n_2}{n_1}\right)\frac{\bar{x}_1^2}{\bar{x}_2^2}\,s_q^2 + \frac{1-f_1}{n_1}\,s^2$$

where the suffices refer to the phases. s_q^2 is calculated as above from the second-phase units, and s^2 is the total variance of y, also calculated from the second-phase units.

The first term of the above formula will be recognized as the sampling variance of $\bar{y}$ due to the second-phase sampling of the first-phase sample (regarded as without error), while the second term is the first-phase sampling variance of $\bar{y}$, i.e. the variance which would be obtained if y were determined for all units of the first-phase sample. This subdivision of the variance provides a general method of obtaining the errors of two-phase sampling. In certain types of sampling the circumstance that values of y are not available for all the first-phase units introduces complications into the estimation of the second component of variance which will be dealt with in Section 8.7.

It frequently happens that the available supplementary information is out of date or otherwise subject to error. If, however, values of x are known for all units of the population these values can be used in the calculation of $\bar{x}$ from the selected units. If this is done bias will be avoided, and the effect of these errors on the final estimates will be correctly assessed, provided the original frame is complete. If the frame is not complete the sampling divides into two parts : that covering units included in the original frame, for which the ratio or regression method of estimation can be used ; and that covering units not so included, for which the appropriate method of estimation without supplementary information will be required.

It may also be noted that if the variance $V_x(r)$ of r for fixed x is constant over the whole range of values of x, and if r itself exhibits no trend over this range, $V(\bar{r})$ may be estimated from formulæ 7.5.h and 7.5.e, substituting x for w and r for y. This method of estimation has the advantage of saving computation in cases in which the values of r are directly available while those of y are not, but it will give a biased estimate of error if the above conditions are not fulfilled. An example of the method is given for a stratified sample in Example 7.17.

If $V_x(r)$ is virtually constant, the sampling error of any ratio estimate can be rapidly calculated once the value of $V_x(r)$ has been established, since only $S(x)$ and $S(x^2)$ require to be known, formula 7.5.e being used. Similarly, if $V_x(r)$ is inversely proportional to x, i.e. equal to λ/x, formula 7.5.f can be used, only $S(x)$ being required. The effective constancy of λ, and its value, can be most simply established by calculating $V(\bar{r})$ in the ordinary manner for various batches of data and calculating the resultant values of λ from formula 7.5.f. This is in general preferable to using formulæ 7.5.i and 7.5.f directly.

Example 7.8.a

Estimate the sampling error of the ratio estimate of the acreage of wheat from the random sample of farms (Example 6.9.a).

We have

$$S(y^2) = 207,261 \qquad S(xy) = 902,958 \qquad S(x^2) = 5,061,734$$
$$\bar{r} = \cdot1522430 \qquad 2\bar{r} = \cdot3044860 \qquad \bar{r}^2 = \cdot0231779$$
$$Q = 49,643 \qquad s_q^2 = 400\cdot35$$

$$\text{S.E. } (\bar{r}) = \frac{1}{15,114} \sqrt{\{(1 - 1/20)\,125 \times 400\cdot35\}} = \pm\, 0\cdot01443$$

$$\text{S.E. } (Y) = 273,074 \times 0\cdot01443 = \pm\, 3,940$$

Example 7.8.b

Estimate the sampling error of the estimates of Example 6.9.b.

We have

$$n = 43 \qquad\qquad f = 43/325 = 0\cdot1323 \qquad g = 7\cdot5581$$
$$S(y) = 799 \qquad \bar{y} = 18\cdot5814 \qquad \bar{x} = 79\cdot6977 \qquad S(x) = 3,427$$
$$S(y^2) = 22,065 \qquad S(xy) = 76,965 \qquad S(x^2) = 328,323$$
$$\bar{y}\,S(y) = 14,846\cdot5 \qquad \bar{y}\,S(x) = 63,678\cdot5 \qquad \bar{x}\,S(x) = 273,124\cdot0$$
$$S(y - \bar{y})^2 = 7218\cdot5 \qquad S(x - \bar{x})(y - \bar{y}) = 13,286\cdot5 \qquad S(x - \bar{x})^2 = 55,199\cdot0$$
$$\bar{r} = 0\cdot233149 \qquad 2\bar{r} = 0\cdot466298 \qquad \bar{r}^2 = 0\cdot0543585$$
$$Q = 4023\cdot6 \qquad s_q^2 = 95\cdot80$$

$$\text{S.E. } (100\,\bar{r}) = \frac{100}{3427} \sqrt{(0\cdot8677 \times 43 \times 95\cdot80)} = \pm\, 1\cdot744$$

Had the sample been a random sample of individuals the formula of Section 7.4 would have been applicable, giving

$$\text{S.E. } (100\,\bar{r}) = 100 \sqrt{(0\cdot8677 \times 0\cdot2331 \times 0\cdot7669/3427)} = \pm\, 0\cdot673$$

The large difference between these two standard errors is an indication of the additional variability between kraals, and illustrates the misleading results that may be obtained by using the formula of Section 7.4 when the sampling units consist of groups of individuals and not single individuals.

Since the total number of persons in the reserve is unknown, the standard errors of the total numbers are derived from the formulæ appropriate to a random sample without supplementary information. We therefore have

$$S(y - \bar{y})^2/(n - 1) = 171\cdot87 \qquad S(x - \bar{x})^2/(n - 1) = 1,314\cdot3$$
$$\text{S.E. } (X) = 7\cdot5581 \times \sqrt{(0\cdot8677 \times 43 \times 1314\cdot3)} = \pm\, 1,673$$
$$\text{S.E. } (Y) = 7\cdot5581 \times \sqrt{(0\cdot8677 \times 43 \times 171\cdot87)} = \pm\, 605\cdot3$$

The standard error of the number present in the reserve can be calculated in the same manner from the sum of the squares of the deviations of $(x - y)$, which in turn can be calculated directly from the separate values of $(x - y)$. In the present case, where the separate values of $(x - y)$ are not tabulated, and where $S(x - \bar{x})(y - \bar{y})$ has already been calculated, it is more convenient to obtain the required sum of squares of deviations from the sums of squares and products already calculated (see Section 7.5). Thus

$$S\{(x-y)-(\bar{x}-\bar{y})\}^2 = 55{,}199{\cdot}0 + 7{,}218{\cdot}5 - 2 \times 13{,}286{\cdot}5 = 35{,}844{\cdot}5$$
$$\text{S.E. } (X-Y) = 7{\cdot}5581 \times \sqrt{\{0{\cdot}8677 \times 43 \times 35{,}844{\cdot}5/42\}} = \pm 1349$$

Note that x and y are not independent, being derived from the same sampling units, and therefore $V(X-Y)$ is not equal to $V(X) + V(Y)$, but to $V(X) + V(Y) - 2 \operatorname{cov}(X,Y)$. Putting $\operatorname{cov}(X,Y)$ equal to $13{,}286{\cdot}5/42$ gives the same result as above.

7.9 Ratio method : stratified sample with uniform sampling fraction

(a) When the ratio is assumed to be the same for all strata :

Instead of taking the sum of squares of deviations from the general ratio line, the deviations from a series of lines parallel to this line and passing through the points representing the strata means must be taken, the divisor $n-1$ being replaced by $n-t$.

Thus

$$Q = \Sigma\, S_i\,\{(y-\bar{y}_i) - \bar{r}\,(x-\bar{x}_i)\}^2$$
$$= \Sigma\, S_i\,(y-\bar{y}_i)^2 - 2\bar{r}\,\Sigma\, S_i\,(y-\bar{y}_i)(x-\bar{x}_i) + \bar{r}^2\,\Sigma\, S_i\,(x-\bar{x}_i)^2$$
$$s_q^2 = Q/(n-t)$$

The sums of squares and products will be recognized as the sums of squares and products within strata, similar to those already obtained for the y variate in the pooled estimate of error for a stratified sample without supplementary information.

(b) When the ratio is permitted to assume different values for the different strata :

The common $\bar{r}$ is replaced by $\bar{r}_i$ corresponding to the different strata, so that

$$Q = \Sigma\, S_i\,(y-\bar{y}_i)^2 - 2\Sigma\, \bar{r}_i\, S_i\,(y-\bar{y}_i)(x-\bar{x}_i) + \Sigma\, \bar{r}_i^2\, S_i\,(x-\bar{x}_i)^2$$

The divisor $n-t$ stands.

In this case the contribution to Q from each stratum is best computed separately. If desired the variances of the contributions to Y from the different strata may also be estimated separately. This course is equivalent to assigning slightly different weights to the different contributions to Q, the situation being analogous to that already discussed in Section 7.6.

Note that if the population totals X_i are not known for the different strata but the total X for the whole population is known, the formulæ for case (a) must be used for calculating the sampling errors of $\bar{r}$ and Y, even if the ratio clearly varies from stratum to stratum, since the method of estimation must be that corresponding to case (a).

Example 7.9

Estimate the sampling errors of the estimate of Example 6.10.

The contributions to Q from the six districts are:

District	Q_i	District	Q_i
1 ..	5,107·59	4 ..	20,566·56
2 ..	1,550·71	5 & 7 ..	3,737·14
3 ..	7,963·98	6 .	1,080·92
		TOTAL	40,006·90

Hence

$$\text{S.E. (Y)} = \frac{273,074}{15,114} \sqrt{\left\{ \left(1 - \frac{1}{20}\right) 125 \frac{40,006\cdot90}{119} \right\}} = 3,610$$

7.10   Ratio method : stratified sample with variable sampling fraction

If the ratio is assigned different values in the different strata the variance of the estimated total is

$$V(Y) = \Sigma \left[\frac{X_i^2}{\{S_i(x)\}^2} (1 - f_i) n_i s_{qi}^2 \right]$$

$$= \Sigma \{g_i^2 (1 - f_i) n_i s_{qi}^2\} \text{ approximately,}$$

$$V(\bar{r}) = V(Y)/X^2$$

Here the s_{qi}^2 are estimated separately for each stratum, using the value of the ratio appropriate to the stratum and the divisor $n_i - 1$ for Q_i.

If the ratio is assumed to be the same for all strata the same formula may be used, with the exception that the Q_i are calculated using the general ratio, with divisors $n_i - 1$ as before.

For an illustration of the application of these formulæ see Example 7.17.

7.11   Sampling errors of means and proportions when the sampling units are aggregates of the data units

In many sampling investigations the sampling units consist of aggregates or " clusters " of the units for which the data are collected. Thus we may require information on individuals in a population, but may take a sample of households and include all the individuals in these households in the sample.

From general considerations it is clear that for items for which the values for members of the same sampling unit are positively correlated, e.g. race or standard of nutrition in a household, the errors of a population estimate relating to individuals will be increased by selection of households instead of individuals. The same is true of the estimate for a domain which contains all or most of the members of some of the households. On the other hand, the error of the difference between two domains each of which contains members of the same household, e.g. standard of nutrition for children in different age-groups, may well be less than that for a sample of individuals.

Example 7.8.b provides a very simple illustration of the effect of such positive correlation. From data from a random sample of kraals the estimated standard error of the percentage of the population absent from their kraal was there found to be 1·74, compared with that of 0·67 which would have been obtained had the sample been assumed to be a random sample of individuals.

In this example the data (Table 6.9) consisted of measurements (actually counts) relating to the sampled kraals. This leads naturally to the use of the formulae for $V(\bar{r})$ of Section 7.8. When, however, the data being analysed relate to individuals, the formulae are more intelligible if they are written in a different form.

It will be sufficient here to consider the case in which the sampling units are selected at random. In such sampling the appropriate estimate of the mean for the population of a quantitative variate y relating to individuals will be the unweighted mean $\bar{y}$ over all individuals. Similarly the unweighted mean $\bar{y}_A$ of all individuals possessing attribute A will be the appropriate estimate for members of the population with attribute A.

The following notation will be adopted :

n number of individuals in the sample ;

n' number of sampling units in the sample ;

n'' number of individuals in a particular sampling unit ;

S summation over the whole sample of quantities relating to individuals ;

S' summation over sampling units of quantities relating to the sampling units ;

S'' summation over a particular sampling unit of quantities relating to individuals.

We then have

$$\bar{y} = Sy/n = S'(S''y)/S'n''$$

Thus y is equivalent to a mean ratio $\bar{r}$ based on separate ratios $r=S''y/n''$ for each of the sampling units. The formula for $V(\bar{y})$ can therefore be obtained from the formula for $V(\bar{r})$ of Section 7.8 by replacing S by S', $\bar{r}$ by $\bar{y}$, y by $S''y$, x by n'', and n by n'. Thus, remembering that $S'n''=n$, we have

$$Q = S'(S''y)^2 - 2\,\bar{y}\,S'(n''\dot{S}''y) + \bar{y}^2\,S'(n''^2)$$

$$V(\bar{y}) = \frac{(1-f)n'}{(n'-1)}\,Q/n^2$$

Alternatively $\bar{y}$ can be regarded as a weighted mean of the $\bar{y}''$, with weights n''. The same result will then be obtained from the general formula 7.5.g by replacing S by S', $\bar{y}_w$ by $\bar{y}$, y by $\bar{y}''$, w by n'', and n by n'.

By including only individuals belonging to a particular domain A, the above formulae can be used to estimate the variance $V(\bar{y}_A)$ of the mean for

217

the domain. Strict application of the formulae would require that n' be replaced by n'_A, the number of sampling units containing at least one data unit belonging to domain A. n' is, however, more convenient ; the difference is trivial when n'_A is large enough to give a reliable estimate of the variance, as only the fraction $n'_A/(n'_A-1)$ is changed.

The expression for the variance of the difference $\bar{y}_A - \bar{y}_B$ of the means for two mutually exclusive domains A and B will contain an additional covariance term. We then have

$$V(\bar{y}_A - \bar{y}_B) = \frac{(1-f)n'}{n'-1}\ [Q_A/n_A^2 + Q_B/n_B^2 - 2Q_{A\cdot B}/n_A n_B]$$

where the covariance term $Q_{A\cdot B}$ is given by

$$Q_{A\cdot B} = S'(S''y_A\ S''y_B) - \bar{y}_B\ S'(n''_B\ S''\bar{y}_A) - y_A S'(n''_A\ S''y_B) + \\ \bar{y}_A \bar{y}_B S'(n''_A n''_B)$$

In addition to means of quantitative variates, interest often attaches to the proportion of individuals possessing a given attribute in the population, or in a particular domain, and to the difference between such proportions in a pair of domains. Variances of such estimates can be derived from the above formulae by scoring each individual possessing the attribute as 1, and other individuals as 0.

It will be seen that sampling of the above type is equivalent to two-stage sampling with complete sampling at the second stage. The formulae given above are therefore applicable both to single-stage sampling in which the units of analysis differ from the units of sampling, and to the estimation of the first component of variance (Section 7.17) in multi-stage sampling.

Example 7.11

Compare the sampling errors of estimates of fertilizer dressings that would be obtained from a random sample of farms with those calculated on the assumption that the data were obtained from a random sample of fields.

Data giving the dressings of nitrogen to winter wheat and spring barley to fields in the South Eastern Region were extracted from the 1976 Survey of Fertilizer Practice (Table 7.11.a). The sample was actually one of farms stratified by size and type with variable sampling fractions, but apart from the differing proportions of farms in the different strata the data will be similar to that from a random sample of farms. All fields on the selected farms were included.

The values of Q_W/n^2_W, Q_B/n^2_B and $Q_{W\cdot B}/n_W n_B$ are 12·27, 5·20 and +0·48 respectively. With $n' = 171$, these give the sampling standard errors shown in the first line of Table 7.11.b. Calculation of Q_W, Q_B and $Q_{W\cdot B}$ would be tedious on a desk calculator, but presents no difficulty given a suitable computer program.

218

TABLE 7.11.a—SAMPLE OF 171 FARMS FROM THE 1976 SURVEY OF FERTILIZER
PRACTICE (SOUTH EASTERN REGION) GROWING WINTER WHEAT OR SPRING BARLEY

	Winter wheat	Spring barley
No. of farms growing crop	162	146
No. of fields growing crop	534	353
Mean dressing of N (kg per ha)	115·9	74·8

TABLE 7.11.b—CONTRAST BETWEEN THE ESTIMATED SAMPLING STANDARD ERRORS
TAKING INTO ACCOUNT AND IGNORING THE HIERARCHICAL STRUCTURE OF THE DATA

Assumed sampling unit	Standard errors		
	$\bar{y}_W$	$\bar{y}_B$	$\bar{y}_W - \bar{y}_B$
Farms (correct)	3·5	2·3	4·1
Fields (incorrect)	1·8	1·5	2·3

The overall variances per field of the wheat and barley dressings on the individual fields were 1691 and 751 respectively. These give the sampling standard errors, shown in the second line of Table 7.11.b, that result from the assumption that the sample is a random sample of fields.

As will be seen, failure to take account of the fact that the units of analysis do not correspond to the sampling units leads to very serious underestimates of the sampling standard errors. This is a consequence of the close correlation between fertilizer applications to fields of the same crop on a farm. The applications to wheat and barley might be expected to show a similar correlation, but in fact the covariance term $Q_{W \cdot B}$ is small.

It should be noted that the difference in the average nitrogen dressings to wheat and barley does not provide a direct measure of the assessment by individual farmers of the difference between the nitrogen requirements of their wheat and barley crops. If, for instance, farmers who grow a large proportion of wheat give substantially higher dressings of nitrogen to both their wheat and barley than do farmers who grow a large proportion of barley, the difference in the average dressings will be greater than the average of the differences on the individual farms.

To determine the within-farm differences we must take the difference of the mean dressing for wheat and that for barley on each individual farm which grows both wheat and barley. The unweighted mean of these differences is 33·6±3·3, compared with the overall average difference of 41·1±4·1.

7.12   Regression method : random sample

The estimation of error in the regression method follows much the same lines as in the ratio method. The sum of squares of deviations from the ratio

219

line is replaced by the sum of squares of deviations from the regression line, and the divisor $n - 2$ is used instead of $n - 1$, since an additional degree of freedom is accounted for by the fact that the regression line not only passes through the mean point, but has its slope determined independently from the data.

The sum of the squares of the deviations from the regression line is given by the equation

$$Q = S(y - y_l)^2$$
$$= S(y - \bar{y})^2 - b S(x - \bar{x})(y - \bar{y})$$
$$= S(y - \bar{y})^2 - \frac{\{S(x - \bar{x})(y - \bar{y})\}^2}{S(x - \bar{x})^2}$$

so that

$$s_l^2 = Q/(n - 2)$$

and, if errors in b are neglected

$$V(\bar{y}) = \frac{1 - f}{n} s_l^2$$

The error variance of b, if the variance of y for fixed x is constant, is

$$V(b) = \frac{s_l^2}{S(x - \bar{x})^2}$$

so that if the regression is truly linear the error variance of a standardized value y_0 of y for the value x_0 of x is

$$V(y_0) = \left\{ \frac{1}{n} + \frac{(x_0 - \bar{x})^2}{S(x - \bar{x})^2} \right\} s_l^2$$

The correction for finite population is here omitted since standardized values are ordinarily used for comparative purposes.

Allowance for errors in b can be made in $V(\bar{y})$ in the same way, but such errors will always be small relative to the other component of error. They will on the average increase the error variance approximately in the ratio $n/(n - 1)$.

In the case of two-phase sampling the sampling variance of $\bar{x}_1$ will introduce the additional term $b^2 V(\bar{x}_1)$ into the above expression for $V(\bar{y})$. The general approach is given in Section 8.7.

If an arbitrary value b_0 of the regression coefficient is used, instead of the value b calculated from the data, the formula for the sum of the squares of the deviations from the arbitrary regression line becomes

$$Q = S(y - \bar{y})^2 - 2b_0 S(x - \bar{x})(y - \bar{y}) + b_0^2 S(x - \bar{x})^2$$

and

$$s_l^2 = Q/(n - 1)$$

This procedure is equivalent to the analysis of the values $y - b_0 x$ from the individual sampling units. Use of the above expression for Q saves the trouble of calculating the values of $y - b_0 x$ for the individual sampling units, at the expense of calculating the sums of squares and products of x and y instead of the sums of squares of $y - b_0 x$.

No allowance has to be made for errors in b_0, but an arbitrary value b_0 should not be used for standardization unless it is known that b_0 approximates closely to b, so that the error $(b_0 - b)(x_0 - \bar{x})$ introduced into the standardization correction is small.

Example 7.12.a

Estimate the errors of the estimates of Example 6.12.a.

We have

$$Q = 164,904 - 0 \cdot 19316 \times 624,739 = 44,229$$
$$s_l^2 = Q/123 = 359 \cdot 59$$
$$V(\bar{y}) = (1 - 1/20) \, 359 \cdot 59 \, / \, 125 = 2 \cdot 733 = (\pm 1 \cdot 653)^2$$
$$\text{S.E. } (Y) = 2496 \times 1 \cdot 653 = \pm 4126$$
$$V(b) = \frac{359 \cdot 59}{3,234,270} = 0 \cdot 0001112 = (\pm 0 \cdot 01054)^2$$

Example 7.12.b

Estimate the error of the estimates of total volume of timber of Example 6.12.b.

Except for the estimate derived from the arbitrary value of the regression coefficient $b_0 = 1$ the computations follow the lines already given and are left as an exercise for the reader.

When $b_0 = 1$ we have

$$Q = 115,266 - 2 \times 52,069 + 82,296 = 93,424$$
$$s_l^2 = 93,424/24 = 3893$$

The values of the error variance per unit, and the resultant standard errors of the various estimates, are as follows:

	Variance per unit	S.E. (total volume)	Relative efficiency
Sample plots only	4,803	$\pm$ 710,000 cu. ft.	72
Ratio method	4,230	$\pm$ 602,000 cu. ft.	82
Regression, $b_0 = 1$	3,893	$\pm$ 639,000 cu. ft.	89
Regression, $b = \cdot 6327$	3,579	$\pm$ 613,000 cu. ft.	96
Regression, $b_0 = \cdot 55$	3,454	$\pm$ 603,000 cu. ft.	100

The relative efficiency of the various methods of estimation is inversely proportional to the value of the variance per unit. Setting the last value at 100 we obtain the relative efficiencies of the last column. The relative efficiencies fall in the order given. If the information from the sample plots only is used, neglecting the eye estimates, about 40 per cent. more sample plots will be required to give results of the same accuracy as those obtained by using a regression of 0·55 on the eye estimates.

The values for $b_0 = 0·55$ have been included to illustrate the fact that any value of the regression coefficient near to the value derived from the data will give results which are of about the same accuracy. Here there is an apparent small gain in accuracy owing to change in the degrees of freedom from 23 to 24. There is therefore no point in attempting to take account of small differences in the regression coefficient for different parts of the data, or to determine b very exactly. Any value reasonably near the correct value will give a satisfactory adjustment.

The ratio method has given an estimate of the standard error which is relatively low because $S(x)$ for the sample is high. The average performance of the ratio method may best be judged by the variance per unit.

Limits of error can be assigned to the possible bias in the eye estimates by calculating the standard error of the difference $\bar{x} - \bar{y}$, or of the ratio $\bar{r}$. We have

$$\text{S.E.} (\bar{x} - \bar{y}) = \sqrt{(3893/25)} = \pm\, 12·5.$$

The actual difference, $-15·3$, is $1·22$ times its standard error, and these data therefore do not by themselves furnish conclusive evidence of the existence of bias in the eye estimates. Taking limits of error of $\pm$ twice the standard error gives limits to the bias of $-40·3$ and $+9·7$, *i.e.* -27 per cent. and $+7$ per cent.

Similarly S.E. $(\bar{r}) = \pm\, 0·098$, and since $\bar{r} = 1·116$ the ratio of the deviation of $\bar{r}$ from unity to its standard error is $1·19$, which compares with the value of $1·22$ for $\bar{x} - \bar{y}$.

As mentioned in Example 6.12.b, the more extensive data of the full survey confirmed the existence of bias, giving at the same time a more accurate determination of its average magnitude and variation for different types of woodland.

7.13 Regression method : stratified and balanced samples

(*a*) Uniform sampling fraction, regression coefficient the same for all strata :

As for a random sample, except that

$$Q = \Sigma\, S_i\, (y - \bar{y}_i)^2 - b\, \Sigma\, S_i\, (y - \bar{y}_i)\, (x - \bar{x}_i)$$
$$s_t{}^2 = Q/(n - t - 1)$$
$$\text{V}\,(b) = s_t{}^2/\Sigma\, S_i\, (x - \bar{x}_i)^2$$

If an arbitrary value b_0 is taken, the formula for Q must be rewritten in the same manner as in an unstratified sample, the divisor being $n - t$.

(*b*) Uniform sampling fraction, regression coefficients different for the different strata:

$$b_i = S_i \, (y - \bar{y}_i) \, (x - \bar{x}_i) / S_i \, (x - \bar{x}_i)^2$$

If a pooled estimate of error is used,

$$Q_l = \Sigma \, S_i \, (y - \bar{y}_i)^2 - \Sigma \, b_i \, S_i \, (y - \bar{y}_i) \, (x - x_i)$$
$$s_l^2 = Q/(n - 2t)$$
$$\mathsf{V} \, (b_i) = s_l^2 / S_i \, (x - \bar{x}_i)^2$$

If the variances from the regression lines in the different strata are likely to be different, separate estimates of σ_i^2 should be used when evaluating $\mathsf{V}\,(b_i)$.

(*c*) Variable sampling fraction:

The method is similar to that outlined for the ratio method.

(*d*) Balanced sample:

As far as the estimation of error is concerned, a balanced sample must be treated as if it were an unbalanced sample from which estimates have been derived by the use of regression on the balanced variate. There will be no regression adjustment to the estimates of the population values, since $\bar{\mathbf{x}} - \bar{x}$ will be zero because of the balancing.

7.14 Calibration of eye estimates

When a regression is used to calibrate eye estimates, as described in Section 6.15, the sampling variance of $\bar{y}$ can be split into three parts, that due to errors in b', that due to the sampling variance of $\bar{x}_1$ arising from the main sampling process, and that due to the variance of $\bar{x}_1 - \bar{x}$ arising from the variance about the regression line.

The component of variance due to errors in b' is usually sufficiently small to be neglected. To a first approximation it equals

$$(\bar{x}_1 - \bar{x})^2 \, \mathsf{V} \, (b') / b'^4$$

where $\mathsf{V}\,(b')$ is calculated in the ordinary manner from the regression.

The variance of $\bar{x}_1$ is calculated from the values of x for all the selected units in the manner appropriate to the method of sampling adopted. The contribution to $\mathsf{V}\,(\bar{y})$ from this source is approximately

$$\mathsf{V} \, (\bar{x}_1) / b'^2$$

A closer approximation is obtained by multiplying this variance by $\{\mathsf{V}\,(x) - \mathsf{V}_l\,(x)\}/\mathsf{V}\,(x)$, where $\mathsf{V}\,(x)$ is that part of the variance of x which contributes to $\mathsf{V}\,(\bar{x}_1)$ and $\mathsf{V}_l\,(x)$ is the residual variance of x about the regression line.

The variance of $\bar{x}_1 - \bar{x}$ due to variance about the regression line is calculated from the residual variance $\mathsf{V}_l\,(x)$ of x about this line. If n_1 and

n represent the numbers of units in the original sample and the sub-sample for eye estimates, the contribution to $V(\bar{y})$ when all units are given equal weight in the mean is

$$(n_1 - n) V_l(x)/b'^2 nn_1$$

If the x's are weighted according to area a or other weights, the last expression becomes

$$\frac{V_l(x)}{b'^2} \left(\frac{S'(a)}{S_1(a)}\right)^2 \left[\frac{S(a^2)}{\{S(a)\}^2} + \frac{S'(a^2)}{\{S'(a)\}^2}\right]$$

where S_1, S and S' indicate summation over the whole sample, over the sub-sample, and over the part of the sample not included in the sub-sample, respectively.

Example 7.14

Estimate the variance of the mean yield per acre obtained in Example 6.15.

The component of variance due to errors in b' is negligible, since $\bar{x}_1 - \bar{x}$ is nearly zero.

Since $\bar{x}_1$ was calculated by weighting the weighted mean eye estimates of the individual farms by the wheat acreages of these farms, the ratio method of calculating the sampling error at the first stage is applicable. A table was therefore prepared giving for each farm, (1) the total wheat acreage, (2) the weighted mean yield per acre based on the eye estimates of all the chosen fields on that farm, and (3) the product of these two numbers. Columns (1) and (3) constitute, in the ordinary ratio notation, the x and y values. Using these tabulated values, and the ordinary formula for the variance of a ratio, with the inclusion of the factor $(1 - f)$, we find $V(\bar{x}_1) = 0 \cdot 8200$,* and consequently the corresponding component of variance is $0 \cdot 8200/0 \cdot 6926^2$ or $1 \cdot 7095$. The factor $(1 - f)$ can properly be included here since for the majority of farms all the fields were taken. On the other hand, although the variance per field of the eye estimates is probably reasonably constant, the alternative approach outlined in Section 7.5, making use of this fact, would present difficulties, since the sampling units at the first stage are farms and not fields. The direct approach is therefore simpler.

The residual variance about the regression line was found to be, from an analysis of the unweighted data for fields, $V_l(x) = 7 \cdot 038$. The sums and sums of squares of the areas of the individual fields for which actual yields are and are not available, and of all fields, are

$$S(a) = \quad 610 \qquad S'(a) = \quad 1279 \qquad S_1(a) = \quad 1889$$
$$S(a^2) = 15{,}172 \qquad S'(a^2) = 33{,}899 \qquad S_1(a^2) = 49{,}071$$

Substitution in the formula above gives a component of variance of $0 \cdot 4137$.

* The closer approximation gives the value $0 \cdot 733$.

Fields and not farms can reasonably be used here, since errors in the eye estimates may be expected to be reasonably independent from field to field. This is not the case with $V(\bar{x}_1)$, since the yields of fields on the same farm often show considerable correlation.

The standard error of the adjusted mean yield is therefore $\sqrt{(1 \cdot 7095 + 0 \cdot 4137)} = \pm 1 \cdot 46$ bushels per acre. The main source of error is that due to sampling errors introduced by the variation in yields from farm to farm. The eye estimates are shown to be sufficiently consistent and to give adequate differentiation between differing yields. Regarded as an estimate of the mean yields of the fields actually sampled, the adjusted mean yield has a standard error of only $\sqrt{0 \cdot 4137} = \pm 0 \cdot 64$ bushels per acre.

7.15 Sampling with probabilities proportional to size of unit

In this case unbiased estimates of the sampling variances are those based on the mean square deviation of r. When units selected more than once are included the number of times they are selected, no correction for finite population is required. If s_r^2 is the estimated variance of r we then have, for a random sample,

$$s_r^2 = S\,(r - \bar{r})^2/(n - 1)$$

$$V(\bar{r}) = \frac{1}{n} s_r^2$$

If the size of the population is known we therefore have

$$V(\bar{y}) = \bar{x}^2\, s_r^2/n$$
$$V(Y) = X^2\, s_r^2/n$$

If the size of the population is not known, the sampling variance of the estimate of total size X is derived from the formulæ for a qualitative variate (Section 7.4). We have, for a random sample, following the notation of Section 6.16,

$$V\left(\frac{n}{n_0}\right) = \frac{n}{n_0}\left(1 - \frac{n}{n_0}\right)\Big/n_{\bullet}$$

Hence, if A is known exactly,

$$V(X) = A^2 \frac{n}{n_0}\left(1 - \frac{n}{n_0}\right)\Big/n_0 = \frac{X^2}{n} \cdot \frac{n_0 - n}{n_0}$$
$$V(Y) = X^2\, V(\bar{r}) + \bar{r}^2\, V(X)$$
$$= \frac{X^2}{n}\left(s_r^2 + \frac{n_0 - n}{n_0}\bar{r}^2\right)$$

If A is not known exactly its estimation will contribute some slight additional variance, the amount of which depends on the precise method of location of the points. This, however, will in general be sufficiently small to be neglected. Substituting $A = n_0/d$ for A, we have

$$V(X) = \frac{n(n_0 - n)}{d^2 n_0}.$$

Example 7.15

Estimate the sampling errors of the estimates of Examples 6.16.a and 6.16.b, given that the standard deviation per field of the yield per acre is 3·5 cwt. per acre, and that the distribution of points can be taken as random.

The standard error of the mean yield per acre is

$$\text{S.E. } (\bar{F}) = 3·5/\sqrt{529} = \pm 0·152 \text{ cwt.}$$

For the estimates of Example 6.16.a,

$$V(X) = 2560^2 \times \frac{529 \times 7788}{8317} \text{ acres}^2$$

$$\text{S.E. } (X) = \pm 57,000 \text{ acres}$$

$$V(Y) = \frac{1,354,000^2}{529} \left(3·5^2 + \frac{7788}{8317} 15·7^2 \right) = 84·24 \times 10^{10} \text{ cwt.}^2$$

$$\text{S.E. } (Y) = \pm 45,900 \text{ tons}$$

For the estimates of Example 6.16.b,

$$V(X) = 640^2 \times \frac{2202 \times 31,053}{33,255} = 842·2 \times 10^6 \text{ acres}^2$$

$$\text{S.E. } (X) = \pm 29,000 \text{ acres}$$

$$V(Y) = 1,409,000^2 \times 3·5^2/529 + 15·7^2 \times 842·2 \times 10^6 = 2536 \times 10^8 \text{ cwt.}^2$$

$$\text{S.E. } (Y) = \pm 25,200 \text{ tons}$$

Note that if the total area of crop were known accurately from other sources we should have

$$V(Y) = \frac{1,354,000^2}{529} 3·5^2 \text{ cwt.}^2$$

$$\text{S.E. } (Y) = \pm 10,300 \text{ tons}$$

If the acreage is not known a survey of this type will clearly be more efficient if sample harvesting is carried out at a proportion only of the sample points, the presence or absence of the crop being determined at the remaining points. This point is discussed in more detail in Section 8.17.

If the sample is such that it can be regarded as stratified by districts it might at first sight appear that the between-districts component should be eliminated from the variance of r. Unless the crop areas of the different districts are accurately known, however, this must not be done, since the proportion of sampling points falling in the crop in a district will not be accurately proportional to the area of the crop in that district.

If the sample points are confined to some localities only by a two-stage sampling process, with localities as first-stage sampling units, the above variances will represent the second-stage components of variance only. The full sampling errors must be determined from the first-stage units as explained in Section 7.17.

7.16 Sampling from within strata with probabilities proportional to size of unit

Since the number of units within each stratum is in general small, a pooled estimate s_r^2 of the within-strata variance of r based on the mean square deviations of r within strata will be required. Consequently

$$s_r^2 = \frac{\Sigma\, S_i\, (r - \bar{r}_i)^2}{n - t}$$

We then have

$$\mathsf{V}\,(\bar{r}_i) = s_r^2\, (1 - f_i)/n_i$$

and thus

$$\mathsf{V}\,(\mathsf{Y}) = s_r^2\, \Sigma\, \{\mathbf{X}_i^2\, (1 - f_i)/n_i\}$$
$$\mathsf{V}\,(\bar{r}) = \mathsf{V}\,(\mathsf{Y})/\mathbf{X}^2$$

If the variation in the within-strata variances is large it may be necessary to introduce weights when forming the pooled estimate of error, in order to avoid bias.

The correction for finite sampling is not required if the units selected more than once are included the number of times they are selected. In the more usual case in which each unit is included once only, additional units being selected by the method of Section 3.10, the correction for finite sampling should be included. In this case, since probability of selection is not strictly proportional to size, the formulæ are approximate only.

Example 7.16

Estimate the sampling errors of the estimate of Example 6.17

The analysis of variance of the values of the ratio for the individual " combined " parishes is given in Table 7.16. It follows the lines of Example 7.7.

TABLE 7.16—ANALYSIS OF VARIANCE OF THE VALUES OF THE RATIO

	Degrees of freedom	Sum of squares	Mean square
Between districts . .	6	·04952	·008253
Within districts . .	10	·02649	·002649
TOTAL	16	·07601	·004751

This gives a value of $s_r{}^2$ of 0·002649. We then have

$$\Sigma \{X_i{}^2 (1 - f_i)/n_i\} = 22{,}932^2 \times \tfrac{8}{9} + 43{,}591^2 \times \tfrac{13}{16}/3 + \ldots = 37{\cdot}188 \times 10^8$$
$$\text{S.E. (Y)} = \sqrt{(0{\cdot}002649 \times 37{\cdot}188 \times 10^8)} = \pm\ 3140$$

It will be noted that the variance of r within districts is less than the overall variance. Consequently stratification by districts appreciably reduces the sampling error.

7.17 Multi-stage sampling

If the sampling fraction at the first stage is small, the total sampling error of multi-stage sampling is obtained from the first-stage unit values, estimating each unit value from the results of the sampling at the second and following stages, and using the method of estimation appropriate to the method of sampling at the first stage. The additional variability contributed by the second and following stages is automatically included in this estimate of error.

If, on the other hand, the sampling fraction f' at the first stage is not sufficiently small for the factor $(1-f')$ to be neglected, the sampling error will be increased on account of the fact that the selected first-stage unit values are themselves subject to sampling error, instead of being known exactly, as in single-stage sampling. The increase in the sampling variance of whatever estimate is under consideration will be equal to f' times the variance in this estimate resulting from the sampling at the second and following stages. This variance can be calculated by regarding the selected first-stage units as strata which are sampled by the sampling at the second and following stages.

Thus, for example, in two-stage random sampling, with n' selected first-stage units, and n'' second-stage units selected from each first-stage unit, if s'^2 is the estimate of the sampling variance of the first-stage unit means, *i.e.* the means of the second-stage values for the separate selected first-stage units, s''^2 is the variance of the second-stage units about the first-stage unit means, and f' and f'' are the first- and second-stage sampling fractions, the sampling variance of the mean of the selected first-stage units will be $(1 - f'')\, s''^2/n'n''$, since this mean is based on $n'n''$ second-stage units, and therefore the sampling variance of the mean of the population is

$$V\,(\bar{y}) = \frac{1 - f'}{n'}\, s'^2 + f'\, \frac{1 - f''}{n'n''}\, s''^2.$$

Example 7.17

Calculate the sampling errors of the estimates of the mean dressing of nitrogen per acre obtained in Example 6.19.

The y's and x's entering into the ratio method estimate at the first stage are the successive terms of the expressions for $S\,(g''y)$ and $S\,(g''x)$, already given for small farms, namely 1·36, 3·78, 3·30, . . . and 2, 6, 6, . . . respectively.

The calculation of the sampling error follows the lines indicated in Section 7.10, the ratio (cwt. nitrogen per acre) being assumed, for the reason given at the end of Section 7.9, to be the same for all strata.

The values of s_{qi}^2 are found to be

$$\text{Small farms:} \qquad 27{\cdot}519/21 = \quad 1{\cdot}3104$$
$$\text{Medium farms:} \quad 583{\cdot}81/35 \quad = \ 16{\cdot}680$$
$$\text{Large farms:} \quad 2779{\cdot}1/8 \qquad = 347{\cdot}39$$

We then have, neglecting the factors $(1 - f_i)$,

$$\mathsf{V}\,(\bar{r}) = (105^2 \times 22 \times 1{\cdot}3104 + 59^2 \times 36 \times 16{\cdot}680 + 30^2 \times 9 \times 347{\cdot}39)/58{,}229^2$$
$$= (\pm\, 0{\cdot}0392)^2$$

The sampling fractions are here all small and the variance at the second stage therefore need not be considered. With the present material this variance could not in any case be estimated since only one field per farm was selected. In such cases, when the f_i are not small, it is still best to neglect them. The sampling error will then be slightly overestimated.

Farms without sugar beet have been excluded from the above calculation. Their inclusion, though substantially decreasing the values of s_{qi}^2, would make little difference to the final estimate of error, since the values of n_i in the formula for $\mathsf{V}\,(\mathsf{Y})$ would be correspondingly increased.

From inspection of the data, and from the nature of the material, we may expect that the variance of the mean dressing per acre r will be substantially constant, irrespective of the size of field to which it is applied. The alternative procedure of calculating the variance of r directly without any weighting may therefore be followed without serious risk of introducing any marked bias into the estimate of error.

TABLE 7.17—ANALYSIS OF VARIANCE OF DRESSINGS OF NITROGEN PER ACRE

	Degrees of freedom	Sum of squares	Mean square
Small farms . . .	21	1·1304	0·0538
Medium farms . . .	35	1·3500	0·0386
Large farms . . .	8	0·9836	0·1230
TOTAL	64	3·4640	0·0541

The within-size-groups sums of squares and mean squares of r are shown in Table 7.17. There is no marked difference between the different size-groups, and in the following calculations we will therefore use the pooled estimate of the mean square, $s_r^2 = 0{\cdot}0541$.

From the formulæ of Section 7.5 we have, for a single stratum,

$$\mathsf{V}\,(\bar{r}_i) = s_{ri}^2\,(1 - f_i)\,S_i\,(x^2)/\{S_i\,(x)\}^2$$

and

$$\mathsf{V}\,(\mathsf{Y}_i) = s_{ri}^2\,(1 - f_i)\,S_i\,(x^2)$$

where the x's are those entering into the first-stage sampling, *i.e.* $g''x$ in the full notation. The sums of squares of $g''x$ will be found to be 580, 15,245 and 39,993 for small, medium and large farms respectively. Consequently, summing over all strata as before, omitting the $(1 - f_i)$, and taking the variable sampling fraction into account, we have

$$V(\bar{r}) = 0 \cdot 0541 \,(105^2 \times 580 + 59^2 \times 15,245 + 30^2 \times 39,993)/58,229^2$$
$$= (\pm 0 \cdot 0391)^2$$

This is almost identical with the value previously obtained.

The comparison between the two methods of calculating the variance may be taken a stage further by estimating the values of s_{ri}^2 from those of s_{qi}^2 and comparing them with those obtained directly. Equating the two expressions for $V(\bar{r}_i)$, we find $s_{ri}^2 = n_i\, s_{qi}^2/S_i\,(g''x)^2$. Using the values of s_{qi}^2 already given we obtain for s_{ri}^2 the values 0·0497, 0·0394, 0·0782 respectively. These show no consistent divergence from the values of Table 7.17, and we may therefore conclude that the bias in the estimation of error by the second method is likely to be small. A more thorough investigation could be made by tabulating a number of comparisons of the above type from various batches of similar data.

The value of s_r^2 given by Table 7.17 is directly appropriate for the calculation of the variance of the estimate from the unweighted means (Table 6.19.c).

The sampling standard error of the mean dressing over all fields is $\sqrt{(0 \cdot 0541/67)} = \pm 0 \cdot 0284$. This standard error does not include any errors due to bias, but will be appropriate, or at least approximately so, to comparisons of such a nature that the major part of the bias is eliminated. If there were large differences between size-groups the question of whether the pooled within-size-groups variance or the overall variance is appropriate to the comparison in question would have to be considered—this, however, involves other problems, such as how far the differences observed are due to differences in size-group proportions (see examples 7.5 and 7.7.b).

It will be noted that the standard error of the properly weighted ratio estimate is considerably greater than the standard error of the straight mean. The ratio of the squares is $0 \cdot 0392^2/0 \cdot 0284^2 = 1 \cdot 91$. Thus about double the number of farms, excluding those without sugar beet, are required to attain the same accuracy when unbiased estimates are required. This is inevitable in a survey of this kind where the sampling fractions cannot be adjusted so as to be proportional to the areas of the crop being sampled, a course which is impossible when a number of crops are covered in the same survey, even if the necessary information is available.

7.18   Systematic samples

No fully valid estimate of the sampling error of a systematic sample is possible, since the units are not located at random within defined strata. Approximate estimates can be made in various ways. The simplest, which will suffice for most census and survey work, is to divide the material arbitrarily into strata, and calculate the sampling error as if the units were selected at random from these strata.

In the case of a systematic sample from a list it will usually be sufficient to take account of the major groupings of the list, treating these as strata, and to ignore any minor and ill-defined groupings. An example of this has already been given (Example 7.7.a).

In the case of one-dimensional systematic sampling, e.g. equally spaced points on a line, or equally spaced lines covering an area, the strata may be taken to contain pairs of successive units, so that the error variance is estimated from the differences between the members of the pairs. Each difference contributes one degree of freedom. If there are n' such differences d, the error variance per unit is therefore

$$s^2 = \tfrac{1}{2} S(d^2)/n'$$

Since the pairing is arbitrary, instead of taking alternate differences between successive units all differences may be taken. This is equivalent to taking two sets of overlapping strata. The accuracy of the estimate of s^2 is thereby somewhat increased, though it is not doubled, owing to lack of independence between the successive differences.

In the case of two-dimensional sampling on a square or rectangular pattern the strata should consist of sets of four units in a 2×2 pattern. By this means variability in both directions will be taken into account. There is no point in taking overlapping strata. Since each such stratum contributes 3 degrees of freedom the formula for the error variance per unit is

$$s^2 = [S(y^2) - \tfrac{1}{4} \Sigma \{S_i(y)\}^2]/3\, n'$$

where n' is the number of strata.

In the case of line sampling a complication arises if all the lines are not of approximately equal length. If the total area covered by the sample is known, the most accurate estimate of the quantity under consideration will be obtained by the ratio method. In this case the calculation of the sampling error should strictly follow the method given in Section 7.9 for a stratified sample estimated by means of a constant ratio. With strata of two units the formula for Q becomes

$$Q = \tfrac{1}{2} S(d_y^2) - 2\bar{r} \cdot \tfrac{1}{2} S(d_x d_y) + \bar{r}^2 \tfrac{1}{2} S(d_x^2)$$

This will eliminate the variability due to variation in length of line. If the total area is not known, so that the final estimate is obtained by multiplication of the total over all the lines by the appropriate raising factor, the difference method given above, and not the ratio method, must be used.

These methods of estimation of the sampling error are also applicable to line samples in which the lines are randomly located in pairs within blocks and thus form a proper stratified random sample. In this case the estimate of error will be fully valid.

In either systematic or stratified random line sampling, the variation in the length between neighbouring lines will not be large unless the boundaries of the area covered are very irregular. Consequently the approximate method based on the direct differences can be used in most cases without serious inaccuracy.

The above methods of estimation of error for systematic samples will give overestimates of the sampling error, provided there are no periodic features in the material, and provided in two-dimensional sampling that there are no marked strip effects running in straight lines across the material in such a manner that the whole of one line of sample points falls on the same strip. If a closer estimate is required, an alternative, but rather more complicated, procedure is available. In one-dimensional sampling, instead of taking successive differences, differences of the type

$$d = \tfrac{1}{2}y_1 - y_2 + y_3 - y_4 + y_5 - y_6 + y_7 - y_8 + \tfrac{1}{2}y_9$$

can be taken. Such differences may be called *balanced differences*. Most of the systematic component of variation is thus eliminated. The number of terms included in each difference is to a certain extent arbitrary, but 9 is chosen as a convenient compromise. With extensive material there will be no need to take overlapping differences, the best procedure being to have overlap of the end terms only, so that the y_9 of the first difference is taken as the y_1 of the second. With this convention the sum of all the differences is equal to one-half the first and last included terms plus the sum of all the remaining odd terms minus the sum of all the even terms. The square of each difference contributes one degree of freedom, the divisor being given by the sum of the squares of the coefficients, i.e. 7·5. Consequently $s^2 = S(d^2)/7 \cdot 5\, n'$.

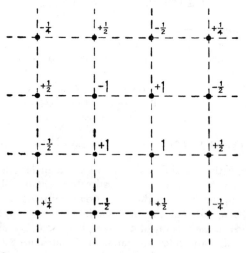

FIG. 7.18—COEFFICIENTS FOR CALCULATING THE ERROR OF A SYSTEMATIC
TWO-DIMENSIONAL SAMPLE

ESTIMATION OF THE SAMPLING ERROR

A similar procedure can be followed in the case of two-dimensional systematic sampling, the most convenient type of difference being that given by the coefficients shown in Fig. 7.18. Here again, the margins of the square covering one difference may be taken as the margins of neighbouring squares. The divisor in this case will be $6\frac{1}{4}$.

The estimates provided by balanced differences will also in general be overestimates of the sampling error, but may be expected to be closer than those based on ordinary differences. If there is no wide discrepancy between the two types of estimate it may be concluded that the degree of overestimation is not likely to be great. More exact estimates can only be obtained by taking supplementary observations at intermediate points allocated either at random or systematically. The one-dimensional case has been discussed in detail by Yates (1948, A).

The above methods of estimation can be applied both to quantitative and qualitative data, but in the case of qualitative data, based on either one- or two-dimensional point sampling, a rapid estimate of the sampling error can be made by using the formulæ for a random sample, as in Example 7.15. This will tend to give greater overestimation of the sampling error than the above methods, but if the parts of the line or area possessing the attribute are small and irregularly distributed, with no great variation in density in different parts of the line or area, the estimate will be sufficiently good for most practical purposes.

The estimation of the number of units in an aggregate of short lists when a systematic sample is taken from each of the lists presents special problems. With a sampling fraction of 1 per cent. for example, the estimated length of a list of length $100h+k$, where k is an integer between 0 and 99, will be $100h$ if the starting point is greater than k, and $100(h+1)$ if less than or equal to k. The corresponding errors will therefore be $-k$ and $100-k$.

Table 7.18.a sets out the numerical values of these errors for various starting points. If the length is an exact multiple of 100 there is no error,

TABLE 7.18.a—ERRORS IN THE ESTIMATED LENGTH OF A LIST WHEN A SYSTEMATIC SAMPLE OF 1 IN 100 IS TAKEN WITH VARIOUS STARTING POINTS, SHOWING BIAS WHEN k IS UNIFORMLY DISTRIBUTED

Starting point	\multicolumn{13}{c}{Last two digits of last number in the list (k)}	Bias
	00	01	02	..	24	25	..	49	50	..	75	76	..	99	
1	0	+99	+98	..	+76	+75	..	+51	+50	..	+25	+24	..	+1	+49·5
2	0	−1	+98	..	+76	+75	..	+51	+50	..	+25	+24	..	+1	+48·5
..	..	..	..	..	..	..	..	..	..	..	..	..	..	..	
25	0	−1	−2	..	−24	+75	..	+51	+50	..	+25	+24	..	+1	+25·5
..	..	..	..	..	..	..	..	..	..	..	..	..	..	
50	0	−1	−2	..	−24	−25	..	−49	+50	..	+25	+24	..	+1	+0·5
..	..	..	..	..	..	..	..	..	..	..	..	..	..	
76	0	−1	−2	..	−24	−25	..	−49	−50	..	−75	+24	..	+1	−25·5
..	..	..	..	..	..	..	..	..	..	..	..	..	..	
100	0	−1	−2	..	−24	−25	..	−49	−50	..	−75	−76	..	−99	−49·5
Mean 25, 76	0	−1	−2	..	−24	+25	..	+1	0	..	−25	+24	..	+1	0

whatever the starting point ; for a list of length 300, for example, 3 units will always be selected.   For a list of length 301 there will be selection of 4 units if the starting point is 1, giving an error of $+99$, but only 3 units if the starting point is 2 or more, giving an error of -1.   These errors are shown in the first two columns of the table.

The means of the lines of the table are shown in the last column.   The sets of deviations from these means are all the same, ranging from $-49 \cdot 5$ to $+49 \cdot 5$, giving a variance of $0 \cdot 083325 \times 100^2$ or very nearly $\frac{1}{12} \times 100^2$ for the values in each row about the row mean.

The row means give the bias in the estimated mean list length when samples with the same starting point are taken from a number of lists and the values of k for the various lists are uniformly distributed over the range 0 to 99.   The bias is large for starting points in the neighbourhood of 1 or 100, and is least for a starting point of 50 or 51, as common sense would indicate.

If a random starting point is taken for each list the mean bias will be zero, but for any particular list the bias for the selected starting point will contribute to the sampling error.   The variance of the biases is the same as the variance within rows, and the total sampling variance per list of the number of units in the sample will therefore be $0 \cdot 083325 + 0 \cdot 083325 = 0 \cdot 16665$, or very nearly $\frac{1}{6}$.   (The general formula for a sampling interval of q is $\frac{1}{6}(1 - 1/q^2)$.)

If instead of taking each starting point at random independently we take a pair at random such that the sum of the starting points for each pair is 101, the biases of each such complementary pair will cancel one another and will therefore not contribute to sampling error.   The sampling variance will therefore be halved, i.e. will be nearly $\frac{1}{12}$ per list.

The above properties only hold if the distribution of k is uniform.   If it is not, both the biases and the sampling variances will be altered.   If the distribution of k is known, the value of the biases and sampling variances can be determined by weighting the columns of Table 7.21.a by the relative frequencies.

If the k are not uniformly distributed, the bias even for a starting point of 50 may be quite substantial.   If all k are greater than 50 and are distributed symmetrically about a mean of 75, for example, the bias will be $+25$.   The mean bias will, however, be zero for each complete set of starting points from 1 to 100 whatever the distribution of k.   This follows from the fact that the sum of each column of Table 7.18.a is zero.   For each complete set of 100 lists the sampling variance will always be less than the value of $\frac{1}{12}$ per list given above.   The contribution to error of the biases of the residue of the lists, not forming a complete set, will be minimized by taking complementary pairs.

Example 7.18.a

In the 1942 **Census of Woodlands the** total area of woodland **shown on the maps was** determined **for each** county by estimating the area of land coloured green on the 1-inch O.S. maps.   This was done by measuring the total length of the E–W kilometre grid lines which fell in green areas.   The results for O.S. sheet No. 115 covering part of Kent are given in Table 7.18.b. Estimate the sampling error of this process.

TABLE 7.18.b—WOODLAND AREAS FROM LINE INTERCEPTS (*cm.*)

Grid line	Length of line, x	Length coloured green, y	Successive differences, d_y	Grid line	Length of line, x	Length coloured green, y	Successive differences, d_y
98	3·5	0·0	—	83	30·0	3·8	+ 1·4
97	4·2	0·9	+ 0·9	82	29·4	4·1	+ 0·3
96	9·2	0·0	− 0·9	81	29·1	4·9	+ 0·8
95	12·6	0·0	0·0	80	28·8	6·0	+ 1·1
94	15·5	0·3	+ 0·3	79	28·6	5·4	− 0·6
93	21·2	0·1	− 0·2	78	28·2	2·3	− 3·1
92	25·2	0·5	+ 0·4	77	27·2	2·9	+ 0·6
91	25·4	3·1	+ 2·6	76	26·3	2·1	− 0·8
90	31·2	2·8	− 0·3	75	25·4	6·3	+ 4·2
89	34·2	2·7	− 0·1	74	25·5	8·2	+ 1·9
88	34·1	2·8	+ 0·1	73	25·2	5·4	− 2·8
87	33·0	2·6	− 0·2	72	24·9	6·6	+ 1·2
86	31·4	2·3	− 0·3	71	24·6	6·6	0·0
85	31·0	3·5	+ 1·2	70	20·8	4·1	− 2·5
84	30·7	2·4	− 1·1				
					716·4	92·7	

The successive differences d_y of the lengths coloured green, y, are shown in the fourth and eighth columns.   We find $S(d_y^2) = 63·21$ and consequently, since there are 29 lines,

$$s^2 = \tfrac{1}{2} \, 63·21/28 = 1·1288$$
$$\text{S.E.} \{ S(y) \} = \sqrt{(29 \, s^2)} = \pm \, 5·72$$

A length corresponding to 1 km. represents an area of 1 sq. km. Consequently the raising factor to be applied to the total length measured in cm. to give the estimated area in acres is $63,360 \times 247·11/100,000 = 156·57$. The total area of woodland is therefore

$$156·57 \times S(y) = 156·57 \times 92·7 = 14,514 \text{ acres}$$

and the standard error of this area is $156·57 \times 5·72 = \pm \, 896$ acres, i.e. 6·2 per cent.

The same procedure can be followed for the other maps covering the county, and the square root of the sum of the squares of the resultant standard errors will give the standard error for the whole county, since the errors of the different maps are virtually independent.   The results for Kent gave a percentage standard error of 3·4 per cent.

If the ratio method is used the corresponding successive differences d_x of the total lengths of the grid lines, and the sums of squares and products $S(d_x^2)$ and $S(d_x d_y)$ are required. The latter are found to be 159·23 and $+2·62$ respectively for the map in question. Using the ratio $\bar{r} = 0·12940$ derived from this map, we find

$$s_q^2 = Q/28 = 1·1643$$

As expected, there is no appreciable difference in the error calculated by the two methods. The simpler method is consequently all that is really required, even when the total area of woodland is calculated from the total area of the county and the ratio of the length coloured green to the total length of the grid lines. If, however, the first grid line is much shorter than the rest, owing to its cutting the map boundary at a small angle, it should be omitted, or the length made up by taking the relevant part of the line on the neighbouring map. This trouble will only arise if the error is estimated separately for each map and the grid lines are not exactly parallel to the map boundary.

On the other hand, the calculation of the error from the total variance of y, i.e. without stratification, would give very misleading results.

Example 7.18.b

Evaluate the sampling errors of the numbers of households from the 1 per cent. sample from the 1951 Population Census of Great Britain (Section 5.17) on the assumption that k is uniformly distributed, and relate these to the discrepancies shown in Table 7.18.a.

Since there were 49,318 O.E.D.s in England and Wales the sampling standard error with a uniform distribution of k, excluding the bias due to the excess of odd-numbered districts, will be $\sqrt{(0·083\,325 \times 49,318)} = \pm64$. The similar standard error for Scotland is $\sqrt{(0·083\,325 \times 9,730)} = \pm28$. The actual discrepancies, after correcting for the above bias, again on the assumption of a uniform distribution of k, were evaluated in Section 5.17. That for Scotland was only -14, i.e. 0·5 times the sampling standard error, but that for England and Wales was -178, i.e. 2·8 times the sampling standard error.

If k is not uniformly distributed, the bias due to the excess of odd O.E.D.s may be altered. In addition, pairs of odds and evens may also be biased. If the distribution is known, the bias for excess odds may be calculated from line 25 of Table 7.18.a, and that for pairs by the mean of lines 25 and 76.

The mean number of households per O.E.D. for England and Wales is 268. We may therefore expect some increase in the bias for excess odds, but this is not likely to be very large. If we take the value as $+200$ instead of $+153$, there will be a discrepancy of -225 to be accounted for. The number of O.E.D.s in pairs is approximately $49,316 - 612 = 48,703$. The expected bias per O.E.D. is therefore $-225/48,703 = -0·0046$. From the last line of Table 7.18.a it is apparent that with a mean k of 68 a negative bias is to be expected.   We may therefore surmise that in England and Wales the expected bias due to the excess of odds has been masked by an unsuspected bias in the matched pairs.

The apparent success of the adjustment for excess odds in accounting for the whole of the Scottish discrepancy may be due to a much wider range in the number of households per O.E.D. and consequently a more even distribution of k for Scotland ; the low mean number of households per O.E.D. is clearly caused by the sparsely populated areas, but numbers of households per O.E.D. similar to those of England and Wales may be expected in the more densely populated areas.

7.18.1   Lattice samples

A general paper on the estimation of the sampling errors of lattice samples has been published by H. D. Patterson (1954, A''). Estimation from complete data (case (d) below) is discussed by Hansen, Hurwitz and Madow in their book.

(a) *Square lattice*

No valid estimate of error is possible for a sample containing p units. With a sample of $2p$ units, however, a valid estimate is possible. The simplest procedure is to divide the lattice into p mutually exclusive sets of p units. Any Latin square effects such a subdivision, the letters defining the sets. These sets may themselves be regarded as complex sampling units. If two sets are selected at random from all p sets the contrasts between the sets will therefore provide an estimate of the sampling error with one degree of freedom. This estimate of error will not be of much value if only one square is sampled. If there are a number of squares each square will contribute one degree of freedom to the estimate of error.

When p is even, however, there is an alternative procedure by which a sample of $2p$ units can be made to yield $\frac{1}{2}p$ degrees of freedom for error. We start with a basic pattern made up of 2×2 squares, such as that shown for an 8×8 square on the left side of Table 7.18.1.a. We then rearrange first, the rows in random order amongst themselves, and secondly the columns in random order amongst themselves. If the two orders are 21748365 and 13524678 the arrangement on the right of the table will be obtained.

The symbols may now be taken to indicate the values actually observed. An estimate of the error variance per unit with 4 degrees of freedom will then be provided by

$$s^2 = \tfrac{1}{16} \left[(a_1 - a_2 - a_3 + a_4)^2 + (b_1 - b_2 - b_3 + b_4)^2 \right. $$
$$\left. + (c_1 - c_2 - c_3 + c_4)^2 + (d_1 - d_2 - d_3 + d_4)^2 \right]$$

In the general case the divisor will be $2p$. Allowing for finite sampling, the sampling variance of the mean will be $(1 - 2/p)s^2/2p$. With the specified randomization process it can be shown that this estimate of error is unbiased.

237

TABLE 7.18.1.a—8 × 8 ROTATIONAL SCHEMES FOR TWO TYPES OF SUBDIVISION

	Basic pattern									Actual arrangement							
	1	2	3	4	5	6	7	8		1	2	3	4	5	6	7	8
1	a_1	a_2							1	a_3			a_4				
2	a_3	a_4							2	a_1			a_2				
3			b_1	b_2					3							d_1	d_2
4			b_3	b_4					4		b_3			b_4			
5					c_1	c_2			5							d_3	d_4
6					c_3	c_4			6		b_1			b_2			
7							d_1	d_2	7			c_3			c_4		
8							d_3	d_4	8			c_1			c_2		

(b) *Cubic lattice, sample of 2p units*

In the case of a cubic lattice no unbiased estimate of error of the above type appears to be possible with a sample of $2p$ units. The best that can be done is to obtain one degree of freedom by selecting and contrasting two out of p^2 mutually exclusive sets of p units. Such a group of sets may be constructed as follows. Let $(a_1, a_2, \ldots a_p)$, $(b_1, b_2, \ldots b_p)$, $(c_1, c_2, \ldots c_p)$ be three random permutations of the numbers 1 to p. Also let β and γ be two random numbers, the same or different, but not both 1, between 1 and p. Then the lattice co-ordinates of the two sets are

$$(a_1, b_1, c_1), (a_2, b_2, c_2), \ldots.$$
$$(a_1, b_\beta, c_\gamma), (a_2, b_{\beta + 1}, c_{\gamma + 1}) \ldots.$$

with the proviso that when any suffix is greater than p, it is reduced by p. For example, if for a 6^3 lattice the random permutations are (5 6 1 2 4 3), (2 4 1 3 6 5), (4 3 6 2 5 1) and β, γ are 6, 4, the two sets are

(5, 2, 4), (6, 4, 3), (1, 1, 6), (2, 3, 2), (4, 6, 5), (3, 5, 1),
(5, 5, 2), (6, 2, 5), (1, 4, 1), (2, 1, 4), (4, 3, 3), (3, 6, 6).

With a sample of $4p$ units an estimate similar to that of the square lattice sample is possible when p is even, using a basic pattern of $\frac{1}{2}p$ $2\times2\times2$ cubes. Unfortunately, however, the different variance components enter in different proportions into this estimate. Investigation shows that the unbiased estimate involves a difference of the mean squares corresponding to the two-factor and three-factor interactions of the $2 \times 2 \times 2$ cubes. It will rarely be more accurate than the estimate based on the three degrees of freedom obtained by taking four sets of p units at random from p^2 mutually exclusive sets.

(c) *Cubic lattice, sample of $2p^2$ units*

When p is even, an unbiased estimate of error with $\frac{1}{4}p^2$ degrees of freedom, similar to that for the square lattice, can be obtained. The basic pattern is made up of $2 \times 2 \times 2$ cubes. It can be represented by a Latin square made up of 2×2 squares, and a second Latin square obtained by reversing all the 2×2 squares of the first square. The letters then represent the third co-ordinate. Table 7.18.1.b shows a suitable pair of squares for $p = 6$. Larger squares can be constructed in a similar manner.

TABLE 7.18.1.b—BASIC PATTERN FOR A SAMPLE OF $2p^2$ UNITS FROM A CUBIC LATTICE

A	B	C	D	E	F		B	A	D	C	F	E
B	A	D	C	F	E		A	B	C	D	E	F
C	D	E	F	A	B		D	C	F	E	B	A
D	C	F	E	B	A		C	D	E	F	A	B
E	F	A	B	C	D		F	E	B	A	D	C
F	E	B	A	D	C		E	F	A	B	C	D

As before, the rows, columns, and letters must be randomized, the same randomization being used for each square. The randomization of letters is effected in the same manner as that for rows and columns, writing the letters *A–F* in random order, say *B F A C D E*, and substituting *B* for *A*, *F* for *B*, *A* for *C*, etc.

For the estimate of error we must calculate for each of the original $2 \times 2 \times 2$ cubes the difference between the sum of the four units of the first square and the sum of the four units of the second square belonging to that cube. These differences can be represented geometrically by assigning opposite signs to the points at the two ends of each edge of the relevant cube. The components of the differences can easily be picked out, since (with the letter randomization adopted) *B* goes with *F*, *A* with *C*, and *D* with *E*, and the components of each cube occur at the intersection of two rows and two columns which are the same for each square.

There are $\frac{1}{4}p^2$ such differences. If these are denoted by *d* we have for the variance of a single unit

$$s^2 = \frac{1}{2p^2} \, S\,(d^2)$$

The sampling variance of the mean of the $2p^2$ units is therefore $(1 - 2/p)\, s^2/2p^2$.

(d) *Estimation of error from complete data*

If we have available data on all units of one or more lattices, we can without difficulty estimate what the sampling error of lattice sampling would have been. Such information is necessary in the planning of surveys both for determination of size of sample and for the study of the relative efficiency of lattice sampling and other methods.

The procedure of the analysis of variance can be applied to the complete data of a square lattice in the manner of Table 7.18.1.c.

TABLE 7.18.1.c—ANALYSIS OF VARIANCE FOR A COMPLETE SQUARE LATTICE

	Degrees of freedom
Rows (R)	$p - 1$
Columns (C) . . .	$p - 1$
Remainder ($R \times C$) . .	$(p - 1)^2$
Total	$p^2 - 1$

The sum of squares for rows is given by the sum of the squares of the deviations of the row totals, divided by p, and similarly for the columns. The sum of squares for the remainder (known as the two-factor interaction $R \times C$) is obtained by subtraction. The remainder mean square then gives an estimate s^2 of the variance per unit in lattice sampling. If we require to determine the relative precision of lattice sampling and simple stratification by rows, we calculate a new mean square for columns plus remainder, adding the degrees of freedom and the sums of squares. Similarly, rows plus remainder gives an estimate of the variance for stratification by columns, while the total line gives the estimate for random sampling.

The procedure in the case of a p^3 lattice is similar. Denote the three classifications by R, C and L. Summation over L gives a $p \times p$ table of totals which can be analysed in the manner of Table 7.18.1.c to give R, C, and

TABLE 7.18.1.d—ANALYSIS OF VARIANCE FOR A COMPLETE CUBIC LATTICE

	Degrees of freedom	Mean square
R	$p - 1$	
C	$p - 1$	
L	$p - 1$	
$R \times C$	$(p - 1)^2$	A
$R \times L$	$(p - 1)^2$	B
$C \times L$	$(p - 1)^2$	C
$R \times C \times L$	$(p - 1)^3$	D
Total	$p^3 - 1$	

$R \times C$. The sums of squares must be divided by an additional factor p because we are working with totals of p units. Tables for R and L and C and L can be constructed similarly. These can then be combined into a single table in the manner of Table 7.18.1.d. The three-factor interaction $R \times C \times L$ is then obtained by subtraction.

The estimate of error variance for a lattice sample of p^2 units (based on a Latin square) is given by the mean square D for $R \times C \times L$. The estimate of error for a lattice sample of p units is given by the expression

$$s^2 = \left\{ A + B + C + (p - 2) D \right\} / (p + 1)$$

If p is large this tends to the pooled mean square for all four interactions.

(e) *Multi-stage schemes*

When each cell of the lattice contains a number of second-stage units, of which some only are selected, the estimation of error follows the ordinary lines for multi-stage schemes. If an estimate of the second-stage error is not required the selection of one second-stage unit from each first-stage unit will normally provide an adequate estimate of the total sampling error, since the first-stage sampling fraction is not usually large.

Another type of two-stage sampling arises when the members of one of the lattice classifications are themselves a sample from a larger number of such classes. Rotational schemes for censuses of road traffic (Section 4.31) provide an example of this type. In such a scheme the observation points constitute a random selection from all possible points. Thus, for example, a single Latin square arrangement of observations on 12 points extending over 12 days with 12 periods in each day forms a second-stage sample of one unit out of the 12 units defined by any set of 12 squares which together comprise the whole of the traffic passing these points. The difference between the totals for a pair of Latin squares from the set will therefore only give an estimate of the sampling error at the second stage. To obtain an estimate of the total sampling error, two different sets of 12 points must be taken for the two Latin squares. The square for each set should be independently randomized, but the points of each set can be obtained (with some gain in precision) by random selection of two points from each of 12 strata instead of 24 points from a single stratum. Thus in the simple case of the estimation of traffic along a single main road the road can be divided into 12 equal sections. Two points are then located at random in each section, one of each pair of points being allocated at random to the first square, and the other to the second square. Each pair of squares yields only one degree of freedom. This limitation, however, is not of great importance in schemes which are extensive either in space or time, since many pairs of squares will be required for the whole scheme.

7.19 Sampling on successive occasions

It will be sufficient if we record the variances of the estimates given in Sections 6.21 and 6.22.

(*a*) Two occasions only : sub-sample on the second occasion.

$$V(\bar{y}) = \{V(y) - \mu b^2 V(x)\}/\lambda n = (1 - \mu r^2) V(y)/\lambda n$$
$$V(\bar{y} - \bar{x}) = \{V(y) + (\lambda - 2\lambda b - \mu b^2) V(x)\}/\lambda n$$

If the population value on the second occasion is estimated by adding the estimate of change derived from the sub-sample only to the overall mean on the first occasion, i.e. by $y' - \bar{x}' + \bar{x}$, the variance of this estimate will be

$$V(\bar{y}' - \bar{x}' + \bar{x}) = \{V(y) - \mu(2b - 1) V(x)\}/\lambda n$$

(*b*) Two occasions only : part of the sample replaced on the second occasion.

$$V(\bar{y}_w) = \frac{(1 - \mu r^2) V(y)}{n(1 - \mu^2 r^2)}$$

or, in the case of unequal numbers on the two occasions,

$$V(\bar{y}_w) = \frac{(1 - \mu r^2) V(y)}{n' + n''(1 - \mu r^2)}$$

With equal numbers on the two occasions, the variance of the estimate of change given by formula 6.21.b is approximately

$$V(\text{change}) = \frac{(1 - r)\{V(y) + V(x)\}}{n(1 - \mu r)}$$

The variance of the estimate given by the difference of the means of the units occurring on both occasions is approximately

$$V(\bar{y}' - \bar{x}') = (1 - r)\{V(y) + V(x)\}/\lambda n$$

and of that given by the difference of the overall means is approximately

$$V(\bar{y} - \bar{x}) = (1 - \lambda r)\{V(y) + V(x)\}/n$$

The exact expressions in the last two cases are given by replacing *r* by

$$2 \operatorname{cov}(xy)/\{V(x) + V(y)\}$$

which is equal to *r* when $V(x) = V(y)$.

(*c*) Successive occasions : same fraction replaced on each occasion.

The limiting value of $V(\bar{y}_h)$, subject to the restrictions mentioned in Section 6.22, is as follows :—

$$V(\bar{y}_h) = \varphi V(y)/\mu n$$

The variance of the estimate of change given by $\bar{y}_h - \bar{y}_{h-1}$ is

$$V(\bar{y}_h - \bar{y}_{h-1}) = 2\varphi V(y)\{1 - r(1 - \varphi)\}/\mu n$$

Example 7.19.a

Estimate the sampling errors of the estimates of Example 6.21.

$V(x)$ and $V(y)$ may reasonably be taken as equal. The pooled estimate, based on all the observations on each occasion (22 degrees of freedom) is 0·08767.

We then have

$$V(\bar{y}_w) = \frac{0\cdot08767\,(1 - \tfrac{1}{3} \times 0\cdot847^2)}{12\,(1 - \tfrac{1}{3} \times 0\cdot847^2)} = 0\cdot0777^2$$

This may be compared with the variance of the overall mean $\bar{y}$, which is 0·08767/12, i.e. 0·0855². The ratio of these variances is 1·210. Thus the gain in efficiency by the use of the information provided by the sampling on the first occasion is 21 per cent.

Similarly

$$V(\text{change}) = \frac{2 \times 0\cdot08767\,(1 - 0\cdot847)}{12\,(1 - \tfrac{1}{3} \times 0\cdot847)} = 0\cdot0558^2$$

The variance of the change estimated from the units sampled on both occasions is $2 \times 0\cdot08767\,(1 - 0\cdot847)/8$, i.e. 0·0579². The ratio of these variances is 1·077, and the gain is therefore 8 per cent. If the change is estimated from the difference of the overall means the variance is

$$V(\bar{y} - \bar{x}) = 2 \times 0\cdot08767\,(1 - \tfrac{2}{3} \times 0\cdot847)/12 = 0\cdot0798^2$$

The ratio of this variance to the first variance is 2·042, and the gain is therefore 104 per cent.

It will be noted that when $V(x)$ is taken as equal to $V(y)$ all the above gains depend solely on the values of λ and r.

Example 7.19.b

Estimate the sampling errors of the estimates of Example 6.22.

Owing to variation in the numbers on the different occasions the above formulæ will only give approximate estimates of the sampling errors. Excluding January, the average number of observations per occasion is 9·8 and the average value of λ is 0·664. Since $r = 0\cdot811$, $1 - r^2 = 0\cdot343$, and hence

$$\varphi = \frac{-0\cdot343 + \sqrt{[0\cdot343\,(1 - 0\cdot657 \times 0\cdot108)]}}{2 \times 0\cdot664 \times 0\cdot657} = 0\cdot254$$

$V(y)$ was found to be 0·0871, and hence

$$V(\bar{y}_h) = \frac{0\cdot254 \times 0\cdot0871}{0\cdot336 \times 9\cdot8} = 0\cdot0820^2$$

$$V(\bar{y}_h - \bar{y}_{h-1}) = \frac{2 \times 0\cdot254 \times 0\cdot0871\,(1 - 0\cdot811 \times 0\cdot746)}{0\cdot336 \times 9\cdot8} = 0\cdot0729^2$$

7.20 The error graph

In various instances in the preceding sections pooled estimates of the error variance have been used.   Such pooled estimates are only legitimate if the error variance is reasonably constant over the parts of the population for which the pooling is carried out.   In many types of material such constancy does not exist, and in such cases, when the number of degrees of freedom is too small for accurate determination of the error variances of the different parts of the population, other procedures must be followed if sampling errors are required separately for the different parts.

The simplest and most convenient device for practical use is the error graph.   The estimates of the error variance are plotted against some other characteristic of the parts of the sample which is believed to govern the magnitude of the error variance, and a smooth curve is drawn to fit the points as closely as possible.   This curve gives the variance law from which revised estimates of the variance can be obtained for any given value of the determining characteristic.

Fig. 7.20 shows a graph of this kind obtained in the course of a survey of wireworm infestation in grass fields.   In each field 20 cores were taken and the wireworms counted in each core.   There are thus 19 degrees of freedom for the determination of sampling error in each field.   The estimates of the percentage variance so obtained were plotted against the estimated number of wireworms per acre in the various fields.   The smooth curve so obtained was used to provide a table of errors that might be expected in similar sampling. Table 7.20 gives a small abstract of this table, and also of the inverse table,

TABLE 7.20—DISTRIBUTION AND PROBABLE LIMITS OF ERROR OF SAMPLE ESTIMATES OF WIREWORM POPULATIONS OF GRASS FIELDS SAMPLED BY TWENTY 4 IN. CORES (1 CORE = 1/500,000 ACRES)

(1,000 per acre)

True population	One-eighth of sample estimates		Estimated population	Population which, in one-eighth of cases, would give an estimate	
	less than	greater than		not less than that observed	not greater than that observed
200	105	295	200	128	325
400	260	540	400	284	567
600	428	772	600	451	804
800	597	1,003	800	624	1,040
1,000	766	1,234	1,000	797	1,277

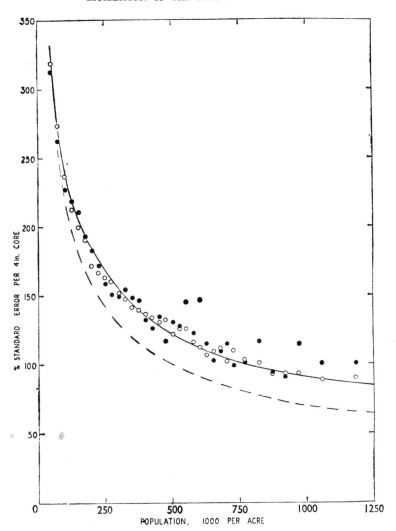

FIG. 7.20—STANDARD ERRORS PER UNIT CORE OF 4 IN. DIAM.
(WIREWORM SURVEY, 1940–1)

o means for 2272 fields grass in 1940   ——— fitted to data from grass fields
● means for 525 fields arable in 1940   – – – Poisson distribution

Reproduced from Yates and Finney (1942, J) with the
permission of the editors of the *Annals of Applied Biology*.

obtained by interpolation from the first table, giving the fiducial limits associated
with any given observed number of wireworms. This procedure is approximate
in a number of respects, but detailed discussion would be out of place here.

There are, of course, various other methods of dealing with problems of this type. In biological work the original variates are often transformed to other variates, such as logarithms or square roots, which may in the material in question be expected to have a more constant variance, and thus permit pooling of the estimates of error. Such procedures introduce a number of complications ; in particular the means of the transformed variates, when transformed back into the original variates, will be biased. They are not generally necessary or advisable in sample censuses and surveys.

7.21 Determination of errors due to bias

As has been pointed out in Chapter 2, bias can arise either in the selection of the sample, or in the estimation process.

Although biased methods of estimation can in general be avoided, occasions arise, such as that discussed in Example 6.19, where biased estimates are considerably more accurate than the corresponding unbiased estimates. The biased estimates are also sometimes considerably simpler to calculate than the unbiased estimates. For these reasons it is sometimes advisable to use biased estimates for comparative purposes. Such estimates can clearly be used with more confidence if the amount of bias that they introduce is not large.

In general it is possible to make an estimate of the expected magnitude of a bias in estimation by a combination of mathematical analysis and detailed numerical analysis of the sampling results. Such methods, however, are complicated and vary with the type of sampling adopted. We shall therefore not describe them here.

An alternative and relatively simple method is to compare the biased estimates with the corresponding unbiased estimates of the same quantity. Any one comparison will of course be affected by random sampling errors in both the estimates, but if the material is sufficiently extensive to provide a number of comparisons, the mean difference will provide an estimate of the average bias whose accuracy can be judged by the variation of the individual differences.

Bias in the selection of the sample can in general only be assessed by comparison with another sample known to be free from bias. If, however, the distribution of some supplementary variate is known, bias in selection can sometimes be assessed, and if necessary eliminated, at least in part, by the use of regression. The calibration of eye estimates by means of regression, described in Section 6.15, provides an example of this procedure. As already pointed out, the procedure is subject to qualifications and cannot be relied on to compensate for all possible sources of bias. The only certain guarantee that bias is absent is the use of methods of selection and observation which are free from bias.

Example 7.21

Assess the evidence for bias in the estimate of the dressing of nitrogen per acre derived from the unweighted mean over all fields of Example 6.19.

The results already given in Tables 6.19.b and 6.19.c show that in each size-group the unweighted mean dressing is less than the weighted mean. The apparent bias in the overall unweighted mean dressing is — 0·052. A larger number of comparisons of the same type may be obtained by dividing the 22 farms of the small-size group into two groups of 11 farms each, and the 36 farms of the medium-size group into four groups of 9 farms each. Division of Table 6.19.a into blocks in this manner gives the comparisons shown in Table 7.21.

Six out of the seven differences are negative. The mean difference is — 0·040, and the standard error of this difference, estimated from the sum of the squares of the deviations of the individual differences from their mean, is ± 0·022. The evidence for bias on this small amount of data is therefore not conclusive.

The procedure has here been adapted to the data given in Table 6.19.a. If the sampling were systematic the division of the groups should be made by a random or systematic process such that each of the sub-groups constitutes a substantially random sample of the whole of the group. The procedure is also subject to the qualification that any bias arising from differential weighting of the different size-groups will be excluded by this method of estimation, since the differences are based on comparisons within size-groups.

TABLE 7.21—ESTIMATION OF BIAS

	Weighted mean	Unweighted mean	Difference
Small farms	·521	·484	— ·037
	·378	·362	— ·016
Medium farms	·584	·537	— ·047
	·417	·446	+ ·029
	·503	·470	— ·033
	·378	·361	— ·017
Large farms	·576	·416	— ·160

An alternative method of subdivision which overcomes this limitation is to form sub-samples in which all the size-groups are represented in the correct proportions. If the data from a number of counties are available, no subdivision will be necessary, since the differences between size-group means and between the overall means for the different counties will provide all necessary comparisons.

7.22   The sampling error of an estimate of bias

In Section 7.21 it was pointed out that an estimate of the bias arising from the use of a biased method of estimation could be obtained by comparing the biased with the unbiased estimates for relevant subdivisions of the sample. In the common case in which the biased estimate under consideration is the unweighted mean of some variate z, whereas the unbiased estimate is provided by some form of weighted mean of z, an alternative approach to the problem is provided by the calculation of the regression of z on the weights w. If B denotes the estimated bias and b the regression coefficient of z on w we have

$$B = \bar{z} - \bar{z}_w = \frac{Sz}{n} - \frac{Swz}{Sw} = -\frac{S(w - \bar{w})(z - \bar{z})}{Sw}$$

$$b = \frac{S(w - \bar{w})(z - \bar{z})}{S(w - \bar{w})^2}$$

Hence

$$B = -\frac{bS(w - \bar{w})^2}{Sw}$$

Thus for a given distribution of weights the magnitude of the bias is proportional to b. In particular, if the value of b for the population is zero there will be no bias, and if b is the same for different domains of study which have similar distributions of the weights the comparisons between the unweighted means for the various domains will be free from bias. The contribution to the sampling error of B due to error in estimation of b can be obtained by calculating the error variance of b from the formula given in Section 7.12. There will be a further contribution due to variation in the distribution of w from sample to sample, but this can ordinarily be neglected.

It should be noted that the calculation of b does not give a more accurate estimate of the bias of the unweighted mean than that provided by the difference of the weighted and unweighted means. The two estimates are in fact identical. The calculation of the standard error of B does, however, provide an indication of the probable limits of the bias.

Example 7.22

Use the above method to assess the evidence for bias in the estimate of the dressing of nitrogen per acre derived from the unweighted mean over all fields of Example 6.19.

In the notation of Example 6.19, $w = g'\,g''\,x$ and $z = r = y/x$. The quantities Sw, Sz, Sw^2, Swz, Sz^2 can best be calculated for each size-group separately, before introduction (where necessary) of the factors g' and g'^2. Sw and Swz have already been given in the formula for $\bar{r}$. Sz is given in Table 6.19.c. The results for the whole sample are:

$n = 67$ $Sw = 58,229$ $Sz = 29{\cdot}36$

$S(w - \bar{w})^2 = 44,849,800$ $S(w - \bar{w})(z - \bar{z}) = 2996$ $S(z - \bar{z})^2 = 3{\cdot}4822$

$b = 2996/44,849,800 = +\,0{\cdot}000,066,80$

$B = -\,0{\cdot}000,066,80 \times 44,849,800/58,229 = -\,0{\cdot}0515$

This agrees with the difference $0{\cdot}438 - 0{\cdot}490 = -0{\cdot}052$ of the unweighted and weighted means. Following the method given in Section 7.12 and illustrated in Example 7.12.a we find

$$\text{S.E. } (b) = \pm\, 0{\cdot}000{,}033{,}6$$

Hence

$$\text{S.E. } (B) = \pm\, 0{\cdot}0258$$

The actual value of B is almost double its standard error. There is therefore some evidence that the unweighted mean is biased, but the magnitude of the bias is not accurately determined. The limits of error given by plus or minus twice the standard error are $+\,0{\cdot}0001$ and $-\,0{\cdot}1031$.

It should be noted that the amount of bias obtained in sampling a population depends not only on the properties of the population, but also on the sampling method followed. Changes in the sampling method will consequently affect the bias. In this example, for instance, the weights are affected by the first-stage sampling fractions. If sufficiently extensive data were available, the biases to be expected with different sampling fractions could be estimated by adjusting the contributions of the different strata to Sw, Sz, Sw^2, Swz, Sz^2 so as to conform to the new sampling fractions.

7.23   Relative precision of biased and unbiased estimates

The use of biased estimates introduces errors (due to the bias) which may be large relative to the random sampling errors. Nevertheless we often have good grounds for believing, either on the basis of previous experience of the material under consideration, or from detailed statistical analysis, that the errors due to bias are actually small, particularly when comparisons between different domains of study are required. If, therefore, a large reduction in the random sampling error is effected by the use of a biased estimate this may be preferable to the unbiased estimate. The relative precision of biased and unbiased estimates (i.e. the reciprocal of the ratio of their respective sampling variances) can be determined by estimating the variances of the two types of estimate by the methods appropriate to the estimates in question. In the particular but important case in which the unbiased estimate consists of a weighted mean of the observed values z with weights w, and the sources of variation are such that all z can be regarded as subject to the same variance, the unweighted mean of the values will provide an estimate of the mean which has minimum variance. The ratio of the variances of the two estimates will then be, from formula 7.5.e,

$$\frac{1}{n} \left| \frac{S\,(w^2)}{\{S\,(w)\}^2} = \frac{(\text{mean of } w)^2}{\text{mean of } w^2} \right.$$

Example 7.23

An estimate of rabbit damage to the wheat crop of a county is made on a random sample of farms. On each selected farm one of the fields growing wheat is selected with probability proportional to the area of the field. The damage is estimated by comparing fenced and unfenced areas. If the distribution of wheat acreages is that given in Table 7.2 what is the relative precision of the weighted and unweighted estimates ?

In this case the components of variation are such that the unweighted mean of the losses per acre on all sampled fields is likely to provide the estimate with approximately minimum error variance. This estimate will, however, be biased if the loss per acre is correlated with the area of wheat on the farm. An unbiased estimate will be provided by a weighted mean of the losses per acre, the weights being proportional to the areas of wheat on the sampled farms. The farms not growing wheat must be excluded from the calculation, since they are automatically excluded from both the weighted and unweighted means, even if they are included in the original sample. From the results given in Example 7.2.b we have

$$n' = 45, \qquad S(y) = 2301, \qquad S(y^2) = 207{,}261$$

where n' is the number of farms growing wheat. Here $w = y$. Hence the relative precision is

$$\frac{1}{45} \left/ \frac{207{,}261}{(2301)^2} \right. = 0{\cdot}568$$

The biased estimate has therefore nearly twice the precision of the unbiased estimate. If it were possible to select farms with probability proportional to area of wheat grown (one field, selected with probability proportional to area, being sampled on each selected farm), the unweighted mean would provide the unbiased estimate. The reciprocal of the above fraction, 1·76, therefore gives the advantage, when an unbiased estimate is required, of using this method of sampling, i.e. of giving each acre an equal chance of being selected for assessment of loss instead of taking a random sample of farms. It may be noted here that the relative precision of the two estimates is very similar to that found in Example 7.17 for the rather similar case of the rate of application of fertilizers. In that case the value was $1/1{\cdot}91 = 0{\cdot}524$.

7.24   Interpenetrating samples : comparison of observers

The error variance of the difference between two observers, estimated from interpenetrating samples, can be obtained by calculating the error variance appropriate to each observer, and adding these variances. This procedure, however, is subject to certain qualifications. In the first place the correction for finite sampling must not be applied. In the second place only those components of variance must be included which affect the comparisons between the observers. Thus, if a two-stage sampling process is adopted and each of the selected primary units is sampled by both observers, only the second-stage sampling error will enter into the comparison between observers.

If the data relevant to the comparison between two observers are at all extensive it is often possible to make a direct estimate of the error of this comparison by subdividing the material so that a number of independent differences are obtained, in the same manner as in the estimation of bias from different methods of estimation.   Thus, with the above two-stage sampling process the difference between the observers might be obtained for each primary unit separately.

7.25   Estimation of the sampling error from duplicate samples

If a survey is carried out in two or more interpenetrating parts and the results are tabulated separately, an estimate of the sampling error can be obtained from the differences of the two samples.   For such an estimate to be of any value there must be a number of independent differences, so that at least a moderate number of degrees of freedom are available.   Even with extensive surveys the number of available differences is likely to be small, so that such estimates are usually rather rough.   Nevertheless they are useful when the detailed results are not available.

If the two samples are distinguished by single and double dashes, and the estimate of the population total is given by the sum of the t parts 1, 2, etc., we have $Y' = Y_1' + Y_2' + \ldots$, and $Y'' = Y_1'' + Y_2'' + \ldots$ If the sizes of the two samples are in the ratio $\lambda : \mu$, where $\lambda + \mu = 1$, the estimate Y of the population total from the two samples is $\lambda Y' + \mu Y''$, with similar expressions for Y_1, etc.   An unbiased estimate of the error of Y is given by

$$V(Y) = \lambda \mu (1 - f) \{(Y_1' - Y_1'')^2 + (Y_2' - Y_2'')^2 + \ldots\}$$

where f is the sampling fraction for the whole survey.   When the parts vary considerably in size this estimate is very inefficient, since excessive weight is given to the larger totals.   If the approximate relation between the variances of the totals of the parts is known, a more efficient estimate can be obtained, though this will be biased if the assumed law of variance is incorrect.   If the variances are proportional to Y_1, Y_2, etc., the efficient estimate is

$$V(Y) = \frac{Y \lambda \mu (1 - f)}{t} \{(Y_1' - Y_1'')^2/Y_1 + (Y_2' - Y_2'')^2/Y_2 + \ldots\}$$

This law of variance is likely to be approximately true for area surveys if the density per unit area of the quantity surveyed does not vary very greatly from part to part.

If the variances are proportional to some other quantity, such as the number of units in each part, these numbers must be substituted for Y_1, Y_2, etc. and their sum for Y in the above formula.

Example 7.25

In the 1942 Census of Woodlands the total volumes of timber for the seven regions of the survey obtained from the first and second 5 per cent. samples, excluding areas surveyed in 1938–9, and with allowance for felling in the interval between the two samples, are shown in Table 7.25. Estimate the sampling error to which the combined estimate of the total volume of timber for the country is subject.

TABLE 7.25—VOLUMES OF TIMBER IN THE DIFFERENT REGIONS ESTIMATED FROM THE FIRST AND SECOND 5 PER CENT. SAMPLES OF THE 1942 CENSUS OF WOODLANDS

Region	Volume, m. cu. ft.			$A - B$	$\dfrac{(A - B)^2}{\text{Mean}}$
	Sample A	Sample B	Mean		
A	253	201	227	$+52$	11·9
B	74	84	79	-10	1·3
C	100	107	104	-7	0·5
D	148	164	156	-16	1·6
E	209	227	218	-18	1·5
F	94	78	86	$+16$	3·0
G	112	119	116	-7	0·4
	990	980	986	$+10$	20·2

The computations are shown in the last three columns. The sum of squares of the differences $A - B$ is 3738, and therefore from the first formula the standard error of the total is

$$\sqrt{(\tfrac{1}{2} \times \tfrac{1}{2} \times \tfrac{9}{10} \times 3738)} = \pm\, 29 \cdot 0 \text{ m. cu. ft.}$$

The sum of the last column is 20·2, and therefore by the second formula the standard error is

$$\sqrt{(986 \times \tfrac{1}{2} \times \tfrac{1}{2} \times \tfrac{9}{10} \times \tfrac{1}{7} \times 20 \cdot 2)} = \pm\, 25 \cdot 3 \text{ m. cu. ft.}$$

It will be noted that an estimate of the standard error of any regional total can be obtained directly from the second formula by substituting this total for the grand total 986.

The above estimates are very rough, since they are based on only 7 degrees of freedom. The estimate obtained by the method described in Section 7.18 is $\pm\, 19 \cdot 7$ m. cu. ft. With a more extensive set of comparisons between the two samples a more accurate estimate could be obtained.

7.26   Presentation of the sampling errors of extensive surveys

Before computers were available the calculation of standard errors for extensive surveys was exceedingly troublesome, and demanded considerable expertise in the use of punched card machines.   Consequently there was no great danger of including an excessive number of errors in reports.   Computers, however, can readily produce standard deviations or standard errors, at least for means and totals derived from random samples, and some restraint must therefore be exercised in their presentation, to avoid cluttering up tabular material with masses of unwanted or irrelevant errors.

When considering what errors should be included and the comments to be made on them, the following questions may be asked :

(a) Do the errors it is proposed to include give at least a general indication of the magnitude of the sampling errors affecting the survey ?

(b) Are an adequate number of key errors included ; are they appropriate to the purposes for which they are likely to be used, and are their limitations in this respect sufficiently explained ?

(c) Are there any known material defects in the way the errors were computed ?

(d) Could there be greater economy of presentation ?

(e) Could some of the errors provisionally included be omitted without loss ?

The answers to these questions depend very much on the nature and purposes of the survey, and the sampling design.   In a census-type survey based on a random sample of individuals there is no need to append standard errors to tables of counts or percentages, as the standard error of any count or percentage can be derived directly from the formulae given in Section 7.4.   These formulae can be taken as known, or quoted directly, or relevant values derived from them can be given in an auxiliary table.   A similar table can be constructed for the contrasts between two percentages based on the same set of units.

If, however, the survey is based on a two-stage or cluster sample these formulae will not be directly applicable.   A concept which has a certain limited usefulness in such cases is the *design effect* (Deff), usually defined as the ratio of the actual sampling variance to the variance of a random sample with the same number of units.   1/Deff is then equivalent to relative efficiency as defined in Section 8.1.   If the design effect were in fact substantially the same for all attributes included in the survey, knowledge of its value would enable the standard error of any percentage value to be determined.   As is to be expected, however, there is considerable variation.   Average values for $\sqrt{(\text{Deff})}$ over a set of six surveys on consumer expectations and finances reported by Kish in his book, for example, range from 0·99 to 1·89.   Nevertheless, to a perceptive reader a table of typical Deffs for attributes of different types may give a fair indication of the likely errors.

The standard errors of quantitative variates cannot in general be summarized so concisely.   Each variate will have a different overall variance, and even in a random or stratified sample the variances within domains and strata may be considerably less than the overall variance, and may vary markedly from domain to domain. Nevertheless, considerable economy of presentation is often possible, though each case must be considered on its merits.

As a simple example, consider a survey of the yields of an agricultural crop in which the method of selection of the sample is such that fields of the crop are effectively selected at random from within regions with probability proportional to size. Suppose also that the varieties grown on the sampled fields are recorded. The means of the yields per unit area, tabulated by variety and region, will then form a table similar to Table 5.21.a for the yields of potatoes. As was shown in Example 7.7.b, the sampling errors of both the regional and varietal marginal means, and also those of the sub-classes, can be reasonably accurately summarized merely by reporting the unit variances given in Table 7.7.f. As, however, the main object of the survey is to esti-mate the mean (or total) yields of the regions, the errors of the regional means and the overall means will rank as key errors which should be reported indi-vidually. It is, however, unnecessary and probably inadvisable to report individually the similar errors for the varietal means; attachment of errors to them may encourage misguided comparisons between them. For varietal comparisons, estimates freed from regional effects, similar to those of Table 5.21.b, are primarily what are required. One set of these should obviously be presented, together with their individual standard errors, similar to those calculated in Example 7.5.a, or for the least-squares estimates in Section 9.8. Probably, however, if the least-squares estimates are given, a full table of the standard errors of the differences of all pairs of varieties similar to Table 9.8 will be judged to be unnecessary, though if the differences of any particular pairs are discussed in the report the exact standard errors of these can be given in the text in preference to those derived from the approximate formula of Section 9.8.

In addition to random sampling errors the results may be biased. Biased methods of estimation may be used deliberately to improve the accuracy of the overall estimates or of comparisons between domains, as in the rabbit damage survey (Example 7.23) and in the survey of fertilizer practice (Example 6.19). Comparisons between the biased and unbiased estimates are then informative.

When reporting errors, it is of course essential to emphasize that sampling errors are not the only source of error, and to give assessments, if any are available, of the likely magnitude of such errors, and also of the degree of non-response, and incompleteness of the frame. Any available comparisons with comparable estimates from other sources are also of great value in coming to a balanced judgement of the accuracy of the results.

CHAPTER 8

EFFICIENCY

8.1 General remarks

The methods described in Chapter 7 enable the sampling error associated with a sample of a given type and size to be calculated from the data furnished by the sample itself. When planning a sample census or survey, we have to solve the more general problem of calculating the sampling errors of samples of various types and sizes from the data furnished by a sample of a particular type and size. We can then determine which method of sampling is likely to be most efficient and the size of sample necessary to give the required accuracy.

The determination of the sample size in the case of a random sample from a large population has already been discussed in Section 4.32. It was there shown that, for qualitative characters which are attributes of the sampling units, the number of units required could be determined without any prior knowledge of the material other than the approximate proportion of units possessing the given attribute in the population ; and that for quantitative characters knowledge of the standard deviation of the character in question per sampling unit was all that was required.

The formulae of Section 4.32 apply when the population is large relative to the size of sample required. If the population is not large a correction must be made to allow for finite sampling. This is most simply done by calculating the number of units n_0 that would be required if the population were large, and the corresponding sampling fraction $f_0 = n_0/N$. The required sampling fraction is then given by

$$f = \frac{f_0}{1 + f_0} \qquad (8.1)$$

In this calculation f_0 may be greater than unity.

The method followed in Section 4.32, i.e. that of taking the appropriate formula for the standard error of a sample of size n and rewriting this formula to give an equation for n, is a general one and can be applied to the more complicated types of sample, using the appropriate formulæ for the standard errors given in Chapter 7. It is apparent, however, that these formulæ can only be used if the relevant variances per sampling unit are known or can be estimated. In certain cases, also, the formulæ cannot conveniently be rearranged so as to give n directly. This, however, is a minor point, since the required solution can always be quickly found by trial once estimates of the relevant variances are available.

In the following sections we will discuss the problems that arise in the estimation of the variances relevant to different types of sample when the basic data consist of a sample of a different type. In certain cases data relating to all the units of a population will be available. This situation does not differ in any essential particulars from that in which data are derived from a random sample of the population.

We will here define the sense in which we shall use various terms in the subsequent discussion.

The *relative accuracy* of two samples which differ in respect of method of sampling or size of sample, or both, may be defined as the reciprocal of the ratio of the sampling variances of the estimates provided by them.

The *relative precision* of two different methods of sampling based on the same type of sampling unit may be defined as the reciprocal of the ratio of the sampling variances of the estimates given by the two methods when the same number of units are taken.

The *relative efficiency* of two different methods of sampling based on the same type of sampling unit may be defined as the reciprocal of the ratio of the numbers of units required to attain a given accuracy with the two methods.

In the case of a random sample from a large population, or a stratified sample with fixed strata from such a population, the relative efficiency is equal to the relative precision. But if the size of the strata depends on the number of units in the sample, or if the population is not large relative to the size of the sample, there is a difference between the two concepts.

The term *efficiency* is already in current use in the theory of estimation. It is there used in an absolute sense. An estimate is *efficient* (i.e. has an efficiency of 100 per cent.) if in large samples it is one of the class of most accurate estimates, i.e. estimates with minimum variance. An estimate has an efficiency of x per cent. if it has $100/x$ times this minimum variance. This use of the term is analogous to *precision* in our terminology. The reason why no distinction has to be made between precision and efficiency in the theory of estimation is that only large populations are normally under consideration, in which case the two concepts are synonymous. Since no confusion is likely to arise, we shall continue to use the term efficiency when discussing the relative accuracy of different estimates derived from the same sample.

The concepts of relative precision and relative efficiency may be extended to cover methods of sampling based on different types of sampling unit, by replacing numbers of units by the amount of material included in the sample. They may be further extended to cover the relative accuracy for a given cost and the relative cost for a given accuracy.

It may be noted here that the relative precision and relative efficiency of different types of sampling should as far as possible be judged from estimates of the sampling variances derived from the same set of data. Comparisons based on estimates derived from independent samples of different types are subject to errors of estimation which are considerably larger, and comparisons based on samples from different aggregates of similar material are even more subject to uncertainty. No very general conclusions should, however, be drawn from a single comparison based on a small amount of data, even when a single set of data is used. The relative precision of stratified and random samples, for instance, will depend on the differences between strata, and these differences may vary considerably even in apparently similar material.

8.2 Qualitative data

If the variates under consideration are attributes of the sampling units, the effect of stratification, with either uniform or variable sampling fraction, can be determined from a knowledge of the proportions of units possessing the given attribute in the different strata. In other cases qualitative variates must be treated similarly to quantitative variates, as in the estimation of sampling errors.

Formulæ for the required size of a stratified random sample with uniform sampling fraction, analogous to those for a random sample given in Section 4.32, can be written down without difficulty. A somewhat simpler approach, however, is to estimate the percentage standard error of a stratified sample of any convenient size (e.g. the size of the sample of which the data are available) on the assumption that the population is large. The size of sample required to give any predetermined percentage standard error is then given, if the population is large, by the formula

$$\frac{\text{Size of sample required}}{\text{Size of actual sample}} = \frac{(\text{Actual percentage standard error})^2}{(\text{Required percentage standard error})^2}$$

Allowance for the effect of finite population size can then be made by formula 8.1.

In the case of a stratified random sample with variable sampling fraction the same procedure can be followed, with the exception that allowance for the effect of finite population size cannot be made in the above manner. If, therefore, any of the correction factors $(1 - f_i)$ are sufficiently large to be of importance, the approximate size of sample required may first be calculated as above and the final size found by trial. Variable sampling fractions, however, are not likely to be much used for qualitative data.

Example 8.2.a

If a large population of individuals is divided into five strata containing equal numbers of people, determine the relative sizes of a stratified and a fully random sample of the same accuracy when the percentages of individuals giving a positive answer to a given question in the different strata are (a) 70, 60, 50, 40 and 30 per cent, (b) 10, $7\frac{1}{2}$, 5, $2\frac{1}{2}$ and 0 per cent.

A sample of 500 people will have 100 in each stratum. The variance of the number in the sample giving positive answers will be, in case (a), for a stratified sample,

$$100 \times \cdot 7 \times \cdot 3 + 100 \times \cdot 6 \times \cdot 4 + 100 \times \cdot 5 \times \cdot 5 + 100 \times \cdot 4 \times \cdot 6 + 100 \times \cdot 3 \times \cdot 7 = 115$$

and for a random sample,

$$500 \times \cdot 5 \times \cdot 5 = 125$$

The ratio of the required sizes is therefore $125/115 = 1 \cdot 087$, *i.e.* the random sample will have to be $8 \cdot 7$ per cent. larger. In case (*b*) a similar calculation shows the random sample will have to be $2 \cdot 7$ per cent. larger.

Example 8.2.b

Determine from the data of Examples 6.5 and 7.6.a the numbers of farms required to give a sampling standard error of 5 per cent. in the estimate of the number of farms growing wheat (*a*) when the sample is random, (*b*) when it is stratified by size-groups.

(*a*) We have $\text{p} = 54/125 = 0 \cdot 432$.   Consequently, from Sections 4.32 and 8.1

$$n_0 = \frac{10{,}000 \times 0 \cdot 568}{0 \cdot 432 \times 5^2} = 526$$

$$f_0 = 526/2496 = 0 \cdot 210$$

$$f = \frac{0 \cdot 210}{1 + 0 \cdot 210} = 0 \cdot 174$$

$$n = 433$$

(*b*) We have already found that for the stratified sample $\mathsf{U} = 1080$ and S.E. $(\mathsf{U}) = \pm 71 \cdot 64$. If the population had been large, therefore, we should have had S.E. $(\mathsf{U}) = 71 \cdot 64/\sqrt{(1 - 1/20)} = 73 \cdot 5$. Consequently in this case S.E. $\% (\mathsf{U}) = 6 \cdot 80$. Hence

$$\frac{n_0}{125} = \frac{6 \cdot 80^2}{5^2}$$

$$n_0 = 231$$

$$f_0 = 231/2496 = 0 \cdot 0927$$

$$f = 0 \cdot 0927/(1 + 0 \cdot 0927) = 0 \cdot 0848$$

$$n = 212$$

The standard error of the total estimated from a random sample of 125 farms is $20\sqrt{(125 \times 0 \cdot 432 \times 0 \cdot 568 \times 19/20)} = 108 \cdot 0$. Consequently the relative precision of the stratified and random samples of 125 units (or indeed of any number of units) is given by $108 \cdot 0^2/71 \cdot 64^2 = 2 \cdot 27$. The relative efficiency, when a 5 per cent. standard error is required, however, is $433/212 = 2 \cdot 05$. The relative efficiency is slightly less than the relative precision because we are sampling from a finite population.

8.3   Random sample and stratified sample with uniform sampling fraction

The general principle to be followed is to construct an analysis of variance which corresponds as closely as possible to that appropriate to the required type of sample. The procedure varies somewhat according to the type of data available.

(*a*) From the data of a stratified sample with uniform sampling fraction:

An analysis of variance within and between strata in the form of Table 7.7.a must be made. The within-strata mean square s_1^2 gives an estimate of the error variance per unit in a stratified sample, and the mean square s^2 from the total line gives a similar estimate for a random sample. If separate estimates of the error variance per unit have been made for the different strata, as in Example 7.6.a, a pooled within-strata sum of squares may be calculated by multiplying the within-strata error variances by the degrees of freedom $n_i - 1$ for each stratum, and summing the products, or by summing the sums of squares directly.

The formula of Section 4.32 can then be used to determine the size of sample, using s_1^2 in place of s^2 for a stratified sample, and correcting for finite population in the same manner as in Section 8.1. Since for a stratified sample $V(\bar{y}) = s_1^2 (1 - f)/n$, and for a random sample $V(\bar{y}) = s^2 (1 - f)/n$, the relative precision of stratified and random sampling will be given by the ratio of s^2/s_1^2. The relative efficiency will be somewhat less than the relative precision when the corrections for finite sampling are appreciable.

This procedure is approximate in two respects. In the first place, if the variances within the different strata are unequal they do not enter into the mean square B with quite the correct weights, as already explained in Section 7.7. In the second place, a stratified sample has a slightly greater overall variance per unit than a random sample from the same population, and consequently C is not the best estimate of the variance per unit of a random sample. Neither of these approximations gives rise to errors of any importance in the comparison of a random and a stratified sample, but it may be noted that the bias in C can be almost completely eliminated by calculating s^2 from the formula

$$s^2 = \{(n - 1) C + B\}/n \qquad (8.3)$$

An extension of this formula is of use in the case of multiple stratification (Section 8.4). Method (*c*) below takes account of both sources of disturbance.

(*b*) From the data of a random sample:

An analysis of variance within and between strata can be made in the same manner as with a stratified sample with uniform sampling fraction, and s_1^2 and s^2 can be estimated as in a stratified sample.

For this procedure it is only necessary that the units of the sample be classified by strata. The numbers of units of the whole population falling in the different strata do not require to be known.

If these numbers are known, Method (*c*) below can be followed. This will give slightly more accurate results at the cost of a little additional computation, since allowance is made for the fact that the numbers in the different strata in the sample will not be exactly proportional to the numbers in the population, owing to the fluctuations of random sampling.

(c) From the data of a stratified sample with a variable sampling fraction (or any arbitrary values of the sampling fractions):

Estimates of the average within-strata mean square s_1^2 and of the overall mean square must be calculated from the proportions $h_i = N_i/N$ of the units of the population in the different strata. The formulæ are

$$s_1^2 = \Sigma\, h_i\, s_i^2$$
$$s^2 = s_1^2 + \Sigma\, h_i\, \bar{y}_i^2 - \bar{y}^2 - \Sigma\, h_i\, (1 - h_i)\, s_i^2/n_i$$

where $\bar{y}$ is the estimate of the population mean derived from the sample, and is consequently equal to $\Sigma\, h_i\, \bar{y}_i$. The relation of these formulæ to the analysis of variance of a stratified sample with uniform sampling fraction will be apparent. The terms involving y in s^2 correspond to the between-strata component of variance, the last term of s^2 being the correction required because the $\bar{y}_i$ are themselves subject to sampling error. This correction will be trivial except when the between-strata component of variance is small and there are a large number of strata with few units from each stratum. If the h_i are put equal to n_i/n (uniform sampling fraction), s^2 will be the same, to order $1/n$, as that given by the mean square C of Method (a), with the exception that in Method (a) $\Sigma\, h_i\, \bar{y}_i^2 - \bar{y}^2$ is multiplied by a factor $n/(n-1)$.

It will be noted that the data need not be derived from a sample in which the sampling fractions are chosen with the object of obtaining the most accurate possible estimates : any set of data in which the sampling is random within strata, and from which the proportions of the units in the different strata, the strata means and the within-strata variances can be determined with sufficient accuracy, will be adequate.

Example 8.3.a

Determine the error variances per unit and the relative precision of a stratified random sample with uniform sampling fraction and a fully random sample from the data on wheat acreages of the stratified random sample of Hertfordshire farms (Examples 6.5 and 7.6.a).

The analysis of variance is given in Table 8.3.a. The within-strata sum of squares is obtained directly from Table 7.6.a by summing the column $S_i\,(y - \bar{y}_i)^2$. The between-strata sum of squares is obtained by summing the products of the columns of Table 6.5.b giving the totals and means, and deducting the product of the general total and the general mean. These means should be taken to two and three decimal places respectively. We thus have $s_1^2 = 349 \cdot 2$ and $s^2 = 797 \cdot 1$. The estimate of the relative precision is therefore $2 \cdot 28$.

TABLE 8.3.a—ANALYSIS OF VARIANCE OF THE STRATIFIED RANDOM SAMPLE OF HERTFORDSHIRE FARMS

	Degrees of freedom	Sum of squares	Mean square
Between size-groups .	5	57,278	
Within size-groups .	119	41,558	349·2
Whole sample . .	124	98,836	797·1

Example 8.3.b

Make similar estimates to those of Example 8.3.a, using the data of the random sample (Examples 6.6, 7.2.b and 7.6.b).

The analysis of variance is given in Table 8.3.b. If the N_i are not known, we have $s_1^2 = 488\cdot5$ and $s^2 = 1329\cdot9$. The estimate of the relative precision is therefore 2·72.

TABLE 8.3.b—ANALYSIS OF VARIANCE OF THE RANDOM SAMPLE OF HERTFORDSHIRE FARMS

	Degrees of freedom	Sum of squares	Mean square
Between size-groups .	5	106,775	
Within size-groups .	119	58,129	488·5
Whole sample . .	124	164,904	1,329·9

If the N_i are known the calculations follow the same lines as those of Example 8.3.c below, and are left to the reader. In this case we find $s_1^2 = 436\cdot5$ and $s^2 = 1189\cdot2$, the estimate of the relative precision being again 2·72.

Example 8.3.c

Make similar estimates to those of Example 8.3.a, using the data of the sample with variable sampling fraction (Examples 6.7 and 7.7.a).

Table 8.3.c shows the calculations. The h_i are calculated from the numbers in the population. These are given in Table 6.6.b, except for the last two size-groups, which have the values 215 and 51 respectively. It will be noted that we are here considering a sample stratified for districts as well as size-groups.

TABLE 8.3.c—CALCULATION OF THE AVERAGE WITHIN-STRATA AND OVERALL MEAN SQUARES FROM THE STRATIFIED SAMPLE WITH VARIABLE SAMPLING FRACTION OF HERTFORDSHIRE FARMS

Size-group	h_i	n_i	$\bar{y}_i$	$\bar{y}_i^2$	s_i^2	$h_i(1-h_i)s_i^2/n_i$
1– 5	·174	0	(0)	(0)	(0)	(0)
6– 20	·208	3	0	0	0	0
21– 50	·143	6	4·5	20	53·5	1·1
51–150	·208	26	8·2	67	159·2	1·0
151–300	·160	40	29·1	847	564·2	1·9
301–500	·086	43	76·6	5,868	1,703	3·1
501–	·020	17	172·1	29,618	2,614	3·0
	·999	135	17·03	1,249·3	329·8	10·1
301–500	·811	43	76·6	5,868	1,703	6·1
501–	·189	17	172·1	29,618	2,614	23·5
	1·000	60	94·65	10,357	1,875	29·6
301–	·106	60	94·6	8,959	3,243	5·1
	·999	135	17·03	1,102·0	474·8	9·1

The sums of the products of h_i with $\bar{y}_i$, $\bar{y}_i^2$ and s_i^2 are shown at the foot of their respective columns. We therefore have, since $17·03^2 = 290·0$,

$$s_1^2 = 329·8$$
$$s^2 = 329·8 + 1249·3 - 290·0 - 10·1 = 1279·0.$$

Hence the relative precision is $1279·0/329·8 = 3·88$. It will be seen that the corrections in the last column are here trivial, and could well be omitted.

Size-groups 301–500 and 501– can be combined in the manner shown in the second part of Table 8.3.c. We have, for these two size-groups combined,

$$s^2 = 1875 + 10,357 - 8959 - 30 = 3243$$

We can now insert a fresh line in Table 8.3.c to replace the lines for the last two size-groups in the first part of the table. The previous computation is then repeated, giving

$$s_1^2 = 474·8$$
$$s^2 = 474·8 + 1102·0 - 290·0 - 9·1 = 1277·7$$

Hence the relative precision is $2·69$.

The amalgamation of the two size-groups containing the largest farms has resulted in a considerable loss of precision, the relative precision being $329·8/474·8 = 0·69$.

8.4 Multiple stratification

The gain in precision due to sub-stratification of a sample which is already stratified into main strata can be estimated by methods similar to that of Section 8.3. An example has already been given in Example 8.3.c, where the gain in precision resulting from the subdivision of the size-group 301– into two groups, 301–500 and 501– was determined.

If the data are derived from a sample with uniform sampling fraction which is itself sub-stratified, the comparisons can be made directly between the relevant mean squares in the analysis of variance, as in Method (a). The structure of the analysis of variance in this case is

Whole sample (s^2) $\begin{cases} \text{Between main strata} \\ \text{Within main strata } (s_1{}^2) \begin{cases} \text{Between sub-strata} \\ \text{Within sub-strata } (s_2{}^2) \end{cases} \end{cases}$

The ratio of the mean squares $s_1{}^2$ and $s_2{}^2$ within main strata and within sub-strata will give the required relative precision.

A similar analysis can be constructed from data derived from other types of sample with uniform sampling fraction (Method (b)).

One case of practical importance is that in which both the main and sub-strata are arbitrary subdivisions of an area, all the main and all the sub-strata being of equal size. If there are t' main strata, and t'' sub-strata per main stratum, with k selected sampling units per sub-stratum, the analysis of variance will be of the form shown in Table 8.4.

TABLE 8.4—STRUCTURE OF THE ANALYSIS OF VARIANCE IN A
DOUBLE STRATIFICATION

	Degrees of freedom	Mean square
Between main strata	$t' - 1$	A
Within main strata $\begin{cases} \text{Between sub-strata} \\ \text{Within sub-strata} \\ \text{Total} \end{cases}$	$t'(t'' - 1)$ $t' t''(k - 1)$ $t'(t'' k - 1)$	D $E = s_2{}^2$ $B = s_1{}^2$
Total for sample	$t' t'' k - 1$	$C = s^2$

If k is small the bias in the estimate $s_1{}^2$ provided by the within-strata mean square may be appreciable. This bias can be almost completely eliminated by using the formula

$$s_1{}^2 = \{(t'' k - 1) B + E\}/t'' k$$

which is derived directly from formula 8.3.

8.5 Stratified sample with variable sampling fraction

In the notation of Section 8.3, we have

$$N \, V (\bar{y}) = \Sigma \, s_i{}^2 \, h_i \, (1 - f_i)/f_i$$

The n_i or f_i required for a given accuracy can only be determined uniquely from this equation if the relations between the different f_i have been decided.

It has already been pointed out (Section 3.5) that for maximum accuracy the f_i should be proportional to σ_i, but that in many types of material stratified by size-groups the f_i may be taken proportional to the mean sizes of the size-groups. If we put $f_i = c\,\lambda_i$, where the λ_i are in the required proportions, the above equation can be written

$$\frac{1}{c} \Sigma\,(s_i^2\,h_i/\lambda_i) = \mathbf{N}\,\mathbf{V}\,(\bar{y}) + \Sigma\,s_i^2\,h_i$$

The value of c for any required accuracy can then be calculated. If, however, a value of c is obtained which makes some of the f_i greater than 1 the calculation must be repeated, omitting the terms for these strata from both sides of the equation.

Alternatively the direct expression for $\mathbf{V}\,(\bar{y})$ can be used and the value of c found by trial. This has the advantage that the effect of adjustments of the final sampling fractions to simple fractions is immediately apparent.

The relative precision of stratified samples with variable and with uniform sampling fractions can be obtained by calculating $\mathbf{V}\,(\bar{y})$ for both samples. It should be noted that if the f_i have been taken proportional to the s_i a slight over-estimate of the relative precision will be obtained, owing to errors in the s_i. This point has been discussed by Sukhatme (1935, A), but is not of great importance in practice.

It will be seen that for these calculations we only require sufficiently accurate estimates of the variances within strata and the proportions of units of the population in the different strata. The procedure is therefore the same whether or not the sample from which the data are obtained is stratified. All that is required is that all strata should be adequately represented.

Example 8.5.a

From the data of Table 8.3.c determine the size of sample required to give a standard error of $\pm\,1500$ acres in the estimate of wheat acreage, when sampling fractions proportional to those of Table 3.7.a are used.

The λ_i can be taken equal to the sampling fractions of Table 3.7.a. Tabulating $s_i^2\,h_i$ and $s_i^2\,h_i/\lambda_i$, we find

$$\Sigma\,(s_i^2\,h_i/\lambda_i) = 2913 \qquad \Sigma\,s_i^2\,h_i = 329{\cdot}77$$

Also $\mathbf{N}\,\mathbf{V}\,(\bar{y}) = \mathbf{V}\,(\mathbf{Y})/\mathbf{N} = 1500^2/2496 = 901{\cdot}44$, and hence

$$c = 2913/\{901{\cdot}44 + 329{\cdot}77\} = 2{\cdot}37$$

The total number required in the sample is therefore $135 \times 2{\cdot}37 = 320$, the number in the largest size-group, for example, being $51 \times 2{\cdot}37/3 = 40$. No sampling fraction is greater than 1, and therefore no further computation is required.

In practice the new sampling fractions may well be rounded off, taking, for example, all of the largest size-group, $\frac{1}{2}$ of the next, etc.

The direct approach illustrates the way in which results of this kind can readily be obtained by interpolation. A first approximation to the number required is given by $135 \times 2550^2/1500^2 = 390$. The standard error corresponding to this sample number, obtained by the ordinary methods, is ± 1302. The squares of the reciprocals of this and of the original standard errors can be plotted against the respective numbers in the samples and a smooth curve drawn through these two points and the origin. This curve gives the general relation between sample size and accuracy, and will be found to give a sample number corresponding to a standard error of ± 1500 of approximately 320.

Example 8.5.b

Determine the relative precision of the sample of Table 3.7.a and the sample with sampling fractions proportional to s_i containing the same number of farms.

The standard error, using these sampling fractions, can be calculated in the ordinary manner, and is found to be ± 2420. The relative precision is therefore $2420^2/2550^2 = 0\cdot90$. There is consequently an apparent loss of precision of approximately 10 per cent., but the real loss is likely to be less than this, owing to errors in the estimates of the standard errors.

This apparent loss refers to a single variate, acreage of wheat. If, for instance, the acreage of some other crop were taken, the σ_i would be different and the sampling fractions required to give minimum variance would therefore also be different. Consequently, if several variates have to be determined, a compromise will in any case be required.

8.5.1 Relative precision of different methods of sampling when domains of study cut across strata

The relative precision of various sampling methods when domains of study cut across strata can be studied by the methods already outlined in this chapter. All that is necessary in any particular case is to estimate and compare the expected sampling errors when different methods are used.

From the results already given it will be apparent that with a uniform sampling fraction the gain in accuracy which results from stratification is largely lost for domains of study that cut across strata. It is therefore important where practicable to use strata which correspond to the expected domains of study. This, however, is not always possible, either because of the resulting increase in complexity or because the information necessary for classification of the sampling units into appropriate domains of study is only obtained in the course of the survey. In the National Farm Survey described in Section 5.17, for example, it would have been impossible to stratify for all the various domains of study into which the data were subsequently broken down. Information on such items as type of occupancy was not known in advance (indeed, collection of this information was one of the objects of the survey) ; but even had it been available the number of different types of domain, all of which cut across one another, was so great that the number of sub-classes thereby created would have been far too numerous to be used as strata.

When a variable sampling fraction is used the situation is somewhat different. Although there is likely to be a large increase in variance when domains of study cut across strata, there will still be substantial gains from the use of a variable sampling fraction in place of a uniform sampling fraction. The optimal values of the sampling fractions will, however, differ from those which are optimal for the population estimates and will indeed depend on what quantities—numbers, totals, means or proportions—require estimation. The sampling will be optimal for population estimates of means or totals when the sampling fractions are such that the values of $f_i \sqrt{c_i}$ are proportional to σ_i (Section 8.17 (a)). For the estimation of a mean of a particular domain the sampling will be approximately optimal when $f_i \sqrt{c_i}$ is proportional (apart from errors of estimation) to the square root of $1/(n_i - 1)$ times the quantity in curly brackets in the formula for $V(\bar{y}_a)$ of Section 7.6.1. Replacing $n_{ia} - 1$ by n_{ia} and $n_i - 1$ by n_i gives $f_i \sqrt{c_i}$ approximately proportional to the square root of

$$p_{ia} \{ q_{ia} (\bar{y}_{ia} - \bar{y}_a)^2 + s_{ia}{}^2 \} \qquad (8.5.1.a)$$

For the corresponding total $f_i \sqrt{c_i}$ must similarly be approximately proportional to the square root of

$$p_{ia} \{ q_{ia} \bar{y}_{ia}{}^2 + s_{ia}{}^2 \} \qquad (8.5.1.b)$$

The expression for a ratio is similar.

If the strata consist of size-groups of the variate y, or of a variate x highly correlated with y, and if we take $f_i \sqrt{c_i}$ proportional to $\bar{y}_i$ or $\bar{x}_i$ for the different size-groups, allocation will be about optimal for the estimation of the totals of domains cutting across strata, and for the proportions Y_a/Y, especially in those cases (which are of frequent occurrence) in which s_{ia} is about proportional to the $\bar{y}_i$ or $\bar{x}_i$ of the size-group. In the estimation of the means of different domains, however, a greater proportion will require to be taken from the extreme size-groups, small as well as large. For estimation of the number of units in the different domains sampling will be optimal when $f_i \sqrt{c_i}$ all have approximately the same value. The best balance between these conflicting requirements will usually be attained by increasing somewhat the sampling fractions for the size-groups with small y. This is in fact what was done in the National Farm Survey.

Example 8.5.1

Using the data of Example 7.6.1.b, determine the relative precision of (a) the sample of that example, (b) a stratified random sample with uniform sampling fraction, and (c) a fully random sample, with regard to numbers of farms in the different domains, and their mean and total acreages.

We will here outline the calculations for $V(Y_a)$. It is best, particularly if relative efficiencies require to be evaluated, to introduce the simplification used in arriving at expression 8.5.1.b. We then have

$$V(Y_a) = \Sigma(g_i - 1) N_{ia}(q_{ia}\bar{y}_{ia}^2 + s_{ia}^2) \qquad (8.5.1.c)$$

The values of N_{ia} can be obtained from Table 7.6.1.a by applying the relevant raising factors. The quantities $q_{ia}\bar{y}_{ia}^2 + s_{ia}^2$ can be calculated from Table 7.6.1.b. Estimates of the proportions h_{ia}, etc., of farms falling in the different size-groups are also required for each domain. All these quantities are tabulated in Table 8.5.1.a.

TABLE 8.5.1.a—VALUES OF N_{ia}, ETC.

Size-group	N_{ia}	h_{ia}	$q_{ia}\bar{y}_{ia}^2 + s_{ia}^2$
2	720	0·677966	2041·6
3	316	0·297552	18346·9
4	26	0·0244821	124205·1
	1062	1·000000	

For a stratified sample with uniform sampling fraction containing the same total number of farms $g = 3296/614 = 5\cdot36808$. $V(Y_a)$ is given by formula 8.5.1.c with all g_i equal to g. Thus by summing the products of the second and fourth columns of Table 8.5.1.a, and multiplying by $g - 1$, we obtain

$$V(Y_a) = 4\cdot36808 \times 10497000 = 45850000.$$

For the random sample the estimated variance s_a^2 within domain A over all size-groups is required. The method of Section 8.3 (c) must be followed. The various terms in the formula for s_a^2 (formula for s^2 of Section 8.3.c) can be determined from the values already given in Table 7.6.1.b, and the values of h_{ia} above. To avoid having to calculate $\bar{y}_{ia}$ and $\bar{y}_a$ to more decimal places than are given in Table 7.6.1.b, $\Sigma h_{ia}\bar{y}_{ia}^2 - \bar{y}_a^2$ may be replaced by

$$\Sigma h_{ia}(\bar{y}_{ia} - \bar{y}_0)^2 - (\bar{y}_a - \bar{y}_0)^2$$

where y_0 is a working mean, say 100, near $\bar{y}_a$. This comes to 3920·50. We also find $s_{ia}^2 = 953\cdot62$ and $\Sigma h_{ia}(1 - h_{ia}) s_{ia}^2/n_{ia} = 11\cdot52$. Note that here the original n_{ia} are to be used. Hence

$$s_a^2 = 953\cdot62 + 3920\cdot50 - 11\cdot52 = 4862\cdot60.$$

The estimated value of q_a is $1 - 1062/3296 = 0\cdot677791$. Hence

$$q_a\bar{y}_a^2 + s_a^2 = 10069\cdot74,$$

and thus from formula 7.5.1.p,

$$V(Y_a) = N_a(g - 1)(q_a\bar{y}_a^2 + s_a^2) = 1062 \times 4\cdot36808 \times 10069\cdot74 = 46710000.$$

For total acreage of domain A, therefore, the precision of the stratified random sample with uniform sampling fraction relative to that with the variable sampling fraction (Table 7.6.1.c) is $3394/4585 = 74\cdot0$ per cent. The precision of a fully random sample relative to a stratified sample with uniform sampling fraction is $4585/4671 = 98 \cdot 2$ per cent.   There is, therefore, considerable gain by the use of the variable sampling fraction but very little gain by stratification.

The full results for number of farms, total acreage and mean acreage are shown in Table 8.5.1.b.   For number of farms the stratified sample with uniform

TABLE 8.5.1.b—RELATIVE PRECISION (PER CENT. OF DIFFERENT TYPES OF SAMPLE

		A	*B*	*C*	Whole county
u.s.f./v.s.f.	No. of farms .	152	136	132	—
	Total acreage .	74	65	62	84
	Mean acreage .	77	101	95	84
Random/u.s.f.	No. of farms .	95	98	99	—
	Total acreage .	98	71	88	21
	Mean acreage .	82	60	80	21

sampling fraction is on the average about 40 per cent. more precise than the sample with variable sampling fraction.   For total acreage, on the other hand, the sample with variable sampling fraction is about 50 per cent. more precise than the sample with uniform sampling fraction, the gain in precision being greater for the separate domains than for the whole county.*   For the mean acreage the relative precision for the different domains is very variable.   There is a gain by use of a variable sampling fraction for domain *A* but not for domains *B* and *C*.   The random sample is always less precise than the stratified sample with uniform sampling fraction, but the gain due to stratification is very variable for the different measures and different domains.   These results are, of course, what would be expected from the nature of the variances.

The relative efficiencies will be somewhat nearer unity than the relative precisions.   A simple method of calculating them is described below.

8.5.2 A simple relation between relative precision and relative efficiency

It is worth noting that when two sampling methods are being compared for efficiency and one of them is random or stratified with uniform sampling fraction, there is a simple relation between the relative precision and relative efficiency of the two methods.   Denote the two methods by S_1 and S_2, S_2 being random or stratified with uniform sampling fraction.   Let the relative precision (S_2/S_1) of the two methods for a given sample size be P, and the relative efficiency when S_1 is of the given sample size be E.   Then the relation in question is

$$E = \bar{f}_1 + (1 - \bar{f}_1)\, P$$

* This is less than the relative efficiency reported in Section 5.17 because the smallest size-group has been omitted and there are no farms in the largest size-group in this county.

where f_1 is the overall or average sampling fraction of S_1, i.e. the number of units in the sample divided by the number in the population. From this relation the relative efficiency can be obtained immediately from the relative precision and vice versa. The specification of the size of S_1 is necessary because both P and E will vary with variations in this size.

The above formula depends for its derivation on the fact that for a random or stratified sample with uniform sampling fraction the variance of an estimate is of the form $\lambda \left(\dfrac{1}{n} - \dfrac{1}{N} \right)$, where λ is independent of n. It is therefore applicable to the case in which domains of study cut across strata, provided we admit the approximation used in Example 8.5.1. Consequently the relative efficiencies of the sampling methods of that example can be calculated directly from Table 8.5.1.b.

Example 8.5.2

Calculate the relative efficiencies of the sampling methods of Example 8.5.1 from the relative precisions given in Table 8.5.1.b.

We have $\overline{f}_1 = 614/3296 = 0.186$. For total acreage of domain B, for example, the efficiency of a stratified sample with a uniform sampling fraction relative to the sample with variable sampling fraction is $0.186 + 0.814 \times 0.65 = 0.72$. The precision of a random sample relative to the sample with variable sampling fraction is $0.65 \times 0.71 = 0.46$, and the corresponding relative efficiency is therefore 0.56.

8.6 Supplementary information

The determination of the number of units required in a sample when supplementary information is available presents no essentially new problems. It has been shown in Chapter 7 that apart from the substitution of s_q^2 or s_l^2 for s^2 the formulæ for the variances of estimates based on supplementary information differ little from those for estimates from similar samples without supplementary information. Consequently it will usually be sufficient to estimate the appropriate variance by the methods given in Chapter 7, using this variance instead of the ordinary variance per unit to determine the size of sample. The factor $\bar{x}^2/\bar{x}^2$ in the variance of the ratio estimate differs from unity only because of sampling fluctuations in $\bar{x}$, and can be omitted.

When the ratio method is to be used and $V_x(r)$ is virtually constant for all x, it will often be advantageous to estimate this variance rather than s_q^2. This will generally lead to somewhat simpler and more straightforward computations. Any slight bias introduced into the estimates of error will be of little consequence, since it will merely result in a slightly larger or smaller sample being taken.

We frequently require an estimate of the gain in precision due to the use of supplementary information. This is needed in planning a sample survey when a decision has to be reached whether supplementary observations should be taken. It is also required in the planning of the computations in order to decide whether the utilization of available supplementary information is worth the additional computational labour.

In the case of the regression method the relative precision is very simply calculated, since it depends only on the value of the correlation coefficient r, being in fact

$$\frac{1}{1 - r^2}$$

In calculating r due regard must be had to any restrictions imposed by stratification, the same sums of squares and products being used as in the calculation of the regression coefficient and the residual error. The above expression is approximate in that the reduction by 1 of the error degrees of freedom with the regression has been ignored, but this correction will be small relative to errors in the estimation of r.

If an arbitrary value b_0 of the regression coefficient is used the relative precision will be

$$\frac{1}{1 - r^2 + r^2 (1 - b_0/b)^2}$$

The corresponding expression for the ratio method is obtained by writing $\bar{r}$ for b_0.

Example 8.6.a

From the data of Example 7.17 calculate (a) the number of farms required to give an unbiased estimate of the mean dressing of nitrogen per acre over the farms of the county with a standard error of ± 0.05 cwt., and (b) the number of farms required in each of two equal groups so that the comparison based on the unweighted means of the dressings per acre of the two groups has a standard error of ± 0.05 cwt.

(a) The required number is $67 \times 0.0392^2/0.05^2 = 41$. The correction for finite sampling is trivial in this example. Note that either s_q^2 or s_r^2 can be used to arrive at this result.

(b) If the required number in each group is n, the variance of the difference of the means is $2 s_r^2/n$. Hence $n = 2 \times 0.0541/0.05^2 = 43$.

Example 8.6.b

Obtain the expressions for the relative efficiencies given in Example 7.12.b from the above formulæ.

We have $b = 0.6327$, $r = 52,069/\sqrt{(115,266 \times 82,296)} = 0.537$ and $\bar{r} = 147.36/132.08 = 1.116$. Hence the relative precision of the regression method, compared with the sample plots only, is $1/(1 - 0.537^2) = 1.40$. Similarly the use of differences ($b_0 = 1$) gives a relative precision of $1/\{1 - 0.537^2 + 0.537^2 (1 - 1/0.6327)^2\} = 1.23$, the ratio method ($b_0 = 1.116$) gives a value of 1.14, and the regression ($b = 0.55$) gives a value of 1.39. These correspond to the relative efficiencies already tabulated except in the case of the regression, for which we have here neglected the correction for degrees of freedom.

Example 8.6.c

Determine the gains in precision in the estimation of wheat acreages from the random sample of Hertfordshire farms due to the use of supplementary information on acreages of crops and grass, (*a*) using the ratio method, and (*b*) using the regression method, without taking account of districts.

The standard errors, already obtained, are $\pm$ 7950 for direct estimation without the use of supplementary information (Example 7.2.b), $\pm$ 3940 for the ratio method (Example 7.8.a), and $\pm$ 4126 for the regression method (Example 7.12.a). The apparent gain in precision due to the ratio method is therefore $7950^2/3940^2 = 4\cdot07$, and that due to the regression method is $7950^2/4126^2 = 3\cdot71$.

The value for the regression appears anomalous, since the formulæ given above indicate that regression may be expected to be at least as efficient (apart from the change in degrees of freedom) as the ratio method. The discrepancy is due to the inclusion of the factor $\bar{x}^2/\bar{x}^2$ in the variance of the ratio estimate. Using the above formulæ with $r = 0\cdot8555$, $b = 0\cdot1932$, $\bar{r} = 0\cdot1522$, we find that the relative precision, compared with direct estimation, is 3·73 for the regression method, and 3·32 for the ratio method. An alternative estimate of the relative precision of the regression and the ratio methods is therefore $3\cdot73/3\cdot32 = 1\cdot12$. This latter value gives a better indication of the average value of the relative precision of the two methods.

8.6.1 Utilization of supplementary information on two or more variates

If supplementary information is available, the way in which it can be used most effectively should be investigated. As was pointed out in Section 6.8, if supplementary information on two or more variates is available, functions of these can sometimes give much greater increases in precision than is obtained by the standard ratio or regression method which utilizes supplementary information on one variate only.

A development of this type, which is termed the *double-ratio estimate*, was devised by N. Keyfitz* for the estimation of the total labour force, wages, salaries, materials used, etc., in the case in which there is an initial complete census of production (denoted here by x_1), and of the labour force (denoted by y_1), etc., and subsequently a further complete census of production, x_2, but a sample only for labour force, y_2, etc. In this case it may be anticipated that the production per worker in a factory, though it may vary from factory to

---

* I am indebted to Dr. Keyfitz for permission to publish an account of this method, which he first described in a lecture given at the London School of Economics.

factory, and from period to period in the same factory, will increase or decrease in much the same ratio from period to period for all factories.

The required estimate of the total labour force on the second occasion is, in the case of a random sample,

$$Y_2 = \frac{S(y_2)}{S(x_2)} \cdot \frac{S(x_1)}{S(y_1)} \cdot \frac{Y_1}{X_1} X_2$$

The variance of Y_2 is given by the approximate formula

$$V(Y_2) = \frac{Y_2{}^2}{n} s_t{}^2$$

where

$$s_t{}^2 = \frac{1}{n-1} S \left(\frac{y_2}{\bar{y}_2} - \frac{x_2}{\bar{x}_2} - \frac{y_1}{\bar{y}_1} + \frac{x_1}{\bar{x}_1} \right)^2$$

The parallelism with the formulæ for the single-ratio estimate already given will be apparent. The extension to the case of stratified sampling is similar to that followed in the case of the single-ratio estimate.

In the case of the labour-force estimate Keyfitz found that the squares of the coefficients of variation, i.e. (S.E. of estimate/estimate)², for the double-ratio estimate and the two possible single-ratio estimates were as follows :—

Double-ratio estimate   ...   . $c^2 = 0 \cdot 0012$

Single-ratio estimate, X_2 $\dfrac{S(y_2)}{S(x_2)}$ ...   ... $c^2 = 0 \cdot 040$

Single-ratio estimate, Y_1 $\dfrac{S(y_2)}{S(y_1)}$ ...   ... $c^2 = 0 \cdot 303$

The double-ratio estimate is therefore in this case much more accurate than either of the single-ratio estimates.

8.7 Two-phase sampling

The only case which presents any new features is that in which the first-phase information is used as supplementary information to improve the accuracy of estimates of the second-phase variate y. It has already been pointed out in Section 7.8 that the variance of a two-phase sample is in this case made up of two parts A and B, where

A = variance due to the first-phase sampling, i.e. the variance which would be obtained if y were determined for all the units of the first-phase sample,

B = variance due to the second-phase sampling of the first-phase sample (regarded as without error).

To determine B the methods given in Chapter 7 for supplementary information are followed, the effective sampling fraction being n_2/n_1. To determine A we must use the methods given in the present chapter for the evaluation of the error of a sample of one size and type from the data of a sample of a different size and possibly different type. Thus, if the first-phase sampling is random, and the second-phase sampling is stratified with a variable sampling fraction, it is necessary to calculate the variance of an unstratified random sample of n_1 units from the data of a stratified sample with variable sampling fraction of n_2 units.

Once A and B have been determined the calculation of the relative precision of different possible sampling methods presents no difficulty. If, for example, we wish to ascertain the increase in precision due to taking a two-phase sample of n_1 and n_2 units instead of a single-phase sample of n_2 units, we calculate what the variance A' of a sample of n_2 units would be if the first-phase sampling procedure were followed for a sample of n_2 units. This calculation will follow the same lines as that of A. The relative precision is then $A'/(A + B)$. Similarly the relative precision resulting from the ascertainment of the second-phase information on the n_2 second-phase units only, instead of on all the n_1 units of the sample, will be $A/(A + B)$.

In the simple but general case in which the population is large, and the methods of sampling and estimation are such that the variances of the estimates at each phase are inversely proportional to the numbers of units, apart from the factor $1 - n_2/n_1$, the above relative precisions are capable of simple expression. If the effective variances per unit are $s_1{}^2$ and $s_2{}^2$, with $s_2/s_1 = \kappa$ and $n_2/n_1 = \lambda$, we have

$$A = \frac{1}{n_1} s_1{}^2 \qquad A' = \frac{1}{n_2} s_1{}^2 \qquad B = \frac{1}{n_2} \left(1 - \frac{n_2}{n_1} \right) s_2{}^2$$

Consequently the relative precision giving the gain due to the inclusion of the additional first-phase units is

$$\frac{A'}{A + B} = \frac{1}{(1 - \lambda)\,\kappa^2 + \lambda}$$

Similarly the loss by not ascertaining the second-phase information over all the first-phase units is given by the relative precision

$$\frac{A}{A + B} = \frac{\lambda}{(1 - \lambda)\,\kappa^2 + \lambda}$$

Representative values of these fractions are given in Table 8.7. If, for example, the effective standard error per unit is halved by the use of the first-phase supplementary information, $\kappa^2 = \frac{1}{4}$. Consequently, if we introduce two-phase sampling and quadruple the size of the sample for first-phase information only, instead of using single-phase sampling, the amount of information derived from a second-phase unit is increased by a factor of 2·29. Similarly by collecting second-phase information on only $\frac{1}{4}$ of the first-phase units instead of all the units the amount of information is reduced by a factor of 0·57.

TABLE 8.7—RELATIVE PRECISION OF TWO-PHASE AND SINGLE-PHASE SAMPLING

Two-phase sample :— Single-phase sample :—	n_1 and n_2 n_2			n_1 and n_2 n_1		
κ^2	$\frac{1}{2}$	$\frac{1}{4}$	$\frac{1}{3}$	$\frac{1}{2}$	$\frac{1}{4}$	$\frac{1}{3}$
$\lambda = \frac{1}{2}$	1·33	1·6	1·78	0·67	0·8	0·89
$\lambda = \frac{1}{4}$	1·6	2·29	2·91	0·4	0·57	0·73
$\lambda = \frac{1}{3}$	1·78	2·91	4·27	0·22	0·36	0·53

8.8   Sampling on successive occasions

The relative efficiency of the various estimates can be calculated from the variances given in Section 7.19. When the variances on the different occasions are the same, the relative efficiency of the various estimates, under the conditions set out in Section 6.22, depends only on μ and the correlation r between the successive occasions.

Table 8.8.a gives the efficiencies, relative to those of the overall mean, of the adjusted estimates of the mean on the last occasion (a) when there is a sub-sample on the second occasion, and (b) with partial replacement, the latter being given for both two and a large number of occasions. Values for $\mu = \frac{1}{2}$ and $\mu = \frac{1}{3}$, and for various values of r, are given. With independent samples or a fixed sample the overall means are fully efficient.

TABLE 8.8.a—SAMPLING ON SUCCESSIVE OCCASIONS : EFFICIENCY, RELATIVE TO
THE OVERALL MEAN, OF THE ADJUSTED ESTIMATES OF THE MEAN ON THE LAST
OCCASION

r	$\mu = \frac{1}{2}$			$\mu = \frac{1}{3}$		
	Sub-sample	Partial replacement		Sub-sample	Partial replacement	
		Two occasions	Large number		Two occasions	Large number
0	1·00	1·00	1·00	1·00	1·00	1·00
·25	1·03	1·02	1·02	1·02	1·02	1·03
·5	1·14	1·07	1·08	1·09	1·06	1·07
·6	1·22	1·11	1·12	1·14	1·09	1·11
·7	1·32	1·16	1·20	1·20	1·13	1·18
·8	1·47	1·24	1·33	1·27	1·18	1·30
·9	1·68	1·34	1·65	1·37	1·25	1·59
·95	1·82	1·41	2·10	1·43	1·29	2·02
1·0	2·00	1·50	Inf.	1·50	1·33	Inf.

TABLE 8.8.b—SAMPLING ON SUCCESSIVE OCCASIONS : EFFICIENCY, RELATIVE TO THE DIFFERENCE OF THE OVERALL MEANS, OR TO INDEPENDENT SAMPLES (VALUES IN BRACKETS), OF ALTERNATIVE ESTIMATES OF CHANGE

r	$\mu = \frac{1}{2}$		$\mu = \frac{1}{3}$		Fixed sample
	$\bar{y}_h - \bar{y}_{h-1}$	From last two occasions	$\bar{y}_h - \bar{y}_{h-1}$	From last two occasions	
0	1·00 (1·00)	1·00 (1·00)	1·00 (1·00)	1·00 (1·00)	(1·00)
·25	1·02 (1·16)	1·02 (1·17)	1·02 (1·22)	1·02 (1·22)	(1·33)
·5	1·10 (1·47)	1·12 (1·50)	1·09 (1·63)	1·11 (1·67)	(2·00)
·6	1·18 (1·69)	1·22 (1·75)	1·15 (1·92)	1·20 (2·00)	(2·50)
·7	1·32 (2·03)	1·41 (2·17)	1·27 (2·37)	1·36 (2·56)	(3·33)
·8	1·60 (2·67)	1·80 (3·00)	1·50 (3·22)	1·71 (3·67)	(5·00)
·9	2·43 (4·41)	3·02 (5·50)	2·21 (5·53)	2·80 (7·00)	(10·00)
·95	3·99 (7·61)	5·51 (10·50)	3·58 (9·76)	5·01 (13·67)	(20·00)

The increase in precision due to the use of partial replacement instead of independent samples or a fixed sample can also be obtained from Table 8.8.a. Thus with a correlation of 0·8 replacement of half the units gives a 24 per cent. increase in precision on the second occasion and a 33 per cent. increase after a number of occasions. With one-third replacement the corresponding percentages are 18 and 30.

Table 8.8.b gives similar efficiencies, relative to the differences of the overall means, or to independent samples (values in brackets), of the estimates of change given by $\bar{y}_h - \bar{y}_{h-1}$ and by the weighted estimate based on the last two occasions only (formula 6.21.b).

In the estimation of change the difference between the overall means of two independent samples is less accurate than the difference of the overall means of a sample with partial replacement. This in its turn is less accurate than the difference between the means of a fixed sample. Thus with a correlation of 0·8 the weighted estimate from the last two occasions, with replacement of half the units, is 3·00 times as efficient as the difference of the means of two independent samples, but only 1·80 times as efficient as the difference of the overall means of the replacement sample. A repeated sample under these circumstances is 5·00 times as precise as a pair of independent samples.

It will be noted that the estimate of change derived from the last two occasions is always somewhat more accurate than the estimate $\bar{y}_h - \bar{y}_{h-1}$. With a correlation of 0·8, for instance, there is a gain in efficiency of 12 per cent. when $\mu = \frac{1}{2}$ and of 14 per cent. when $\mu = \frac{1}{3}$.

Example 8.8

Estimate the relative efficiency of the various estimates of Examples 6.21 and 6.22.

In Example 6.21, $r = 0.847$ and $\mu = \frac{1}{3}$. Consequently, from Table 8.8.a, the relative efficiency of $\bar{y}_w$ and $\bar{y}$ is 1·21. From Table 8.8.b the relative efficiency of the weighted estimate of change and the difference of the overall means is about 2·1. The relative efficiency of the difference of the means of the units common to both occasions and the weighted estimate is given by the weight of the former, namely, 0·929.

The relative efficiency of the estimates of Example 6.22 cannot easily be determined exactly, owing to the variation in the numbers of units from occasion to occasion. With the average value of μ of $\frac{1}{3}$ and a correlation of 0·811, the efficiency of $\bar{y}_h$ relative to the overall mean after a number of occasions will be 1·32 (Table 8.8.a) and that of the estimate $\bar{y}_h - \bar{y}_{h-1}$ of change relative to the difference of the overall means will be about 1·6 (Table 8.8.b).

8.9 Sampling with probability proportional to size of unit

The relative precision of sampling with uniform probability and with probability proportional to size of unit depends on the variance laws to which the material is subject. The case in which the mean r for fixed x is the same for all values of x, and in which the variance of r for fixed x is a function of x, may first be considered.

If the total size of all units is known, we shall be concerned with estimates of $\bar{r}$. If we put $V(x)/\bar{x}^2 = \gamma$ we have the results shown in Table 8.9.a for the three variance laws there given, v being a constant.

TABLE 8.9.a—VARIANCES OF $\bar{r}$

Variance of r for fixed x	Variance of $\bar{r}$	
	Uniform probability	Probability proportional to x
v	$v(1 + \gamma)/n$	v/n
v/x	$v/n\bar{x}$	$v/n\bar{x}$
v/x^2	$v/n\bar{x}^2$	$v(1 + \gamma)/n\bar{x}^2$

In sampling for yield per acre in a crop estimation scheme, for example, the variance of the yield per acre may be expected to be about the same for large and small fields. If in addition there is no marked difference between the mean yields per acre of small and large fields, the precision of sampling with probability proportional to size relative to sampling with uniform probability will be $1 + \gamma$.

If the mean r for fixed x varies with x the variances of Table 8.9.a will be increased, and the precision of either method, or the relative precision of the two methods, may best be judged by direct analysis of actual data.

If the acreage of the crop has to be determined by the sampling of fields, the relative precision of sampling with probability proportional to size, and with uniform probability, will also depend on the variance of the acreages. The simplest case is that in which the sampling is used to determine which of the fields carry the given crop, and in which the values of $\bar{x}$ and $V(x)$ are the same for the fields of the given crop and for the remaining fields, the number of fields being large. The variance of the proportion p of the total area under the given crop when n' fields are taken is in this case pq/n' with sampling with probability proportional to size, and $pq(1+\gamma)/n'$ with uniform probability. The relative precision is therefore $1+\gamma$.

In the case of sampling with probability proportional to size, point sampling will often be used. If the part of the land area which consists of fields cannot be recognized on the map, additional points will have to be visited on the ground, and these must be allowed for in assessing the total number of points required.

In the more complicated cases of sampling with probability proportional to size the same general approach as that adopted in the previous sections must be followed, using the data provided by an actual sample to determine the relevant variances. If the basic data are derived from a sample taken with probability proportional to size, s_r^2 can be calculated from the formulæ of Sections 7.15 or 7.16. The value so obtained may then be used to deduce the size of sample required for a given accuracy.

If the basic data are derived from a sample taken with uniform probability of selection, or if data relating to the whole population are available, the various sizes of unit will occur in proportions which are different from those of a sample taken with probability proportional to size of unit. Consequently a different formula is required for the calculation of s_r^2. The appropriate formula for a random sample is

$$s_r^2 = \frac{1}{(n-1)\,\bar{x}}\left[S\left(\frac{y^2}{x}\right) - \frac{\{S(y)\}^2}{S(x)}\right] = \frac{1}{(n-1)\,\bar{x}}\left[S(ry) - \bar{r}_u\,S(y)\right]$$

where $\bar{r}_u = S(y)/S(x)$. If the individual values of y and r are tabulated the second form of the expression is most convenient for computation.

In the case of a stratified sample the expression within the square brackets must be evaluated for each stratum separately. If the number in each stratum is small and there is no great difference between the $\bar{x}_i$, the separate components can then be aggregated and divided by $(n-t)\,\bar{x}$. If there are considerable differences between the $\bar{x}_i$ it is best to calculate s_{ri}^2 separately for each stratum, using the separate values in the calculation of $V(Y)$.

Example 8.9.a

From the data of Table 6.19.a construct a frequency distribution of the acreages of sugar-beet fields on old arable land in Norfolk, and hence calculate the relative precision of estimates of the mean yield per acre derived from a random sample of fields taken (*a*) with probability proportional to size and (*b*) with uniform probability, on the assumption that the variability of the yield per acre is the same for all sizes of field.

In constructing the frequency distribution, account must be taken of the variable sampling fractions at the two stages of sampling. Since the raising factors at the first stage are nearly proportional to 7, 4, 2, the fields on the small, medium and large farms with a single field of sugar-beet must be counted 7, 4 and 2 times respectively. Similarly a field occurring on a farm with 2 fields of sugar beet must be counted 14, 8 or 4 times, etc.

TABLE 8.9.b—FREQUENCY DISTRIBUTION OF THE ACREAGES OF

SUGAR-BEET FIELDS

Acreage	Raised No. of fields	Acreage	Raised No. of fields
2	92	12	4
3	43	13	8
4	117	—	
5	48	20	8
6	54	—	
7	63	24	6
8	42	—	
9	4	29	12
10	60	—	
11	24	48	2
			587

This procedure gives the frequency distribution shown in Table 8.9.b. Following the method of Example 7.1.a for grouped data (the acreages being taken as the working units), we find

$$\bar{x} = 6\cdot681, \qquad s^2 = V(x) = 30\cdot47, \qquad \gamma = 30\cdot47/6\cdot681^2 = 0\cdot681$$

Consequently the relative precision of methods (*a*) and (*b*) is 1·68.

Example 8.9.b

From the data of the sample of Hertfordshire parishes taken with uniform probability (Sample *A* of Section 3.11) estimate the value of s_r^2 for a sample of parishes, stratified by districts, taken with probability proportional to size. Make a similar estimate from the data for all 91 combined parishes.

The data for the 91 combined parishes are shown in Table 8.9.c, the parishes selected for samples *A* and *B* being indicated in the table.

TABLE 8.9.c—ACREAGES OF CROPS AND GRASS (DIVIDED BY 10), AND OF WHEAT, IN THE 91 COMBINED HERTFORDSHIRE PARISHES

Dist.	C. & G.	Wh.	Dist.	C. & G.	Wh.	Dist.	C. & G.	Wh.	Dist.	C. & G.	Wh.
1	249	316a	3	264	386a	4	363	958a	5	380	491
	335	164b		208	366		220	454		347	818b
	664	652		237	311b		390	907		363	741
	226	192		227	319		251	466		405	582
	256	272		220	238a		210	426		337	586
	314	131		436	54		217	263		371	442
	248	26		214	327		305	779		294	416a
				333	228b		230	558b			
2	283	612		464	1074		227	440a	6	252	225b
	247	624		232	313		337	618		307	284a
	205	356a		210	98a		282	710		374	738b
	304	766b		201	407		443	775b		305	244
	220	362		265	466		250	518a		486	562
	344	701b		634	1264		289	495ab		204	194
	237	567		228	276		213	262		257	236a
	204	503b		229	249ab		242	565ab		249	309
	209	573		293	686b		416	862b		337	294
	336	728a		276	651		340	537		323	246
	305	901		281	503		358	1085		350	390
	330	515a		273	604		246	474			
	344	788					393	776	7	306	290b
	226	434					267	702		380	244
	220	506					259	410		272	237
	345	838					388	862		384	318a
							258	424		251	116

27,304 44,676

The parishes selected for samples A and B of Table 3.11.b are indicated by the letters a and b respectively.

The values of x, y, and r for district 4 (sample A) are as follows:

x	y	r
363	958	2·6391
227	440	1·9383
250	518	2·0720
289	495	1·7128
242	565	2·3347
1371	2976	2·1707

Thus we have $S(ry) - \bar{r}_u S(y) = 958 \times 2 \cdot 6391 + \ldots - 2976 \times 2 \cdot 1707 = 161 \cdot 3$. The corresponding values for districts 2, 3 and 6 are 63·9, 116·8, and 0·0, with a sum of 342·0. The sample mean of x for these four districts is 266·4 and consequently $s_r^2 = 342 \cdot 0/(10 \times 266 \cdot 4) = 0 \cdot 1284$, or in acreage units 0·001284.

This value is considerably less than the value 0·002649 obtained in Example 7.16. Each estimate, however, is based on only 10 degrees of freedom, so that the discrepancy is not exceptionally large. The corresponding value from the data for all 91 combined parishes, calculated in the same manner, is 0·002222. This calculation is left as an exercise for the reader.

Example 8.9.c

Compare the relative precision, in the estimation of wheat acreage, of samples of Hertfordshire parishes taken with uniform probability and with probability proportional to size, by calculating the expected standard errors of samples of types A and B of Table 3.11.b.

The data for all 91 combined parishes give a value of s_q^2 of 23,483 when districts are eliminated and the same ratio is taken for all districts, and a value of 22,427 when different ratios are taken for the different districts.

In calculating the expected standard error the formula of Section 7.10 may be used, so as to allow for the variation in sampling fraction from district to district. The factors $X_i^2/\{S_i(x)\}^2$ may be replaced by $1/f_i^2$ since we are considering the average error to be expected over a series of similar samples. This will lead to a slight underestimation of the average error.

We find $\Sigma(1 - f_i)\, n_i/f_i^2 = 402\cdot83$, and consequently $V(Y) = 9\cdot460 \times 10^6$ when the same ratio is taken for all districts, and $9\cdot034 \times 10^6$ when different ratios are taken.

Similarly, in the case of sampling with probability proportional to size, from the results already given in Example 7.16, and the value of s_r^2 given in Example 8.9.b, we find $V(Y) = 8\cdot263 \times 10^6$.

The standard errors corresponding to these variances have already been given in Table 3.11.b.

The relative precision of sampling with probability proportional to size, and with uniform probability using a single value of the ratio, is therefore $9\cdot460/8\cdot263 = 1\cdot14$. There is thus a gain in precision of 14 per cent., but it must be recognized that sampling with probability proportional to size will result in parishes of larger average size being included in the sample. Neglecting the disturbance due to the probability being only approximately proportional to size, the average size of parish in this case will be given by $S(x^2)/S(x)$, where the summations are taken over the whole population (or a sample selected with uniform probability). This gives an average size of 3244 acres of crops and grass, compared with the arithmetic mean of 3000 acres, i.e. an average size greater by 8 per cent.

8.9.1 Relative efficiency of sampling with probability proportional to size and stratified sampling with variable sampling fraction

A further question that arises when considering sampling with probability proportional to size is how it compares in efficiency with stratification by size and the use of a variable sampling fraction.

For any particular type of material this question can be dealt with by comparing the variances of the two types of sample, calculated by the methods already described. It is worth noting, however, that the analysis of variance procedure provides a rapid and elegant approximate method of making this comparison.

If we have a random sample taken with probability proportional to size x and we stratify this sample (after selection) into size-groups of x, we can perform an analysis of variance between and within size-groups on the values of r obtained from the sample. This can be arranged as in Table 8.9.1.a, where $V_w(r)$ represents the pooled estimate of the variance of r within size-groups.

TABLE 8.9.1.a—ANALYSIS OF VARIANCE OF r

	Degrees of freedom	Mean square
Between size-groups	$t - 1$	
Within size-groups	$\Sigma (n_i - 1)$	$V_w(r)$
Total	$n - 1$	$V(r)$

If the ranges of the size-groups are sufficiently small for the variation in size of x within size-groups to be neglected, the sum of squares of r within size-groups can be taken as equal to $\Sigma (n_i - 1) s_i^2/\bar{x}_i^2$. If in addition there is no great variation in the $V_i(r)$ for the different size-groups, or if all n_i are large, we have approximately

$$V_w(r) = \frac{\Sigma n_i s_i^2/\bar{x}_i^2}{\Sigma n_i} = \frac{\Sigma n_i s_i^2/\bar{x}_i^2}{n} \tag{8.9.1}$$

If the number of units in the ith stratum of the stratified sample is taken as $n_i = n\, X_i/X$ the total number of units in the two samples will be equal, and the sampling fractions will be proportional to $\bar{x}_i$, and therefore about optimal. In this case $f_i = n_i/N_i = n\bar{x}_i/X$, and from formula 7.6.a

$$V(Y) = N^2 V(\bar{y}) = (X^2/n^2) \Sigma n_i s_i^2 (1 - f_i)/\bar{x}_i^2$$

From equation 8.9.1 this approximates to $X^2 V_w(r)/n$, apart from the factors $(1 - f_i)$.

For the sample with probability proportional to size $V(Y) = X^2 V(r)/n$ (Section 7.15). Consequently the ratio of the mean squares $V(r)/V_w(r)$ in the analysis of variance gives an estimate of the efficiency of stratified sampling relative to sampling with probability proportional to size. This estimate is approximate because (a) it has been assumed that the variation in x within a size-group can be neglected, (b) corrections for finite sampling, $1 - f_i$, have been omitted, and (c) the $n_i - 1$ have been replaced by n_i. Allowance for (a) will cause a decrease in the estimate, allowance for (b) will cause an increase, while (c) is not likely to be important. Some further gain may be expected in stratified sampling by taking optimal sampling fractions instead of fractions proportional to mean size.

The inverse of the process can also be used, starting with the data of a stratified sample with variable or uniform sampling fraction, or with the data of a random sample stratified after selection. This process is illustrated in the example below.   Apart from saving computation it has the merit that reference to the original data for the calculation of values of r (which would be necessary if the formula for s_r^2 of Section 8.9 were used) is not necessary.

Example 8.9.1

Using the data on Hertfordshire farms described in Section 3.7, etc., and the within-size-group variances of Table 8.19.a, compare the relative efficiency of sampling from within size-groups with sampling fractions proportional to the mean farm acreages of the size-groups, and unstratified sampling with probability proportional to farm acreage.

TABLE 8.9.1.b—HERTFORDSHIRE FARM DATA : CALCULATIONS FOR THE CONSTRUCTION OF AN ANALYSIS OF VARIANCE OF r

Size-group (1)	$\bar{x}_i$ (2)	$\bar{y}_i$ (3)	$\bar{r}_i$ (4)	s_i^2 (5)	$s_i^2/\bar{x}_i^2$ (6)	n_i (7)	$n_i \bar{r}_i$ (8)
6–	10	0·2	·02	2	·02	5·2	·104
21–	30	1·7	·0567	15	·01667	10·8	·6124
51–	85	11·0	·1294	160	·02215	44·2	5·7195
151–	200	31	·155	650	·01625	80	12·4
301–	365	77	·2110	1700	·01276	76·6	16·1626
501–	600	172	·2867	4500	·0125	30	8·6010
			$\bar{r} = $ ·17666			246·8	43·5995

As before, we may take x to represent farm acreage (acres crops and grass), and y to represent wheat acreage.   We shall require rough estimates of $\bar{x}_i$ and $\bar{y}_i$ for all size-groups.   For $\bar{x}_i$ the size-group means were assumed to be situated at one-third the group interval from the lower limit of the group. For $\bar{y}_i$ weighted means were calculated from Tables 6.5.b, 6.6.b and 6.7.b. The values obtained are shown in Table 8.9.1.b.   Taking sampling fractions of $x_i/1000$ we obtain from Table 8.19.a the values of n_i shown.   The rest of the calculations in the table are self-explanatory ($\bar{r} = \Sigma n_i \bar{r}_i / \Sigma n_i$).

From columns 4 and 8 we find for the sum of squares of r between size-groups

$$\Sigma n_i \bar{r}_i^2 - \bar{r} \Sigma n_i \bar{r}_i = 0\cdot8728$$

and from columns 6 and 7 we find for the estimated sum of squares of r within size-groups

$$\Sigma (n_i - 1) s_i^2/\bar{x}_i^2 = 3\cdot8152.$$

The reconstructed analysis of variance of r therefore takes the form shown in Table 8.9.1.c.

TABLE 8.9.1.c—RECONSTRUCTED ANALYSIS OF VARIANCE OF r WHERE
SAMPLING IS WITH PROBABILITY PROPORTIONAL TO SIZE

	Degrees of freedom	Sum of squares	Mean square
Between size-groups .	5	0·8728	
Within size-groups .	240·8	3·8152	·01584
Total . . .	245·8	4·6880	·01907

The approximate relative efficiency of the two methods of sampling is therefore $\cdot01907/\cdot01584 = 1\cdot20$. Owing to variation of x within size-groups this is a slight overestimate when all sampling fractions are small, as explained above. The correction to the total mean square from this cause is in fact of the order of $-\cdot0015$. Unless all sampling fractions are small, however, corrections for finite sampling are required. This will increase the efficiency of the stratified sampling, as will adjustment of the sampling fractions to their optimal values.

8.10 Multi-stage sampling

The sampling variance of two-stage sampling can be divided into two parts, A and B, where

A = variance due to the first-stage sampling when there is complete ascertainment at the second stage, i.e. when all the second-stage units which go to make up the selected first-stage units are known,

B = variance due to the second-stage sampling of the selected first-stage units.

Thus the formula of Section 7.17 for $V(\bar{y})$ in two-stage random sampling may be rewritten

$$V(\bar{y}) = \frac{1-f'}{n'} s_0'^2 + \frac{1-f''}{n' n''} s''^2 \qquad (8.10.a)$$

where

$$s_0'^2 = s'^2 - \frac{1-f''}{n''} s''^2 \qquad (8.10.b)$$

The first term constitutes part A and the second part B.

The second term will be recognized as $(1-f'')/(1-f)$ times the variance that would be obtained with single-stage sampling of the second-stage units, the same total number of second-stage units being taken, with the first-stage units as strata and uniform sampling fraction f. If f'' is small, therefore, the first term gives the increase in variance due to the adoption of the two-stage process.

The above subdivision is alternative to that given in Section 7.17. Part A is dependent only on the first-stage sampling, being unaffected by the intensity or type of sampling at the second stage. This fact considerably simplifies the problem of determining the sampling errors for different intensities of sampling at the two stages: with the subdivision of Section 7.17 the variation in s'^2

for different intensities of sampling at the second stage has to be taken into account.

The only new point that arises in the estimation of the relevant variances is the determination of part A from the data of a two-stage sample. In general this simply requires that the variance per first-stage unit due to the second-stage sampling of the first-stage units be deducted from the variance per first-stage unit calculated from the sample. Thus for a two-stage random sample formula 8.10.b is used.

Example 8.10.a

Calculate the expected sampling errors of the wheat acreages derived from the two-stage sample B_1 of Hertfordshire farms of Table 3.11.b, and discuss the effects of varying the number of parishes in the sample, with adjustment of the second-stage sampling fraction so as to give the same total number of farms in the sample.

Part A of the variance has already been determined in Example 8.9.c. We have $A = 8\cdot263 \times 10^6$.

The determination of part B requires the evaluation of the variance of the r for individual parishes due to the second-stage sampling of these parishes. These variances were evaluated separately for each of the 17 parishes of the sample, using method (a) of Section 7.9. The mean value of these variances $V''(r)$ was found to be $0\cdot003575$.

The equation of estimation of the total acreage is $Y = \Sigma\, X_i\, \bar{r}_i$. The second-stage variance of $\bar{r}_i$ is $V''(r)/n_i$, and part B of the variance is therefore given by

$$B = V''(r)\, \Sigma\, X_i{}^2/n_i = 0\cdot003575 \times 45\cdot429 \times 10^8 = 16\cdot24 \times 10^6$$

Hence $V(Y) = 24\cdot50 \times 10^6$.

Exact treatment of the effects of varying the number of parishes is complicated by the fact that the first-stage sampling fractions are bound to vary somewhat from district to district, and that the number of farms per parish is also variable. Ignoring these sources of disturbance we may write for any number n' of parishes and n'' of farms per parish

$$V(Y) = \frac{1-f'}{n'}\, a + \frac{1-f''}{n'\, n''}\, \beta$$

The values of a and β can be determined from the values of A and B. The mean number of farms per parish is $N'' = 2496/91 = 27\cdot429$. Putting $n' = 17$, $f' = 17/91$, $f'' = \frac{1}{4}$, and $n'' = \frac{1}{4} \times 27\cdot429 = 6\cdot857$, we have

$$a = \frac{17}{1 - 17/91} \times 8\cdot263 \times 10^6 = 172\cdot7 \times 10^6$$

$$\beta = \frac{17 \times 6\cdot857}{1 - \frac{1}{4}} \times 16\cdot24 \times 10^6 = 2524\cdot2 \times 10^6$$

The effect of any variation in n' and n'' can now be determined from the formula for $V(Y)$. If the total number of farms $n\,(= n'\,n'')$ is to be kept fixed, the formula is best rewritten in the form

$$V(Y) = (a - \beta/N'')/n' + \beta/n - a/N'$$
$$= \{80 \cdot 71/n' + 2524 \cdot 2/n - 1 \cdot 8982\} \times 10^6$$

with the checks that, when $n' = 91$ and $n = 2496$, $V(Y)$ is zero, and when $n' = 17$ and $n = 17 \times 6 \cdot 857 = 116 \cdot 57$, it equals the value given above.

The values of $V(Y)/10^6$ for 5, 10, 20, and 30 parishes and a number of farms, 116·57, approximately the same as that of the actual sample, are 35·9, 27·8, 23·8 and 22·4 respectively. If all 91 parishes were sampled the corresponding variance would be 20·6. There is thus no great gain in taking more than 20 parishes when sampling within parishes is with a uniform sampling fraction.

It should be noted that the use of the above formula when the fraction of parishes sampled is large is unrealistic, in that sampling with probability proportional to size could not be adopted in such cases. It serves, however, to illustrate the use of the similar formulæ which could be developed for sampling with uniform probability at the first stage.

Example 8.10.b

Repeat the analysis of Example 8.10.a for the sample B_2 of Table 3.11.b.

Part A of the total variance will be the same as in sample B_1.

Part B can be calculated in the same manner as in Example 8.10.a, using the method of Section 7.10, with a separate value of the ratio for each parish. This gives $V''(r) = 0 \cdot 0008167$, and $B = 3 \cdot 710 \times 10^6$. Hence $A + B = 11 \cdot 97 \times 10^6$.

The expression of $V(Y)$ in terms of n' and n'' is complicated by the variable sampling fraction at the second stage. As a first approximation we may take an average sampling fraction f'' for the given sample, and use this to obtain a formula of the same form as in Example 8.10.a. This gives $f'' = 0 \cdot 29121$ and we then find

$$V(Y) = \{146 \cdot 83/n' + 710 \cdot 76/n - 1 \cdot 8982\} \times 10^6$$

with checks as before.

This formula gives values of $V(Y)/10^6$ for 5, 10, 20 and 30 parishes and 135·8 farms of 32·7, 18·0, 10·7 and 8·2 respectively. With more accurate sampling of the farms within the selected parishes, therefore, there is a more marked decrease in variance as the number of parishes is increased. The above values underestimate the decrease, as with the reduction in the number of farms per parish the change in the second-stage sampling fractions will result in a somewhat smaller increase in the variance than that given by the formula. More accurate values could be obtained by recalculating the second-stage

285

variances with various intensities of second-stage sampling, using graphical methods for interpolation between the calculated values.

Example 8.10.c

Investigate the relative precision of the determination of the acreages of crops by the measurement of the areas of fields and part fields included in a sample of rectangular areas, and the use of grids of points covering these areas (Section 4.24).

If a sample area has an area a and the proportion of the area occupied by a given crop is $\mathbf{p}$, the area $\mathbf{y}$ occupied by this crop in this sample area equals $a\,\mathbf{p}$. If a random set of n points is taken over the area the variance of the estimate y given by the proportion of points falling in the given crop is

$$\mathbf{V}\,(\mathbf{y}) = \mathbf{pq}\,a^2/n$$

This variance will be additional to the sampling variance of the **y** over the sample areas. This latter variance depends on the sampling method and the variability of the **y** from area to area, and can only be determined from actual sample data.

As an example we may consider the case in which the areas are randomly selected, and the frequency distribution of the areas with proportions $0\cdot0$, $0\cdot1$, . . . of the given crop is as follows:

Proportion, $\mathbf{p}$	0·0	0·1	0·2	0·3	0·4	0·5	0·6	0·7	0·8	0·9	1·0
Frequency, φ	0·05	0·15	0·20	0·15	0·12	0·10	0·07	0·05	0·03	0·03	0·05

The average variance due to the point sampling is given by

$$\Sigma\,\varphi\,\mathbf{V}\,(\mathbf{y}) = (a^2/n)\,\Sigma\,\varphi\mathbf{pq} = 0\cdot1656\,a^2/n$$

The sampling variance of the **y** is given by

$$\mathbf{V}\,(\mathbf{y}) = \Sigma\,\varphi\,\mathbf{p}^2\,a^2 - (\Sigma\,\varphi\mathbf{p}\,a)^2 = 0\cdot069024\,a^2$$

The proportional increase in variance due to point sampling is therefore

$$0\cdot1656/0\cdot069024\,n = 2\cdot40/n$$

With 9 points there is an increase of 27 per cent. in the variance and with 16 points an increase of 15 per cent. In the latter case about one-seventh more areas will be required for the same accuracy. Against this must be set the fact that only fields in which the points fall need be examined and recorded. The occurrence of mixed crops and systematic location of the points on a rectangular grid will also reduce the sampling variance.

8.10.1 Interpretation of the analysis of variance

The analysis of variance can be interpreted in the manner set out below. This interpretation is of particular use when we are concerned with multi-stage sampling, and with the effect of change of size of the sampling units.

If the units fall into groups of any kind, such as strata, the unit values of a variate y can be regarded as made up of the sum of two parts, one, u, which varies from group to group but has a fixed value for all units of a particular group, and the other, v, which varies from unit to unit independently of the groups. The variances of u and v may be denoted by $\mathbf{U}$ and $\mathbf{V}$ respectively. Thus u and v may be random sample values from normal distributions, though the condition of normality is not necessary.   In this hypothetical framework zero mean can be assigned to the parent distribution of v without loss of generality, but even so the mean of the v's for all the units of a finite population, or for all the units of a particular group, will not be exactly zero, and consequently the group means are not exactly equal to the u's.   For this reason the values of u and v cannot be uniquely determined from the values of y.

The mean squares of the analysis of variance provide estimates of $\mathbf{U}$ and $\mathbf{V}$. If A and B are the mean squares between and within groups, C is the overall mean square, k is the number of units in each group, and h the number of groups, we have

$$A = k\,\mathbf{U} + \mathbf{V}$$
$$B = \mathbf{V}$$

Hence
$$\mathbf{U} = (A - B)/k$$

We also have, from the analysis of variance, $(hk - 1)\,C = h\,(k - 1)\,B + (h - 1)\,A$.   Consequently if σ^2 is the overall variance and σ_1^2 the variance within groups we have, from formula 8.3,

$$s^2 = \mathbf{U}\,(h - 1)/h + \mathbf{V}$$
$$s_1^2 = \mathbf{V}$$

The factor $(h - 1)/h$ is analogous to the correction for sampling from a finite population.

The relative precision of stratified and random sampling will be obtained by taking the groups as strata.   We then have, with t strata,

$$\frac{s^2}{s_1^2} = 1 + \frac{t - 1}{t} \cdot \frac{\mathbf{U}}{\mathbf{V}}$$

An alternative formulation is possible in terms of the *intra-class correlation*, i.e. the correlation between members of the same stratum when the strata themselves are regarded as a random sample from an infinite set of similar strata (R. A. Fisher, *Statistical Methods for Research Workers*, Section 40). The estimate r_i of this correlation is given by

$$r_i = \frac{A - B}{A + (k - 1)\,B} = \frac{\mathbf{U}}{\mathbf{U} + \mathbf{V}}$$

and consequently
$$\frac{s^2}{s_1^2} = 1 + \frac{t - 1}{t} \cdot \frac{r_i}{1 - r_i}$$

Looked at from this point of view, the intra-class correlation coefficient may be regarded as a quantitative expression of association which is alternative to the ratio $\mathbf{U}/\mathbf{V}$. In this book we shall use the concept of additive components of variance, since this appears to be more easily capable of generalization, and is otherwise preferable to the concept of intra-class correlation.

When there is compensation between the different units of the same stratum the definition of U as a variance breaks down, and has to be extended (see Yates, *Experimental Design*). Complete compensation occurs when all the strata means (or the first-stage units in a two-stage scheme) are equal. In this case $K U + V = 0$, i.e. $U = - V/K$, where K is the number of units in each group of the population. Negative values of U between 0 and $- V/K$ are therefore admissible.

8.11   Application of the analysis of variance to two-stage samples

It is often helpful to carry out an analysis of variance on data derived from a two-stage sampling process. The situation is simplest when the number of sampled second-stage units n'' in each first-stage unit is the same. Each stage of the analysis then follows the same pattern as the analysis of a single-stage sample of the same type. At the first stage, however, the values entering into the analysis must be either the means or the totals of the second-stage unit values. It is customary (though not essential) to tabulate the sums of squares of the first stage in terms of the second-stage units. If the first-stage unit means are used, therefore, all sums of squares at the first stage must be multiplied by n'', while with totals all sums of squares must be divided by n''.

In the case of two-stage random sampling, for example, the degrees of freedom and mean squares will be

	Degrees of freedom	Mean square
Between first-stage units	$n' - 1$	$n'' s'^2 = V + n'' U$
Within first-stage units between second-stage units	$n' (n'' - 1)$	$s''^2 = V$
TOTAL	$n' n'' - 1$	

We then have

$$V(\bar{y}) = \frac{1 - f'}{n'} U + \frac{1 - f}{n} V \qquad (8.11.a)$$

where n is the total number of second-stage units and f is the overall sampling fraction ($n = n' n''$ and $f = f' f''$). The second term of this subdivision is the estimate of the variance that would be obtained with single-stage sampling of the second-stage units, the same total number of sampling units being taken, with first-stage units as strata, and uniform sampling fraction. The analysis of variance therefore provides a further alternative subdivision of the sampling variance.

The results are similar with stratification with uniform sampling fraction at either or both stages.

An analysis of variance of the above form can still be made if the numbers n'' of second-stage units in each first-stage unit differ. The within first-stage degrees of freedom are then $S'(n'' - 1)$, and the sum of squares between first-stage units is calculated by the " mean $\times$ total " rule, i.e. each first-stage mean is multiplied by the corresponding total. If the additive components-of-variance model defined in Section 8.10.1 correctly represents the variance law, a virtually unbiased estimate of U will still be obtained from the analysis of variance by replacing n'' by $\bar{n}''$. The first term in the formula for $V(\bar{y})$, however, must be multiplied by

$$M = S'(n''^2)/n' \, \bar{n}''^2,$$

i.e. approximately by $1 + V(n'')/\bar{n}''^2$. That this is so is easily seen by expressing $\bar{y}$ in terms of the u's and v's, and applying formula 7.5.a for the variance of a linear function.

With a sample stratified at the first stage, $V + n''U$ will be given by the within-strata mean square ($n' - t$ degrees of freedom). If the n'' vary and the number of first-stage units in each stratum is small, replacement of n'' by $\bar{n}''$ will result in appreciable bias. This can be avoided by using instead

$$\bar{n}''_0 = \frac{\Sigma\{S'n'' - S'(n''^2)/S'n''\}}{n' - t} \tag{8.11.b}$$

where Σ indicates summation over the t strata, and S' summation over the first-stage units of a stratum.

Section 8.12 provides a numerical example of the analysis of variance for a four-stage sampling scheme with stratification at the first stage. In this example, comparable values of $\bar{n}''$ and $\bar{n}''_0$ were found to be 1·44 and 1·29 respectively.

It is, of course, unrealistic to expect that the variance law governing any particular set of data will conform exactly to that postulated by the above model. The variance of the v's may well be greater in the larger first-stage units, and the variance of the u's may also be associated in some manner with the size of the first-stage units. Such differences in variance can be investigated by grouping the first-stage units by size and partitioning the between and within components of the analysis of variance according to these size-groups. Such an investigation may be of value when planning a large-scale survey. Note that if the data have been derived by the use of variable sampling fractions at the second stage the n'' will not be directly proportional to size.

When all n'' are equal, estimates of $V(\bar{y})$ from the analysis of variance are equivalent to ratio estimates, but even if there is little variation in the n'' for the population this is not likely to hold for the numbers n''_A belonging to a particular domain. In general, therefore, the analysis of variance approach should be reserved for general studies of variability.

Note that in the analysis of variance n' must be taken to be the number of first-stage units for which n'' is not zero. In other words, $n' - 1$ is the number of degrees of freedom for the first-stage units. This refinement is not important for ratio estimates, as these contain the factor $n'/(n' - 1)$.

It does not appear profitable to develop an analysis of variance and covariance model for estimating $V(\bar{y}_A - \bar{y}_B)$. Such a model would be complicated and unrealistic.

Example 8.11

Compare the sampling variances obtained in Example 7.11 with those derived from the analysis of variance model.

The analyses of variance are set out in Table 8.11.a, and the necessary computations in Table 8.11.b. With the adjustment given above to allow for the variability of the n'' relative to their means, the variances given by the analysis of variance model do not differ greatly from the ratio estimates.

Direct division of the mean squares between farms by the numbers of fields, which is the appropriate procedure for an analysis of variance with

TABLE 8.11.a—ANALYSIS OF VARIANCE OF NITROGEN DRESSINGS (KG PER HA) ON WINTER WHEAT AND SPRING BARLEY

| | Winter wheat | | Spring barley | |
	D.F.	M.S.	D.F.	M.S.
Between farms	161	3930	145	1303
Within farms	372	722	207	365
Total	533	1691	352	751

TABLE 8.11.b—COMPUTATION OF ADJUSTED ESTIMATES OF $V(\bar{y}_W)$ AND $V(\bar{y}_B)$ FROM THE ANALYSES OF VARIANCE

	Wheat	Barley
No. of farms (n')	162	146
No. of fields (n)	534	353
Fields per farm $(\bar{n}'')$	3·296	2·418
$S'(n''^2)$	3978	1669
$M = S'(n''^2)/n'\bar{n}''^2$	2·260	1·955
$V = B$	722	365
$U = (B - A)/\bar{n}''$	973	388
$V(\bar{y}) = MU/n' + V/n$	14·9	6·2
Ratio estimate	12·3	5·2
Unadjusted estimate		
$\quad U/n' + V/n = A/n$	7·4	3·7

A and B here represent the mean squares between and within farms, as in Table 7.7.a.

equal numbers in the sub-classes, leads to the serious underestimates shown by the last line of the table.

8.12 An example of a pilot sampling scheme for crop estimation

In order to investigate the practicability of obtaining estimates of the yields of cereal crops in the United Kingdom by the harvesting of sample areas, the yields of a number of wheat fields were determined by this method in each of the years 1934–1938 (Cochran, 1939, A). Fields were taken in several districts each year, one or two fields being selected at random from the fields growing wheat on each chosen farm. The selection of farms in each district was not random, the farms being taken in the neighbourhood of the centres at which the investigators were located.

The sampling of the individual fields followed the lines described in Section 4.29, the fields being traversed in the direction of the rows, along two lines selected at random. Two sets of unit areas were taken from each line. Each unit area consisted of $\frac{1}{4}$ metre of each of 6 contiguous rows. For the

TABLE 8.12.a—SAMPLING OF WHEAT FIELDS, 1937: MEAN YIELDS OF GRAIN PER UNIT AREA (0·0000565 ACRES) IN GRAMS

	District I				District II			District III					
1st line { Set 1	47	48	75	105	93	58	76	92	89	89	75	70	80
Set 2	63	51	71	82	84	78	57	83	111	58	72	85	97
2nd line { Set 3	67	45	75	97	75	68	79	93	90	70	82	76	111
Set 4	55	46	85	86	80	83	78	96	115	70	81	102	66
	232	190	306	370	332	287	290	364	405	287	310	333	354
	422		676		619			769		597		687	
Totals	1098				909			2053					

	District IV											District V	
1st line { Set 1	29	45	57	69	78	59	68	97	60	65	81	77	60
Set 2	21	39	63	55	109	59	56	88	53	59	94	74	88
2nd line { Set 3	29	69	46	21	90	58	74	109	43	49	93	44	84
Set 4	31	57	66	40	51	53	61	95	48	71	92	57	97
	110	210	232	185	328	229	259	389	204	244	360	252	329
	320		417									581	
Totals	2750											581	

	District VI												
1st line { Set 1	66	93	55	127	84	80	81	93	21	84	87	79	90
Set 2	73	70	56	106	80	86	107	106	63	51	67	79	117
2nd line { Set 3	64	80	83	84	63	88	135	71	50	82	135	71	112
Set 4	73	67	60	98	89	110	82	83	29	80	114	89	122
	276	310	254	415	316	364	405	353	163	297	403	318	441
	586		669		680								
Total	4315												

most part, sets each contained three unit areas, equally spaced along the line, with a random starting point.

The yields of grain obtained in 1937 are shown in Table 8.12.a. The mean yield of all the unit areas in each set is given. In order to allow for differences in row spacing on the different fields the yields have been reduced to a 6-inch row spacing, and therefore represent the yields in grams of areas of $\frac{1}{4}$ metre $\times$ 3 ft. Fields on the same farm are indicated by brackets. In District III, where three fields from a single farm were sampled, each field was growing two varieties which were sampled separately.

The analysis of variance was carried out in units of the totals of the four sets, i.e. on yields of areas of 1 metre $\times$ 3 ft. or 0·000226 acres. Thus the sum of squares of the sets is multiplied by 4, and the sum of squares of the line totals by 2.

The sums of squares for 1937 can be obtained from Table 8.12.a by calculating the sum of squares for each classification, disregarding the others, and deducting the sum of squares corresponding to the next higher classification. The rule of " mean $\times$ total " or " total²/(number of units) " is followed in each case. Thus the correction for the mean is $11,706^2/39 = 3,513,601$. The sum of squares for districts is

$$\tfrac{1}{4} (1098)^2 + \tfrac{1}{3} (909)^2 + \ldots - 3,513,601 = 54,224$$

The sum of squares for farms is

$$\tfrac{1}{2} (422)^2 + \tfrac{1}{2} (676)^2 + \tfrac{1}{2} (619)^2 + 290^2 + \tfrac{1}{6} (2053)^2 + \ldots$$
$$- 3,513,601 - 54,224 = 132,062$$

The arithmetical work can be simplified by omitting items which are repeated in more than one sum of squares. In particular the sum of squares for varieties is

$$364^2 + 405^2 + 287^2 + \ldots - \tfrac{1}{2} (769)^2 - \tfrac{1}{2} (597)^2 - \tfrac{1}{2} (687)^2 = 1326$$

Furthermore the sum of squares corresponding to the difference between any two totals containing the same number of units can be obtained by squaring the difference and dividing by twice the number of units in either total. Thus the sum of squares due to varieties is also given by $\tfrac{1}{2} (41^2 + 23^2 + 21^2)$. This gives a useful check in cases in which, as here, many of the sums of squares depend on differences of pairs of values. Thus the sum of squares between sets is given by $2 (16^2 + 12^2 + 3^2 + 1^2 + \ldots) = 52,368$, that between lines by $12^2 + 8^2 + \ldots = 33,924$.

The sum of squares between fields within farms (excluding District III) is given by $\tfrac{1}{2} (42^2 + 64^2 + 45^2 + 100^2 + \ldots) = 27,702$. The sum of squares between fields within farms for District III has to be calculated in the ordinary manner, since there are three fields, and is $\tfrac{1}{2} (769)^2 + \tfrac{1}{2} (597)^2 + \tfrac{1}{2} (687)^2 - \tfrac{1}{6} (2053)^2 = 7,401$, giving a corresponding total sum of squares of 35,103.

Those not fully familiar with the analysis of variance technique should recalculate the sums of squares of this example in the various alternative ways indicated.

For general purposes it is best to convert the mean squares into some common units such as (cwt. per acre)^2. The conversion factor is here 0·0075861. This is done in Table 8.12.b, which also shows the results of the similar analyses for the other four years of the investigation.

TABLE 8.12.b—ANALYSIS OF VARIANCE PER FIELD OF YIELDS OF WHEAT GRAIN (CWT. PER ACRE)

	1934		1935		1936		1937		1938	
	d.f.	m.s.	d.f.	m.s.	d.f.	m.s.	d.f.	m.s.	d.f.	m.s.
Between districts .	4	66·5	6	318·4	4	79·4	5	82·3	4	206·8
Within districts be- tween farms .	11	38·9	12	27·1	7	62·2	19	52·7	14	65·3
Within farms be- tween fields .	—	—	15	22·8	8	31·2	11	24·2	8	12·1
Within fields : Sampling error .	16	5·33	40	6·20	22	11·39	39	6·60	28	9·80
Between sets .	32	2·11	80	2·18	45	2·52	78	5·09	55	4·78
Mean yield . .	29·1		23·3		24·3		26·2		30·7	

The mean squares for the same component of variance in the different years are not estimates of precisely the same quantities, owing to variation in the numbers of fields per farm, etc. In view of the small number of degrees of freedom in each year, however, we shall not lose much information by pooling all the years, weighting the mean squares in proportion to the degrees of freedom. This pooled estimate is shown in Table 8.12.c.

From the degrees of freedom we may deduce that there are 91 farms and 133 fields in all in the sample, i.e. a mean of 1·46 fields per farm. Denoting the variance per set by V_1, the additional components of variance per line,

TABLE 8.12.c—COMBINED ANALYSIS OF VARIANCE

	Degrees of freedom	Mean square	Estimate
Between districts . .	23	162·3	
Within districts between farms . . .	63	49·3	$\frac{1}{4} V_1 + \frac{1}{2} V_2 + U_1 + 1·46 U_2$
Within farms between fields . . .	42	22·7	$\frac{1}{4} V_1 + \frac{1}{2} V_2 + U_1$
Within fields between lines . . .	145	7·69	$\frac{1}{4} V_1 + \frac{1}{2} V_2$
Within lines between sets	290	3·50	$\frac{1}{4} V_1$

per field and per farm by V_2, U_1 and U_2 respectively, and ignoring the fact that a few fields have more than two lines and that the number of fields per farm is variable, we have the mean square equivalences shown in Table 8.12.c. Hence $V_1 = 14\cdot0$, $V_2 = 8\cdot4$, $U_1 = 15\cdot0$, $U_2 = 18\cdot2$.

The value of U_2 is an underestimate, firstly because we have used the mean number of fields per farm, instead of calculating the correct value of $\bar{n}_0''$ from formula 8.11.b, and secondly because of the fact that the sample of farms was not random. For 1937 the value of $\bar{n}_0''$ is $1\cdot29$, compared with the value of $\bar{n}''$ of $1\cdot44$.

From the above estimates of the different components of variance we may calculate the variance to be expected with a sample of any given type and size. If the unweighted mean of the yields per acre of the different fields can be taken as the estimate of the mean yield per acre over the country, i.e. if the potential bias due to association of yield per acre with size of field, etc., can be ignored, the variance of the mean yield per acre with a fixed amount of sampling of individual fields and with equal numbers of fields taken from all selected farms will depend solely on the number of farms and the number of fields in the sample. If these are n_1 and n_2 respectively the variance of the mean yield per acre with the same amount of sampling per field as that actually adopted will be

$$U_1'/n_2 + U_2/n_1$$

where $U_1' = \tfrac{1}{4} V_1 + \tfrac{1}{2} V_2 + U_1 = 22\cdot7$. Thus with 200 fields from 100 farms, 2 fields per farm, the variance with the above values of the components of variance is $0\cdot296$. This is equivalent to a standard error of $0\cdot54$ cwt. per acre, or $2\cdot0$ per cent. of the mean yield.

This is an over-simplification of the practical situation. In general the possibility of bias cannot be ignored, and a properly weighted mean must therefore be taken. Any statement in general terms would be difficult, since the weighting depends on the variation in numbers and acreages of the fields on the individual farms and the sampling method adopted. Given the numbers and acreages of the fields on an adequate sample of farms, however, the weighting coefficients for any chosen method of sampling can be determined. If these are denoted by w, and if $[w]$ denotes the sum of the weights for all the sampled fields on a farm, the variance of the weighted mean will be

$$\{U_1' \, S \, (w^2) + U_2 \, S \, ([w]^2)\}/\{ S \, (w)\}^2$$

Thus the relative precision of alternative methods of sampling can be evaluated without difficulty.

The above procedure is approximate in another respect which is not entirely irrelevant to the practical situation. It has been assumed that the component of variance from field to field on the same farm, and that between farms, are independent of the size of the farm. This is not likely to be strictly true, and may introduce appreciable inaccuracy when a variable sampling fraction is used for farms of different sizes.

It will be noted that no corrections for sampling from a finite population are necessary, provided the fraction of the farms in the sample is small. The second-stage sampling fraction of fields from farms may be large, but this fraction does not enter into the formula for the partition of the variance given by the analysis of variance, as for example is shown by formula 8.11.a.

Although the variance of the unbiased estimate depends on the acreages, and therefore cannot be easily formulated in general terms, certain general statements about the relative precision of different types of sample can be made from the above results.

In the first place we may consider the effect of varying the amount of sampling of the selected fields. If the number of sets per line is reduced to one, for example, the sampling variance per field will be $\frac{1}{2} V_1 + \frac{1}{2} V_2 = 11 \cdot 2$, instead of $7 \cdot 7$. If at the same time the number of lines is increased to four, the sampling variance per field will be $\frac{1}{4} V_1 + \frac{1}{4} V_2 = 5 \cdot 6$.

There is little to be gained by increase in the accuracy of the determination of the yields of individual fields, however. With one field per farm the effective variance per field if the yields are determined without error will be $U_1 + U_2 = 33 \cdot 2$, instead of $40 \cdot 9$. Consequently with the intensity of sampling actually adopted the relative accuracy is $0 \cdot 81$. Doubling the number of lines per field with two sets per line would only increase the accuracy by 10 per cent.

The question of whether to sample one or more fields per farm requires more consideration. For farms growing a given number of fields of wheat (greater than one) the variance if two fields are sampled will be $\frac{1}{2} U_1' + U_2 = 29 \cdot 6$, whereas if one field is sampled the variance will be $40 \cdot 9$. The whole question of the methods of sampling farms and fields is bound up with the question of costs, and will be further considered in Section 8.17.

The effect of varying the number of unit areas per set cannot be precisely determined from the above analysis. If the unit areas of each set were randomly and independently located, the variance of the set means would be inversely proportional to the number of units per set, and would be determined from the variance between sets within lines. With the even spacing of units within the set, however, we may expect the reduction in variance with increasing numbers of unit areas to be somewhat greater than in the case of random location. From the basic data giving the yields of the separate unit areas it would be possible to determine the variance per unit area within sets, and from this variance and the variance between sets within lines a fair idea of the departure from the random law could be obtained.

As a first approximation, however, we may assume that the random law holds. In this case, taking two instead of three unit areas per set would multiply the value of V_1 by 3/2, and would therefore raise the value of $U_1' + U_2$ from $40 \cdot 9$ to $42 \cdot 6$. It was in fact recognized after the first year's work that there was little to be gained from having more than a small number of unit areas per set, and the number, which was five in the first year, was then reduced to three.

8.13 A special case of two-stage sampling

The possibility of sampling from within strata with probability proportional to size at the first stage, and with second-stage sampling fractions so chosen that the overall sampling fraction is uniform, has already been mentioned in Section 3.10 and subsequently.

This case is of considerable practical importance, and also provides a useful example of the application of the above methods to the more complicated types of two-stage sampling.

From the results already given in Sections 7.16 and 7.17 we have

$$V(Y) = \Sigma\, s_{ri}'^2 \mathbf{X}_i^2 (1 - f_i')/n_i' + \Sigma\, f_i'\, \mathbf{X}_i^2\, V''(\bar{r}_i) \qquad (8.13)$$

where $V''(\bar{r}_i)$ is the estimated second-stage variance of $\bar{r}_i$.

We will consider the case in which the sampling at the second stage is random (or stratified with uniform sampling fraction) and the number of second-stage units is taken as the measure of size. If n_i' first-stage units are selected from the ith stratum the probability of selection of the jth unit will be $n_i'\, \mathbf{N}_{ij}/\mathbf{N}_i$, where $\mathbf{N}_{ij}$ is the number of second-stage units in the jth first-stage unit, etc. The second-stage sampling fraction for this unit, if selected, will be $f \mathbf{N}_i/n_i'\, \mathbf{N}_{ij}$, where f is the uniform overall sampling fraction. The number of second-stage units selected will be $f \mathbf{N}_i/n_i'$. Thus the same number of second-stage units will be selected from each of the selected first-stage units in a given stratum. If the variance $\sigma_i''^2$ of y per unit at the second stage can be taken as constant for the whole of the ith stratum the estimated variance of $r_{ij}\ (= \bar{y}_{ij})$ will be

$$V''(r_{ij}) = s_i''^2\, \frac{(1 - f_{ij}'')}{f\, \mathbf{N}_i/n_i'}$$

which is constant for all the selected units of the ith stratum except for the factor $(1 - f_{ij}'')$. Provided all the f_{ij}'' are moderately small it will be sufficient to replace them by $f_i'' = f/f_i'$. We then have

$$V''(\bar{r}_i) = s_i''^2 (1 - f_i'')/f \mathbf{N}_i$$

The $\mathbf{X}_i$ of formula 8.13 will be replaced by $\mathbf{N}_i$, and some form of pooling can be adopted to estimate an average value $s_r'^2$ of $s_{ri}'^2$. In cases in which the $s_{ri}'^2$ are likely to vary markedly, weights corresponding to those given by the first term should be used.

We then have

$$V(Y) = s_r'^2 \Sigma\, \mathbf{N}_i^2 (1 - f_i')/n_i' + \Sigma\, f_i'\, s_i''^2\, \mathbf{N}_i (1 - f_i'')/f$$

Following the previous procedure this may be re-written as

$$V(Y) = s_{ro}'^2 \Sigma\, \mathbf{N}_i^2 (1 - f_i')/n_i' + \Sigma\, s_i''^2\, \mathbf{N}_i (1 - f_i'')/f$$

where

$$s_{ro}'^2 = s_r'^2 - \frac{\Sigma\, (1 - f_i')\, s_i''^2\, \mathbf{N}_i (1 - f_i'')}{f\, \Sigma\, \mathbf{N}_i^2 (1 - f_i')/n_i'}$$

If the f_i' are approximately equal, as will usually be the case, and the $s_i''^2$ are the same for all strata, we have

$$V(Y) = s_{ro}'^2 (1 - f') \Sigma N_i^2/n_i' + s''^2 N (1 - f'')/f$$

$$s_{ro}'^2 = s_r'^2 - \frac{s''^2 N (1 - f'')}{f \Sigma N_i^2/n_i'}$$

The second term of $V(Y)$ will be recognized as $(1 - f'')/(1 - f)$ times the variance which would be obtained with single-stage sampling of the second-stage units with uniform sampling fraction and the first-stage units as strata. If f'' is small, therefore, the first term gives the increase in variance due to the adoption of the two-stage process, as in the case of two-stage random sampling.

Instead of considering the variability of the values of r_{ij} between the different first-stage units of a stratum, we may consider the variability of the corresponding totals of y for the different first-stage units.

This leads to a somewhat more elegant form of the above formulae. If the sample total of y for the jth unit (presumed selected) of the ith stratum is denoted by Y_{ij} ($= S_{ij}(y)$), the estimate of the variance of these totals within stratum i will be given by

$$s_{ti}'^2 = \frac{1}{n_i' - 1} S (Y_{ij} - \bar{Y}_{ij})^2$$

Since there will usually only be very few (often two) first-stage units per stratum, some form of pooled estimate will be required, derived from the estimates of the various strata. This can be denoted by $s_t'^2$. The formula for the variance then becomes

$$V(Y) = g^2 \{ s_t'^2 \Sigma n_i' (1 - f_i') + \Sigma s_i''^2 n_i f_i' (1 - f_i'') \}$$

where n_i is the total number of selected second-stage units in the ith stratum. This formula is exactly similar in form to that for $V(Y)$ in a two-stage sample with equal probability of selection at the first stage. This latter formula can be easily derived from the formula for $V(\bar{y})$ given in Section 7.17.

If the number of second-stage units in each first-stage unit is taken as the measure of size, the same number, n_i'' ($= f N_i/n_i'$), of second-stage units will require to be selected from each first-stage unit in a given stratum, and $r_{ij} = \bar{y}_{ij}$. In this case we can of course work equally well with the totals Y_{ij} or the means $\bar{y}_{ij}$. If, however, the preliminary data on the number of second-stage units in each first-stage unit are not exact, as for example may occur when the second-stage frame for the selected first-stage units is constructed after selection of the first-stage units, the formula given above, based on totals, still holds. An actual case in which this contingency occurred is described in Section 4.16.

8.14 Effect of change in size of the sampling units

If the population is divided into N large units, each of which is subdivided into K small units, and if a sample of k small units from each of n large units is taken, an analysis of variance between and within large units can be made, and the components of variance U and V estimated as in Section 8.10.1.

The estimate of the overall variance between small units will be given, as before, by $V + U(N - 1)/N$. That between large-unit means of k small units will be given by $1/k$ times the expectation of the mean square between large units, i.e. by $V/k + U$. The variance between large-unit means when all K small units of each large unit are included will be $V/K + U$.

These results enable us to determine the effect on the sampling error of the alternatives of using the large or the small units as sampling units. It will be noted that for this determination it is not necessary to have data in which all the small units that go to make up the selected large units are observed. The analogy with two-stage sampling will be apparent.

Various extensions of these results are of interest. If the large units are stratified, with N_t units per stratum in the population and n_t in the sample, there will be a further between-strata component of variance U_t, and the mean squares in the analysis of variance will provide estimates as follows :—

Between strata	$V + kU + k\,n_t\,U_t$
Within strata between large units	$V + kU$
Within large units between small units ..	V

The estimates of the different variances will then be :

Small units within strata	$V + U(N_t - 1)/N_t$
Large units (means) within strata	$V/K + U$
Large units (means) overall	$V/K + U + U_t(t - 1)/t$

The first two variances are the same as previously, except that N is replaced by N_t.

The above approach enables us to determine the effect of simultaneous change of size of unit and size of strata. This is relevant when the strata can be of any size, and the size is therefore chosen to contain two units (or one unit) per stratum. If the size of unit is halved and the amount of material in the sample remains the same, for example, there will be twice as many units, and the size of the strata can therefore be halved. We shall then require a four-fold analysis of variance into whole strata, half-strata, whole units, and half-units. The minimum amount of data required for this purpose will be two whole units (of which each half-unit is separately recorded) in each half-stratum.

The expressions for mean squares and variances will be similar to those given above, N_t being the number of whole units per half-stratum in the population, and t the number (two) of half-strata per stratum. These expressions are as follows :—

Mean squares :

Within whole strata between half-strata ..	$V + 2U + 4U_t$
Within half-strata between whole units ..	$V + 2U$
Within whole units between half-units ..	V

Variances :

Half-units within half-strata	..	..	$V + U (N_t - 1)/N_t$
Whole units within half-strata	..	..	$\frac{1}{2} V + U$
Whole units within whole strata	..	..	$\frac{1}{2} V + U + \frac{1}{2} U_t$

Since there will be half as many whole units as half-units the relative precision of the two methods of sampling will be given by

$$\frac{V + 2 U + U_t}{V + U (N_t - 1)/N_t}$$

8.15   Variation in size of strata

When the strata boundaries are arbitrary, the size of the strata may be varied in such a manner that a fixed number of units require to be selected from each stratum, whatever the size of the sample. The strata will naturally be taken as small as possible, i.e. so as to contain two units if a rigorous estimate of error is required, or one unit otherwise.

In order to determine the size of sample required for a given accuracy under these conditions it is necessary to know the relation between the size of the strata and the within-strata variance. The simplest way in which this relation can be determined is to obtain data for all units of a representative sample of the largest strata that are of interest. Strata of any smaller size can then be constructed, and the within-strata variances calculated.

A minor difficulty in this construction is that the original strata will only be exactly subdivisible into strata of smaller size if these contain numbers of units which are integral fractions of the numbers in the original strata. If in area sampling the strata are also to be of the same shape, only squares of integral fractions, i.e. $\frac{1}{4}$, $\frac{1}{9}$, . . ., will give exact subdivision. For strata of intermediate size there will therefore be a certain amount of arbitrariness in the location of the strata boundaries. Some objective rule must therefore be followed. If it appears desirable, overlapping strata may be used. Thus in a case in which the data cover a set of isolated squares, four sets of smaller squares may be taken within each large square, each set having a corner point coincident with one corner of the large square.

If the smallest strata likely to be of interest each contain a large number of units, the collection of data in full for all the units of these basic strata is likely to be laborious. Instead a random sample of such units may be taken. In this case the within-strata variances can be estimated by means of an analysis of variance similar to that used for change in size of sampling units (Section 8.14). If the small units of that section are taken as equivalent to the units of the present case, the large units as equivalent to the basic strata, and the strata as equivalent to the larger strata, the same expressions hold.

This procedure has the disadvantage that variances can only be obtained for strata which contain an integral number of the basic strata—if the strata are all to be of the same shape the number must be a square. This disadvantage can be overcome by sub-stratifying the basic strata, with random selection of units from within these sub-strata. Thus in area sampling with square strata, if each basic stratum is subdivided into nine square sub-strata, with a minimum of two selected units per sub-stratum, square strata can be constructed with areas of 1, $1\frac{7}{9}$, $2\frac{7}{9}$, 4, $5\frac{4}{9}$, $7\frac{1}{9}$, 9, . . . times the area of the basic stratum. Separate analyses of variance will be required for the different sizes of strata, but these have certain elements in common.

When the variances have been calculated for certain sizes of strata an approximate variance–size relationship can be constructed by graphical means.

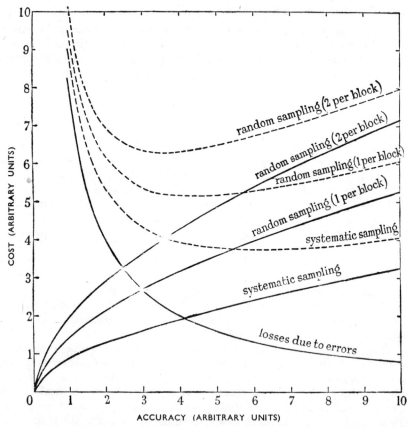

FIG. 8.15—RELATIONS BETWEEN COST AND ACCURACY IN SAMPLING FOR MEAN SOIL TEMPERATURE OVER A PERIOD, WITH STRATA OF VARYING SIZE

The mean temperature is estimated from temperatures taken (a) on two days selected at random from each stratum (block of days), (b) on one day selected at random from each stratum, (c) on days equally spaced throughout the period.

Reproduced by permission of the Royal Society (Yates, 1948, A).

It is often advantageous to plot the log-variance against the log-size. A straight line on this graph represents a variance law of the type $\sigma_z^2 = a z^b$, where z is the number of units per stratum and a and b are constants.

For the purpose of determining the size of sample required for a given accuracy it is better to plot $z\sigma_z^2$ against z. If N is the total number of units in the population the number of units in a sample with two units per stratum will be $2N/z$, and we shall have $z\sigma_z^2 = 2 N V(\bar{y})$. Thus the required size of the strata can be read off from the graph. With one unit per stratum the factor 2 is omitted.

It has been pointed out in Section 3.14 that systematic sampling, when used on the type of material for which it is suitable, is likely to have an error variance which is somewhat less than random sampling with one unit per stratum. In neither type of sampling can the sampling error be estimated with any certainty from the results of a single sample. In random sampling with one unit per stratum, however, an objective estimate of error is possible if additional randomly located units are taken in certain of the strata, but in a systematic sample much more elaborate methods have to be used (Yates, 1948, A), and even then the estimates obtained are not fully objective.

It may be noted here that the common practice of estimating the error of a sample with one unit per stratum by combining the strata in pairs, will give an estimate of error which will generally be somewhat greater than the true error with strata of double the size and two units per stratum.

An example of the relation between the accuracy of sampling with two units per stratum, sampling with one unit per stratum, and systematic sampling, is given in Fig. 8.15. The curves (full lines) are based on the variances found for daily soil temperatures at 1 foot depth, each daily reading constituting a sampling unit. The cost scale is proportional to the number of units, and the accuracy scale gives the accuracy of the sample estimate of the mean soil temperature over a period. The curves themselves are based on relations for σ_z^2 of the type given above. The curve of losses due to errors and the broken curves will be referred to in Section 8.18.

The material is of the type in which the reduction in variance with reduction in size of strata may be expected to be considerable. This is brought out by the curves. The relative precisions of the three types of sampling are given by the intercepts of horizontal lines, which are in the ratio $1 : 1 \cdot 75 : 4 \cdot 24$. The relative efficiencies are given by the reciprocals of the intercepts of vertical lines, which are in the ratio $1 : 1 \cdot 36 : 2 \cdot 22$. This provides an illustration of the marked difference between relative precision and relative efficiency when reduction in the size of the strata results in a considerable reduction in variance per unit.

8.16 Efficiency in terms of cost

In the previous sections we have described how to determine the size of sample necessary to attain results of a given accuracy when various methods of sampling are used. We have also indicated how the relative efficiency (in terms of numbers of sampling units) of different sampling methods and variations in a given method may be judged. Minimization of the number of sampling units or amount of material included in the sample will not in general, however, give maximum efficiency in terms of cost. To attain this the sampling method must be so chosen that the total cost of the survey is minimized.

To minimize the total cost it is necessary to know the relative costs of the different operations. Exact evaluation of these costs is usually troublesome, and is only worth while if an extensive survey has to be undertaken, or if a series of surveys on similar material is contemplated. The matter is complicated by the fact that for many purposes it is the marginal cost of an additional unit, rather than the average cost per unit that is required. Nevertheless it is not difficult in the course of survey operations, or even in the course of a pilot survey, to obtain data which will serve to give rough estimates of the main components of the costs. With the aid of such estimates the efficiency of further surveys of similar material can often be substantially improved.

When information on the costs of different types of operation is available it is possible to determine the values of the sampling fractions, etc., which for a given sampling method will give results of the required accuracy for the least cost. Such values may be termed the *optimal values*. It is also possible to determine which of two methods, each employed in the most efficient manner, will be the least costly.

The determination of optimal values of the sampling fractions, etc., requires minimization of the cost function, and will be dealt with in the next section. The choice between different methods when the optimal values of the sampling fractions, etc., are known, or when there are no variants of this type, can be obtained directly from the results of the previous sections.

Thus in the case in which there is the possibility of using supplementary information, if c_s represents the cost per unit of obtaining the supplementary information, c_0 the marginal cost per unit when no supplementary information is obtained (these costs being taken to include the marginal costs of abstraction and computation), and C_1 represents the additional computational cost of utilizing the supplementary information (which apart from the above marginal cost per unit may be taken as broadly independent of the size of the sample), the total cost of a sample of n_s units with supplementary information, excluding elements of cost which are fixed for both methods, will be

$$C_s = C_1 + n_s \, (c_0 + c_s)$$

and that for a sample of n_0 units without supplementary information will be

$$C_0 = n_0 \, c_0$$

Under conditions in which the error variance is inversely proportional to the number of units in the sample, the two samples will be of equal accuracy when the numbers of units are in inverse ratio to the relative precision of the two methods with equal numbers. If the regression method of adjustment is used, therefore, and the sample is random,

$$n_s/n_0 = 1 - \rho^2$$

where ρ is the true correlation coefficient between the main and the supplementary variates (estimate r).

Hence the use of supplementary information will be more efficient if

$$n_0\, c_0 > C_1 + n_0\,(1 - \rho^2)(c_0 + c_s)$$

i.e. if

$$n_0\,(c_0 + c_s)\,\rho^2 > n_0\,c_s + C_1$$

If the cost of adjustment C_1 can be ignored this inequality becomes

$$c_s/c_0 < \rho^2/(1 - \rho^2)$$

which is independent of n_0, and therefore of the accuracy required. Thus, for example, under these conditions, if $\rho = \frac{1}{2}$ the use of supplementary information will be worth while if the cost of collection is less than one-third the cost of taking an equal number of additional units.

With $\rho = \frac{1}{2}$, however, the gain will not be marked unless the ratio of the costs is considerably less than $\frac{1}{3}$. With a ratio of $\frac{1}{6}$ the total costs will be in the ratio of $7:8$ (minimum value, with zero value of the cost ratio, $3:4$). With higher values of ρ the gains are more marked. With $\rho = \frac{3}{4}$ the two methods have equal cost when $c_s/c_0 = 9/7$. In this case, when the ratio has the values $\frac{1}{2}$ and $\frac{1}{4}$ the ratio of the total costs will be $21:32$ and $35:64$ respectively (minimum value $7:16$).

8.17 Minimization of the cost function

When the sampling fractions, etc., of a method of sampling are not fully determined by the accuracy required in the results, the optimal values can be determined by minimizing the cost function.

The total cost can usually be expressed as a linear function of the numbers of sampling units n_1, n_2, . . . in the various strata, etc., at least to a first approximation, using marginal costs. The simplest procedure is then to add a multiple K of this linear function to the expression in terms of n_1, n_2, . . . for the variance of the required estimate, and differentiate the resultant expression with respect to n_1, n_2, . . . in turn. This minimizes the variance for fixed cost, which is equivalent to minimizing the cost for fixed variance. The exact procedure will be apparent from the first of the cases treated below.

(a) Variable sampling fraction

If the marginal cost of taking an additional unit of the ith stratum is c_i, the total cost C, omitting constant elements, is given by

$$C = \Sigma\, c_i\, n_i$$

Hence

$$\mathbf{V}\,(\Upsilon) = \Sigma\, \sigma_i^2\,(1 - f_i)\,\mathbf{N}_i^2/n_i \qquad (8.17.a)$$
$$= \Sigma\, \sigma_i^2\,(1/n_i - 1/\mathbf{N}_i)\,\mathbf{N}_i^2 + K\,(\Sigma\, c_i\, n_i - C)$$

Differentiating with respect to the n_i and equating to zero, we have the t equations $(i = 1, 2, \ldots t)$

$$- \sigma_i{}^2 N_i{}^2/n_i{}^2 + K c_i = 0$$

Hence, since $n_i/N_i = f_i$,

$$\frac{f_1}{\sigma_1/\sqrt{c_1}} = \frac{f_2}{\sigma_2/\sqrt{c_2}} = \ldots = \frac{1}{\sqrt{K}} \qquad (8.17.b)$$

Thus the optimal sampling fractions are proportional to $\sigma_i/\sqrt{c_i}$. This is an extension of the formula already given in Section 3.5.

The actual values of the sampling fractions required to attain a given accuracy can be obtained by substituting for the f_i in equation 8.17.a and solving for K.* This gives

$$(\sqrt{K}) \Sigma N_i \sigma_i \sqrt{c_i} = V(Y) + \Sigma N_i \sigma_i{}^2 \qquad (8.17.c)$$

If any of the f_i are equal to unity the corresponding terms must be omitted from both sides of the equation. This may require a trial solution.

(b) Two-phase sampling

If c_1 represents the cost per unit of obtaining the first-phase information, c_2 the additional cost per unit of obtaining the second-phase information, and there are n_1 first-phase units, of which n_2 are included in the second phase, the total cost, apart from constant elements, will be given by

$$C = n_1 c_1 + n_2 c_2$$

When the methods of sampling and estimation are such that the effective variances of the estimates at each phase (apart from corrections for finite sampling) are inversely proportional to the numbers of units, from the results of Section 8.7 we have

$$V(\bar{y}) = \frac{1 - f_1}{n_1} \sigma_1{}^2 + \frac{1}{n_2} \left(1 - \frac{n_2}{n_1} \right) \sigma_2{}^2$$

Following the above procedure, we find

$$\frac{n_2{}^2}{n_1{}^2} = \frac{c_1}{c_2} \cdot \frac{\sigma_2{}^2}{\sigma_1{}^2 - \sigma_2{}^2} = \frac{c_1}{c_2} \cdot \frac{\kappa^2}{1 - \kappa^2} \qquad (8.17.d)$$

where $\kappa = \sigma_2/\sigma_1$. The values of n_1 and n_2 required for a given accuracy can be obtained by substituting for n_2 in terms of n_1 in $V(\bar{y})$.

(c) Two-phase point sampling

We will only consider the special case arising in crop estimation (Examples 6.16.b and 7.15).

If the acreages are determined from $n_0{}'$ points and the yields per acre are determined on fields of the crop in question in which n of these points fall, and if c' is the cost of visiting the field to determine the nature of the crop,

---

* An example of the calculations will be found in Example 8.19.

and c the additional cost of a yield determination, we have, when a proportion p of the area is under the crop and the mean yield per acre is $\bar{r}$,

$$C = c' \, n_0' + c \, n$$
$$\mathbf{V}(Y)/Y^2 = q/p \, n_0' + \mathbf{V}(r)/n \, \bar{r}^2 \qquad (8.17.e)$$

Hence

$$\frac{n^2}{n_0'^2} = \frac{p \, \mathbf{V}(r) \, c'}{q \, \bar{r}^2 \, c} \qquad (8.17.f)$$

The values of n_0' and n are best obtained by substitution for n in terms of n_0' in the equation for $\mathbf{V}(Y)$.

(d) Two-stage sampling

If c' is the cost per first-stage unit, and c'' the additional cost per second-stage unit, the total cost is given by

$$C = n' \, c' + n \, c''$$

With a random or stratified random sample with uniform sampling fraction and equal numbers of second-stage units per first-stage unit, $\mathbf{V}(\bar{y})$ is given by formula 8.11.a. Thus

$$\mathbf{V}(\bar{y}) = \mathbf{U}/n' + \mathbf{V}/n + \text{const.},$$

where

$$\mathbf{U} = \sigma_0'^2 - \sigma''^2/N'', \quad \text{and} \quad \mathbf{V} = \sigma''^2.$$

Following the previous procedure, we find

$$n^2/n'^2 = n''^2 = \mathbf{V} \, c'/\mathbf{U} \, c'' \qquad (8.17.g)$$

In other words the number of second-stage units per first-stage unit is independent of the accuracy required. The values of n' and n required for any given accuracy can be obtained by substitution in the equation for $\mathbf{V}(\bar{y})$.

The same formulæ hold for any form of two-stage sampling in which $\mathbf{V}(\bar{y})$ can be written in the above form. Thus stratification with a variable sampling fraction at the second stage is covered, provided none of the sampling fractions are unity.

(e) Two-stage sampling with probability proportional to size at the first stage

The solution of the case of sampling from within strata with probability proportional to size of unit at the first stage follows similar lines. We find that if the cost per second-stage unit is the same for all first-stage units in all strata, one condition for minimum cost is that the second-stage sampling fractions are so chosen that the overall sampling fraction is uniform. Thus the use of a uniform overall sampling fraction, which is computationally convenient, is justified on grounds of minimum cost. The assumption of constant cost per second-stage unit will not in fact hold for the component of cost due to travel, since the same number of second-stage units will be taken from any selected unit of a stratum, and consequently the travel cost per unit will be greater for the larger units. This, however, is not likely to reduce the efficiency greatly unless travel costs at the second stage are very large.

The relation between the first-stage and second-stage sampling can also be very simply expressed. In the case considered in Section 8.13, in which the size of the first-stage units is represented by the number of second-stage units, if all the second-stage variances are equal we may put $\sigma_{ri_0}'^2 - \sigma''^2 N_i'/N_i = U_i$ and $\sigma''^2 = V$, the costs per first and second-stage unit being taken as c_i' and c''. We then have

$$f = \frac{V N + \sqrt{(V/c'')} \, \Sigma \, N_i \, \sqrt{(U_i \, c_i')}}{V \, (Y) + \Sigma \, N_i^2 \, U_i/N_i'}$$

$$n_i'^2 = f^2 \, N_i^2 \, U_i \, c''/V \, c_i'$$

Since the number of second-stage units n_i'' per first-stage unit in stratum i is the same for all selected units, and independent of which particular units are selected, the above equations give $n_i''^2 = V \, c_i'/U_i \, c''$, as before.

(f) Two-stage sampling of farms and fields

Any fully general treatment is difficult owing to the fact that the numbers of fields per farm carrying a given crop are usually small, and consequently what would otherwise be the optimal values of the second-stage sampling fractions will give numbers of fields per farm which are not only non-integral, but which will in many cases be less than unity.

If the numbers of fields per farm are sufficiently large for this source of disturbance to be neglected, the optimal values of the sampling fractions can be simply expressed.

In order to standardize the notation we may replace the between-farms component of variance U_2, as defined in Section 8.12, by U', and the between-fields within-farms component U_1' by U''. The cost of visiting a farm may be taken as c', and that of sampling a field as c''.

We will consider the case in which the farms are divided into size-groups with fields which have mean areas $\bar{a}_1, \bar{a}_2, \ldots$. We will further assume that the mean acreages per field within a size-group of farms of 1, 2, 3, ... fields are the same, and that $V(a_i)/\bar{a}_i^2$ is constant for all size-groups and for all numbers of fields within a size-group.

In the first place we find that in this case the second-stage sampling fractions within a size-group should all have the same value, which is given by

$$f_i''^2 = \frac{U''}{U'} \, \frac{c'}{c''} \, \frac{N_i'}{[N_i']_2} \tag{8.17.h}$$

where $N_{i_1}', N_{i_2}', N_{i_3}', \ldots$ are the numbers of farms in the group with 1, 2, 3, ... fields respectively, and $[N_i']_2 = N_{i_1}' + 4 \, N_{i_2}' + 9 \, N_{i_3}' + \ldots$. The ratios of the first-stage sampling fractions are given by

$$\frac{f_1'^2}{\bar{a}_1^2 \, [N_1']_2/N_1'} = \frac{f_2'^2}{\bar{a}_2^2 \, [N_2']_2/N_2'} = \cdots\cdots = k^2 \text{ say} \tag{8.17.i}$$

These equations will serve to give first approximations to the relative sampling fractions. The relative efficiency of different variants which are practically applicable can then be tested by use of the expression for the variance of the weighted mean given in Section 8.12. It will usually be sufficient to use the mean acreage for each size-group in evaluating the weights, but in evaluating the actual size of sample required the factors $\bar{a}_i^2$ should be replaced by $\bar{a}_i^2 + \mathbf{V}(a_i)$.

Example 8.17.a

Determine, from the data of Examples 6.12.b and 7.12.b, the optimal proportion of sample plots to eye estimates on conifer stands in a two-phase sampling scheme in which eye estimates only are made at the first phase, when the cost of visiting a stand and making an eye estimate is $\frac{1}{10}$ the additional cost of measuring a sample plot.

Consideration of this problem would be relevant if it were possible to demarcate and classify the stands into conifers of over 20 years of age, etc., from aerial photographs. In this case, if the sampling of stands is with probability proportional to size, the variances of the volumes per acre, given in Example 7.12.b, will be required. We then have, with the regression method of estimation,

$$s_1^2 = 4803 \qquad s_2^2 = 3579$$
$$\kappa^2 = 0.745 \qquad \kappa^2/(1 - \kappa^2) = 2.92 \qquad c_1/c_2 = 1/10$$
$$n_2/n_1 = \sqrt{(2.92/10)} = 0.540$$

Thus sample plots should be taken on about one-half the stands which are visited.

There is, however, in this case no appreciable gain by the use of two-phase sampling. If n_2' is the number of sample plots required if no eye estimates are made we have, for equal variance,

$$\frac{1}{n_2'}\sigma_1^2 = \frac{1}{n_1}\sigma_1^2 + \frac{1}{n_2}\left(1 - \frac{n_2}{n_1}\right)\sigma_2^2$$

which gives

$$n_2/n_2' = n_2/n_1 + (1 - n_2/n_1)\kappa^2$$
$$= 0.540 + 0.460 \times 0.745 = 0.883$$
$$n_1/n_2' = 1.64$$

Thus with two-phase sampling

$$C/n_2' c_1 = 1.64 + 10 \times 0.883 = 10.47$$

and with single-phase sampling $C/n_2' c_1$ lies between 10 and 11, depending on the saving due to the omission of the eye estimates on the stands that are visited.

One of the reasons why the use of eye estimates is here of little value is that the determination of the volumes of individual stands by means of a single sample plot per stand is very inaccurate. If more sample plots per stand were taken the overall variance of y would be reduced, while the covariance of y and x and the variance of x would remain unaltered. Under these circumstances two-phase sampling would be more advantageous. Given information on the within-stand variance of the sample plots and the cost of taking different numbers of sample plots from a stand, the optimal number of sample plots per stand could be determined.

Example 8.17.b

If in the crop survey of Examples 6.16.b and 7.15 the additional cost of crop-cutting in order to obtain an estimate of yield is 20 times the cost of visiting a sampling point to ascertain the nature of the crop, calculate the optimal ratio of the number of yield determinations to total sample points, and the number of points required to give an estimate of the total yield with a standard error of 5 per cent.

From the results already given we have

$$p = 2{,}202/33{,}255 = 0 \cdot 0662 \qquad q = 0 \cdot 9338$$
$$\bar{r} = 15 \cdot 7 \qquad V(r) = 3 \cdot 5^2 \qquad V(r)/\bar{r}^2 = 0 \cdot 0497$$

Hence, from equation 8.17.f,

$$\frac{n}{n_0{'}} = \sqrt{\frac{\cdot 0662 \times \cdot 0497}{\cdot 9338 \times 20}} = 0 \cdot 0133$$

Substituting in equation 8.17.e,

$$(\cdot 05)^2 \, n_0{'} = \cdot 9338/\cdot 0662 + \cdot 0497/\cdot 0133$$
$$n_0{'} = 7140 \qquad n = 95 \qquad n' = 473$$

The large number of points that have to be visited to ascertain the crop is accounted for by the small fraction of the total land area under crop. If the crop is an important one it will occupy a considerably larger fraction of the cultivated area, and if, therefore, the non-cultivated areas can be excluded, the total number of points required will be considerably reduced. Alternatively sparsely cultivated areas may be sampled with a lower intensity.

It is also worth noting that if several crops have to be surveyed it will probably be possible to make the acreage determinations of all crops simultaneously prior to the crop-cutting work. This will alter the above cost relationships. The general case can be dealt with by minimization of the combined cost function. In the simple case in which there are a number of crops each occupying the same area and having the same variance and cost relationships, and in which the same accuracy is required for each crop, the above solution holds, the cost of the acreage determinations being spread equally over all the crops, and adjustment of c being made for the cost of revisits. Thus in the above example, with 5 crops and a cost of revisit per point of double the original cost (owing to wider dispersion), all that is necessary is to put $c/c' = 110$. We then find $n_0' = 9160$, $n = 52$, $n' = 606$.

Example 8.17.c

If county lists of farms are not available, and if the cost of the construction of a list of the farms of a parish is 10 times the cost of visiting a single farm within the parish and ascertaining the wheat acreage, determine the optimal sampling fractions at the first and second stage which will give estimates of the acreage of wheat having a standard error of ± 4500 (i.e. approximately 10 per cent.), using the methods of sampling followed in samples B_1 and B_2 of Hertfordshire farms.

From the equation for $V(Y)$ in terms of n' and n given in Example 8.10.a and equation 8.17.g, we have

$$n'' = \sqrt{(2524 \cdot 2 \times 10/80 \cdot 71)} = 17 \cdot 7$$

Hence, for the required accuracy,

$$4 \cdot 5^2 = 80 \cdot 71/n' + 2524 \cdot 2/17 \cdot 7\ n' - 1 \cdot 8982$$
$$n' = 10 \cdot 1$$
$$n = 178 \cdot 8$$

Similarly, from Example 8.10.b,

$$n'' = 7 \cdot 0 \qquad n' = 11 \cdot 2 \qquad n = 78 \cdot 4$$

For the reasons already given these latter values are approximate. When proper allowance is made for the changes in sampling fractions at the second stage a somewhat smaller sample will be found to be necessary.

Example 8.17.d

From the data of Table 6.19.a determine suitable sampling fractions for a crop-estimation scheme for sugar beet, on the assumption that variances between fields on the same farm, and between farms, are the same as those found for wheat in Section 8.12, and that the cost of visiting a farm is (a) equal to, and (b) twice that of sampling a field.

From Table 6.19.a, including farms not growing sugar beet on old arable land, we obtain the following values:

			n_i'	$[n_i']_2$	$\bar{a}_i$
Small farms	...	...	30	51	3·1
Medium farms	...	...	47	224	8·4
Large farms	...	...	14	121	12·3

Taking the estimates of $N_i'/[N_i']_2$ given by $n_i'/[n_i']_2$, formulæ 8.17.h and 8.17.i give the following values for f_i'' and f_i' :

	f_i''	f_i''	f_i'/k
	$c_1 = c_2$	$c_1 = \frac{1}{2} c_2$	All c
Small farms	0·86	1·21	4·0
Medium farms	0·51	0·72	18·3
Large farms	0·38	0·54	36·2

Thus when $c_1 = c_2$ we may consider variants on the following scheme. For small farms sample all fields ; for medium farms sample one field from farms growing 1 or 2 fields, two fields from farms growing 3 or 4 fields, etc. ; for large farms sample one field from farms growing 1, 2, 3 or 4 fields, two fields from farms growing 5, 6 or 7 fields, etc. ; sample farms in the proportions 1 : 4 : 8.

A similar scheme can be drawn up for the case when $c_1 = \frac{1}{2} c_2$. In this case, since f_1'' is greater than unity, some increase in the proportion of small farms included may be advisable.

Further investigation is left to the reader.

8.18 Losses due to errors

We have so far considered the minimization of the cost of a survey when a given accuracy is required in the results. Since, however, errors in the results themselves give rise to losses when these results are used as a basis for further action, the accuracy should itself be determined in such a manner that the sum of the cost of the survey and the expected losses due to the resultant errors is minimized.

If the loss due to an error Z in an estimate Y is equal to $a Z^2$, where a is a constant, the average loss in a series of samples of the same size and type in which the estimates are free from bias will be $a \mathbf{V}(Y)$, whatever the actual form of the distribution function of the errors.

When the loss due to an error is proportional to the square of the error, therefore, minimization of the sum of the cost C of a survey and the average loss due to errors requires minimization of the function

$$C + a \mathbf{V}(Y)$$

the sampling method and size of sample being so chosen that $\mathbf{V}(Y)$ has its minimum possible value for the cost C.

Under these circumstances $\mathbf{V}(Y)$ will always be expressible as a function of C. In many cases, as we have seen, this function is of the form

$$\mathbf{V}(Y) = \frac{h}{C - C_o} - k$$

where h and k are constants depending on the population which is being surveyed, and C_o is the overhead or constant component of cost which is independent of the size of the survey.

In this case the minimum value will be attained when

$$(C - C_0)^2 = a\,h$$

We then have

$$V\,(Y) = \frac{\sqrt{h}}{\sqrt{a}} - k$$

This implies that the more accurate the results that can be obtained with a given cost, i.e. the smaller the value of h, the higher should be the accuracy aimed at with a given loss function. The value of the cost-plus-loss function at the minimum is in fact $2\,C - C_0 - a\,k$. Any saving due to increased accuracy should therefore be divided equally between reduction in the cost of the survey and reduction in the loss due to errors. Equally if the loss due to a given error is multiplied by a factor λ, the funds devoted to the survey (excluding overhead) should be multiplied by $\sqrt{\lambda}$.

The same general conclusions hold when the variance–cost relation is of a more complicated form than that given above. A case of this type is illustrated in Fig. 8.15, in which a loss curve of the form $a\,V\,(Y)$ has been inserted. We see that with the more accurate methods of sampling the minima of the cost-plus-loss functions (shown by broken lines) are attained when both the cost of the sampling and the loss due to errors are less than with the less accurate methods.

Other loss functions will lead to more complicated expressions for the average loss. The most general loss function which is capable of relatively simple expression in terms of $V\,(Y)$ is that in which the loss due to a positive error is equal to $a\,Z^b$, and that due to a negative error is $a'\,(-Z)^b$, a, a' and b being constants. Provided the distribution of errors has the same form for all values of $V\,(Y)$, the average loss is then equal to $a''\,\sigma_0{}^b$, where $\sigma_0{}^2 = V\,(Y)$ and a'' has a value which is a linear function of a and a'. The actual linear function can only be determined if the distribution function of the errors is known. In general terms, if the distribution function of the errors of Y is $f_1\,(z)\,dz$, where $z = Z/\sigma_0$, we have

$$a'' = a \int_0^{+\infty} z^b f_1\,(z)\,dz + a' \int_{-\infty}^{0} (-z)^b f_1\,(z)\,dz$$

If the distribution of the errors is normal, $f_1\,(z)\,dz$ will be of the form given in Section 7.3, with $\sigma = 1$. The two integrals will in this case (as in any symmetrical distribution) be equal. Their values for any value of b can be obtained from existing tables*, those for $b = 1\cdot0$, $1\cdot25$, $1\cdot5$, $1\cdot75$, $2\cdot0$ being $0\cdot3989$, $0\cdot4097$, $0\cdot4300$, $0\cdot4599$, $0\cdot5$ respectively. It must be emphasized, however, that the distribution of sampling errors is frequently not sufficiently normal for the use of these values to be justified. In such cases, also, the form of the distribution may be expected to change with change in the size of the sample.

* Tables of the Gamma function.

With this more general loss function we require to minimize the function

$$C + a'' \left(\frac{h}{C - C_0} - k \right)^{b/2}$$

which will be minimum when

$$(C - C_0)^2 \left(\frac{h}{C - C_0} - k \right)^{1 - b/2} = \tfrac{1}{2} a'' b h$$

This equation can easily be solved by trial and error, or directly if k can be neglected, as will be the case when all sampling fractions are small. The same general conclusion that the accuracy should be increased with a more accurate method of survey still holds.

When $\mathbf{V}(Y)$ is a more complicated function of C (possibly only determined in numerical form) the minimum of the cost-plus-loss function can itself be determined by trial and error.

8.19   Optimal allocation for more than one variate

When an estimate of the population total or mean of a single variate is required, the optimal sampling fractions for a stratified random sample with variable sampling fraction are given by the equations of Section 8.17 (a). If estimates for two or more variates with different within-strata variances are required, these equations will give different values of the sampling fractions. It is often asked how the sampling fractions should be chosen in such circumstances.

Suppose there are two variates which are denoted by y and y' with within-strata variances σ_i^2 and $\sigma_i'^2$. Three cases arise. Firstly, if sampling fractions are chosen which are optimal for Y it may be found that Y' is estimated with more than the required accuracy. Secondly, Y may be estimated with more than the required accuracy when the sampling fractions are optimal for Y'. Thirdly, neither of these conditions may hold. In the first two cases no problem arises, since we choose the sampling fractions which are optimal for Y or Y' respectively. In the third case, as mentioned in Section 3.5, sampling fractions which are sufficiently nearly optimal can often be determined by compromise, without any exact investigation, but a formal solution is possible on the lines of Section 8.17.

Without imposing additional conditions we cannot minimize the variance for a given cost, since two variances are involved. We can, however, minimize the cost for given values of $\mathbf{V}(Y)$ and $\mathbf{V}(Y')$. If these values are denoted by A and A' and the cost is C, we have

$$\mathbf{V}(Y) = \Sigma \, \sigma_i^2 \, (1 - f_i) \, \mathbf{N}_i^2 / n_i = \Sigma \, \mathbf{N}_i \, \sigma_i^2 \, (g_i - 1) = A \qquad (8.19.a)$$
$$\mathbf{V}(Y') = \Sigma \, \sigma_i'^2 \, (1 - f_i) \, \mathbf{N}_i^2 / n_i = \Sigma \, \mathbf{N}_i \, \sigma_i'^2 \, (g_i - 1) = A' \qquad (8.19.b)$$
$$C = \Sigma \, c_i \, n_i$$

We require to minimize C subject to conditions 8.19.a and 8.19.b. Following the procedure of Section 8.17 we may add multiples L and L' of these conditions and differentiate with respect to each n_i in turn. Equating the differentials to zero we obtain the equations

$$c_i - L\,\sigma_i{}^2/f_i{}^2 - L'\,\sigma_i{}'^2/f_i{}^2 = 0$$

Hence

$$f_i = \frac{\sqrt{(L\,\sigma_i{}^2 + L'\,\sigma_i{}'^2)}}{\sqrt{c_i}} \qquad (8.19.c)$$

Extension to three or more variates merely requires the insertion of additional terms $L''\,\sigma_i{}''^2$, etc., in the numerator.

Substitution for the $g_i\,(=1/f_i)$ in equations 8.19.a and 8.19.b will give two simultaneous equations for L and L'. Unfortunately these equations cannot be easily solved. A solution by successive approximation on the lines of the example below is therefore necessary.

If we put $L = \lambda/K$ and $L' = \lambda'/K$ with $\lambda + \lambda' = 1$, so that $L + L' = 1/K$, we have the alternative form

$$f_i = \frac{1}{\sqrt{K}}\;\frac{\sqrt{(\lambda\,\sigma_i{}^2 + \lambda'\,\sigma_i{}'^2)}}{\sqrt{c_i}} \qquad (8.19.d)$$

The numerator is now a weighted mean of $\sigma_i{}^2$ and $\sigma_i{}'^2$, and the analogy with equation 8.17.b will be apparent.

Putting $\sqrt{c_i}/\sqrt{(\lambda\,\sigma_i{}^2 + \lambda'\sigma_i{}'^2)} = g_{i0}$ we have $g_i = g_{i0}\sqrt{K}$, and therefore, from equations 8.19.a and 8.19.b,

$$\sqrt{K}\,.\,\Sigma\,\mathbf{N}_i\,\sigma_i{}^2\,g_{i0} = \mathbf{V}\,(\mathbf{Y}) + \Sigma\,\mathbf{N}_i\,\sigma_i{}^2 \qquad (8.19.e)$$

$$\sqrt{K}\,.\,\Sigma\,\mathbf{N}_i\,\sigma_i{}'^2\,g_{i0} = \mathbf{V}\,(\mathbf{Y}') + \Sigma\,\mathbf{N}_i\,\sigma_i{}'^2 \qquad (8.19.f)$$

For any assumed value of λ the value of $\sqrt{K}$ which gives $\mathbf{V}\,(\mathbf{Y}) = A$ can therefore be determined from equation 8.19.e. The value of $\mathbf{V}\,(\mathbf{Y}')$ for this value of $\sqrt{K}$ can then be found from equation 8.19.f. The value of λ which gives $\mathbf{V}\,(\mathbf{Y}') = A'$ under these circumstances can thus be determined by successive approximation, taking different trial values of λ. This provides an alternative method of solution. If desired the roles of y and y' can be interchanged in this process. With more than two variates this alternative method of solution is preferable, since one dimension is thereby eliminated.

Equations 8.19.e and 8.19.f are also of use when trial values of L and L' are taken directly. With such trial values neither $\mathbf{V}\,(\mathbf{Y})$ nor $\mathbf{V}\,(\mathbf{Y}')$ will have the required value. We can, however, adjust both L and L' by a factor θ, so chosen that neither $\mathbf{V}\,(\mathbf{Y})$ nor $\mathbf{V}\,(\mathbf{Y}')$ exceeds its required value. The sampling fractions will then all be multiplied by a factor $\sqrt{\theta}$. The value of $\sqrt{\theta}$ is given by equation 8.19.e or 8.19.f (the larger value being taken) with $\sqrt{K}$ replaced by $1/\sqrt{\theta}$ and g_{i0} by the raising factors calculated from the trial values of L and L'.

If any of the f_i are found to be greater than unity the solution must be revised, as indicated in Section 8.17 (a).

The problem is also capable of re-formulation in terms of the minimization of costs plus losses and in this form has a direct solution. If, following Section 8.18, the expected losses are taken to be $a\,\mathbf{V}\,(\mathbf{Y})$ and $a'\,\mathbf{V}\,(\mathbf{Y}')$ and we minimize the costs plus losses, we shall obtain the equations 8.19.c with L and L' replaced by a and a'. If a and a' are known, these equations can be solved immediately.

If a and a' are not known but the ratio between a and a' can be assumed, we have $\lambda = a/(a + a')$ and $\lambda' = a'/(a + a')$, and a value of $\sqrt{K}$ can be chosen which gives the required general level of accuracy, using equations 8.19.e and 8.19.f. In a survey to determine the amount of unemployment in various industries stratified by districts, for example, it might be reasonable to attach equal importance to errors in the unemployment totals proportional to the square roots of the numbers employed in the industries, in which case a, a', etc., would be taken inversely proportional to these square roots. It should be noted, however, that this does not imply that the actual standard errors will be in the ratio of these square roots. The ratio obtained will be given by equations 8.19.e and 8.19.f and may differ substantially from the ratio of the square roots.

Example 8.19

Using the data on Hertfordshire farms described in Section 3.7, etc., and the size-groups of Example 6.7, determine the sampling fractions which will be optimal for simultaneously estimating the number of farms growing wheat and the total wheat acreage, each with a standard error of 5 per cent. The cost per farm is to be taken to be the same for all strata, i.e. all $c_i = 1$.

We require the numbers of farms in each stratum, the proportion of farms growing wheat in each stratum, and the within-strata variances of the acreages of wheat of individual farms. Composite estimates from the various tables given in the preceding examples have been made. (The actual values could, of course, have been calculated from the original data, but these were not

TABLE 8.19.a—HERTFORDSHIRE FARM DATA

Stratum	Size-group	N_i	p_i	s_i^2	$s_i'^2$	$N_i s_i^2$	$N_i s_i'^2$	s_i	s_i'
1	1 —	440	0	0	0	0	0	0	0
2	6 —	520	·04	·0384	2	20·0	1040	·19596	1·414
3	21 —	360	·2	·16	15	57·6	5400	·4	3·873
4	51 —	520	·6	·24	160	124·8	83200	·48990	12·649
5	151 —	400	·8	·16	650	64	260000	·4	25·495
6	301 —	210	·9	·09	1700	18·9	357000	·3	41·231
7	501 —	50	1·0	0	4500	0	225000	0	67·082
		2500				285·3	931640		

readily available.) These are shown in Table 8.19.a, where p_i denotes the proportion of farms growing wheat and $s_i'^2$ represents the within-strata variances of the wheat acreages. There is no need to describe how the estimates were obtained.

From the table the estimate of the number of farms growing wheat, denoted for convenience by Y, is

$$Y = \Sigma\, N_i\, p_i = 963 \cdot 8.$$

A standard error of 5 per cent. will therefore be equivalent to a variance of $48 \cdot 19^2$ or 2322. The total wheat acreage is 44,676 acres (Section 3.7) and the corresponding variance is therefore 4,990,000. The values of s_i^2 for numbers of farms are given by the equation

$$s_i^2 = p_i\, q_i$$

These values are tabulated in Table 8.19.a.

As a first step the sampling fractions (or raising factors) which are optimal for each variate separately should be calculated. Following the method of Section 8.17 (a) we require to calculate $\Sigma\, N_i\, s_i\, \sqrt{c_i}$ and $\Sigma\, N_i\, s_i^2$ and the similar functions for wheat acreage. $N_i\, s_i^2$ and $N_i\, s_i'^2$, which are required in the subsequent calculations, are tabulated in Table 8.19.a, as are s_i and s_i'. We then find

$$\Sigma\, N_i\, s_i\, \sqrt{c_i} = 723 \cdot 65 \qquad \Sigma\, N_i\, s_i^2 = 285 \cdot 3$$
$$\Sigma\, N_i\, s_i'\sqrt{c_i} = 30918 \qquad \Sigma\, N_i\, s_i'^2 = 931640$$

Hence from equation 8.17.c we have, for the optimal sampling fractions for number of farms,

$$1/\sqrt{K} = 723 \cdot 65/(2322 + 285 \cdot 3) = 0 \cdot 2775$$
$$L_0 = 1/K = \cdot 07701$$

The values of f_i may now be calculated from equation 8.17.b, using the values for s_i given in Table 8.19.a. These values are shown in Table 8.19.b.

TABLE 8.19.b—OPTIMAL SAMPLING FRACTIONS FOR NUMBER OF FARMS, FOR ACREAGE, AND FOR BOTH VARIATES SIMULTANEOUSLY

Stratum	No. of farms	Acreage	Both variates
2	·0544	·0074	·0478
3	·1110	·0202	·0981
4	·1359	·0660	·1268
5	·1110	·1331	·1313
6	·0832	·2153	·1603
7	0	·3502	·2324

Similarly for the optimal sampling fractions for acreage

$$1/\sqrt{K'} = \cdot005221$$
$$L_0' = \cdot00002726$$

giving the values of f_i for acreage of Table 8.19.b.

It is clear that neither of these two sets of sampling fractions gives the requisite accuracy in the other variate. Something intermediate is therefore required. As a first trial (a) we may take $L = \cdot04$ and $L' = \cdot000013$, i.e. roughly half L_0 and L_0'. The intermediate calculations for $\text{V}(Y)$ and $\text{V}(Y')$ are given in Table 8.19.c. The values for g_i can be calculated on the slide

TABLE 8.19.c—VALUES OF f_i AND g_i FOR $L = \cdot04$, $L' = \cdot000013$

Stratum	$Ls_i^2 + L's_i'^2$	f_i	g_i
2	·001562	·0395	25·317
3	·006595	·0812	12·315
4	·011680	·1081	9·251
5	·014850	·1219	8·203
6	·025700	·1603	6·238
7	·058500	·2419	4·134
	·118887		

rule using equation 8.19.c. From equations 8.19.a and 8.19.b we then find

$$\text{V}(Y) = 2728$$
$$\text{V}(Y') = 5220000$$

Both of these values are too high. Further trial values (b) $L = \cdot06$, $L' = \cdot000013$ and (c) $L = \cdot06$, $L' = \cdot000010$ were therefore taken. The values of $\text{V}(Y)$ and $\text{V}(Y')$ obtained were

(b) $L = \cdot06$, $L' = \cdot000013$ (c) $L = \cdot06$, $L' = \cdot000010$
$$\text{V}(Y) = 2274 \qquad\qquad \text{V}(Y) = 2332$$
$$\text{V}(Y') = 4812000 \qquad\qquad \text{V}(Y') = 5301000$$

If three points A, B and C with coordinates corresponding to the values (a), (b) and (c) of L and L' are plotted, the points P and Q on the lines AB and BC where $\text{V}(Y)$ is estimated to have the required value 2322 can be determined by linear interpolation. The line PQ then represents an approximation to the curve of values of L and L' for which $\text{V}(Y)$ has the required value. A similar line can be drawn for $\text{V}(Y')$. These two lines are found to intersect at the point $L = \cdot059$, $L' = \cdot0000120$. A check computation of the values of $\text{V}(Y)$ and $\text{V}(Y')$ for these values of L and L' gives

$$\text{V}(Y) = 2310 \qquad\qquad \text{V}(Y') = 4972000$$

Hence L and L' have been determined with all necessary accuracy. The resultant optimal values of f_i are included in Table 8.19.b.

The alternative method of solution using λ and λ' gives similar results and is left as an exercise to the reader.

It is worth noting that all three sets of trial values give sampling fractions which, after adjustment so that neither variance exceeds its required value, are reasonably efficient. In order to compare the efficiencies the sampling fractions given by the chosen values of L and L' must be adjusted by multiplication by $\sqrt{\theta}$, calculated as explained above. After these adjustments the numbers of farms in the samples and the relative efficiencies are found to be:

	λ'	No. of farms	Relative efficiency
(a)	·000325	231·7	96·4
(b)	·000217	223·9	99·8
(c)	·000167	230·5	96·9
Optimal	·000203	223·4	100

8.20 Choice of probability function in sampling with variable probability

Sampling with probability proportional to size is one form of what may be termed *sampling with variable probability*. The essence of such sampling is that the probability of selection of the different individual units is made proportional to some known quantitative characteristic of the units themselves. The same general theory will apply whatever the measure adopted. In two-stage sampling the number of second-stage units often provides a convenient measure of the size of the first-stage units (Section 8.13), but in some cases greater efficiency will be obtained if the probabilities are taken proportional to some function of the number of second-stage units, instead of proportional to the actual number.

This question has been discussed by Hansen and Hurwitz (1949, A'). They consider the case of two-stage sampling with uniform overall probability (self-weighting) and a cost function of the form

$$C = c_1 n_1 + c_2 n_2 + c_3 n_3$$

where n_1 is the number of selected first-stage (primary) units, n_2 is the expected number of second-stage units included in the selected first-stage units, and n_3 is the number of second-stage units included in the sample. The first and third terms of this cost function correspond to the terms of the cost function of Section 8.17 (*d*). The second term arises when a frame has to be constructed for each selected first-stage unit, e.g. by listing or ground survey, and there are other costs, e.g. of travel, which are proportional to the total number of second-stage units in the selected first-stage units. Taking an illustrative example from the 1940 Census of Housing, Hansen and Hurwitz tabulate the relative efficiencies of selecting first-stage units with (a) equal probability, (b) probability proportional to the square root of the number of second-stage units, and (c) probability proportional to the number of second-stage units, for various ratios between c_1, c_2 and c_3. For most of the chosen ratios, selection with probability proportional to the square root of the number of second-stage units is most efficient. Even when c_2 is zero, method (b) is only slightly less efficient than method (c).

8.21   Concluding remarks

The preceding sections give an indication of the ways in which the efficiencies of different sampling methods can be compared, and the techniques of determining the optimal sampling fractions, size of sample required for a given accuracy, etc. It has been further shown that the accuracy which should be aimed at is itself related to the losses resulting from errors in the survey results.

Since determination of the optimal accuracy from the expected losses due to errors demands knowledge of the loss-function, it will chiefly be of relevance when action in the economic sphere has to be based on the results of the survey. An error in the estimate of the yield of a crop, for instance, may require changes in an import programme, or may lead to wastage, and the resultant additional costs may be assessable, at least roughly. The losses due to errors in estimates provided by surveys of the research and investigational type can scarcely be assessed. Indeed, it is usually impossible to give any quantitative estimate in monetary terms of the value of the information provided by such surveys. The decision to undertake the survey, and the accuracy aimed at, must then be a matter of judgment on the part of those who require the information, and those who are concerned with the allocation of resources.

Even if the optimal accuracy cannot be quantitatively determined, arbitrary decisions on the accuracy required should as far as possible be avoided. Before any decision as to accuracy is taken, estimates should be prepared of the costs of obtaining results of differing degrees of accuracy, and these estimates should be considered in relation to the purposes for which the results are required.

Minimization of costs can of course be carried out whether or not a loss-function is available. In this chapter we have only considered this minimization when a single quantity requires estimation.* In most censuses and surveys such treatment would be an over-simplification. A number of quantities will require to be estimated, frequently for many domains of study. It may then be necessary to carry out a more elaborate investigation, minimizing the cost for defined accuracies of all the estimated quantities. Alternatively, if loss-functions are available for all of these quantities, the combined cost-plus-loss function can be minimized. Frequently, however, one of the quantities is of dominant importance, and the situation is such that when adequate accuracy is attained on this quantity the remaining quantities are determined with more than the required accuracy. In this case minimization can be conducted solely with reference to this quantity.

* Except in Section 8.19.

Many of the examples worked out in this chapter are based on very small amounts of data, and the conclusions reached on the relative efficiencies of different methods, even in the particular circumstances of the chosen examples, must therefore be treated with reserve. These examples are, in fact, merely intended to illustrate the computational procedures, and bring out the various points that have to be taken into account when making calculations of relative efficiencies, optimal sampling fractions and size of sample. They are in no way intended as general investigations into the relative efficiency of the different methods.

On the other hand, it should be borne in mind that no very exact determinations of the optimal sampling fractions and size of sample are required in the practical planning of surveys. If the values adopted are somewhere near the optimal the total cost, or the total of costs-plus-losses, will be very near the minimum.

We must also not be deterred from undertaking a survey by the fact that there is little information on which to base exact planning of the sampling methods. As we have seen, surveys themselves provide information which will enable future surveys on similar material to be more efficiently planned. In surveys on relatively unknown material one of the points to be kept in mind in the planning is that information will be required both on variances and on costs. Equally, if preliminary rough estimates are required, pilot investigations can be designed so as to provide such estimates, as well as information on which to base the planning of a larger survey.

The study of the relative efficiency of different sampling methods depends not so much on having a large amount of data as on having data which are relevant to the methods concerned. Thus the small pilot investigation on the estimation of wheat yields by sampling methods described in Section 8.12 was of sufficient size to give estimates of both the field-to-field and farm-to-farm components of variance with all necessary accuracy. On the other hand, in the early years of the Survey of Fertilizer Practice—although a large amount of data was accumulated—it was impossible to determine the field-to-field components of variation of the fertilizer dressings, since only one old- and one new-arable field of each crop was taken on each farm. This must be regarded as a defect in the planning of this survey, which could have been remedied had a pair of fields been taken for the various crops on a small proportion of the farms. Incidentally, lack of this information also prevented any consideration of the question of the extent to which individual farmers vary their fertilizer practice from field to field of the same crop.*

Given the necessary data, the increase in the efficiency of survey methods requires proper statistical investigations of the types outlined in this chapter. The need for thorough investigation of the efficiencies of different sampling methods in different circumstances is great, and it is to be hoped that many more will be made and reported in the future. Such investigations are often

* This was remedied shortly after the above was written—see footnote on p. 86.

neglected because, once a survey has been completed, the question of whether it could have been carried out more efficiently is largely historical as far as that survey is concerned. One of the reasons why both the theory and practice of sample censuses and surveys has made rapid advances in recent years is that permanent organizations—often part of, or attached to, research statistical institutes—have been set up in a number of countries. These organizations have been actively engaged both in the planning and the execution of surveys covering various fields of enquiry. Consequently they have not only had access to the necessary data, or the means of collecting it, but they have also had a continuing interest in investigations of efficiency, and a body of workers who have both the training and experience to carry out these investigations.

Further progress may be expected on the same lines. In particular the problems that arise in censuses and surveys of undeveloped areas will be likely to receive very much more thorough investigation when more centres which are actively concerned with the planning and execution of surveys in these areas are developed. Only in this way will a body of experience be built up which is relevant to the special problems of such surveys.

CHAPTER 9

CRITICAL ANALYSIS OF SURVEY DATA

9.1 Introduction

As has been pointed out at various places in the preceding chapters, surveys fall into two main classes : those which have as their object the assessment of the characteristics of the population or different parts of it, and those which are investigational in character. In the census type of survey, estimates of the characteristics, quantitative and qualitative, of the whole population and usually also of various previously defined subdivisions of it are required. These estimates form the basis of administrative action, either directly or after incorporation with information from other sources. The accuracy to be aimed at is determined by the nature of the administrative action that is envisaged. In the investigational type of survey we are more concerned with the study of relationships between different variates, and with contrasts between different domains. In such surveys estimates appertaining to the whole population are usually of relatively minor interest.

The critical analysis of the results of an investigational survey is a much more difficult task than is the calculation of estimates and their errors in a survey of the census type. The matter has already been briefly discussed in Sections 5.19, 5.20 and 5.21. The present chapter continues this discussion at a more advanced level, and includes some additional examples. It must be emphasized, however, that the chapter is not intended to be an exhaustive treatise on the analysis of investigational surveys : this would require much more space than is available here.

9.2 Inferences on causal relations

At the outset a clear distinction must be made between the types of deduction that can be made with certainty from survey data and the types that are speculative. If in a nutrition survey, for example, we find that children of large families are worse fed than children of small families we can draw the definite conclusion that size of family is associated with malnutrition of the children, and we can give quantitative estimates of the degree of malnutrition actually existing amongst children of families of different sizes. We cannot, however, assert with certainty that size of family is the cause of this malnutrition, though the fact that in large families the income per head is automatically less if there is a fixed total income would lead us to expect an underlying causal relationship.

Even in situations where a definite causal relationship is known to exist, deductions as to the magnitudes of the effects of given factors can never be made with certainty from survey data. We may, for instance, find that fields receiving fertilizers give higher yields per acre than fields without fertilizers. Yet we cannot attribute the observed differences solely to differences in fertilizers. The farmers using the fertilizers may be farming better land, they may be growing higher-yielding varieties, and they may be carrying out their farming operations with greater skill.

Clearly definable extraneous factors which may influence the effects of other factors on some quantity of interest such as the yield of a crop can be determined in the course of the survey.   But there will always be other undetermined and possibly unascertainable factors which cannot be taken into account.

Even in cases in which there is reasonable confidence that values of all factors of importance have been ascertained, it may not be possible to determine and separate their effects with any degree of confidence, because some or all of them are closely correlated.   Furthermore, if one of two correlated factors is imperfectly ascertained, not only will its apparent effect be too small, but the apparent effect of the factor with which it is correlated will be too large.   Thus, for example, the standard of nutrition of the children of a family is likely to depend both on income and on knowledge of what constitutes good nutrition. The better educated parents may be expected to be more knowledgeable in this respect, but general educational attainment will certainly be an imperfect measure of this knowledge.   Degree of educational attainment will also be correlated with income.   Consequently, if steps are taken to increase the income of large families, e.g. by child allowances, the improvement in the nutrition of the children of these families resulting from this increase may well be markedly less than the estimate obtained from a survey in which general educational attainment is taken as a measure of knowledge of nutrition.

In order to determine with certainty the magnitude in the causal sense of the effect of any given factor, experiments must be undertaken.   Surveys cannot be regarded as equivalent to experiments.   Nevertheless they are of value in situations in which experiments are difficult or impossible, though in such cases all conclusions must be tentative.   They are also of value as a preliminary to experimental work, since they frequently indicate the factors that are likely to be most worth investigation.

9.3   What constitutes an experiment?

The essence of an experiment is that the effect of some treatment or change of condition which is of interest is measured by actually imposing the treatment on some of the units and observing the resulting difference between these units and those on which the treatment is not imposed.   In a survey, on the other hand, the effect of the treatment is assessed by observing the difference between the units which happened to have had the treatment and those which have not, eliminating as far as possible differential effects of other factors.

The basic requirements of a good experiment are that each of the treatments that are to be compared should be applied to several experimental units, i.e.

there should be replication, and that the units chosen for the experiment should be assigned at random to the different treatments. Random assignment of the units ensures that no treatment is unduly favoured. Replication, besides its obvious effect in increasing the accuracy of the results, provides an estimate of the experimental error; this is fully valid (in the statistical sense) if random assignment has been adhered to.

Accuracy can be further increased by various other devices. The simplest, the so-called randomized block design, is to form groups or " blocks " of experimental units each of which contains as many units as there are treatments, in such a manner that the units within a group are as alike as possible. The treatments are then randomly assigned to the units of a group, so that each group contains one replicate of each treatment. Only the differences between experimental units in the same group then contribute to experimental error.

Choice of experimental unit depends on the nature of the experimental material and the problems that are to be investigated. For some agricultural problems fields or even farms might be used, but for experiments on fertilizers, crop varieties, etc., which can be tested on small plots, a set of plots on a single field results in much greater accuracy, and has other advantages. The results of an experiment on a single field relate, of course, only to that field in a particular year. Such experiments have therefore to be repeated on different fields over a run of years to obtain results of general validity.

For many sociological and economic problems small experimental units are impracticable. In our child nutrition example, direct supplementation of the diet of children of single families might be feasible, but if the effect of supplementation by school meals is to be compared with supplementation of income by child allowances the school is probably the smallest practicable unit, income being supplemented (nominally as a compensating measure) for children not attending these schools. If, in addition, we wish to obtain a reliable estimate of the improvement resulting from these measures, we require some control experimental units with no supplementation. For these, local authority areas are probably the smallest that are practicable. We can then with advantage use a two-level hierarchy of experimental units as in Table 9.3. This is analogous to the use of split plots in agricultural field trials. To satisfy the randomization requirements, local authorities must be selected who are willing to take part in a genuine experiment and agree to abide by the

TABLE 9.3—A HIERARCHICAL EXPERIMENTAL DESIGN

Level	Units	Treatments		
1	Local authorities	(a) No benefit	(b) Benefit	
2	Schools	—	(b₁) School meals	(b₂) Supplementary income

luck of the draw as to whether they do or do not provide benefit (which of course may be financed from outside sources), and also be able and willing to provide facilities for school meals at such schools as are selected for this and not for the others.

Various other devices may be used to increase the accuracy and reduce the cost of an experiment of this kind. We may, for example, form like pairs of local authorities at level 1, and of schools at level 2. Sub-samples of families may be selected for detailed medical observations, and the selection of these may be based on a larger sample of families on which preliminary medical observations are made.

9.4   Use of ratios and regressions in investigational work

Ratios and regressions have many uses in investigational work. Whenever the nature of a pair of variates is such that the ratio between them may be expected to be relatively constant then the replacement of the variates by some estimate of the ratio is likely to simplify considerably the interpretation of the data. The choice of estimate depends on the nature of the variability. If x and y do not vary very widely then the unweighted means of the individual ratios $r\ (=y/x)$ calculated from the pairs of values are likely to give the most accurate results. Under such circumstances the biases of the unweighted means are usually of relatively little importance in comparative work. Once the individual ratios have been calculated the unweighted means are less trouble to compute than any form of weighted mean, particularly when many alternative groupings of the data are required. If on the other hand x and y vary widely (and particularly if there are some very small values) the estimates $r = S(y)/S(x)$ may be subject to smaller errors, as well as being unbiased.

Several examples where ratios have proved of use have already been given. In Example 6.19 both types of estimate were examined.

It is important to recognize that the use of a ratio does not imply that its value is necessarily constant over the range of x. If it appears desirable the relation between r and x (or r and y) can be examined, e.g. by the use of a regression of r on x (or r on y). It is sometimes objected that this procedure is inadmissible since random errors in x will give rise to a negative correlation between r and x. This objection is, however, only valid if x and y are in fact subject to random errors which are independent. In a study of the relation of earnings to factory size in an industry, for example, the size, as represented by the number of workers, will usually be correctly ascertained, and the relation between the average earnings per worker and size will therefore not have any spurious component of correlation ; by taking earnings per worker and size instead of total wage-bill and size as the variates for analysis we obtain the data in a form which is considerably easier to study.

When the value of the ratio changes considerably over the range of x it may be more appropriate to use the regression of y on x. Either a linear or a curved regression may be used. The calculation of a linear regression has been described in Section 6.12.

In investigational work we may be interested not only in the relation of y to x over the whole population, but also in differences in this relationship for different domains of study. In such cases ratios or regression lines can be calculated for each domain separately. For comparative purposes slight departure from the assumed law is often of little consequence. Thus *provided the mean of x is similar for the different domains* linear regressions may be used when some degree of curvature in the lines is apparent from the data. If, however, the means of x differ considerably the slopes of the linear regressions will differ, although the whole of the data may in fact be adequately described by a single curved regression line.

A full discussion of the use and interpretation of regression analysis is not possible here. One or two points may, however, be mentioned.

If regressions are calculated for different domains of study which are, for example, parts of a random sample, or strata of a stratified sample, we shall obtain regression equations of the form:

$$\text{Domain } A: \; y_l = \bar{y}_a + b_a\,(x - \bar{x}_a)$$
$$\text{Domain } B: \; y_l = \bar{y}_b + b_b\,(x - \bar{x}_b)$$

etc. To enable these equations to be compared directly $\bar{x}_a$, $\bar{x}_b$, etc., must be replaced by some standard value $\mathbf{x}_0$ which should be chosen conveniently near the general mean. The equations will then become

$$y_l = y_{a0} + b_a\,(x - \mathbf{x}_0)$$
$$y_l = y_{b0} + b_b\,(x - \mathbf{x}_0)$$

etc., with $y_{a0} = \bar{y}_a + b_a\,(\mathbf{x}_0 - \bar{x}_a)$, etc. The formula for the error of y_{a0}, etc., has been given in Section 7.12. The differences between the standardised values y_{a0}, etc., will represent the differences between the domain y's for the standard value $\mathbf{x}_0$ of x. If b_a, b_b, etc., differ substantially, the differences between y_{a0}, y_{b0}, etc., will depend on the value chosen for $\mathbf{x}_0$. If there is no evidence of differences between the b's then a mean b can be taken (in which case all the regression lines will be parallel). The mean b can be calculated from the formula of Section 6.13. The question of whether the b's differ can be examined by means of the standard errors of b_a, b_b, etc., calculated from the formula of Section 7.12. The analysis of variance can also be used in this connection, but the procedure is rather complicated for those not familiar with the technique. (See, for example, Quenouille, *Associated Measurements*.)

A common tactical error made by those unfamiliar with regression work is to take $\mathbf{x}_0$ as zero, and re-write the regression equations in the apparently simpler form

$$y_l = a_a + b_a\,x$$
$$y_l = a_b + b_b\,x$$

etc.   Unless the means of x are near to zero these equations are unsuitable for comparison, for although a_a, a_b, etc., give the estimated values of y for $x = 0$ they are subject to large errors, both because of errors in b_a, b_b, etc., and because although the assumption that the regressions are linear may be reasonably correct over the range of x actually covered, it may be by no means correct if the range is extended to zero.

When comparing regression lines it is often useful to plot all the lines on the same graph. Relations which are implicit in the equations will then be immediately apparent.

Regressions can easily be calculated from grouped data. If the data are grouped for x only and the mean value of y is calculated for each group, the plot of these values (apart from sampling errors and a small error introduced by the grouping) represents the regression of y on x. With a large sample this is quite a good way of examining the regression. At the same time the plot will reveal whether the regression is truly linear. It is often advisable, however, to make an exact calculation of the regression from the group means, rather than to draw a line by eye, since it is difficult otherwise to make proper allowance for the varying numbers of observations in the different groups. If n_1, n_2, $\ldots$ represent the numbers in the groups, x_1, x_2, $\ldots$ the values of x for the midpoints of the groups, and $\bar{y}_1$, $\bar{y}_2$, $\ldots$ the means of y, the formula for the regression coefficient will be

$$ b = \frac{S\,(n_r\,x_r\,\bar{y}_r) - n\,\bar{x}\,\bar{y}}{S\,(n_r\,x_r{}^2) - n\,\bar{x}^2} $$

where $n = S\,(n_r)$, $\bar{x} = S\,(n_r\,x_r)/n$, $\bar{y} = S\,(n_r\,\bar{y}_r)/n$.

With two variates there are two regression lines, that of y on x and that of x on y. The data themselves will not tell us which, if either, of these regression lines is appropriate for our purposes. If we wish to estimate the values of y for individual units for which only the value of x is known, or the mean y for a set of such units, then the regression of y on x will give the appropriate estimation equation, *provided the units can be regarded as further random samples of one unit each from the population of which the data are a sample*. This is the justification for using the regression method in the manner outlined in Chapter 6 for improving the accuracy of a sample for which supplementary information is available either from the whole population or from the first phase of a two-phase sample. For estimation of this type, errors in x can be ignored in calculating the regression. If x is subject to large errors the numerical value of the regression coefficient will be reduced. We shall consequently estimate the unobserved y to be closer to the mean of the already observed y than would be the case if x were free from error. That this is as it should be is obvious if we consider the limiting case where x is subject to such large errors as to be worthless. In this case the mean value of the observed y provides the best estimate of all unobserved y.

If, on the other hand, we are concerned with the estimation of the mean of the y's of a new sample which, although possessing certain features in common with the original sample on which both x and y were measured, cannot be regarded as a random sample from the same population, the whole situation is altered. We then have to consider in much more detail the nature of the measured variates and the errors affecting them. This situation is discussed in more detail in Section 9.16.

If the regressions are being used in investigational work we shall not usually be concerned with problems of the estimation of further values of y from observed values of x. We shall rather be concerned to evaluate the underlying laws which govern the relationships between various variates. In this case if y is believed to depend causally, in part at least, on x we shall normally require the regression of y on x. This will give the relation between x and the mean value of y for given x. The variation of the actual y's about the mean value of the y's for a given x may then be attributed to the influence of other variates on y and to random errors in y (of observation, etc.). If x is subject to error a correction to the regression coefficient will in this case be required, as is explained in Section 9.16.

9.5 Multiple regression

Instead of taking a regression on a single variate it is possible to take a multiple regression on two or more variates. The formulæ will be given for two variates ; they can easily be extended to more variates when required.

To shorten the formulæ we may write S_{11} for $S(x_1 - \bar{x}_1)^2$, S_{12} for $S(x_1 - \bar{x}_1)(x_2 - \bar{x}_2)$, S_{1y} for $S(x_1 - \bar{x}_1)(y - \bar{y})$, etc. If x_1 and x_2 are the two independent variates the regression can be written in the form

$$y_l = \bar{y} + b_1(x_1 - \bar{x}_1) + b_2(x_2 - \bar{x}_2) \qquad (9.5.a)$$

The regression coefficients b_1 and b_2 are given by the solution of the two simultaneous linear equations (known as the *normal* equations) :

$$\left. \begin{array}{l} b_1 S_{11} + b_2 S_{12} = S_{1y} \\ b_1 S_{12} + b_2 S_{22} = S_{2y} \end{array} \right\} \qquad (9.5.b)$$

In order to obtain the standard errors of b_1 and b_2, or any linear function of b_1 and b_2, it is necessary to perform an operation known as inverting the matrix given by the coefficients of the b's in the above equations. This is equivalent to solving the two pairs of equations of which the left-hand sides are those of the normal equations, but which have as numerical terms 1, 0 and 0, 1 respectively, instead of S_{1y}, S_{2y}. The values of b_1 and b_2 given by these equations are commonly denoted by c_{11}, c_{12}, and c_{21}, c_{22}. The c's form the inverse matrix, which has diagonal symmetry, so that $c_{12} = c_{21}$.

When the c's have been obtained the b's can be calculated from the formulæ

$$\left. \begin{array}{l} b_1 = c_{11} S_{1y} + c_{12} S_{2y} \\ b_2 = c_{12} S_{1y} + c_{22} S_{2y} \end{array} \right\} \qquad (9.5.c)$$

The sum of squares of the deviations from the regression line will be

$$Q = S (y - y_l)^2 = S_{yy} - b_1 S_{1y} - b_2 S_{2y} \qquad (9.5.\text{d})$$

Since two degrees of freedom have been absorbed by the regression line, and one degree of freedom because deviations from the mean have been taken, the total number of degrees of freedom remaining will be $n - 3$. The residual variance of a single observation after fitting the regression is therefore

$$s_l^2 = Q/(n - 3)$$

We then have

$$\mathsf{V} (b_1) = c_{11} s_l^2 \qquad \mathsf{V} (b_2) = c_{22} s_l^2 \qquad \mathsf{cov} (b_1, b_2) = c_{12} s_l^2$$

Hence the variance of any linear function is given by

$$\mathsf{V} (l_1 b_1 + l_2 b_2) = (l_1^2 c_{11} + 2 l_1 l_2 c_{12} + l_2^2 c_{22}^2) s_l^2 \qquad (9.5.\text{e})$$

The reader should verify that if b_2 is omitted from the above formulæ the formulae already given in Sections 6.12 and 7.12 for a regression on a single variate are obtained.

The above formulæ can be adapted to the fitting of curvilinear regression lines. If, for example, we require to fit a quadratic

$$y = a + bx + cx^2$$

a multiple regression on x and x^2 can be calculated. Since x and x^2 are highly correlated if all x are of the same sign it is better for purposes of computation to take $(x - x_0)^2$ for the second variate, x_0 being a convenient value of x near $\bar{x}$.

The amount of computation required to evaluate even a single regression equation with more than two or three independent variates is very substantial, but computer programs for regression analysis are now widely available. These enable much more extensive use to be made of regression techniques, but some knowledge of the basic structure and associated formulae, and of the various pitfalls, is required if they are to be used to good effect. Many reported results of regression analysis serve to confuse rather than to enlighten.

9.6 Models

As was pointed out in Section 5.20, the estimates of the separate effects of the different factors in a table classified by a number of factors can be most efficiently obtained by maximum likelihood, using a regression type model.

If the value y of a variate depends on the values $x_1, x_2, x_3, \ldots$ of some other variates this can be expressed as a functional relationship

$$y = f(x_1, x_2, x_3, \ldots, \beta_1, \beta_2, \beta_3, \ldots) \qquad (9.6.\text{a})$$

in which the β's represent numerical parameters appearing in the function. Occasionally we may have knowledge of the precise form of the functional relationship, but may require estimates of the values of the β's. Usually, however, we do not know the precise form. Instead, some reasonably simple empirical functional relationship is assumed and is then tested against the data to see whether it provides an adequate representation. Any functional relationship of this type is termed a *model*.

A simple example of the use of models on a two-factor table is provided by the data of Table 5.21.a. As already indicated in Section 5.21, it is apparent from inspection of the table that the yields of the varieties differ and that these yields are also affected by the regions. The simplest type of model is to assume that apart from random variation the differences between varieties are the same for all the regions, which implies also that the differences between regions are the same for all varieties ; in other words, to assume that the effects of the varieties and regions are additive. This can be represented by the functional relation

$$\bar{y}_{vr} = \mu + \beta_v + \gamma_r + \varepsilon_{vr} \qquad (9.6.\text{b})$$

where $\bar{y}_{vr}$ is the yield of the sub-class mean for variety v in region r. The β_v are five parameters $(\beta_1, \beta_2, \ldots, \beta_5)$ whose differences represent the differences between the yields of the five varieties. Similarly, the differences between $\gamma_1, \gamma_2, \ldots, \gamma_5$ represent the differences between the five regions. μ is an additive term which can be functionally taken to represent a general mean. The ε term represents random errors and properties of the data not accounted for by the model.

The problems here are:

(a) Given the data, to estimate as accurately as possible the values of the β's and γ's; we will denote these estimates by $b_1, \ldots, b_5$, and $c_1, \ldots, c_5$.

(b) To assess the errors of estimation of these estimates.

(c) To see whether there is any evidence that the model can be improved, i.e. whether the model fits the data within the expected limits of error, and whether any unnecessary terms have been included.

As written, there are two redundant parameters in this model, since we are using five β's to represent differences between the five varieties, and five γ's for the five regions. These redundancies can be removed by deleting β_1 and γ_1.

Equation 9.6.b, with β_1 and γ_1 deleted, may be written in the form

$$\bar{y}_{vr} = \mu' + \beta'_2 x_2 + \ldots + \gamma'_2 x'_2 + \ldots + \varepsilon_{vr} \qquad (9.6.\text{c})$$

where x_2 has the value 1 if the yield relates to variety 2, and 0 otherwise, and similarly for the other x's and the x''s. The model is thus equivalent to a multiple linear regression containing a general constant, four terms for varieties and four for regions. $\beta'_2, \ldots, \beta'_5$ will then represent differences of varieties 2 to 5 from variety 1, and similarly for the γ''s. This is convenient arithmetically, but has the consequence that μ' is not equal to the general mean.

The general method of estimating the values of the parameters of a model is by maximum likelihood. This is equivalent, for regression type models fitted to quantitative data, to the method of least squares.   Denoting the estimates of μ', β'_2, . . . , γ'_2, . . ., by m', b'_2, . . ., c'_2, . . ., we obtain a set of equations

$$\bar{y}_{vr} = m' + b'_v + c'_r + e_{vr} \qquad (9.6.\text{d})$$

where e_{vr} are the *residuals*, i.e. the differences of the observed values from those of the fitted model.   The method of least squares determines the values of m' and of the b''s and c''s which minimize the sum of squares of the residuals.

If, as here, the observed values $\bar{y}_{rs}$ are of differing accuracy, weights inversely proportional to their variances must be assigned to the equations.   The weighted sum of squares of the residuals will then be minimized.   Here the numbers of fields in the sub-classes give appropriate weights.

Although in this simple example the values of m' and the b''s and c''s, though not their estimated errors, can be determined by relatively simple arithmetical operations (as was illustrated in earlier editions of this book), the task can be much more expeditiously effected by the use of a computer program for multiple regression, especially one such as GENSTAT or GLIM which enables separate parameters for the different levels of qualitative variates (here variety and region) to be included without having to specify the values of the dummy x variates of equation 9.6.c.

The fitted values for each cell of the table can be set out in tabular form as in Table 9.6.a, but there is usually little point in presenting the full table of

TABLE 9.6.a—PARAMETRIC REPRESENTATION OF THE CELLS OF A TWO-FACTOR TABLE WHEN REDUNDANCIES ARE REMOVED

Variety	Scotland	Regions North	E. Mid.	...	No. of fields
Majestic	m'	$m' + c'_2$	$m' + c'_3$	...	$393(n_1.)$
King Edward	$m' + b'_2$	$m' + b'_2 + c'_2$	$m' + b'_2 + c'_3$	...	$250(n_2.)$
Great Scot	$m' + b'_3$	$m' + b'_3 + c'_2$	$m' + b'_3 + c'_3$	...	$84(n_3.)$
...	...	...	...	...	...
No. of fields	$174(n_{.1})$	$177(n_{.2})$	$189(n_{.3})$	...	$901(n_{..})$

25 values, as the varietal differences will be the same in all regions, and the regional differences will be the same for all varieties.   Any column of the table, or any combination of the columns, will represent the estimated varietal differences.   If the columns are weighted proportionally to the column totals of the numbers of fields, and the rows are weighted similarly, the weighted means of the resultant varietal and regional estimates will equal the general mean of the whole sample.   The addition of

$$(n_{.2} c'_2 + n_{.3} c'_3 + \ldots)/n_{..} \qquad (9.6.\text{e})$$

to m', $m' + b'_2$, . . . , and of

$$(n_{2.} b'_2 + n_{3.} b'_3 + \ldots)/n_{..}$$

to m', $m' + c'_2$, . . . will give the required values.

The deviations of these estimates from the general mean are in fact the estimates $b_1, b_2, \ldots, c_1, c_2, \ldots$ of the β's and γ's of the original symmetrical formulation of the model, subject to the conditions

$$
\left.
\begin{aligned}
n_1.\, \beta_1 + n_2.\, \beta_2 + \ldots &= 0 \\
n_{.1}\, \gamma_1 + n_{.2}\, \beta_2 + \ldots &= 0
\end{aligned}
\right\}
\tag{9.6.f}
$$

The results produced by GLIM are given in Table 9.6.b. The quantity to be added to m', $m'+b'_2,\ldots$ is $(177 \times -1 \cdot 177 + 189 \times 1 \cdot 002 + \ldots)/901 = -0 \cdot 993$.

TABLE 9.6.b—ESTIMATES OF m'_2, b'_2, ..., c'_2, ..., AND THEIR STANDARD ERRORS PRODUCED BY GLIM

$$m' \quad 9 \cdot 490 \pm 0 \cdot 253$$

b'_2	$-2 \cdot 226 \pm 0 \cdot 210$	$c'_2 \quad -1 \cdot 177 \pm 0 \cdot 285$
b'_3	$-0 \cdot 386 \pm 0 \cdot 376$	$c'_3 \quad -1 \cdot 002 \pm 0 \cdot 300$
b'_4	$0 \cdot 771 \pm 0 \cdot 321$	$c'_4 \quad -1 \cdot 434 \pm 0 \cdot 298$
b'_5	$-1 \cdot 087 \pm 0 \cdot 303$	$c'_5 \quad -1 \cdot 320 \pm 0 \cdot 289$

The required estimate for Majestic is therefore $9 \cdot 490 - 0 \cdot 993 = 8 \cdot 50$, that for King Edward is $9 \cdot 490 - 2 \cdot 226 - 0 \cdot 993 = 6 \cdot 27$, etc. These adjustments give the estimates already presented in Table 5.21.b.

There is, of course, nothing sacrosanct about using the weights proportional to n_1. etc. as the reference level for reporting the results. In the present example, for instance, weighting the regions according to the areas of potatoes grown would give a set of varietal estimates which would provide estimates of the average yields to be expected from the different varieties over the whole of the United Kingdom if the varieties were grown in the same proportions in all regions.

The derivation of the standard errors of the estimates is described in Section 9.8 below. The standard errors shown in Table 9.6.b are those for the difference of the estimate for variety 1 from the other varieties, and similarly for regions. They therefore give only a partial picture of the accuracy of the results.

9.7   Structure of the analysis of variance for a fitted model

The general adequacy of the fit of a model can best be judged by the associated analysis of variance. That for the above example is set out in Table 9.7.a.

Lines 5, 6 and 7 show the amount of variation in the sub-class means that is accounted for by the fitted model. There are here, apart from μ', eight independent parameters in the model. There are thus 12 degrees of freedom for deviations from the values of the sub-class means given by the model. The weighted sum of squares of these deviations (the residuals) is $78 \cdot 7$, giving a mean square of $6 \cdot 56$, compared with the mean square of $58 \cdot 07$ for deviations

TABLE 9.7.a—POTATO SURVEY : ANALYSIS OF VARIANCE, ADDITIVE MODEL FOR
EFFECTS OF VARIETIES AND REGIONS

	Degrees of freedom	Sum of squares	Mean square
(1) Regions only	4	173·7	43·42
(2) Varieties (regions fitted)	4	909·0	227·25
(3) Varieties only	4	887·3	221·82
(4) Regions (varieties fitted)	4	195·4	48·85
(5) Varieties and regions	8	1082·7	135·34
(6) Remainder	12	78·7	6·56
(7) Between sub-classes	20	1161·4	58·07
(8) Within sub-classes	880	3713·6	4·22
(9) Total	900	4875·0	5·42

from the general mean.   Thus an estimated 89 per cent. of the variance of
the sub-class means has been accounted for.*

Note, however, that if we require to test whether there has been a significant reduction in the variance by fitting the model, the appropriate ratio for
the F-test is that between lines 5 and 6, not 7 and 6.   Many computer programs give only the residual mean squares and associated degrees of freedom
at each stage of the fitting process, i.e. in this case those for lines 7 and 6.   The
required F ratio can then be obtained by reconstructing the complete analysis
of variance table, or directly from the formula

$$F = (n_1 F' - n_2)/(n_1 - n_2) \qquad (9.7)$$

where F' is the ratio between the residual mean squares at the two stages, and
n_1 and n_2 are the associated degrees of freedom.   The degrees of freedom for
the F-test are $n_1 - n_2$ and n_2.

Part of the variation in the sub-class means is attributable to random
sampling variation in the selection of fields.   The estimated variance due to
this cause is given by the mean square in line 8.   This is a pooled estimate of
the variance within variety–region sub-classes.   Comparison with line 6 by
means of an F-test gives $F = 6 \cdot 56/4 \cdot 22 = 1 \cdot 55$, $P = 0 \cdot 10$.   There is therefore some slight indication that there are some residual differences between
the sub-class means which are not accounted for by the model.   This, of
course, is not surprising; it merely indicates that the assumption that the
differences between varieties are the same for all regions does not hold exactly.

We have here used a pooled estimate of the variance within sub-classes,

---

* The proportion of the variance accounted for, which may be denoted by R'^2, is
a better general measure of the goodness of fit than the proportional reduction in the
sum of squares, usually denoted by R^2. R is the so-called multiple correlation coefficient,
i.e. the correlation between the observed and fitted values; it will always be increased by
the inclusion of additional parameters, however irrelevant.

as it is reasonable to expect that the variability from field to field within a sub-class is approximately the same for each sub-class. In other types of data, there may well be clear evidence of differences in variability. In such cases the differences can be taken into account by adjusting the weights assigned to the sub-class means, with corresponding weighting of the sums of squares within sub-classes before pooling.

There is no doubt here that terms for both regions and varieties are required in the model. Sometimes, however, we are doubtful whether a set of parameters is worth including. If a test for a particular set is required, this must be made after fitting the other set or sets.* Thus the test for the regional set of parameters is obtained by omitting this set from the model and taking the difference of the sum of squares thus accounted for from that accounted for by both varieties and regions. This is shown in lines 3 and 4 of Table 9.7.a. Lines 1 and 2 give the similar test for regions. Here there is little difference between the mean squares for lines 2 and 3, or for those between lines 1 and 4. If, however, each variety had been mainly confined to a single region, the mean squares for lines 2 and 4 would have been much less than those for lines 3 and 1. This would serve as a warning that the differences due to varieties could not be accurately separated from those due to regions, as common sense would indicate.

Once the principles are understood there is no need to set out an analysis of variance of this type in full. The importance of any set of parameters or of a particular combination of sets can be judged from the reduction in the residual mean square resulting from their inclusion *after* the other sets have been fitted. If a formal test of significance is required formula 9.7 can be used.

Table 9.7.b shows the residual sums of squares and mean squares reported by GLIM. The importance of regions, for example, can be judged by the

TABLE 9.7.b—POTATO SURVEY : RESIDUAL DEGREES OF FREEDOM, SUMS OF SQUARES
AND MEAN SQUARES REPORTED BY GLIM

Effects fitted	D.F.	S.S.	M.S.
(1) —	20	1161	(58·07)
(2) Varieties only	16	274·1	17·13
(3) Regions only	16	987·7	61·73
(4) Varieties and regions	12	78·67	6·556

comparison of line 2 (varieties only) and line 4 (varieties and regions), *not* by lines 1 and 3 (which actually show a slight increase in the residual mean square). The value of F' is $17·13/6·556 = 2·613$, and the formal test of significance is therefore given by

$$F = (16F' - 12)/4 = 7·45$$

* Note, however, that important parameters should not be excluded merely on the ground that they are not statistically significant. Section 9.14 gives an example.

with 4 and 12 degrees of freedom.   Note that lines 1 to 7 of Table 9.7.a can be constructed from Table 9.7.b.

In the above analysis the model was fitted to the sub-class means.   It could alternatively have been fitted directly to the data for the individual fields—indeed many computer programs for survey analysis contain regression programs which only make provision for this form of analysis.

A regression based on the unit data will give the same values for the parameters, but the residuals from the fitted model will be those for the individual fields, leading to the analysis of variance shown in Table 9.7.c.   The remainder term here is the sum of lines 6 and 8 of Table 9.7.a.

TABLE 9.7.c—POTATO SURVEY : ANALYSIS OF VARIANCE WHEN DATA ON INDIVIDUAL FIELDS ARE USED FOR CALCULATING ADDITIVE EFFECTS OF VARIETIES AND REGIONS

	Degrees of freedom	Sum of squares	Mean square
Varieties and regions	8	1082·7	135·34
Remainder	892	3792·3	4·25
Total	900	4875·0	5·42

Only 22 per cent. of the variance of the individual fields is accounted for by varieties and regions.   This contrasts with the 89 per cent. accounted for when the model was fitted to the variety–region sub-class means.   The latter indicates that the additive model for varieties and regions accounts for most of the variability that can be attributed to these two factors.

Explanation of the variability of individual fields might be sought by adding additional terms to the model to represent effects of further factors for which data were collected.   Information was in fact available on fertilizer application, previous crop, etc.   The effects of fertilizers were investigated in detail (at considerable labour—this was in the pre-computer era) by including terms $b_N x_N$, $b_P x_P$, $b_K x_K$ in the regression, where x_N, x_P, x_K represent the amounts of nitrogen, phosphate, and potash applied.   Suffice to say that the apparent responses were much less than those to be expected from the results of fertilizer experiments.   This did not greatly surprise us—indeed the main object of the exercise was to convince those who believed otherwise that the existence of survey data, however accurate and extensive, and however elaborate the analysis, does not obviate the need for experiments.

Note that the full analysis represented by Table 9.7.a can, if required, be obtained from the individual yields by supplementary analyses which include parameters for (a) regions only, (b) varieties only, (c) varieties, regions and variety–region interactions.   The interaction terms can be specified by adding 16 additional parameters i'_{vr} ($v = 2 \ldots 5$, $r = 2 \ldots 5$) to the cells of Table

9.6.a, excluding those in the first row and column. (This specification is implied by GLIM by the instruction FIT V*R.)

9.8 Standard errors of the least-squares estimates

The variances and covariances of all the parameters in a model can be obtained from the inverse matrix given by the least-squares solution. For sets of additive parameters fitted to multifactor tables, contrasts between the parameters in the same set are often the only ones of interest. The standard error of any such contrast can be simply derived from the standard errors of the differences between all pairs in the set. If, for example, in the symmetrical formulation of the model, s_{12}, s_{13}, s_{23} are the standard errors of the differences of b_1, b_2, b_3, the variance of a linear function $l_1b_1 + l_2b_2 + l_3b_3$, where $l_1 + l_2 + l_3 = 0$, is given by

$$-l_1l_2s_{12}{}^2 - l_1l_3s_{13}{}^2 - l_2l_3s_{23}{}^2 \qquad (9.8.\text{a})$$

Thus the difference between variety 3 and the unweighted mean of varieties 1 and 2 is equal to $\frac{1}{2}(b_1+b_2) - b_3$, and its variance is therefore $-\frac{1}{4}s_{12}{}^2 + \frac{1}{2}s_{13}{}^2 + \frac{1}{2}s_{23}{}^2$.

The standard error s'_r of the difference of any parameter b_r of a set from a weighted mean of all the parameters in the set can be obtained from the above formula. For weights w_1, $w_2 \ldots (S(w) = 1)$, s'_1, for example, is given by

$$s'_1{}^2 = (w_2s_{12}{}^2 + w_3s_{13}{}^2 + \ldots) - Sw_r w_s s_{rs}{}^2 = S_1 - T, \text{ say} \qquad (9.8.\text{b})$$

where the second summation includes all pairs in the set. If S_1, S_2, $\ldots$ are first calculated, T is given by

$$T = \tfrac{1}{2}(w_1S_1 + w_2S_2 + \ldots) \qquad (9.8.\text{c})$$

These standard errors may be appended to the parameter values. They provide a useful indication of their relative accuracy, and they direct attention to those values that differ significantly from the weighted mean. Furthermore, if the set contains k parameters $\sqrt{[(s'_r{}^2 + s'_s{}^2)k/(k-1)]}$ provides a rough estimate of the standard error of the difference of parameters r and s. The approximations given by this rule may be judged sufficiently accurate to dispense with an auxiliary table of the standard errors of all differences of pairs within sets.

The standard errors of differences between the parameters for varieties in the potato example are shown in Table 9.8. These can be abstracted from the full results of a GLIM analysis. The table also gives the standard errors of the deviations of the parameter values from their weighted mean, calculated as above.

All these standard errors are based on the mean square, 6·56, of the deviations from the fitted model (Table 9.7.b). This is based on only 12 degrees of freedom, and is therefore subject to considerable uncertainty. Note that if the model had been fitted directly to the data for the individual fields, standard errors based on the pooled mean square, 4·25, of Table 9.7.c would have been obtained. Note also that the standard errors of means of sub-class

TABLE 9.8—STANDARD ERRORS RELATING TO THE VARIETAL PARAMETERS

| | S.E.s of differences of pairs | | | | Deviations from weighted mean |
	Maj.	K.E.	G.S.	A.B.	
Majestic	–				$+0.72 \pm 0.10$
King Edward	0.210	–		.	-1.51 ± 0.14
Great Scot	0.376	0.392	–		$+0.33 \pm 0.34$
Arran Banner	0.321	0.344	0.446	–	$+1.49 \pm 0.28$
Kerr's Pink	0.303	0.320	0.422	0.385	-0.37 ± 0.25

means calculated in Example 7.5.a are based on a mean square of 4·22. This must be allowed for when comparing these errors with those of the least-squares estimates. The standard error of the difference of the unweighted means of sub-class means for the Scottish and Northern regions was there found to be 0·277. The efficiency of this estimate is therefore $(0.285^2 \times 4.22/6.56)/0.277^2 = 68$ per cent.

9.9   Examination of residuals

Examination of the residuals for the individual data units sometimes reveals inadequacies in the postulated model. There are two approaches. The records appertaining to units with large residuals may be examined to see whether they possess any exceptional features which might explain the discrepancies ; or units may be selected because they possess some specific attribute not included in the model to see whether their residuals are outstanding.

If the units used in the fitting have different weights, it is to be expected that on the average the residuals of those with smaller weights will be larger. If, therefore, any examination is to be made, the residuals should first be standardized so as to equalize expected random errors. This can be done by multiplying each residual by the square root of its associated weight. It should also be borne in mind that the residuals are subject to certain restrictions. Thus in the potato example the weighted sum of the residuals in any row or column will be zero. A large positive residual will therefore be associated with one or more negative residuals.

The potato survey provides an illustrative example of the second procedure. It was suggested in Example 7.5.a that varieties which yielded particularly poorly in a region might for this reason be grown less frequently there than in other regions. If this is so, we should expect the residuals for regions containing a small proportion of a given variety to be markedly negative.

The values of the standardized residuals are shown in Table 9.9. Those associated with regions in which a particular variety is grown much less frequently than in other regions are marked with an asterisk. (Great Scot has not been marked because the number of fields is small in all regions.) All the values so marked are negative, four of them large. A rough test of whether

TABLE 9.9—POTATO DATA : STANDARDIZED RESIDUALS

	Scot.	North	E. Mid.	South	West
Majestic	−0·19*	−1·10	−0·66	+0·92	+0·93
King Edward	+2·45	−1·40*	+0·73	+0·27	−2·99
Great Scot	+0·50	−1·33	—	+1·21	−0·03
Arran Banner	−3·21*	+0·94	—	−4·81*	+3·29
Kerr's Pink	−0·94	+2·31	—	—	−1·68*

they deviate significantly from the remaining residuals can be made by the
t-test. This gives the value of $t = 3·20$, $P<0·01$. There is thus evidence
that the slight discrepancy of the data from the fitted model mentioned above
can be attributed to the fact that the proportions of the different varieties
grown in a region are influenced by their suitability for the region.

This point could of course be tested formally by introducing an extra
term into the model, but this is scarcely worth doing.

9.10 Factors at two levels

If all the factors of a multifactorial table are at two levels, the average or *main*
effect of each factor and of any interactions between two or more factors
can be expressed as simple differences, one for each effect. The analysis and
presentation of the results is therefore particularly simple.

An estimate of the average effect of each factor can be obtained by taking
the weighted mean of the differences of each pair of sub-classes which differ
only in the levels of that factor. The procedure is that illustrated in Section
5.21. The same procedure can be used, more tediously, to evaluate interactions. If, however, there are more than two factors, weighted means of
differences will not give quite the most accurate estimates. These can be
obtained by the method of least squares, which is in any case preferable if
computer facilities are available, as less desk computation is required, particularly when interactions are investigated.

Facilities are provided by most computer programs for investigating interactions between any number of factors. In general, however, only interactions
between factors which show clear average effects are likely to be of sufficient
magnitude—having regard to random sources of error—to be worth examination. The best strategy, therefore, is usually first to fit main effects only, and
then add two-factor interactions between pairs of factors which show significant main effects. Only if some of these are substantial is it likely to be worth
proceeding further.

Apart from the question of deciding which interactions, if any, are of
importance, there are various other points that arise in the interpretation of a
least-squares analysis of multifactorial tables. These can best be made clear
from a specific example, using the data of Table 9.10.a, in which all the factors
are at two levels. The main results that emerged were first presented in a
lecture at the London School of Economics given by Professor N. Keyfitz,

who kindly provided me with details of the investigation in advance of its appearance in published form (Keyfitz, 1953, C').

TABLE 9.10.a—1941 CENSUS OF CANADA : AVERAGE NUMBERS OF CHILDREN AND NUMBERS OF FAMILIES IN A SMALL SAMPLE CLASSIFIED IN SIX WAYS

	Present age							
	45–54				55–74			
	Age at marriage							
	15–19		20–24		15–19		20–24	
	Years of schooling							
	0–6	7+	0–6	7+	0–6	7+	0–6	7+
Average number of children								
Low income, French area :								
Far from city . . .	9·4	10·7	10·3	9·8	10·1	14·5	10·4	9·8
Near city . . .	7·4	12·9	8·3	6·7	10·0	11·0	7·6	8·6
Low income, mixed area :								
Far from city . . .	12·9	10·9	8·9	9·8	8·3	12·8	8·4	9·6
Near city . . .	9·7	11·3	9·4	7·1	9·0	9·9	8·6	8·6
High income, French area :								
Far from city . . .	10·9	12·9	10·6	9·8	12·1	12·5	9·0	11·3
Near city . . .	8·3	8·7	7·1	10·3	10·8	13·2	10·9	9·9
High income, mixed area :								
Far from city . . .	12·8	14·3	9·4	11·2	10·6	12·0	9·9	9·0
Near city . . .	10·5	12·2	7·6	8·8	11·0	11·0	8·6	8·4
Number of families								
Low income, French area :								
Far from city . . .	15	14	35	20	18	6	34	12
Near city . . .	5	8	10	37	9	8	15	22
Low income, mixed area :								
Far from city . .	14	11	15	21	16	9	16	17
Near city . . .	3	7	14	49	12	8	17	29
High income, French area :								
Far from city . . .	35	29	24	29	31	15	22	27
Near city . . .	6	15	7	28	14	18	14	30
High income, mixed area :								
Far from city . . .	9	10	14	13	14	2	9	4
Near city . . .	15	6	25	12	14	3	26	10

An analysis for average effects, and indeed also for two-factor interactions, using weighted means of differences, was included in the second and third editions of this book. The average effects obtained from weighted means of differences are shown in Table 9.10.b, together with the overall effects ignoring other factors. Four of the six factors show substantial effects relative to their

TABLE 9.10.b—EFFECTS OF THE SIX FACTORS

	Eliminating effects of other factors	Ignoring other factors
Present age	$+ 0.38 \pm 0.27$	$+ 0.30 \pm 0.28$
Age at marriage	$- 1.77 \pm 0.28$	$- 2.02 \pm 0.28$
Years of schooling . . .	$+ 0.72 \pm 0.28$	$+ 0.16 \pm 0.28$
Income	$+ 0.90 \pm 0.28$	$+ 1.20 \pm 0.27$
Relation to city (near — far) . .	$- 1.28 \pm 0.28$	$- 1.58 \pm 0.27$
Type of area (mixed — French) .	$- 0.15 \pm 0.28$	$- 0.74 \pm 0.28$

standard errors. There are large discrepancies between the average and overall effects for two of the factors.

The GLIM results for the least-squares analysis, fitting main effects only, are shown in the first column of Table 9.10.c. These, as is to be expected,

TABLE 9.10.c—LEAST-SQUARES ANALYSIS FOR MAIN EFFECTS ONLY, AND FOR MAIN EFFECTS AND TWO-FACTOR INTERACTIONS BETWEEN A, Y, I, R

	Main effects only	Main effects and interactions		
	From GLIM	From GLIM	Unweighted adjusted values	Weighted values from regression
M'	10·76	10·36	10·14	9·97
P	$+0.40 \pm 0.265$	$+0.39 \pm 0.266$	Unchanged	Unchanged
A	-1.74 ± 0.277	-0.82 ± 0.527	-1.80 ± 0.282	-1.79 ± 0.280
Y	$+0.68 \pm 0.270$	$+1.33 \pm 0.981$	$+0.84 \pm 0.282$	$+0.70 \pm 0.272$
I	$+0.94 \pm 0.268$	$+0.68 \pm 0.522$	$+0.97 \pm 0.280$	$+0.92 \pm 0.270$
R	-1.41 ± 0.273	-1.17 ± 0.609	-1.32 ± 0.283	-1.36 ± 0.274
T	-0.24 ± 0.273	-0.19 ± 0.279	Unchanged	Unchanged
$A.Y$	—	-1.28 ± 0.559		
$A.I$	—	-0.36 ± 0.558		
$A.R$	—	-0.32 ± 0.565		
$Y.I$	—	$+0.61 \pm 0.549$	Unchanged	Unchanged
$Y.R$	—	-0.31 ± 0.548		
$I.R$	—	$+0.33 \pm 0.546$		
Res. M.S.	21·93 (57 D.F.)	20·96 (51 D.F.)		

The standard errors shown are adjusted to the within-cell M.S., 18·15, to make them comparable within the table and with those in Table 9.10.b.

agree well with the average effects of Table 9.10.b, and have slightly smaller standard errors.

When, however, the two-factor interactions between the four factors showing significant main effects are included, the estimates of the apparent main effects of these factors show little relation to those previously obtained and have much larger standard errors.

TABLE 9.10.d—PARAMETERS FOR A 2×2 TABLE, INCLUDING INTERACTION

	b_0	b_1	Effect of b	No. of units
a_0	m'	$m' + b'$	b'	$q_a n$
a_1	$m' + a'$	$m' + a' + b' + i'$	$b' + i'$	$p_a n$
Effect of a	a'	$a' + i'$		
No. of units	$q_b n$	$p_b n$		n

The reason for this is apparent from Table 9.10.d, which gives the parameters for two factors, a and b, including a parameter i' for interaction. This corresponds to the top left-hand corner of Table 9.6.a. Defining the main effect A of a as the unweighted mean of the effects a' and $a' + i'$ at levels b_0 and b_1 of b, we have

$$A = a' + \tfrac{1}{2} i'$$

Similarly,

$$B = b' + \tfrac{1}{2} i'$$

If the interaction $A.B$ is defined as the difference of the A (or B) effects,* we have

$$A.B = i'$$

The unweighted mean M is given by

$$M = m' + \tfrac{1}{2} a' + \tfrac{1}{2} b' + \tfrac{1}{4} i'$$

Thus m', a' and b' require adjustment, but the interaction is unchanged.

If the interaction term is omitted, only m' requires adjustment. M then represents the adjusted additive constant, but will differ somewhat from the unweighted mean if the frequencies are not proportionate.

The formulae for three or more factors are similar. Thus for three factors, with a parameter for the three-factor interaction included,

$$\left.\begin{aligned}
M &= m' + \tfrac{1}{2} a' + \tfrac{1}{2} b' + \tfrac{1}{2} c' + \tfrac{1}{4} i'_{ab} + \tfrac{1}{4} i'_{ac} + \tfrac{1}{4} i'_{bc} + \tfrac{1}{8} i'_{abc} \\
A &= a' + \tfrac{1}{2} i'_{ab} + \tfrac{1}{2} i'_{ac} + \tfrac{1}{4} i'_{abc} \\
A.B &= i'_{ab} + \tfrac{1}{2} i'_{abc} \\
A.B.C &= i'_{abc}
\end{aligned}\right\} \qquad (9.10.\text{a})$$

The standard errors of the main effects also require adjustment. With three factors, including only two-factor interactions, the variance of A, for example, is

$$\begin{aligned}
V(A) &= V(a') + \tfrac{1}{4} V(i'_{ab}) + \tfrac{1}{4} V(i'_{ac}) \\
&\quad + \operatorname{cov}(a', b'_{ab}) + \operatorname{cov}(a', i'_{ac}) + \tfrac{1}{2} \operatorname{cov}(i'_{ab}, i'_{ac})
\end{aligned} \qquad (9.10.\text{b})$$

* In the analysis of factorial experiments with factors at two levels there are advantages in defining the interactions between two factors as *one half* the above difference, with similar factors of $\tfrac{1}{4}$, $\tfrac{1}{8}$, etc., for interactions between three or more factors. If this is done, the main (average) effects and all interactions will have the same standard error, and the effect of a factor at any combination of upper and lower levels of other factors is the sum of the main effect $\pm$ each of the relevant interactions, without multipliers of $\tfrac{1}{2}$, $\tfrac{1}{4}$, etc.

The values of the variances and covariances are provided by the inverse matrix.

The computation of the adjusted standard errors, although straightforward, is tiresome. If all two-factor interactions were included in the model, then with six factors, as here, A would contain six terms. Thus the relevant 21 values for each of the six factors would have to be extracted from an inverse matrix containing 253 values. Even here, four sets of 6 values have to be extracted from a matrix of 91 values.

The adjusted values are shown in the third column of Table 9.10.c. The main effects are now similar to those in the first column. The standard errors are also similar ; the inclusion of interactions has led to slight increases, as is to be expected with non-orthogonal data.

Only one, $A.Y$, of the six interactions is significant, and its actual magnitude is not at all accurately determined. If the estimate is accepted, the estimate of the effect of schooling at the lower age of marriage will be

$$+0 \cdot 84 - \tfrac{1}{2}(-1 \cdot 28) = 1 \cdot 48$$

and at the higher age will be

$$+0 \cdot 84 + \tfrac{1}{2}(-1 \cdot 28) = 0 \cdot 20$$

Without knowing more about the data it is difficult to offer an explanation of this apparent difference. It may well be due to chance.

The definition of the main effect of a factor a as the unweighted mean of its effects at all combinations of levels of the other factors is conceptually and arithmetically simple, but is in a sense arbitrary. Reverting to Table 9.10.d, we see that instead of taking the unweighted mean A of a' and $a'+i'$ as an estimate of the effect of a we might more appropriately assign weights q_b and p_b to give a weighted estimate A_w. If the numbers in the separate cells are proportionate, i.e. if the number in cell 0, 0 is $q_a q_b n$, etc., A_w will equal the average effect of a for the whole sample, and will be equal to the estimate of the a effect obtained when the interaction term is not included in the model.

It will also have minimum variance. With proportionate frequencies and a within-cell variance of σ^2 the variance is $\sigma^2/(n p_a q_a)$, whereas the variance of the unweighted estimate A is $\sigma^2/(4n p_a q_a p_b q_b)$. Thus if $p_b = 0 \cdot 2$ or $0 \cdot 8$, for example, the ratio of the variances is 16/25.

The inclusion of the interaction term in the model implies that the effects of a at the two levels of b may differ, and that consequently the true values of A_w and A may also differ ; its exclusion implies that the true values are the same. A least-squares solution excluding the interaction term automatically assigns weights to the effects of a at the two levels of b which give the most accurate estimate. If the frequencies are not proportionate, these weights will differ from q_b and p_b, but if p_b differs substantially from $0 \cdot 5$ and the frequencies are not too disproportionate, the estimate will be closer to the value of A_w obtained when the interaction term is included than to A.

Estimates of the weighted effects and their standard errors could be obtained from the second column of Table 9.10.c by a procedure similar to that given above for the unweighted effects, but this is even more tiresome arithmetically. We can, however, simplify the whole process by abandoning the specification of the model in terms of factor levels, using instead regression type specification. A solution in the required form can then be obtained directly for any chosen set of weights, either equal or unequal, as explained in the next section.

9.11 Factors at two levels : specification in terms of regression

If a variate x_a is used to denote the level of factor a, with $x_a = 0$ for level 0 and $x_a = 1$ for level 1, then for two factors the regression equation

$$Y = m' + a' x_a + b' x_b + i' x_a x_b \qquad (9.11.a)$$

will give values of Y equal to the cell values shown in Table 9.10.d when the appropriate values of 0 or 1 are assigned to x'_a and x'_b, and is thus equivalent to the factorial specification of main effects and the two-factor interaction.

Instead of regressing on x_a and x_b we may replace them by new variates $x'_a = x_a - \mathbf{x}_{a0}$ and $x'_b = x_b - \mathbf{x}_{b0}$, where $\mathbf{x}_{a0}$ and $\mathbf{x}_{b0}$ are any constants. This will give a regression equation which is equivalent to the one above, but which will have different numerical coefficients. It may be written

$$Y = M' + (x_a - \mathbf{x}_{a0})A' + (x_b - \mathbf{x}_{b0})B' + (x_a - \mathbf{x}_{a0})(x_b - \mathbf{x}_{b0})A.B \qquad (9.11.b)$$

A' and B' will then give estimates of the main effects with weights which depend on the values assigned to $\mathbf{x}_{a0}$ and $\mathbf{x}_{b0}$. If values of 0.5 are assigned, the unweighted estimates A and B will be obtained. If values of p_a and p_b are assigned, the weighted estimates A_w and B_w defined above will be obtained.

To obtain a computer solution with the required weighting it is of course necessary to express the factor levels of the individual cells in terms of the new variates. Most regression programs make provision for preliminary transformations of this kind.

The fourth column of Table 9.10.c shows the weighted estimates provided directly by GLIM by the above procedure, taking $p_p = 0.47$, $p_a = 0.62$, $p_y = p_i = 0.50$, $p_r = 0.47$, $p_t = 0.42$, corresponding to the marginal frequencies, and including the same interactions as before. It will be seen that the weighted estimates agree more closely than do the unweighted estimates with those obtained by fitting main effects only. The standard errors of the weighted estimates are also less than those of the unweighted estimates, and are only trivially greater than those in column (a).

The residual discrepancies in the main effects are attributable to disproportionate frequencies in the table of counts. Only if components added to a model are orthogonal with those already included will the values of the latter remain unaltered. The standard errors of estimates of parameters already included in the model will also be increased by the addition of other non-orthogonal parameters.

The regression equation 9.11.b may be used to derive formulae for M_w', A_w', B_w', ... for a set of weights p'_a, p'_b, ... in terms of the corresponding estimates M_w, A_w, B_w, ... for weights p_a, p_b, ... by equating the coefficients of x_a, x_b, $x_a x_b$, and the absolute terms in the equations for the two sets of weights. We then obtain (for two factors)

$$\left.\begin{aligned}
M_{w'} &= M_w + (p'_a - p_a) A_w + (p'_b - p_b) B_w + (p'_a - p_a)(p'_b - p_b) A.B_w \\
A_{w'} &= A_w + (p'_b - p_b) A.B_w \\
B_{w'} &= B_w + (p'_a - p_a) A.B_w \\
A.B_{w'} &= A.B_w
\end{aligned}\right\} \quad (9.11.\text{c})$$

These formulae can easily be extended to include three or more factors. If three-factor interactions are included, additional terms for these must be added.

The above formulae can be used to obtain expressions for the effect of a given factor at some defined level of another factor (or factors). Thus the effect of a at level 0 of b ($p'_b = 0$) is $A - \frac{1}{2}A.B$ if calculated from unweighted effects, and $A_w - p_b A.B_w$ if calculated from weighted effects. Similarly the appropriate expression for the mean for cell 0,1 of the $a \times b$ table, for example, is $M - \frac{1}{2}A + \frac{1}{2}B - \frac{1}{4}A.B$ or $M_w - p_a A_w + q_b B_w - p_a q_b A.B_w$.

These expressions can be used to form a table of age at marriage $\times$ schooling from the third or fourth column of Table 9.10.c.

The parallelism of the factorial and regression forms of specification will now be apparent. The variates in the above example were in fact all quantitative, and apart from computational problems (much lessened now that computers are available) a multiple regression could have been calculated directly from the basic data. Apart from changes of scale the values of the main effects and interactions would then be mirrored by the values of the corresponding regression coefficients. In many respects direct regressions for quantitative data are preferable if the relations are approximately linear; in particular they give direct estimates of the effect of unit change of an independent variate. If quantitative data are divided into two classes this can only be deduced from knowledge of the class means of the independent variate, and then with substantially lower accuracy. Moreover, differences in the class means in different domains may give rise to spurious differences in the apparent effects. On the other hand, tabular presentation with grouping into three or more classes enables the results to be more easily comprehended, and directs attention to non-linear relations and to the existence of interactions.

One point that is brought out very clearly by the above example is the need to use equations of the type 9.11.b with $\mathbf{x}_{a0}$, etc. taken as overall means of x_{a0}, etc., instead of the apparently simpler type 9.11.a. Nothing is more confusing or disconcerting than to find, on the insertion of some apparently innocuous terms for possible interactions, that all the regression coefficients already obtained have been radically altered.

9.12 Factors at three or more levels

With factors at more than two levels the main (average) effect of each factor separately cannot be expressed as a single difference, since more than one contrast is involved. The natural form of presentation is therefore as a set of means for each factor, which are estimates of the means that would be expected if the cell frequencies were equal or proportionate to some chosen marginal frequencies.

As was shown in Section 9.6, if main effects only are fitted, these means can be obtained without difficulty from the parameters provided by the least-squares solution. If, however, two-factor interactions are included there will be distortions in the apparent main effects similar to those observed for factors at two levels. Adjustments similar to those given by formulae 9.10.a are therefore required to make them comparable with the values obtained when main effects only are fitted.

Table 9.12.a shows the parameters for two factors a and b at three levels. Factor levels are numbered from 1 upwards. The interaction parameters are

TABLE 9.12.a—PARAMETERS FOR A 3×3 TABLE, INCLUDING INTERACTIONS

	b_1	b_2	b_3	No. of units
a_1	m'	$m' + b'_2$	$m' + b'_3$	$p_a n$
a_2	$m' + a'_2$	$m' + a'_2 + b'_2 + [a_2 b_2 \cdot]$	$m' + a'_2 + b'_3 + [a_2 b_3]$	$q_a n$
a_3	$m' + a'_3$	$m' + a'_3 + b'_2 + [a_3 b_2]$	$m' + a'_3 + b'_3 + [a_3 b_3]$	$r_a n$
No. of units	$p_b n$	$q_b n$	$r_b n$	n

denoted by $[a_2 b_2]$, etc., instead of i''_2, etc., to indicate the factors and levels with which they are associated.

With this notation the formula for A given above (formulae 9.10.a) for equal weights can be extended. That for A_2 for three factors, a and b at three levels, c at four levels, for example, is

$$A_2 = a'_2 + \tfrac{1}{3}([a_2 b_2] + [a_2 b_3]) + \tfrac{1}{4}([a_2 c_2] + [a_2 c_3] + [a_2 c_4]) \qquad (9.12.a)$$

Similarly that for M is

$$M = m' + \tfrac{1}{3}(a'_2 + a'_3) + \tfrac{1}{3}(b'_2 + b'_3) + \tfrac{1}{4}(c'_2 + c'_3 + c'_4)$$
$$+ \tfrac{1}{9}([a_2 b_2] + \ldots) + \tfrac{1}{12}([a_2 c_2] + \ldots) + \tfrac{1}{12}([b_2 c_2] + \ldots) \qquad (9.12.b)$$

To construct a two-factor table of means which exhibits the effect of two interacting factors, e.g. a and b, the parameters $m', a'_2, a'_3, b'_2, b'_3$ of Table 9.12.a can be replaced by m'', a_2, etc., where m'' is given by the above formula for M, including only m' and terms which contain letters relating to factors other than a and b, i.e. here terms 1, 4, 6 and 7. Similarly a''_2 is given by the formula for A_2 omitting the second term. Alternatively the table can be obtained by taking marginal means of the fitted values for the individual cells.

The weighted formulae can be obtained from the corresponding formulae for equal weights. If the weights for factor a at levels 1, 2, 3 are p_a, q_a, r_a, etc., with a sum of unity for each factor, the general rule is to omit the numerical fractions, attach weights to each term corresponding to its factor symbols, and delete a weight occurring in all terms.   Thus

$$A_{2w} = a_2' + q_b[a_2\ b_2] + r_b[a_2\ b_3] + q_c[a_2\ c_2] + r_c[a_2\ c_3] + r_c[a_2\ c_4]$$

the common weight q_a being deleted.

The above formulae are tiresome computationally if adjustments have to be made for several pairs of interacting factors, and the computation of the associated standard errors from the inverse matrix would be very onerous. Unfortunately, with factors at more than two levels the model cannot be specified simply in regression form; consequently the alternative approach adopted for factors at two levels is not feasible.

Probably the best long-term solution of this difficulty would be to provide additional facilities in computer programs to enable them to provide results in the required form.   This does not appear to be a major problem.

There remains the question of determining which interactions, if any, should be included in the final model.   As there are, for example, four parameters for the interactions between two factors each at three levels, the significance of an interaction component may be obscured by being spread over these.   Usually the question can be resolved by the pattern of values.   If, for example, the main effects of the two factors a and b are both substantially linear, we may expect the main component of the interaction to be linear × linear.   Apart from errors of estimation the values of $[a_2\ b_2]$, $[a_2\ b_3]$, $[a_3\ b_2]$, $[a_3\ b_3]$, will then be in the ratio of 1, 2, 2, 4.   If the two higher levels of each of the factors are equal, the interaction parameters may be expected to be substantially equal.

The second contingency is equivalent to taking the main interaction component as that between the lowest level and the mean of the two higher levels of each factor.   The remaining three degrees of freedom can be excluded from the model in this or the linear × linear case, or, more generally, by replacing the interaction parameters of the model by a product term, as was done in the two-level case.

For a linear × linear interaction, regression on a product term $(x_a - 2)$ $(x_b - 2)$ will suffice, but for the more general form auxiliary variates z_a, z_b, ... must be defined, with regressions on the products of these.   Any multiples of the deviations of the values for the levels of a factor given by the fit of main effects only will suffice for the z values of that factor.   Thus if values 2, 6, 7 of the main effects are obtained for factor a we can assign values of -3, $+1$, $+2$, or alternatively -1, $+\frac{1}{3}$, $+\frac{2}{3}$, to z_a.   With the latter, z_a represents the contrast between the lowest level of a and a weighted mean, with weights $\frac{1}{3}$ and $\frac{2}{3}$, of the two upper levels.   A further refinement, when the marginal frequencies for a factor differ greatly, is to adjust the weights to take account of the differing accuracy of the values of the main effects.

It should be noted that the isolation of a single degree of freedom for interactions in this manner should only be used to decide which sets of interactions are to be included in the final model.   If, for example, factors a and b appear to interact, the ordinary factorial specification for $A.B$, which automatically includes all $a \times b$ interaction parameters, should be used for the final model. There is no guarantee that the actual form of the interaction conforms exactly to the component isolated by the above analysis. Omission of the remaining components would give a spurious and misleading appearance of regularity to the two-factor table of adjusted yields without any countervailing economy of presentation, and might conceal components of interaction which were in fact of importance.

9.13   Percentage data : the logit transformation

For qualitative data direct representation of the effects by additive parameters is unrealistic. This is apparent in the percentage data for milk fever given in Table 5.20.b. In these data seasonal differences in the percentages for the different lactations are roughly proportional to the average incidence for each lactation. Each sub-class percentage consequently approximates to the product of two parameters, the value of one being dependent on lactation, the other on season.

A multiplicative model of this type can be converted into an additive model by replacing the percentages $(100 \, p)$ by their logarithms $(z = \log_e p)$. This will be satisfactory if all the percentages are small, but clearly cannot hold over the whole of the percentage scale, as this has an upper limit of 100. Instead, what is known as the *logit* transformation can be used. This is defined by

$$z = \tfrac{1}{2} \log_e \frac{p}{q} \tag{9.13.a}$$

or alternatively by

$$z = \log_e \frac{p}{q} \tag{9.13.b}$$

The first definition was commonly used in statistical work before the introduction of electronic desk and pocket calculators. The $\tfrac{1}{2}$ is included in *Statistical Tables* and in many other published tables of logits, but with the spread of calculators provided with the functions $\log_e x$ and e^x there are obvious advantages in dropping it. The second definition is that adopted by GLIM, and will be used here.

For small percentages the logit transformation (without the $\tfrac{1}{2}$) is almost equivalent to the log transformation, and consequently effects which are additive on the logit scale are multiplicative on the percentage scale. Logits have a range of $-\infty$ to $+\infty$ over the whole of the percentage scale, with a value of 0 at 50 per cent. For a random sample their variance is given by

$$\mathbf{V} \, (\text{logit } \mathsf{p}) = 1/n\mathbf{pq} \tag{9.13.c}$$

If the values of a table are converted to logits, estimates of the effects of the factors in terms of logits could be obtained from weighted or unweighted means of the logits for the sub-classes in the same manner as in Section 5.20. For factors at two levels, weighted differences of the logits could be used.

346

These procedures, however, will not be satisfactory if any of the percentages are small or based on few units. To obtain a maximum likelihood solution we require a computer program which embodies the iterative procedure that was originally developed for probit analysis. The analysis then follows the same general lines as would a similar analysis on untransformed quantitative data. The values obtained for the parameters, however, will relate to the transformed variate, and their transformation back to percentages presents certain subsidiary problems which are not encountered with untransformed data.

As a simple example we may take the milk-fever data of Table 5.20.b, already discussed in Section 5.20. A model with additive parameters in the logit scale was postulated. The parameter values given by GLIM are shown in Table 9.13.a.

TABLE 9.13.a—MILK-FEVER DATA : PARAMETER VALUES FROM GLIM

$$m' \quad -6{\cdot}108 \pm 0{\cdot}1440$$

b'_2	$0{\cdot}6393 \pm 0{\cdot}0920$	c'_2	$2{\cdot}010 \pm 0{\cdot}1531$
b'_3	$0{\cdot}9851 \pm 0{\cdot}0914$	c'_3	$2{\cdot}785 \pm 0{\cdot}1468$
b'_4	$0{\cdot}6494 \pm 0{\cdot}0824$	c'_4	$3{\cdot}197 \pm 0{\cdot}1483$
		c'_5	$3{\cdot}392 \pm 0{\cdot}1413$

The fitted logit values for all cells can be obtained from the parameter values, using the expressions given in Table 9.6.a ; they can then be transformed back to percentages to give a table of estimated percentages for each of the cells. Alternatively, the percentages can be calculated directly from the ancillary information provided by GLIM when residuals are requested.

Marginal means can be calculated for this percentage table, using weights proportional to the marginal frequencies, or other chosen weights. This will give a full representation of the fitted model in percentage form. Often, however, a single set of percentages for each factor, with a weighted mean for each set equal to the overall percentage mean of the table, may be considered sufficient to indicate their effects.

In untransformed data such sets could be obtained directly by the use of formula 9.6.e. This is not possible with transformed data, as the appropriate weights for use in this formula cannot be readily determined. If, however, all the w of the formula are taken to be equal to $1/k$, where k is the number of classes in the second set, a set of logit values will be obtained whose unweighted mean is equal to that of all the logit values for the individual cells. A similar set can be constructed for the second factor. The logit for any cell can then be obtained, if required, by taking the sum of the appropriate pair of logits from the two sets and subtracting the overall logit mean. The two sets are in fact the unweighted marginal means of the expressions given in Table 9.6.a.

Table 9.13.b gives the set of logits for the seasonal factor of the milk-fever data derived from Table 9.13.a, together with the corresponding percentages.

TABLE 9.13.b—COMPUTATION OF ADJUSTED PERCENTAGE VALUES

	Logit	Percentages From logit	Percentages Adjusted
Jan.	−3·831	2·12	2·03 ± 0·107
May	−3·192	3·95	3·78 ± 0·230
Aug.	−2·846	5·49	5·27 ± 0·274
Oct.	−3·182	3·98	3·82 ± 0·166
Mean	−3·263	3·61 (wtd.)	3·46 (wtd.)

The weighted mean, $\bar{p}_w$, of these percentages does not agree exactly with the overall percentage mean, $\bar{p}'_w$, here 3·46. If, as here, the differences are small, the percentages can be adjusted by calculating an adjusting factor

$$\lambda = (\bar{p}'_w - \bar{p}_w)/\bar{p}_w \bar{q}_w$$

and using the formula

$$p' = p(1 + \lambda q) \qquad (9.13.d)$$

to adjust each percentage. Thus for the seasonal percentages

$$\lambda = (0·0346 - 0·0361)/(0·0361 \times 0·9639) = -0·0431$$

A similar computation for the lactation factor gives an unadjusted weighted mean percentage of 3·74 and adjusted percentages of 0·36, 2·63, 5·54, 8·14, 9·72.

It should be noted that these percentages will not agree exactly with the weighted marginal means of the fitted cell values. The difference of the logits of the means of any two sets of percentages whose logits differ by a constant amount d will always be somewhat less than d. The percentages in fact represent those that would be obtained for the (hypothetical) level of the other factor giving a marginal mean of 3·46.

The above adjustment formula is not very accurate for large adjustments. In such cases preliminary adjustment of the logits is advisable. If, for example, we require percentages with an overall mean of 5 per cent. for comparison of the effects from surveys in other years, the logits may be increased by logit (0·05) − logit (0·0346).

In presenting the results it is good practice to give the basic logit values as well as the adjusted percentages (columns 1 and 3 of Table 9.13.b).

Standard errors are given by GLIM in logits, and are directly applicable to logit comparisons. To obtain the standard error of the corresponding proportion p the following formula may be used :

$$\text{S.E. of proportion} = pq \times \text{logit S.E.} \qquad (9.13.e)$$

If the logit standard error relates to the difference of two logits, the standard error of the difference of the corresponding percentages p_1 and p_2 may be obtained to all necessary accuracy by replacing pq by $\frac{1}{2}(p_1 q_1 + p_2 q_2)$ in the above formula. The standard errors shown in Table 9.13.b were calculated in the same manner as in Section 9.8 after conversion of the logit standard errors of the differences of pairs to percentage standard errors by the above formula.

The fit of the model can be judged by the residual deviance given by GLIM. This is 17·61 with 12 degrees of freedom, and has a χ^2 distribution for random incidence within cells corresponding to a P value between 0·1 and 0·2. As the data were collected from herds, some departure from independence is to be expected. The residuals did not show any systematic deviations from the model.

9.14   A multifactorial example of disease incidence

The milk-fever example of the last section provides a simple illustration of the value of some suitable transformation, such as logits, for the separation of the effects of a pair of factors operating on percentage data. The method can be applied to similar problems, such as studies of the probable causes of human diseases, in which many factors may be suspected of playing a part in the causative process. In such studies the overall correlation between incidence and a particular factor may well be very misleading, because of high correlations with other factors.

A study by Higgins and Koch (1977) of the incidence of byssinosis—a form of pneumoconiosis to which workers exposed to cotton dust are subject— will serve as an example. This example is of particular interest as untransformed data were analysed in the original study. The results by the two methods may therefore be contrasted.

The data on which the study is based recorded the incidence of byssinosis amongst 5419 workers classified by five factors :

Type of work	Workplace 1, 2, 3
Employment (yrs)	<10, 10–, 20–
Smoking	Smoker, non-smoker in the last five years
Sex	Male, female
Race	White, other

The data therefore constitute a $3 \times 3 \times 2 \times 2 \times 2$ contingency table, but this is better thought of as a $3 \times 3 \times 2 \times 2 \times 2$ table of percentage incidences, together with a table of the numbers on which the percentages are based.

The overall incidences for each factor separately are shown in Table 9.14.a. From these it might be concluded that all the factors, including sex and race, were affecting the incidence.

As a first step in sorting out the effects of the different factors main effects of all factors were fitted. The complementary loglog transformation

$$u = \log_e \{-\log_e(1-p)\} \qquad (9.14.a)$$

was used for the analysis. For small percentages this is very similar to the log and logit transformations, but it has some merit in studies of persistent disease in which the probability of acquiring the disease in a short time δt is $\beta \delta t$, so that the probability of acquiring it in a time t is

$$p = 1 - e^{-\beta t}$$

TABLE 9.14.a—BYSSINOSIS : OVERALL INCIDENCE FOR EACH FACTOR SEPARATELY

| | Employment (years) | | | Workplace | | |
	<10	10–	20–	1	2	3
Incidence, per cent.	2·31	3·65	3·84	15·70	1·38	1·22
Employees	2729	712	1978	669	1300	3450

| | Sex | | Race | | Smoking | |
	Male	Female	White	Others	Yes	No
Incidence, per cent.	4·39	1·48	2·62	3·84	3·92	1·79
Employees	2916	2503	3516	1903	3189	2230

Thus for two types of exposure to risk operating for the same time, $u_1 - u_2$ gives an estimate of $\log_e \beta_1 - \log_e \beta_2$, i.e. of $\log_e \beta_1/\beta_2$.

The results are shown in Table 9.14.b. Workplaces, as is to be expected from Table 9.14.a, show very large differences. Duration of employment and smoking also show large effects, but the effects of race and sex are negligible—the estimated values are only about half their standard errors. Inspection of the complete contingency table (not reproduced here) confirms that the differences in incidence for sex and race shown in Table 9.14.a are due to the very uneven distribution of the sexes and races amongst the different combinations of the other factors.

TABLE 9.14.b—BYSSINOSIS : GLIM RESULTS, USING THE COMPLEMENTARY LOGLOG TRANSFORMATION

All main effects				Workplace 1 only, sex and race pooled	
m'	$-2·012 \pm 0·218$	SEX	$0·103 \pm 0·223$	m'	$-1·982 \pm 0·172$
W2	$-2·488 \pm 0·285$	E2	$0·538 \pm 0·244$	E2	$0·732 \pm 0·284$
W3	$-2·639 \pm 0·207$	E3	$0·711 \pm 0·202$	E3	$0·814 \pm 0·218$
RACE	$0·098 \pm 0·192$	SMOK	$-0·618 \pm 0·186$	SMOK	$-0·877 \pm 0·260$
Res. deviance 42·47 (57 d.f.), $P = 0·92$				$0·385$ (2 d.f.), $P = 0·82$	

As 8 parameters were fitted and 7 of the 72 factor combinations contained no individuals, the residual deviance has 57 degrees of freedom. If there are no interactions and the expectations in the remaining cells of the contingency table are not too small, the deviance should be distributed approximately as χ^2. Its low value here may be due, at least in part, to small expectations in many of the cells.

To test whether there were any significant deviations from the additive model, the two-factor interactions were added to the model after removing race and sex. Removal of race and sex increased the deviance by 0·48 (2 d.f., $P = 0·79$), and the addition of the interactions decreased it by 10·00 (8 d.f.,

$P = 0.27$). Although there are clearly no large interaction effects, the values of the interaction parameters suggested that workplaces 2 and 3 might differ somewhat from workplace 1 : the two workplace $\times$ smoking parameters were both positive, the first being double its standard error, and all four workplace $\times$ employment parameters were negative.

To get rid of the disturbance in the deviances due to very small expectations the data for race and sex can be pooled. This will have no effect on the parameter values obtained for models which do not include race or sex, but will affect the residual deviances and consequently the nominal standard errors.

The pooled data are shown in Table 9.14.c. It will be seen that workplaces 2 and 3 do not differ in any consistent manner, and may therefore

TABLE 9.14.c—BYSSINOSIS : INCIDENCE (PER CENT.) AND NUMBERS OF EMPLOYEES (TOTALS OVER SEX AND RACE)

Work-place	Years of employment						All classes
	<10		10–19		20–		
	Smoking						
	Yes	No	Yes	No	Yes	No	
Incidence (per cent.)							
1	12·88	5·56	23·88	15·00	27·15	11·11	15·70
2	0·74	1·77	2·12	1·96	1·69	1·29	1·38
3	1·16	0·95	0·94	0·62	2·05	0·90	1·22
2 & 3	1·03	1·18	1·21	0·94	1·96	1·02	1·26
Number of employees							
1	233	126	67	20	151	72	669
2	403	283	94	51	237	232	1300
3	951	733	320	160	733	553	3450

reasonably be combined for tests of interactions between workplace and the other factors. Fitting W.S and W.E to the pooled data in addition to the main effects gave a reduction in deviance of 5·88 (3 d.f., $P = 0.12$) and fitting W.S. only gave a reduction of 3·04 (1 d.f., $P = 0.081$).

This confirms the indications obtained in the first analysis that the same additive model (main effects only) on the complementary loglog scale does not well summarize all three workplaces. It does, however, adequately summarize the effects of smoking and employment for workplace 1. The GLIM results for this are also shown in Table 9.14.b.

In spite of the pooling of sex and race the residual deviance is still surprisingly small. To check that this was a chance effect deviances were obtained for a series of 75 randomly generated sets of values. These agreed satisfactorily with the χ^2 distribution.

TABLE 9.14.d—COMPARISON OF THE PERCENTAGE INCIDENCES FOR THE TWO
MODELS FOR WORKPLACE 1 WITH THE OBSERVED DATA

Employment	<10		10–19		20–	
Smoking	S	—	S	—	S	—
Observed incidence (%)	12·9	5·6	23·9	15·0	27·2	11·1
No. of employees	233	126	67	20	151	72
Loglog model	12·9	5·6	24·9	11·2	26·7	12·1
Additive model	14·0	4·6	24·0	14·5	24·0	14·5

Table 9.14.d gives a comparison of the observed and expected percentages for workplace 1, and also the expected percentages given by the final model of the original paper. Except for the fourth cell, which contains only 20 individuals, the percentages given by the loglog model agree closely with those observed.

Those of the alternative model do not agree so well. The conclusion that the data indicate that smoking produces a constant percentage difference of 9·5 per cent. whatever the duration of employment is particularly questionable. Since it is clear that smoking substantially increases susceptibility, a larger absolute increase in percentage incidence at the higher levels of incidence may be anticipated. The data indicate an increase by a *factor* of about 2·5 for both the short and long durations of employment. The difference in the percentage differences is $8·8 \pm 6·0$, giving a single tail $P = 0·071$. Furthermore, as there is a large increase in incidence between the short and medium durations of employment, some further increase in the longest duration is to be expected, and this is indicated by the data. It should not be concealed, even though the difference is not in itself significant.

The conclusion in the original investigation that there were no effects of any importance in workplaces 2 and 3—an expectation of 1·2 per cent. was assigned to all cells—is also questionable. Table 9.14.c shows that the incidence is virtually constant for non-smokers, but suggests that the incidence for smokers does increase with length of employment : the contrast (for smokers only) between the first and third employment categories gives $\chi^2 = 2·82$ (with continuity correction), i.e. $P = 0·047$ (single tail). One may suspect that workplaces 2 and 3 are sufficiently clean for cotton dust to have no appreciable effect, but that there is an underlying initial level of pneumoconiosis in the population, which is enhanced by prolonged smoking.

The contrast between these two analyses illustrates two points. First, that when necessary an appropriate transformation should be used ; if this is not done, effects that may be expected to be approximately additive in the transformed scale will be complicated by interactions, which may, as for smoking $\times$ employment here, get lost in the analysis. Second, that there is no virtue in attempting to reduce the number of parameters in the model to such an extent that differences in the data that may be of interest are concealed.

9.15   Contingency tables

If of inherent interest, contingency tables are usually best presented as tables of percentages based on the marginal totals of some suitably chosen factor, or combination of two or more factors; Table 5.20.c is a simple illustration of this. If the chosen factor is at two levels, only the percentages for one of the levels are required, as in Table 5.20.b. The totals on which the percentages are based should also be shown, as this enables the accuracy of the individual percentages to be assessed, with exactitude for random samples, and approximately for more complex samples.

When the factor on which the percentages are based—the " base factor "— is at two levels only, the cross-effects between the remaining factors can be eliminated by the use of the logit or loglog transformation, as illustrated in Sections 9.13 and 9.14. If the base factor is at more than two levels these transformations are not applicable to the full table of percentages, but they can be used if the set of contrasts for the base factor can be subdivided in a meaningful manner into single contrasts. Thus, for example, if disease incidence is recorded as severely affected, mildly affected, not affected, this may be dealt with by two separate analyses : (a) of the percentages affected, grouping the severe and mild cases, and (b) of the percentages severely affected out of all those affected, excluding those not affected.

If such a subdivision is not meaningful, a log transformation of the frequencies of the original contingency table must be used. Suppose, for example, that a survey of car accidents is made, classified into, say, five types of accident (T), with the drivers concerned classified into four age-groups (A), and by sex (S). The results will constitute a $5 \times 4 \times 2$ contingency table. To exhibit the relation between types of accident and age and sex of the driver the results are best presented as percentages of each type of accident in each age–sex class, together with similar percentages for type $\times$ age, type $\times$ sex, and for the whole sample, calculated from the corresponding marginal totals in the contingency table. The points of interest may well emerge from inspection of such a table. Note, however, that if there are variations in the sex ratio for the different ages the marginal distribution for type $\times$ age will be affected by sex differences, and vice versa.

The elimination of such cross-effects depends primarily on decisions as to which interactions (i.e. associations between the different factors) are to be retained in the model. An analysis of the contingency table frequencies, using the log transformation, will effect this. There is a minimal model, which typically will comprise all factors other than the base factor and the interactions between these factors. The effects which comprise the minimal model must always be included in any model that is being fitted. Thus, in the present example, A, S, A.S constitute the minimal model. This leaves T, T.A, T.S, T.A.S. The importance of T.A.S, for example, can be assessed by testing the residual deviance against the χ^2 distribution after fitting all main effects and two-factor interactions. If it is decided to omit T.A.S, the resultant smoothed contingency table can be converted to percentages on the base factor, and these can be combined in any suitable manner. Thus, to exhibit the age effects freed from sex effects, we must take the same proportion of males and females at each age.

Note that if T.A.S is included the " full " model is obtained. In this case there will be no change in the percentages calculated directly from the contingency table. They can be combined as above to exhibit age effects freed from sex effects, and sex effects freed from age effects.

The reader may wonder how an analysis based on the log transformation relates to the conventional tests for association using the Pearsonian χ^2 statistic and to other statistics such as the contingency coefficient, often applied to two-factor contingency tables.

The value obtained for the residual deviance for any particular effect and the corresponding Pearsonian χ^2 are not identical, but are very similar, and the distribution of both tends to the χ^2 distribution in large samples. The general formula for the Pearsonian χ^2 is

$$\chi^2 = \Sigma \ \frac{(\text{observed no.} - \text{expected no.})^{\,2}}{\text{expected no.}} \qquad (9.15.\text{a})$$

with summation over all cells. For a two-factor $p \times q$ table the expected number for the r, s cell, taking the margins as fixed, is

$$n_{r.} \, . \, n_{.s}/n_{..}$$

where $n_{r.}$, $n_{.s}$, $n_{..}$, represent the corresponding marginal totals and the grand total. The resultant χ^2 will have $(p - 1)\,(q - 1)$ degrees of freedom. This corresponds to the deviance for the $P.Q$ interaction, with P and Q previously fitted, in the log transformation model.

For a two-factor table of any size, the Pearsonian χ^2 is tiresome to compute on a desk calculator, but is provided either routinely or on request by many computer survey programs. The corresponding expression for the three-factor interaction in a three-factor table is even more tiresome to compute and is much more rarely provided by computer programs.

Consequently, if a program such as GLIM is available the log transformation provides a much more general approach to the problem of determining which associations are significant, and has the added advantage that it will furnish the standardized deviations of the individual observed values from the corresponding expected values given by any chosen model. Standardized deviations are of use (a) for detecting anomalous values, and (b) for indicating a systematic pattern of deviations from the postulated model, such as a linear component of interaction when one or both the classifying variates are basically quantitative.

A χ^2 or residual deviance test which embraces a set of degrees of freedom must be used with caution. As in the quantitative case, there is always the danger that the significance of a component which is of importance will be obscured by being combined with numerous other small or non-existent components, or, when certain main classes contain very few observations, by components on which there is trivial information.

Particular components of an interaction can be tested formally by including one or more parameters to represent them. Thus a linear by linear component for a 3×4 table can be represented by a term x in the log model, where $x = uv$ with $u = -1, 0, +1$ and $v = -1.5, -0.5, +0.5, +1.5$.

Frequently, however, more can be learnt—and learnt more speedily—by isolating what appear to be salient features of a contingency table and applying simple tests on a desk calculator, instead of making a formal analysis of the full table on a computer. Thus in a table classified by two quantitative factors formation of a 2×2 sub-table by grouping rows and columns, possibly with omission of the central rows and columns, will provide a quick test for the most likely type of interaction.

If a 2×2 table and its marginal totals is denoted by

$$
\begin{array}{cc|c}
n_{11} & n_{21} & n_{\cdot 1} \\
n_{12} & n_{22} & n_{\cdot 2} \\
\hline
n_{1\cdot} & n_{2\cdot} & n_{\cdot\cdot}
\end{array}
$$

the formula for χ^2 (1 d.f.) is

$$
\chi^2 = \frac{(n_{11} n_{22} - n_{12} n_{21})^2 \, n_{\cdot\cdot}}{n_{1\cdot} \, n_{2\cdot} \, n_{\cdot 1} \, n_{\cdot 2}} \tag{9.15.b}
$$

which can be very rapidly computed on a modern desk calculator.

As a refinement, the *continuity correction*, which makes allowance for the discontinuity of the distribution for one degree of freedom, can be applied by subtracting half the total $n_{\cdot\cdot}$ from the cross-product, regarded as positive, before squaring. This is equivalent to adjusting the cell values by $\pm\frac{1}{2}$, keeping the margins fixed.

There is also a simple formula for χ^2 for a $2 \times s$ table ($s - 1$ d.f.). If the numbers in one of the rows are n_{1r} ($r = 1 \ldots s$) and $p_r = n_{1r}/n_{\cdot r}$, $p = n_{1\cdot}/n_{\cdot\cdot}$, $q = 1 - p$, then

$$
\chi^2 = \frac{1}{pq} \left(\Sigma p_r n_{1r} - p n_{1\cdot} \right) \tag{9.15.c}
$$

Note the similarity to the mean $\times$ total rule for the sum of squares for a quantitative variate.

Quick tests for interactions between three or more factors can be made similarly by forming an appropriate $2 \times 2 \times 2 \times \ldots$ table. There is, however, no convenient algebraic formula for computing χ^2 directly for such tables. Instead, the logarithm of the ratio of the cross-products may be used. For a 2^3 table this is

$$
L = \log_e \frac{n_{111} \, n_{122} \, n_{212} \, n_{221}}{n_{211} \, n_{121} \, n_{112} \, n_{222}}
$$

We then have

$$
\chi^2 = L^2 / V(L) \tag{9.15.d}
$$

where $V(L)$ equals the sum of the reciprocals of n_{111}, n_{122}, etc.   A continuity correction can be made by adding $\frac{1}{2}$ to all the frequencies in the numerator and subtracting $\frac{1}{2}$ from those in the denominator, or vice versa.   The extension to 2^4, etc. tables is obvious.

This formula should not be used for very small n, and breaks down entirely if any $n = 0$.

It should be noted that tests based on χ^2 or the residual deviance assume that the sample is random, and that the data refer directly to the sampling units.   If this is not the case, e.g. if raising factors have been applied to parts or the whole of the table, or if the data are derived from a lower level of a multi-stage or cluster sample, the significance is likely to be greatly exaggerated.

Little need be said about the many other "measures of association" which have been proposed at various times and are embodied in some computer programs.   They are likely to confuse rather than enlighten, and are best ignored.

Example 9.15.a

In a pilot survey of the effect of home conditions on the quality of the homework by schoolchildren the results shown in Table 9.15.a were obtained

TABLE 9.15.a—RELATION BETWEEN HOMEWORK CONDITIONS AND TEACHER'S RATINGS

Teacher's rating	Homework conditions					Total
	A'	B'	C'	D'	E'	
	Number of children					
A	141	67	114	79	39	440
B	131	66	143	72	35	447
C	36	14	38	28	16	132
Total	308	147	295	179	90	1019
	Percentages of column totals					
A	46	46	39	44	43	43
B	42	45	48	40	39	44
C	12	9	13	16	18	13

Yates, 1947).   A to C and A' to E' are graded scales, A and A' being the highest ratings.   Is there any evidence that the quality of the homework is affected by the conditions?

The percentages of the column totals indicate that there may be some slight association in the expected direction, but the value of χ^2 for the whole table is only 9·16 (8 d.f.), $P = 0.33$.   If, however, the central grades B and C' are omitted and A' and B', and D' and E', are combined we obtain the 2 × 2 table

	A' + B'	D' + E'	Total
A	208	118	326
C	50	44	94
Total	258	162	420

Formula 9.15.b, with the continuity correction, gives $\chi^2 = 3\cdot03$ (1 d.f.), P (single tail) $= 0\cdot041$, indicating that the observed effects are unlikely to be wholly due to chance.

It may be noted that the smoothed values derived from a formal linear $\times$ linear regression model, although certainly not deviating significantly from the data, would have indicated a progressive reduction in the percentage of A, and a corresponding increase in C, as conditions of work deteriorate, leaving B unchanged, whereas the data indicate a reduction of both A and B. Combining B with A, instead of omitting B, would have given $\chi^2 = 3\cdot85$, P (single tail) $= 0\cdot025$. This illustrates the point made in Section 9.12, that it is often better to show actual rather than smoothed values if a full table is to be presented.

Example 9.15.b

In an experiment on mice (Whittaker and Aitkin, 1978, and earlier papers) pairs of sires and dams were irradiated (treatment A) or not (treatment B). The first three columns of Table 9.15.b give the numbers of litters in which

TABLE 9.15.b—LITTERS OF MICE : EFFECT OF IRRADIATION ON THE MORTALITY OF THE YOUNG

Litter size	Treat-ment	Numbers of litters				Percentages			Mortality of young (%)*	
		No. of deaths				No. of deaths				Difference
		0	1	2+	Total	0	1	2+	Mean	(A − B)
7	A	58	11	5	74	78	15	7	4·1	−0·6 ± 1·30
	B	75	19	7	101	74	19	7	4·7	
8	A	49	14	10	73	67	19	14	5·8	+0·8 ± 1·40
	B	58	17	8	83	70	20	10	5·0	
9	A	33	18	15	66	50	27	23	8·1	+2·0 ± 1·44
	B	45	22	10	77	58	29	18	6·1	
10	A	15	13	15	43	35	30	35	10·0	+2·7 ± 1·58
	B	39	22	18	79	49	28	23	7·3	
11	A	4	12	17	33	12	36	52	12·7	+2·6 ± 1·62
	B	5	15	8	28	18	54	29	10·1	

* Taking the 2+ class as 2 for all litter sizes.

there were 0, 1, 2 or more deaths of the young before weaning, classified by litter size. Is there any evidence of variation in the treatment effect for litters of different sizes ?

The data constitute a $5 \times 2 \times 3$ contingency table. Inspection of the percentages of litters with 0, 1, 2+ deaths, also shown in the table, suggests that there is indeed a differential effect: the percentages for litters of 7 are almost the same for the two treatments, but those for 0 and 2+ steadily diverge as litter size increases.

If there is a differential effect, this should appear as the linear $\times$ linear $\times$ linear component of the three-factor interaction in the contingency table. A rough test can be made, as in the previous example, by condensing the table to $2 \times 2 \times 2$ form by omission and combination of litter levels and death numbers ; 7 $v.$ $(10 + 11)$ for litters, and 0 $v.$ $(1 + 2+)$ for deaths, gives reasonable numbers in all eight classes. From formula 9.15.d we then find (with the continuity correction) $L = 0.8485$, $V(L) = 0.2382$, $\chi^2 = 3.02$, $P = 0.082$.*

A somewhat different story emerges from the GLIM analysis of the contingency table. Denoting deaths by D, litter size by S, and treatment by T, the relevant contributions to the deviance, with elimination in the order shown, are

	d.f.	Deviance
D.T.	2	6.60
D.S.T. {Linear	1	2.343 } 3.159
D.S.T. {Remainder	7	0.816 }

The pooled 8 d.f. for D.S.T. is low ($P = 0.924$) and no 1 d.f. component of this could attain significance at the 5 per cent. level, but the linear $\times$ linear $\times$ linear component absorbs most of the deviance ($P = 0.126$), leaving a residue with a P value of 0.997. The significance of the linear component is less than that given by the rough test above. It is, however, well above expectation, and does confirm that there is some evidence for the three-factor interaction on the log scale.

The value of P for the residue seems at first sight to be improbably near to unity. However, the values in a contingency table must be integral, and any set of values consistent with the marginal constraints must therefore have a finite probability of occurrence, whereas the χ^2 associated with the set of values conforming most closely to expectation may be very small or even zero. Thus, for example, in a 2×2 contingency table with all four marginal totals equal to 50 the probability when there is no association of getting 25 in each cell is 0.158. The value of χ^2 will therefore be zero in about a sixth of all samples.

It would appear, however, that the residue after including a parameter for the linear component of D.S.T may well be subnormal, which implies that the random variations in the data are smaller than would be expected on

*The value of χ^2 trom Bartlett's exact test (Bartlett, 1935), also with the continuity correction, is 3.04.

the average. In such cases any real effects that exist will be particularly noticeable on visual inspection.

One may however ask : why not treat the data quantitatively ? As the experimenter was concerned with the effect of the treatment on the mortality of the young it is better, and also conceptually simpler, to examine the mean mortality rates. These cannot be computed exactly from the data given, but an approximation to them can be made by taking $2+$ as 2 throughout. The values obtained are shown in the last section of Table 9.15.b. The differences between the treatments, and their standard errors calculated from the numbers of 0's, 1's, and 2's in each group, are also shown.

There is a clear trend in these differences. A weighted regression, with weights inversely proportional to the squares of their standard errors, gives*

$$b = 0 \cdot 867 \pm 0 \cdot 463 \qquad \chi^2 = 3 \cdot 507 \qquad P = 0 \cdot 061$$

This, although not quite reaching the mystic 5 per cent. significance level, agrees with the tests above in indicating that the effect is very probably real. Note that if the percentage in the $2 +$ group is large it is likely that it contains a substantial proportion of 3 or more ; this particularly affects the 11 litter-size difference. Consequently the trend would have been increased, and would probably also have been more regular, if the full figures for deaths had been used, though the standard errors would also be larger.†

There is a further point. The contingency table analysis is based on the log transformation ; this implies that if there is no interaction the treatment differences in mortality rates for different litter sizes will be more or less proportional to the mean mortality rates for those litter sizes. The expected mortality rates can be calculated from the fitted values obtained when the interaction term is omitted. For the 7 and 11 litter sizes they are 5·0, 4·0 and 12·2, 10·6 for treatments A and B respectively. Thus 0·6 of the observed 3·2 difference is included in the main effect by the log transformation.

Two other features of the data may be noted. Not surprisingly, the mortality of the young increases markedly with litter size, even for treatment B. Also the total number of litters with treatment A (289) is substantially less than that for treatment B (368). Assuming that equal numbers of pairs of mice were assigned to each treatment this indicates a higher mortality or sterility rate for the irradiated parents ; the proportion of A litters, 0·44, differs significantly from 0·5.

9.16 Regression : effect of random errors in the variates

Random errors in the dependent variate make the regression coefficients less precise, but no consistent error is introduced : as the sample size is increased

* Formulae for a weighted regression can be obtained from those given in Sections 6.12 and 7.12 by using weighted means and including a factor w in all summations.

† The quantitative approach, with estimation of the numbers of litters with 3 or more deaths, was adopted by Koch *et al.* (1976). They reported a significant linear component in the interaction.

the coefficients tend to the underlying population values. If, on the other hand, there are random errors in the independent variates the estimated coefficients, regarded as coefficients of an underlying regression law, are subject to consistent errors which do not decrease as the size of the sample is increased. In the case of a single independent variate with errors of x uncorrelated with those of y, a consistent estimate b' of the coefficient β of the underlying regression law is

$$b' = b/(1 - h) \qquad (9.16.a)$$

where h is the ratio of the error variance of x to the total variance of x (including the error variance). Thus the regression coefficient b calculated in the ordinary manner is on the average too small in absolute magnitude, and is said to be *attenuated*.

In the case of a multiple regression with independent errors in the different x's the estimation of the coefficients of the underlying regression law will be obtained by replacing S_{11}, S_{22}, etc., in the normal equations by $S_{11}(1 - h_1)$, $S_{22}(1 - h_2)$, etc. If the errors of the x's are correlated similar corrections are required for S_{12}, etc. If the errors of the x's are correlated with those of y the terms S_{1y}, etc., must also be corrected.

These adjustments can only be made if the relevant error variances and covariances can be estimated. If they arise solely from sampling errors this will often be possible. Unfortunately errors in the independent variates are not confined to sampling errors. Errors of observation and measurement, and failure of the chosen measures to represent what is really required, will produce similar disturbances. Errors of observation and measurement frequently require supplementary investigations if they are to be assessed, though in some cases, as in the results described below, they will be included in the sampling errors as ordinarily calculated. Failure of the chosen measures to represent what is really required is much more troublesome and the amount of the disturbance cannot ordinarily be assessed. (An example where this point may arise is considered in the next section.)

An example of a case in which adjustments for attenuation were necessary is provided by a survey on potatoes carried out in England and Wales in 1948/50. In this survey the yields were estimated by taking a sample of about 35 fields per county and lifting and weighing small sample lengths of row from the selected fields (Boyd and Dyke, 1950, H'). The means of the sample yields and the official estimates (tons per acre) for the surveyed counties for the three years are shown in Table 9.16, and Fig. 9.16 shows the Ministry of Agriculture's estimates of the yields of the surveyed counties plotted against the sample estimates. The top dotted line is the line on which all points should fall if there were no errors in either the official or the sample estimates. The sample estimates may be taken to be virtually free from bias, but they are subject to random sampling errors owing to selection of fields within a county, and to the sampling of the selected fields.

It appears from the figure that there was a tendency to underestimate counties with high yields. This tendency can be evaluated quantitatively by

TABLE 9.16—YIELDS (TONS PER ACRE) AND REGRESSION COEFFICIENTS IN THE POTATO SURVEY

	Sample yields (x)	Official estimates (y)	Regression coefficients unadjusted	adjusted
1948	9·35	7·69	0·365	0·457
1949	7·52	6·30	0·520	0·606
1950	9·48	7·59	0·415	0·530

calculating the regressions of the official estimates on the sample estimates. The thin lines on the diagram give these regressions for the three years. The coefficients are given in Table 9.16. Since, however, the sample estimates are subject to random errors these regressions require adjustment if they are to represent the average values of the official estimates for given values of the *true* county yields.

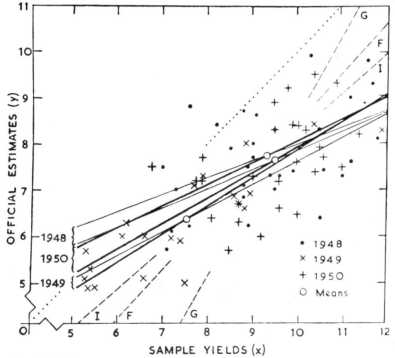

FIG. 9.16—POTATO SURVEY : THE RELATION BETWEEN OFFICIAL ESTIMATES AND SAMPLE YIELDS OF COUNTIES FOR 1948, 1949 AND 1950 (TONS PER ACRE)

The thick lines on the diagram and the coefficients in Table 9.16 give these adjusted regressions. Apart from random errors of estimation they represent the lines that would have been obtained if the yields of all the fields in each county had been determined without error. The calculated regression line for 1948, for example, was

$$y = 7 \cdot 69 + 0 \cdot 365 \,(x - 9 \cdot 35) \qquad (9.16.\text{b})$$

The average error variance per county was $0 \cdot 398$. This includes both the first-stage component due to selection of farms, the second-stage component due to the selection (where necessary) of fields on the selected farms, and the third-stage component due to the sampling of the selected fields. It is calculated from the within-counties variance of the mean sample yield per farm. The total variance of the county sample estimates was $1 \cdot 968$. Hence $h = 0 \cdot 398/1 \cdot 968 = 0 \cdot 202$, and $b' = 0 \cdot 365/(1 - 0 \cdot 202) = 0 \cdot 457$. The adjusted regression line is therefore

$$y = 7 \cdot 69 + 0 \cdot 457 \,(x - 9 \cdot 35) \qquad (9.16.\text{c})$$

It will be seen that part, but by no means the whole, of the apparent under-estimation of high yields can be attributed to random errors in the sample estimates. The lines for the three years have also been brought into somewhat closer agreement by the adjustments, for although the adjustments to the three regression coefficients are very similar, the lower mean yield of 1949 has raised this line relative to the others.

In 1948 the estimates were provided by the Ministry's Crop Reporters, but in 1949 and 1950 the duty of making estimates was transferred to the National Agricultural Advisory Service. The close agreement of the adjusted lines—the differences are no greater than would be expected from random errors—demonstrates the very similar behaviour of the two different groups of reporters. The line obtained by taking a weighted mean of these lines (by a procedure we need not describe) is

$$y = 7 \cdot 19 + 0 \cdot 563 \,(x - 8 \cdot 78) \qquad (9.16.\text{d})$$

These results may be used to establish a formula for correcting official estimates in future years. This is not so simple a problem as it appears at first sight. If both official and sample estimates were available for all counties over a number of years, the regression of the sample yearly means $\bar{x}_t$ on the official yearly means $\bar{y}_t$ would provide the appropriate equation of estimation, at least in so far as future years could be regarded as a random sample from the same population as the years in which the samples were taken. No adjustment for errors in x would be required, since $\bar{x}_t$ is here the dependent variate. The existing results, however, only provide data for a selection of counties for three years, and the regression of $\bar{x}_t$ on $\bar{y}_t$ will therefore be too ill-determined to be of any value. Consequently we have to consider whether it is possible to establish this regression line indirectly.

In the light of the results obtained we may tentatively assume :—

(1) The official county estimates are distributed about the mean adjusted

regression line of y on x given by equation 9.16.d with a residual variance estimated at 0·624.*

(2) There may be an additional common component of error affecting all the official estimates of a particular year, but in view of the closeness of the adjusted regression lines of y on x for the three years this component is likely to be small.

We shall also assume, in order to simplify the discussion, that the errors of a particular county about the regression line are independent from year to year. This is not likely to be wholly correct, since official estimates for a particular county are for the most part made by the same reporters in successive years.

If $\bar{x}_t'$ is the true mean yield of all counties for the year t, and if the common component of error under assumption (2) is negligible, the points $(\bar{x}_t', \bar{y}_t)$ representing the yearly means will deviate from the regression line only by an amount due to the random errors arising from assumption (1). The official estimate for the country is a weighted mean of the county estimates, with weights w proportional to the county potato acreages. Using the county acreages for 1942 (which had a similar total potato acreage to 1950) we find

$$S\,(w^2)/\{\,S\,(w)\}^2 = 0\cdot 0302$$

and consequently from formula 7.5.e the variance of $\bar{y}_t$ about the regression line is given by

$$V_r\,(\bar{y}_t) = 0\cdot 0302 \times 0\cdot 624 = 0\cdot 0188.$$

We also require estimates of the mean and variance of the yearly means of either the true yields or the official estimates. No reliable estimate can be obtained from the present data, since only three years are available, but one can be obtained from the values of the official estimates over past years. Taking the 20-year period for which data are readily available (1930–1949) we obtain

$$\bar{\bar{y}}_t = 6\cdot 80 \qquad V\,(\bar{y}_t) = 0\cdot 333.$$

From the adjusted mean regression line (9.16.d) the corresponding mean x_t' for the true yields is 8·09.

If $\bar{y}_t'$ represents the value of y for the point on the regression line corresponding to $\bar{x}_t'$ we have

$$V\,(\bar{y}_t) = V\,(\bar{y}_t') + V_r\,(\bar{y}_t) = b'^2\,V\,(\bar{x}_t') + V_r\,(\bar{y}_t)$$

and hence

$$\mathrm{cov}\,(\bar{x}_t', \bar{y}_t) = b'\,V\,(\bar{x}_t') = \frac{1}{b'}\Big\{\,V\,(\bar{y}_t) - V_r\,(\bar{y}_t)\Big\}$$

$$= \frac{1}{b'}\,V\,(\bar{y}_t)\,(1-k)$$

where $k = V_r\,(\bar{y}_t)/V\,(\bar{y}_t) = 0\cdot 0188/0\cdot 333 = 0\cdot 0565.$

* This is calculated in the ordinary manner except that b is replaced by b' in the formula for Q (Section 7.12).

The regression coefficient of $\bar{x}_t{}'$ on $\bar{y}_t$ (referred to the y axis) is therefore

$$\frac{\text{cov}\,(\bar{x}_t{}',\,\bar{y}_t)}{\text{V}\,(\bar{y}_t)} = \frac{1}{b'}(1 - k)$$

$$= \frac{1 - 0{\cdot}0565}{0{\cdot}563}$$

$$= 1{\cdot}676$$

The regression equation passing through the point $(\bar{\bar{x}}_t,\ \bar{\bar{y}}_t)$ will be

$$\bar{x}_t{}' = 8{\cdot}09 + 1{\cdot}676\,(\bar{y}_t - 6{\cdot}80)$$
$$= 8{\cdot}43 + 1{\cdot}676\,(\bar{y}_t - 7{\cdot}00) \tag{9.16.e}$$

This is not shown in Fig. 9.16 but is almost the same as the 1949 adjusted regression.

If we had taken the regression of all the observed x on the observed y (disregarding the year classification) we should have obtained the equation (line F in the figure)

$$\bar{x}_t{}' = 8{\cdot}88 + 0{\cdot}966\,(\bar{y}_t - 7{\cdot}27)$$
$$= 8{\cdot}62 + 0{\cdot}966\,(\bar{y}_t - 7{\cdot}00) \tag{9.16.f}$$

This differs considerably in slope from the regression given above. If only the data for the years 1948 and 1950 had been available the difference would have been much greater, the line (line G in the figure) being

$$\bar{x}_t{}' = 9{\cdot}41 + 0{\cdot}593\,(\bar{y}_t - 7{\cdot}65)$$
$$= 9{\cdot}02 + 0{\cdot}593\,(\bar{y}_t - 7{\cdot}00) \tag{9.16.g}$$

whereas the procedure giving equation 9.16.e would have given the line

$$\bar{x}_t{}' = 7{\cdot}64 + 1{\cdot}953\,(\bar{y}_t - 6{\cdot}80)$$
$$= 8{\cdot}03 + 1{\cdot}953\,(\bar{y}_t - 7{\cdot}00) \tag{9.16.h}$$

This line is also not shown in the figure but is nearly the same as the 1950 adjusted regression. The procedure therefore gives a relatively stable line.

If there is an additional common component of error under assumption (2), the slope of the regression equation (relative to the y-axis) should be decreased, but since we have no means of estimating this component the amount of the decrease cannot be assessed. We might, however, adopt the simple compromise of taking the regression of $\bar{x}_t$ on $\bar{y}_t$ passing through the origin. For this regression

$$b = S\,(\bar{x}_t\,\bar{y}_t)/S\,(\bar{y}_t{}^2) = 1{\cdot}222$$

Hence the line (line I in the figure) is

$$\bar{x}_t = 1{\cdot}222\,\bar{y}_t$$
$$= 8{\cdot}55 + 1{\cdot}222\,(\bar{y}_t - 7{\cdot}00) \tag{9.16.i}$$

Apart from any additional common component of error, the success of equation 9.16.e in future years will depend on how far the reporters continue to make the same type of error as they have in the past. If they become aware of their present tendency to underestimate high yields they may endeavour to improve their estimates. This will, of course, vitiate any adjustments based on equation 9.16.e.

The reader will find it instructive to calculate the predicted values of $\bar{x}_t{'}$ for the three years for which data are available.

9.17   The interpretation of multiple regression

The interpretation of the results of multiple regression analysis requires the greatest care. Nothing is easier than to reach false conclusions. The first point to remember is that all regression and correlation analysis merely deals with associations. By itself it tells us little of the causative factors that are operating. Fortunately we are frequently in a position to make at least tentative assumptions about the actual causative system. When this is possible a regression analysis can, under favourable circumstances, confirm or disprove our assumptions, and provide estimates in quantitative terms of the effects of the different factors.

As a specific example of the types of problem involved we may take the case of a survey of housing conditions conducted with the object of finding the influence of such conditions on the health of the occupants. It was observed by M'Gonigle and Kirby (1936) that the health of the inhabitants of " an unhealthy area " of Stockton-on-Tees deteriorated when they were rehoused in a self-contained municipal housing estate, owing to the fact that families moving to better houses had to spend a greater part of their total income on rent, etc., to the detriment of their general living standards, and particularly of their nutrition. On the other hand, in an investigation in Newcastle-upon-Tyne during the depression, which revealed the alarming difference in health between children of working-class (largely unemployed) parents and those of middle-class parents, Dr. J. C. Spence (reported by M'Gonigle and Kirby) came to the conclusion that the main factors responsible for the difference were

    (a) The housing conditions, which permitted mass-infections of young children at susceptible ages.

    (b) Improper and inadequate diet, which prevented satisfactory recovery from their illnesses.

He further stated: "It is probable that these two factors are of equal importance; but I would suggest that opinion on this matter should be reserved until a full inquiry, carried out by competent observers in a scientific manner, has studied the problem more closely."

We will consider how this situation is likely to be reflected in the results of a survey of a group of the population subject to different housing conditions but otherwise relatively homogeneous. For purposes of discussion we will assume that information is available on income as well as on housing conditions and health, but that no information is available on standard of nutrition, etc. It is reasonable to suppose that income affects housing conditions in that those with larger incomes will be in a position to obtain better housing, but that housing conditions do not exert any appreciable influence on income. Both income and housing conditions may be expected to affect health, income operating through housing conditions which are observed, and through other factors, such as nutrition, which are not observed. If U represents total income, V housing conditions and Y health this causative system can be represented by the following diagram :

The arrow between U and Y here represents the " net " effect of income on health, i.e. the effect of income other than that due to change in housing conditions.

This leads to the concept of net income, i.e. income after deduction of rent and other charges associated with a given type of housing. This is the part of the income which is available to produce the net effect of income on health.

To simplify the discussion consider only families of a given size and composition. Take u to represent the total income of such a family, u_n the net income, v the index of housing conditions, and y the health index. If, within the ranges covered by the variates, the causative relations are linear with a superimposed random component, the equations representing the above causative system may be written

$$v - v_0 = \gamma (u - u_0) + e_1 \qquad (9.17.a)$$
$$y - y_0 = a (u_n - u_{no}) + \beta (v - v_0) + e_2 \qquad (9.17.b)$$

where e_1 and e_2 are random components, the Greek letters are numerical coefficients, and u_0, u_{no}, v_0, y_0 represent a set of values of u, u_n, v, y near their means which conform to the linear relationship defining the causative system.

The coefficient a represents the average increase in health index that may be expected if incomes are raised one unit but people are prevented from spending any of this additional income on improvement of housing conditions. Similarly the coefficient β represents the average increase in health index that would result from an improvement of housing conditions if this improvement entailed no additional charges either direct or indirect on the occupier. The coefficient γ represents the average increase in the housing-condition index that may be expected to result from unit increase in income.

If information has been collected for the individual families on all the necessary points, the net income of each family can be calculated directly. In this case this variate should be used. Here, however, we wish to consider the case in which detailed data of this type are not available, but sufficient information is available to make an estimate of the average charge k on the total income resulting from a unit increase in the housing-condition index.

We then have

$$u_n - u_{no} = (u - u_0) - k (v - v_0) + e_3 \qquad (9.17.c)$$

The equation for y may now be written

$$y - y_0 = a\,(u - u_0) + (\beta - ka)\,(v - v_0) + e_2 + a\,e_3 \qquad (9.17.\text{d})$$

The basic data will be in the form of observations on the three variates, u, v and y. For purposes of analysis such data may well be condensed by grouping over income classes and over housing conditions so as to form a two-way table with income and housing condition as the two classifications, the entries in the table being the mean health indexes for all the families belonging to that cell. An auxiliary table giving the number of families in each cell will also be required.

If, now, we calculate the multiple regression of y on u and v we shall have an estimate of the constants of equation 9.17.d. If this regression is

$$y = \bar{y} + a\,(u - \bar{u}) + b\,(v - \bar{v}) \qquad (9.17.\text{e})$$

a provides an estimate of α and b of $\beta - ka$. Consequently the direct effect of housing β is estimated by $b + ka$.

If also, the regression of v on u is calculated, and found to be

$$v = \bar{v} + c\,(u - \bar{u}) \qquad (9.17.\text{f})$$

c provides an estimate of γ.

The total regression of y on u,

$$y = \bar{y} + a'\,(u - \bar{u}) \qquad (9.17.\text{g})$$

is also of interest, as will be explained later.

When this causative system is operating, therefore, the partial regression coefficient b of health on housing conditions, with total income as the second independent variate, does not estimate the direct effect of change of housing conditions on health. It represents the net effect, which is the difference in this direct effect and the effect of the reduction of other aspects of the standard of living due to having to spend more on housing. If there is compulsory improvement of housing, by slum clearance schemes and the like, of amount δv, without improvement of income either direct or indirect (e.g. by rent subsidies), an improvement of health of $b\delta v$ may be expected. On the other hand, if the full additional cost of the housing to the occupiers is covered by subsidies or other means, an improvement of $(b + ka)\,\delta v$ may be expected.

If incomes are raised by an amount δu and the situation is such that housing conditions and rents cannot change, an improvement in health of $a\delta u$ may be expected. If, however, the increased incomes are allowed to produce their natural effect in improving housing conditions, this improvement may from equation 9.17.f be expected to amount to $c\delta u$. The expected improvement in health, using equation 9.17.d, will then be

$$a \left(\delta u - kc \, \delta u \right) + \left(b + ka \right) c \delta u = \left(a + bc \right) \delta u$$

$$= \left\{ a + b \frac{S \left(u - \bar{u} \right) \left(v - \bar{v} \right)}{S \left(u - \bar{u} \right)^2} \right\} \delta u$$

$$= \frac{a \, S \left(u - \bar{u} \right)^2 + b \, S \left(u - \bar{u} \right) \left(v - \bar{v} \right)}{S \left(u - \bar{u} \right)^2} \delta u$$

$$= \frac{S \left(u - \bar{u} \right) \left(y - \bar{y} \right)}{S \left(u - \bar{u} \right)^2} \delta u = a' \, \delta u$$

the last line being derived from equation 9.5.b. The required increase is therefore given by the total regression of y on u, as might be expected.

The above interpretation can be accepted without qualification only if the causative system really conforms to the postulated model. In practice there are likely to be departures from such a simple model, many of which will introduce serious disturbances which may entirely vitiate the conclusions.

In the first place there will usually be external causative agents, not included in our regression system, which affect the various variates. In the above example, for instance, the level of the education of the adult members of the household may be expected to affect income. It may also, to a less extent, affect housing conditions (apart from influence due to income). Provided education does not affect health directly, these influences will not disturb the partial regression coefficients of y on u and v or their interpretation, but they will affect the regression coefficient of v on u. This latter will now represent the sum of the effects of a direct increase in income and of the associated average increase in educational level. The coefficient will therefore no longer give an estimate of the increase in housing conditions that may be expected from a rise in income level of an individual whose educational level remains unchanged.

If, on the other hand, educational level affects health directly, as well may be the case, the partial regression coefficients of health on income and housing conditions will be similarly disturbed.

These disturbances can theoretically be eliminated and the effects of education measured if an appropriate measure of education is available. All that is necessary is to include a term for education in the regression system. In practice, however, it is not possible to eliminate all disturbances of this kind, because of the number of variates that may be involved, because some of them may not be measured or may be unmeasurable, and because correlations between them prevent the separation of their effects.

A further complication which affects interpretation of regression equations arises when there is a two-way causal relationship. In the example considered, ill-health, if continued over any long period of time, may undoubtedly be expected to depress income. Consequently the association between income and health arises not only from the influence of income on health but also from the influence of health on income. The only way of attempting to assess the magnitude of disturbances of this kind and to eliminate their effects is by more detailed observations of different parts of the material. If, for example, the health index is calculated from the health of the occupants of the house other than the wage-earner it may be hoped that the components of this index which affect income will be much reduced.

Errors in the independent variates can also affect the results. In so far as these are random and their variances and covariances are known they can be allowed for by the method of Section 9.16. But this will not be the case in the problem we are considering. There will in fact be many aspects of housing which may affect health, often in different ways. These may be inadequately recorded (and indeed exact assessment of some aspects may be almost impossible) or the index may be imperfectly chosen.* Our regression will then not measure the full effect of housing conditions, but only the effect of such conditions as are correctly summarized in the index.

This complicates the issue in another way. If income is accurately measured but housing conditions are inaccurately measured and the causative system shown above is operating, some of the effect that should be attributed to housing conditions will appear as a direct income effect. In the extreme case, for example, where the chosen housing index bears no relation to the housing conditions which affect health, and is uncorrelated with income, the partial regression coefficient of y on v will be zero, and that of y on u will be equal (except for random errors) to the total regression coefficient of y on u.

When the chosen index measures some aspect of housing which is closely correlated with income but which does not affect health, the total effect of income will be divided between the partial coefficients of y on u and v in a manner which depends to a large extent on the chance distribution of error.

From the above discussion it can be seen that the use of the regression method in the interpretation of survey data is fraught with hazards. In part these hazards arise from the fundamental weaknesses of observational material stressed at the beginning of Section 9.2, but in part they can be attributed to an over-simplified approach to the problem. Housing conditions can vary in many ways, and ill-health can take many forms. The more precise and detailed the observations, the more relevant the quantities observed to the causal systems believed to be operating, and the greater our knowledge of the causal systems themselves, the more confidence can we have in our conclusions. Thus, any detailed analysis of the effect of general housing conditions on health generally must be extremely tentative, and may well, in the light of the above discussion, be judged to be not worth while. On the other hand, if we are dealing with a specific disease, such as dysentery, known to be spread under insanitary conditions, and if we can get a direct measure of these insanitary conditions and can show that the incidence of dysentery is closely related to them, then we shall feel ourselves on much safer ground in drawing

---

* Given ample data, statistical procedures are available for choosing the best index. In fact, all that is necessary is to include the different components of the index as separate terms in the regression equation, but the additional computational work involved in such a procedure will not ordinarily be justified.

the conclusion that if these conditions are remedied we may expect to see a considerable fall in the incidence of dysentery.

If, at the same time, our investigation is extended to other diseases, and these are shown to be related to the types of condition that are known to favour them, our confidence in the validity of all our conclusions will thereby be strengthened. Measuring a single association and drawing an isolated conclusion from it does not, in fact, constitute good investigational work. Such work is much more of the nature of a detective inquiry, in which all the separate pieces of evidence are assembled and fitted together. If and only if they form a coherent picture are we entitled to have confidence in our conclusions. Statistically, therefore, such work is much more difficult, and requires much more critical ability, than the analysis of experimental data, where the effects of separate factors are deliberately isolated in the planning of the experiment.

In the above discussion we have considered the simple case of a regression analysis with two independent variates. If the data are extensive the regression of health on housing conditions can be calculated separately for each income level. The partial regression coefficient b will be a weighted average of these regression coefficients. Similarly the partial regression coefficient a is a weighted average of the coefficients of the regression of health on income for different levels of housing conditions. The multiple regression method, therefore, provides an automatic averaging of the separate regression lines for different parts of the data. If examination of the data indicates that there are real differences of a meaningful nature in the separate regression lines, such averaging will be inappropriate.

I have thought it worth while to consider a specific case of regression analysis in some detail because of the very real dangers of misinterpretation in investigational work. In the particular example chosen the underlying causal relationship may be considered to be of bivariate linear form. The same considerations and qualifications apply when some or all of the variates are qualitative. Indeed the analysis of such data by the methods exemplified in Sections 9.6 to 9.15 is very similar in principle to multiple regression analysis. If the data, whether quantitative or qualitative, are extensive the necessary examination can frequently be made by comparisons of the means of various groups instead of formally calculating regressions. Whatever the method employed, however, the interpretation of the results is governed by the same principles.

CHAPTER 10

COMPUTER PROGRAMS FOR SURVEY ANALYSIS

10.1 Introduction

To perform any operation on a computer a suitable computer program is required. By introducing a sufficient degree of generality the same program can serve for a variety of jobs of the same type. This, aided by the rapid advances of computer technology, has led to the development of general programs and program packages. Many such packages for survey analysis are now in existence. Their contents and effectiveness vary considerably, depending on the skill of the authors and the applications they had in mind when the programs were written. Some are only concerned with the production and printing of tabular matter, others approximate to more general statistical packages with provision for regression and other forms of multivariate analysis, etc.

The extent to which the details of these operations are under the control of the user depends on the design of the program or program package, and this in turn depends on the types of job for which the program is intended, the value set on ease of use, and the skills expected of the user. There is an inherent trade-off here between ease of use and flexibility and power. Clearly, also, the simpler the program the easier it is to design, write, maintain and document, and the less are the likely demands on machine space.

In Chapters 5 and 9 an account has already been given in general terms of the ways in which computers can be used for validation and tabulation of survey data, and for critical analysis. The present chapter has two objectives. The first is to give some idea of the characteristics of existing packages and their strengths and weaknesses. This is illustrated by brief descriptions of four contrasting packages, all of which are in current use, and for one of which I am primarily responsible. The second is to examine what are likely to be the best lines for future development of such packages.

10.2 Types of program

For those not familiar with computers, it may be well to begin with a brief outline of the various types of program.

All modern computers are provided with one or more general high-level languages such as Fortran, Algol, and Cobol. These enable computations to be programmed without resorting to machine code. Modern high-level languages can be used without difficulty for simple computations, but if many computations for the same type of job are required much labour can be saved by writing a more general program or a package of programs. Such programs may be roughly classified as follows:

(1) Straightforward general programs for a well-defined class of job.

371

(2) Introduction of greater flexibility by the use of a problem-oriented language.

(3) A package of subroutines* which the user incorporates in his own steering program for each particular job ; the steering program is written in a general high-level language.

(4) Provision of an auxiliary compiler for use with a package of subroutines.

For the expert programmer, the provision of a package of standard subroutines which can be incorporated in a specially written steering program according to the requirements of each specific job or for a restricted class of jobs gives greater flexibility for some purposes than a problem-oriented language, as the full power of the general high-level language can be utilized at any point when anything exceptional is required. Additional subroutines can also be added to such packages to meet special requirements.

The restrictions attached to subroutines in most general high-level languages are, however, rather limiting for many purposes. If an auxiliary compiler is provided which permits the user to state his particular requirements for the job in hand in a form that is convenient to him, subroutines can be provided which are much more comprehensive and which can be more simply incorporated in his steering program. Thus, for example, in a program for survey analysis the tables required for a particular analysis can be specified in a single list, with all the necessary particulars for each table—variate to be tabulated, classifying variates, etc. This can be read once and for all by the auxiliary compiler, and suitably coded. A standard subroutine can then be provided which will make the necessary entries for the current unit in all the tables of the list by a single call.

A program written in a high-level language has first to be compiled, i.e. translated into machine code. This is done by the computer, using a compiler program. Each type of machine must of course have its own compiler. For any program that is used repeatedly without change the program will be held on a computer file in machine code so that it can be used without recompilation.

The above list covers the various types of single integrated program. Instead of including all the required operations in a single program, packages of such programs may be constructed which are so designed that the results from one program can be readily transmitted to another, i.e. they can " interface " with one another. In addition or alternatively programs or packages may be so designed that they can communicate with other programs or packages which have been constructed independently. By suitable " job control " instructions a job requiring a sequence of programs can be run in one operation without intervention.

Large modern computers are capable of running many jobs simultaneously, sharing the available resources. This is termed *time-sharing*. They can also maintain simultaneous communication with a number of computer

---

* A subroutine is essentially a sub-program which can be invoked at any required point in a program by (in Fortran) the prefix CALL.

terminals which can be linked to the computer by telephone lines, or even by satellite.   Basically a terminal consists of a typewriter keyboard by which the user can transmit messages to the computer.   These messages will be printed or displayed on a video screen as they are typed ; typing errors can be corrected by the equivalent of backspacing and overtyping, or a whole line can be deleted and retyped.   Return messages from the computer will also be printed or appear on the video screen at the terminal.

If a suitable program is available in compiled form on the computer files the instructions for a job can be typed at the terminal, and if only a few results are expected these can be sent direct to the terminal and will appear as they are produced.   If the amount of output is expected to be large the results can be sent to a file, which can then be printed (possibly after preliminary inspection at the terminal) on a line-printer, situated either at a central point, or in the neighbourhood of the terminal.   Similarly card readers and paper or magnetic tape readers can be installed at remote locations for the transmission of large amounts of data to the computer.

The ability to exchange messages with the machine with only trivial delay permits the computer to be used interactively.   An *interactive program* is one which, when it has executed whatever instructions have been given to it initially, asks the user in essence " What next ? ", possibly with guidance on available alternatives.   The user can then issue further instructions in the light of the results so far obtained, thus proceeding with his analysis step by step.

10.3   Operations required for survey analysis

These may be broadly classified as follows :

(a) Validation and editing of the data
(b) Formation of basic tables from the unit data
(c) Formation and printing of the finished tables
(d) Graphical presentation of tabular matter
(e) More advanced statistical processes required for critical analysis

Before describing any actual programs some general points concerning the formation of tables on a computer and their structure will be discussed.   The preparation of tables constitutes the central core of most survey analysis, and even when more advanced statistical processes are required, tabulation of the data in various ways is usually an essential preliminary.

10.4   Formation of tables

To produce a table of counts classified in any manner on a computer the data for each survey unit must be taken in turn, and the values of the classifying variates for the current unit must be determined.   These will define the cell into which the entry is to be made.   The address of this cell in the computer store must then be calculated, and 1 must be added to the existing number in this cell.   For tables of totals the process is similar, but instead of adding 1 the value of the variate of which the totals are being formed must be added. For tables of means, two basic tables, one of totals and one of counts, must be formed.   If there is any possibility of missing or out-of-range values a separate table of *associated counts* must be provided for each table of totals.

Before the entries for a unit are made in the tables various preliminary computations may have to be made on the data for the unit.   These generate new *derived* variates such as ratios, group values for classification, etc.   Variates recorded in literal codes also have to be decoded, unless special provision is made for handling literal values.

Entries in tables may require to be made only when one or more variates assume particular values (*selection* or *restriction*).   If a variable sampling fraction has been used or allowance is to be made for incomplete response rates, the values (counts and variates) may require to be multiplied (*weighted* or *raised*) before entry by the value for the unit of some variate representing the raising factor.   Multiple entries of the values of several variates into different cells of a single table may also be required.   Provision also has to be made for the exclusion of values which are coded as missing or which are out of range.

If a large data file is being handled there will be obvious economies if a whole set of basic tables is constructed simultaneously, as the data then only have to be read once, and much repetition of derived variate computations is avoided.   Further economies can be effected by skilful internal organization of the program.

When the basic tables have been formed, the computer can proceed to the conversion of these tables to the form required for printing and to the actual printing.   Some computations on the values contained in the basic tables will usually be required to obtain the values that are to be printed, e.g. percentages of marginal totals for tables of counts, and calculation of the means for tables of means from the basic tables of totals and associated counts.

10.5   Table structures

A printed table is essentially a collection of cells, which are necessarily set out on paper in a two-dimensional pattern, with the rows and columns suitably labelled.   The contents of each cell may be a count, a total, a mean, or some other defined quantity.   Different groups of cells may contain different quantities.   A large table may be divided into blocks, each with its own label. These can be arranged on paper in various ways, possibly on separate pages. Notionally they can be regarded as forming a third dimension.

Table structures may be quite complicated.   When a complicated table is required there are two possibilities : it may be specified in its final form in all necessary detail, or the component parts may be specified separately in the form of tables each of which has a simple structure.   If the latter alternative is adopted, the parts have then to be assembled, either manually or by a separate set of instructions to the computer.

(a)   Crossed classifications

Much the most generally useful simple structure is a table classified by a single variate or cross-classified by two or more variates.   The classifying variates may be termed *factors* and the classes of each factor may be termed the *levels* of that factor.   A three-factor $A \times B \times C$ table for factors A, B, C with 2, 3 and 4 levels respectively, for example, will contain 24 cells, which can be regarded as forming a three-dimensional $2 \times 3 \times 4$ structure ; this may or may not be the same as the structure used for printing.

It is convenient to recognize also tables classified by no factors.   Such a table will have only a single cell, containing a count, total, etc.   Tables classified by one or no factors may be regarded as degenerate cases of crossed structures.

Margins may be included in a one-factor or cross-classified table.   If all margins are included, an additional marginal level will be added to each factor. The marginal cells constitute tables of lower order : a two-factor $A \times B$ table has two one-factor margins A and B, and a no-factor margin (a single cell) ; a three-factor table has three two-factor margins, three one-factor margins and a no-factor margin.

There is of course no logical reason why margins should be included for all the factors of a table, but it is computationally simpler, if the manipulation of tables is separated from the printing, to include all margins and omit those not required when the table is printed.   When the basic tables are being formed, however, there are considerable advantages in omitting all margins. This saves space in the working store, and considerably reduces the number of entries required.   If this is done marginal totals must be inserted after the basic tables have been formed.

(b)   Nested classifications

Another type of classification that is of common occurrence is the nested classification.   In a nested classification, which may be denoted by A/B, the levels of B associated with the different levels of A are unrelated, and may differ in number.   A simple example is provided by a survey of towns in a country divided into regions.   Towns cannot meaningfully be crossed with regions, but if towns/regions is regarded as a single factor, this can then be crossed with other factors.   Note, however, that marginal values (sub-totals or means) may be required for all towns in a region.   These cannot then be generated by adding a single marginal level to towns.

(c)   Composite tables

Composite tables may be defined as the combination of parts or the whole of crossed and nested tables for single variates as defined above.   Thus we may wish to provide a printed table with a crossed classification containing interleaved rows of means, standard deviations and associated counts instead of three separate tables; or to report standard errors for the column means only of a two-factor table.   There are many variants.   One that is particularly useful and simple is to combine tables classified by the same factor or factors by printing them in parallel columns.

10.6   Tables relating to hierarchical data

If the data are hierarchical there will usually be data for more than one type of unit. Consequently the unit-by-unit operations for the construction of basic tables set out above require elaboration.

As an example, consider an educational survey for which a sample of schools is taken and data are collected for the selected schools and for some or all of the children in them. Six types of basic table may be required. These are shown in Table 10.6.

TABLE 10.6—FORMATION OF TABLES FROM HIERARCHICAL DATA

Type	Relating to	Involving data on
(a)	Schools	Schools
(b)	Children	Children
(c)	Children	Schools + Children
(d)	Schools	Schools + Children (aggr.)
(e)	Children	Children + Children (aggr.)
(f)	Children	Schools + Children + Children (aggr.)

Types (e) and (f) require the creation of a file containing the computed aggregate data, and a concurrent re-read of this file and the original data file.

If all the data are on a single file and the record for a school immediately precedes the records for the children in that school, tables of type (a), (b) and (c) present no major problems. All that is necessary is to provide for entries to be made in tables of type (a) as each school record is encountered, and in tables of type (b) and (c) as each child record is encountered. Two sets of derived variate instructions may be required, one involving school data only, performed after each school record is read, the other involving child data and possibly school data also, performed after each child record is read. The school data will still be available, as the data will have been read into different storage locations.

Tables of type (d) require provision for the calculation, for each school separately, of school attributes based on the child data for the school, e.g. mean I.Q. of all the children in the school. For the most part such attributes are likely to be derivable as simple arithmetic functions of values obtained from subsidiary tabulations (or *aggregations*) (including simple totals of variates) of the child data for the school.

Tables of types (e) and (f) require a rescan of the data after the required school attributes have been calculated. The simplest procedure is to create a file of these attributes. If there are facilities for reading two files simultaneously the required tables can then be constructed immediately. Alternatively the two files can be merged. If the data for the two hierarchical levels are on separate files, merging or reading simultaneously from these two files will also be necessary for tables of type (c) and (d).

When data are read from two files simultaneously it is, of course, essential that the associated records should be in the same order on both files ; if not, a sort of one or other of the files will be required.

If tables of type (b) only are required, recognition of the hierarchical structure will only be necessary for the calculation of standard errors.

The above procedures can be extended to hierarchical structures at more than two levels.

10.7 Relationally structured data

The tabulation of hierarchically structured data is particularly simple because, for a unit at any given level, there is only a single associated unit at each higher level ; the data can thus be arranged on a file in an order that gives easy simultaneous access to associated units. Data structures can, however, be more complicated. Thus in an educational survey information may be obtained from schools on their general characteristics and on the individual characteristics of their pupils, and from parents on the details of the family background, which latter may of course relate to several children.

To handle extensive data of the above type efficiently requires a relational system of file management. Discussion of such systems is beyond the scope of this book. It may be noted, however, that for small surveys sub-files arranged in different ways can be constructed by direct search, starting with separate files of the different types of data unit. Thus in the above example a file of relevant family data appertaining to each child can be written in the order that the children are recorded in the school file, and similarly a file of relevant school particulars can be written in the order of the children on the family file.

10.8 Table files

It will be apparent from Section 10.4 that when the unit-by-unit operations have been completed, the basic tables will exist in the computer store in the form in which they were constructed. Whether they are also accessible to the user depends on the computer program. There is, however, no computer difficulty in making these tables available. This can be done by storing all or a selection of them, as requested, in a file.

Whether the final tables that are to be printed can be similarly written to a file is more dependent on the structure of the program. If the values of a table to be printed are assembled in tabular form in the computer before printing there will be no difficulty. If any necessary computations are done concurrently with the printing, writing them away to a file may be less simple.

For table files to be generally useful the tables in them must have a definite and simple structure.   For survey work the crossed structure is clearly appropriate.   If basic tables are formed without margins it is convenient to permit tables to be filed either with or without margins, and also in packed or unpacked form if packing is permitted in the program.

Provided suitable facilities are available for further processing of tables stored on files, much retabulation from the original data files can be saved. Thus in surveys covering different regions the analysis can be done region by region as the data become available, and summaries for the whole survey can then be produced from the regional tables.   In surveys repeated at intervals, comparisons between the current and previous results can readily be made, and up-to-date summaries can be maintained.

With large data files, particularly if built up over time, e.g. data on road accidents based on police reports, tabulations are often requested which require selection of a small proportion of the units only, e.g. accidents at weekends or at particular times of the day.   If files of tables containing more comprehensive breakdowns are constructed, e.g. by day of the week and time of day, requests for partial information can be met without retabulation.

Table files encourage a step-by-step approach to the tabulation of data. In particular, decisions on the form and content of the final presentation of the results can be made after preliminary examination of the full tables.   They are also useful if graphical output is required.   To obtain a satisfactory graphical representation of tabular matter often requires several attempts. It is clearly advantageous to work directly from table files rather than to have to reconstitute the required values at every attempt.

If such uses are envisaged, it is important that at least a limited amount of information relating to the individual tables be included in the file.   There are obvious advantages in including also general particulars relating to the file, such as lists of variate and level names.

Table files also facilitate the construction of interfaces with other statistical programs that can operate on tabular data.

10.9   Examples of some actual program packages

The following sections contain brief descriptions of four program packages which between them illustrate the very varied approaches that can be adopted to the design of such packages.   Two of these (RGSP and TPL) are limited to the production of tables, with interfaces to other packages for more advanced statistical processes; the third (SPSS) is a comprehensive statistical package oriented to survey analysis ; the fourth (GENSTAT) is a powerful general statistical program which has its own problem-oriented language and is primarily oriented to advanced multivariate techniques and to the analysis of data from planned experiments.   SPSS was selected from a number of similar packages included in the comparative test of tabulating ability described in Section 10.14 because of its wide use in British universities.

None of these programs are interactive, but an interactive version of SPSS (SCSS) is now operational. This is also described. Some other packages are briefly mentioned in later sections.

For the most part the descriptions are based on the program manuals listed in the bibliography. Many program packages are of course improved from time to time by the addition of new features and correction of defects. Some of the criticisms expressed below may therefore have been met by the time this book appears.

10.10 RGSP (Rothamsted General Survey Program)

The Rothamsted General Survey Program (RGSP) has evolved from the scheme for a general survey analysis program which I outlined in the third edition of this book. A primitive program on these lines was implemented on our first very small computer (Yates and Simpson, 1960, 1961). This program was considerably elaborated for our Orion computer and was subsequently transferred with further enhancement to its successor, an ICL 4-70 (Yates, 1973, 1975). It differs from most tabulation programs in that it is divided into two independent parts, that dealing with the unit-by-unit operations, and that dealing with table manipulation and printing. These parts can be used separately or in sequence. Communication between the two parts is by means of table files, which are a distinctive feature of the package. This division was historical in origin, due to the severe storage limitations of our early computers, but once made, it was apparent that it had many advantages over the more conventional approach, at least for our type of work.

Fig. 10.10 illustrates the structure. The continuous lines show the most simple and fundamental uses ; the broken lines indicate other facilities that are available.

(a) The Part 1 compiler

Part 1 comprises a package of standard subroutines with an auxiliary compiler (Type 4 of Section 10.2). The lists of the required basic tables, limits (which must be specified for variates used for classification and may be specified for other variates to exclude gross errors), literal codes, and grouping instructions are supplied to the compiler. Derived-variate instructions other than groupings are written in Fortran and included in a Fortran steering program written by the user for the job in hand.

Variates are referred to by number, but lists of variate names and also level names can be supplied for the simple table prints available in Part 1. Tables are automatically numbered consecutively as listed, starting from 1 or other specified number. The compiler supplies an expanded list of the particulars relating to the specified tables, including their locations in the working store; also the minimum working store requirements for the steering program.

Tables of counts, totals (with, if required, associated counts), maxima or minima can be formed. Each table may be cross-classified simultaneously

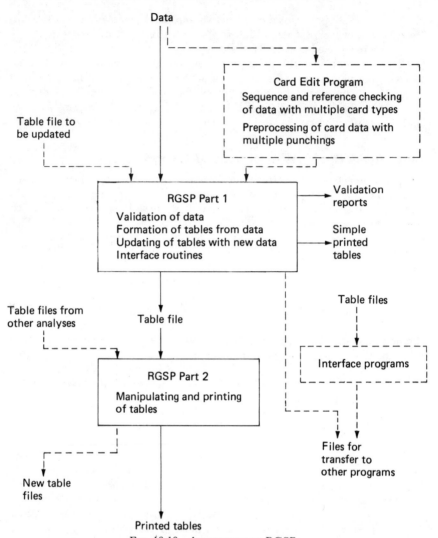

FIG. 10.10—AN OUTLINE OF RGSP
Continuous lines show the simplest and most fundamental uses; dotted lines show other available facilities.

by up to five variates. Table entries may be weighted by the current value of any variate and entries can be restricted, i.e. made only if the value of some variate satisfies a defined condition. All these requirements are specified in compact form in the compiler tables list.

If sets of data for several units or sub-units are included on the same record (" parallel " sets), different variate numbers must be assigned to each set, but such sets can all be entered in the same table or tables. The " parallel " device also enables data for multiple non-exclusive responses coded as separate variates or data for a set of quantitative variates—e.g. areas of different crops in a farm survey—to be entered in the different classes of a table.

Tables are formed without margins, and may be in integer, short (or packed) integer, real or double-length real form. These options reduce storage requirements without sacrificing necessary accuracy. For most jobs, control of form can be left to the computer, but overriding control is available when needed. To further economize working store the compiler lists in both Parts 1 and 2 are packed end-to-end after compilation.

If the value of a classification variate for a table transgresses its limits or is unknown, an entry is made in a single " unclassified " cell unless an extra " unclassified " class is specified for that variate.

(b) *Part 1 steering programs*

The basic structure of a straightforward steering program is illustrated in Table 10.10.1. The standard subroutine START reads the file of particulars

TABLE 10.10.1—OUTLINE OF A SIMPLE RGSP PART 1 STEERING PROGRAM

	Call START (...)
1	Read data for next unit
	If end of file go to 2
	Call CODES (1)
	Derived variate instructions
	Call GROUPS (1)
	Call TABLES (1)
	Go to 1
2	Call STORE
	Stop

produced by the compiler, checks that sufficient working store has been declared, and initializes all tables. For each unit in turn, CODES and GROUPS perform any required de-coding and grouping, and TABLES makes the necessary entries in the specified tables. When the last unit has been processed, STORE writes the completed tables to a table file in the form in which they were constructed, i.e. without margins, as integers, short integers, reals, etc.

The parts of the program that have to be provided in detail by the user are the instructions for reading the data and those for forming derived variates. For reading the data Fortran READ can be used, but is liable to give trouble if there are errors in the data, particularly if there is more than one card type, or the cards contain literal data. A standard subroutine CARD is therefore provided as an alternative. This handles character and unsigned integer fields and permits the use in numerical fields of literal unknown-value symbols, e.g. X and Y, recoding these, blank fields, and illegitimate punching by chosen numerical values, e.g. -1, -2, -3, -4. If there is more than one card type the card-type number of the next record is returned.

The subsidiary program Card Edit performs the same functions as CARD, but in greater generality, and also deals with multi-punched data and over-punching in the first column of a numerical field. Multi-punched non-exclusive responses are retained in binary form on the edited file and are tabulated directly by TABLES. Card Edit also checks for sequence and reference errors in files which contain more than one card type ; this can be done without any numerical editing.

The main advantage of writing derived variate instructions directly in Fortran is that the full power of the language is available when complicated functions are required. There is, however, the penalty that the program must be so written that variates derived from variates any one of which is unknown or transgresses its limits are themselves coded as unknown. The simplest procedure is to set to the unknown value (-10^{15}), after the data for each unit are read, (a) all derived variates (a simple Fortran loop) and (b) all the basic variates from which they are to be derived and which transgress their limits. A standard subroutine TEST is provided for (b). Insertion of conditional jumps at appropriate points is then all that is required.

TEST also returns the number of variates for which the test fails, and thus provides a simple method of excluding completely from the analysis units in which any variate (or any one of a set of key variates) is unknown or transgresses its limits.

Tables will usually be filed for manipulation and printing by Part 2, but a simple standard subroutine PRINT is provided which prints all or a selection of the basic tables in full, without margins, in very compact form ; tables of totals with associated counts are printed as means with interleaved counts. Variate names are included if specified. PRINT has proved useful for program checks, using a small sample of data, and also for preliminary assessments of the data such as are provided by univariate frequency distributions.

The use of a steering program provides a simple and flexible means of dealing with hierarchical data of any degree of complexity. The card-type number of the next card returned by CARD provides overall control. The codes, groups, and tables lists submitted to the compiler can be divided into separate sequences, the sequence required at a particular point in the steering program being indicated by the subroutine parameter, e.g. TABLES (2). Thus tables of types (a) to (d) of Table 10.6 can be formed in a single run. For tables of types (e) and (f) the file of aggregated variates constructed in the first run can be read without merging by Fortran READ.

It may be noted, also, that identical sets of tables for batches of data, e.g. regions of a regional survey, can be formed in a single run by a simple loop in the steering program, with calls to STORE and INIT (which re-initializes tables) at the end of each batch. The tables for each batch will then be assembled in the same table file. Corresponding tables will have the same table numbers, but can be distinguished by a subsidiary index.

(c)   Updating tables

As indicated in Fig. 10.10, a Part 1 table file can be read back by the steering program for updating. This enables a supplementary file of additional data, e.g. late returns, to be added to a set of basic tables without complete retabulation. It also enables corrections to be made to tables for errors in the data discovered after the tables have been formed.

Control of this read-back is very simple. All that is necessary is to replace the Stop of Table 10.10.1 by

Call RESTRT

Go to 1

with minor changes in the START parameters. If this is done before the first set of data is tabulated, no change in Part 1 or in associated Part 2 programs is required for re-runs. The action required in any run is determined by the files that are presented.

(d)   Validation and correction of data files

As mentioned above, card-type sequence and reference checks can be made by the Card Edit program. Numerical and logical tests on variate values of the types described in Section 5.14 can be made by calls to a standard subroutine VALID in a suitably written Part 1 steering program. There are no provisions for scatter diagrams.

VALID can test for the unknown value and for up to four other nominated numerical values, typically those specified in CARD for non-responses, blanks, and illegitimate punching ; also for transgression of the limits (upper and lower separately) specified in the compiler list. The controls enable each type of error to be monitored separately for each variate (basic or derived), the options being (a) test, count, and report, (b) test and count only, (c) do not test. If reported, the actual value, record number, and a unit reference are given, and also the image of the current card if requested. Logical tests, written in Fortran, have similar options.

An edit program is provided for the amendment of data files. This substitutes, inserts and deletes complete records, and can be used even on files which are not accessible to normal text-editing facilities, e.g. magnetic tape. Alternatively the original data file can be divided into (a) a file containing all the records belonging to units (including all associated units of hierarchical data) in which no record is in error, and (b) a file containing all the records for units in which there are one or more errors. The latter file can then be listed, and if on disc the text-editing facilities available on most machines can be used to make the necessary corrections. The two files (a) and (b) can then be tabulated consecutively in the same run, thus obviating the need to amend the original data file.

(e)   *Standard errors, regression estimates, etc.*

Tables of sums of squares and products are required for the calculation of regression coefficients and for standard deviations and standard errors. Tabulation of a variate with weighting by the same or a different variate provides a convenient way of forming such tables without the formation of auxiliary derived variates. The required estimates can then be obtained by Part 2 operations.

For hierarchical data, and data in which the unit of sampling differs from the unit of analysis, tables involving summation over both the levels of the hierarchy are required for standard errors. These cannot be formed directly by TABLES. A standard subroutine STAGE has therefore been provided ; this forms tables $S'(S''y)^2$, $S'(n''S''y)$ and $S'(n''^2)$ required in the formula for Q of Section 7.11. These additional tables are defined in a separate sequence, which is not called, in the compiler tables list. STAGE does not at present form the additional sums of products required for $Q_{A.B}$, but this can be added.*

An alternative mode of STAGE provides a table of $S'[(S''x)^2/n'']$ required for the analysis of variance of Section 8.11. Both modes provide a table of n'.

(f)   *Data-file management*

RGSP lacks provisions for data-file management such as are contained in SPSS. However, as variates are denoted by number, Fortran instructions to write a new file can be included without difficulty in a steering program. Such a file can be written in formatted (character) or unformatted (binary) form, with addition and deletion of variates, and, if required, renumbering. Such operations would be more complicated if variates were denoted by name, as amended lists of variate and level names and limits would have to be transferred.

(g)   *Part 2 processing instructions*

Part 2 is a self-contained program with its own problem-oriented language (Type 2 of Section 10.2). The required operations are controlled by *processing instructions*. These are written in a simple Fortran-like language which permits the use of indexes, loops, conditional and unconditional jumps (Go to's) and subroutines of a simple type. Lists can also be used in various contexts. These facilities save much tedious repetition when similar operations have to be performed on several tables.

The processing instructions comprise : (i) instructions for reading tables from a table file or from an extraneous source, and for storing tables on a new file ; (ii) instructions for performing arithmetical operations on tables and

---

* In Example 7.11 tables of these were formed by a simple piece of Fortran, supplementary to STAGE, in the steering program—a good illustration of the way tables can be defined and accessed in Part 1 for special operations not provided by the standard subroutines.

manipulating them in various ways ; (iii) instructions for printing ; (iv) miscellaneous instructions for setting indexes and numerical constants, controlling the. flow of the program, clearing tables from the working store, etc.

In Part 2 all tables are in real form, with margins. When tables are read from Part 1 files, they are automatically converted to real form and marginal totals are inserted. Tables stored on a new file retain their margins.

The provision in Part 2 for reading tables from several files and for storing tables on one or more new files gives the required flexibility for summarizing the results of surveys covering different regions or repeated at intervals, and for creating a new table file of the summarized results ; also for combining several table files or a selection of tables from them into a single new file. Tables can also be stored temporarily on a back-up file.

(h) Table operations

Many of the arithmetical operations that are required on tables can be conveniently represented in algebraic notation. This notation is adopted for the processing instructions. Thus a table of means T3 can be formed from a table of totals T1 and the corresponding table of associated counts T2 by writing

$$T3 = T1/T2$$

(The table number of an associated count is always one greater than that of the corresponding table of totals.)

Overwriting a table by another with the same factors is permitted. Thus

$$T1 = T1/T2$$

will replace the totals by means.

Only one or two table operands may appear on the right-hand side, but subject to some restrictions (which, however, can be circumvented by using more than one instruction) the tables in such equations may be classified by different factors. If, for example, T2 contains some only of the factors of T1, T3 can be classified by all the factors of T1 (default action) or by the factors of T2 only. Thus if T1 is classified by variates 20 and 21, and T2 by 20 only, we may write

$$T3 = T1 * T2 (20)$$

T3 will then be the cell-by-cell product of the 20 margin of T1 with T2. Similarly

$$T3 = T1 (20)$$

will give a table classified by variate 20, containing the 20 margin of T1, and

$$T3 = T2 (20, 21)$$

will give a two-factor table containing identical columns which are copies of T2.

Tables may also be initialized to a chosen numerical value, e.g. 0 or 1, or multiplied, etc., by such a value.

Many of the calculations exemplified in the preceding chapters can be specified in terms of a table algebra of this type. The formation of a table of standardized rents per acre for the 50 or so counties included in the National Farm Survey by the procedure of Example 6.8 provides a simple example.

The required instructions are shown in Table 10.10.2. These take as their
starting point basic tables T2 and T3 of the total rents and areas (unraised)
of the rented farms in the sample, classified by county × size-group. The
raising factors, 20, 10, 4, 2, 1, applicable to the five size-groups (Table 5.17.a)
are held in table T1.

TABLE 10.10.2—FORMATION OF A TABLE OF STANDARDIZED RENTS BY RGSP
TABLE OPERATIONS

Processing instructions	Contents of new table	Classifying factors		
		New table	Source table(s)	
T4 = T2 * T1	Raised rent totals	(11,12)	(11,12)	(12)
T5 = T3 * T1	Raised area totals	(11,12)	(11,12)	(12)
MAR T4, T5	Recalculate marginal totals			
T6 = T4/T5	Rents per acre	(11,12)	(11,12)	(11,12)
T7 = T5 (12)	Size-group totals of raised areas	(12)	(11,12)	
T8 = T7 (0)	Raised area (grand total of T5)	—	(11,12)	
T7 = T7/T8	Weighting factors for standardization	(12)	(12)	—
T9 = T6 * T7	Rents per acre × weighting factors	(11,12)	(11,12)	(12)
MAR T9	Recalculate marginal totals of T9			
T10 = T9 (11)	Standardized rents per acre	(11)	(11,12)	

T1 : raising factors ; T2 : rent totals ; T3 : area totals.
Variate nos. of classifying factors : County = 11, Size-group = 12.
(0) indicates classification by no factors, i.e. T8 is a single value.
***** denotes multiplication.

The table of rents per acre (T6) and the table of standardized rents per
acre (T10) can be printed as a single composite table, with rents and standardized rents side by side in parallel columns, by the procedure described in
the next sub-section.

The calculation of the standard deviations of a table of means provides a
further simple example which illustrates the use of indexes (30 are available)
and subroutines. A subroutine for this is shown in Table 10.10.3. L, L+1,
L + 2 are the table numbers of the basic tables of totals, associated counts
and sums of squares. Entry is by CALL 40 with L set. T(M) is a working
table. POS gives 1 for positive cell values, 0 otherwise. At exit T(L) contains
means and T(L + 2) standard deviations. T(L + 1) is unchanged.

A particular point to note about this form of subroutine is the equating
of the working table, T(M), to T(L) at the start of the subroutine and its
clearance at the end. This has the effect of assigning the classifying factors
of T(L) to T(M) during the passage of the subroutine and cancelling this
assignment and the associated storage at the end, so that the subroutine can
be entered repeatedly for tables with different factors.

Once written, subroutines for commonly required procedures such as
standard errors, χ^2, ratio and regression estimates can be stored on computer

TABLE 10.10.3—AN RGSP SUBROUTINE FOR STANDARD DEVIATIONS

40 T(M) = T(L)	$S(y)$
T(L) = T(L)/T(L + 1)	$\bar{y}$
T(M) = T(L) * T(M)	c.f.m.
T(L + 2) = T(L + 2) − T(M)	$S(y - \bar{y})^2$
T(M) = POS T(L + 1)	
T(M) = T(L + 1) − T(M)	d.f.
T(L + 2) = T(L + 2)/T(M)	m.s.
T(L + 2) = SQRT T(L + 2)	s.d.
CLEAR T(M)	
RETURN	

files or sets of punched cards, and incorporated when required in Part 2 programs. This has the advantage of flexibility, as such subroutines can easily be modified to meet different requirements. Thus an additional division by T(L + 1) before taking the square root will give standard errors instead of standard deviations. If the sample is random, these will be the sampling standard errors of the various means. If the sample is stratified, the sampling standard error of the population mean can be obtained by insertion of additional instructions to provide a pooled sum of squares within strata.

Although the limitation to two operands on the right-hand side results in somewhat numerous table operation instructions in this type of computation, there is no great difficulty in splitting up a formula containing more than two operands into its component parts ; nor if the required formula is appended as a comment, in checking that this has been correctly done. The limitation simplifies the Part 2 compiler, and also, when operations are done table by table, gives the user better control over temporary working space, an important consideration when large tables are being handled.

In addition to the instructions illustrated above, various instructions covering other types of operation are provided, of which the following may be mentioned :

(1) Percentages based on any defined margin (including margins for two or more factors). The base numbers are retained in the relevant marginal cells in a form that causes them to be printed as integers.

(2) Summation of a set of tables classified by the same factors.

(3) Combination of a set of tables classified by the same factors into a single table with an additional factor.

(4) Combination of levels of a defined factor of a table, e.g. rows or columns of a two-factor table. This is a very general instruction, and provides for combination, rearrangement, deletion, differencing and insertion of subtotals. It can also be used to copy a single level to form another table with one less factor, or, by means of a loop, to split a table into a set of such tables (i.e. the converse of 3 above).

(5) Combination of defined parts of two or more tables into a larger table.

387

(*i*)   *Printing facilities*

The Part 2 printing facilities are designed to print complete tables in an acceptable and compact layout merely by giving a print instruction with a list of table numbers, while permitting more detailed control when this is required for tables intended for direct reproduction in reports.

Tables classified by more than two factors are printed in blocks classified by the last two factors.   If space permits, successive blocks are printed side by side.   Any selected one or more factor margin of a table may be printed, and rows, columns, etc., may be interchanged.   Particular margins may be excluded from the printed table.

Names of any reasonable length may be used for column, side and block headings and for variate names.   If level names are not specified, levels are numbered from their lower limits ; similarly, unnamed variates are referred to by number.

Side headings may be printed on two or more lines.   Column headings are automatically centred over columns, and if too long are printed alternately on two lines ; or they may be divided into two parts, and the first part may be used for a heading extending over several columns ; alternatively the whole of a table heading, including the column headings, can be formatted, allowing the user complete control over layout and content.   Column spacing and number of decimal places can be specified or left to control by the computer. If the latter, each table is printed so that its largest value contains four (or other specified number) of significant figures.

In addition to the provisions for printing tables one by one, sets of tables classified by the same factors can be printed in parallel columns.   The column width and number of decimal places can be specified independently for each table ; the options available for column headings are the same as those for single tables.   There is also provision in this form of printing for grouping rows based on less than a defined minimum value in the first table as " OTHERS ".   Totals or weighted means of the grouped values can be specified for printing for each table independently ; if there is more than one factor the " others " values for each set of levels of the last factor are grouped separately.

This form of printing can be used to append one or more additional columns to, say, a two-factor table, by first splitting the two-factor table into a set of one-factor tables each of one column.   Thus in the National Farm Survey example above, the standardized rents (T10) can be printed alongside the counties × size-group rents table (T6) after splitting T6.   An additional row or rows can similarly be appended to a two-factor table by printing the row as a second table with suppressed column headings.

Tables selected for a report can be automatically numbered consecutively in the printed copy without alteration of their file numbers.   Page width can be controlled, and can be varied from table to table.   There should be no difficulty in providing for output of literal matter in upper and lower case for installations that have the required printing facilities.

388

10.11   TPL (Table Producing Language)

The distinctive feature of TPL* is its provision for direct specification of composite tables of virtually any degree of complexity in the form in which they are to be printed. The required content and layout of a table is defined by its column (*heading*) labels, its row (*stub*) labels, and its block (*wafer*) labels. Thus the skeleton form of a table is one-dimensional (a single row of values), or a two- or three-dimensional crossed structure of values. The structure of each set of labels may be complex, as explained below. The labels do not have to be listed in full; a single reference to two classification variates, A and B, for example, will give separate juxtaposed or interleaved sub-tables (counts, totals, means, or any combination of them) for A and B separately, or all combinations of A and B, as required.

The package is particularly designed to enable users to access and tabulate selected items of data from census-type surveys, including surveys containing hierarchical data with any number of levels, without direct knowledge of the file structure of the basic data file, and without having to prepare lists of variate and class names. To effect this a *code book* of the basic file, or of selected variates from it, is prepared. This is done by a separate program. Variate and class names are assigned for use in table specification and for printing ; optionally longer labels can be specified for printing. The code book program produces a working data file and gives a list of all the variates in it, distinguishing between variates used for classification (*control variables*) and variates which are to be tabulated (*observation variables*). The associated names and any comments that are considered necessary are also included. If it is thought that the same variable may be required both for classification and tabulation it can be duplicated in control and observation form under different names in the code book. The working data file can then be made available to the tabulation program by its code-book name, and variables and classes can be referred to by their names. Further observation variables (derived variates), can be formed from other observation variables by COMPUTE statements. These are limited to algebraic expressions containing $+$, $-$, $*$, $/$, square root. In the latest version (V4), COMPUTE statements can be made conditional on values of either control or observation variables. Further control variables can be generated by DEFINE statements. These provide for selection and grouping of values of both observation and control variables ; the groups need not be mutually exclusive or exhaustive of all values.

TPL does not have a general editing routine. However, if a control variable used in a request has invalid variable values in the data file, any affected processing unit will be omitted from the tabulation and the record in error will be listed. Limits can be imposed on an observation variable by a SELECT statement or a DEFINE statement. Missing values of control variables can be

---

* TPL was developed by the Division of General Systems of the U.S. Department of Labor, Bureau of Labor Statistics, under the guidance of Rudolph C. Mendelssohn (see Bibliography). This description refers to Version 4.0, 1978.

excluded by DEFINE, i.e. DEFINE can screen units from particular tabulations without excluding the entire unit as SELECT does.

For statistical computations not provided by TPL, interfaces with SOUPAC and, in the V4 version, with SAS are provided. These transmit the values in each column of a table as a SAS or SOUPAC variable. Several tables can be so transmitted. There is, however, no reverse interface : any subsequent table printing would have to be done by SAS or SOUPAC.

Table specification

Wafer, stub and heading specifications are written in that order, with commas as separators. Each contains a string of variable names separated by BY (nesting) or THEN (concatenation, i.e. followed by), with brackets if necessary. If only headings are specified the table produced will be on a single line. If headings and stubs the table will be two-dimensional, and if wafers also three-dimensional.

The content of each cell of a table is defined by the variables relating to that cell in the wafer, stub and heading labels. If they are all control variables a count will be formed, or if required a count can be specified explicitly by COUNT or by a record name assigned to the records of the survey. If one of them is an observation variable a total of that variable over all selected units will be formed. For obvious reasons not more than one observation variable may relate to any one cell. Hence observation variables may occur in one only of the wafer, stub and heading specifications.

Marginal totals must be specified explicitly where required as separate items by the word TOTAL (preceded or followed by THEN).

The following examples illustrate the flexibility of these specification rules:

A and B are control variables with 2 and 3 levels respectively. REGIONS and DISTRICTS are control variables defining a nested Regions/Districts classification. X is an observation variable.

(1)   TOTAL THEN A THEN B
will give the column headings

$$\text{TOTAL} \quad \text{A1} \quad \text{A2} \quad \text{B1} \quad \text{B2} \quad \text{B3}$$

This table will therefore contain the grand total and the A and B marginal totals of a full $A \times B$ table of counts, printed on a single line.

(2)   (A THEN TOTAL) BY (B THEN TOTAL)
will give

A1				A2				Total			
B1	B2	B3	Total	B1	B2	B3	Total	B1	B2	B3	Total

i.e. the full $A \times B$ table, with marginal totals, printed on a single line.

(3)   A THEN TOTAL, B THEN TOTAL
will give the conventional $A \times B$ cross-tabulated layout.

(4)   X BY (TOTAL THEN A THEN B)
will give a layout as in (1) but with totals of X instead of counts.

(5) X THEN COUNT, TOTAL THEN A THEN B

will give the totals of X as in (4) with the corresponding counts printed beneath them.

(6) (TOTAL THEN A THEN B) BY (X THEN COUNT)

will give a layout as in (1) with each total of X followed by the corresponding count.

(7) REGIONS BY (DISTRICTS THEN TOTAL) THEN TOTAL,
      (TOTAL THEN A THEN B) BY (X THEN COUNT)

will give rows of totals and counts as in (6), with a side classification by districts within regions, regional totals and a grand total.

Specifications of the above type generate tables of counts and totals, corresponding to the basic tables of RGSP, but usually of more complex structure. If means, etc. are required, additional computations must be made after the tables are formed. These are specified by POST COMPUTE statements.

If, for example, a table classified by region and containing totals of income, average income per person, and counts of persons is required, this will be given by

POST COMPUTE

AVR_INCOME = INCOME/PERSONS

TABLE A: REGION, INCOME THEN AVR_INCOME THEN
            PERSONS

Essentially the post compute statement creates an additional column of cells for which the values are calculated after the tabulations are completed. If the totals of income are not required in the printed table, INCOME is omitted from the table specification.

From the above it will be seen that post compute statements are equivalent to the RGSP arithmetic operations on tables, but are limited to tables or subtables having the same pattern. They are not designed for operations such as those exemplified in Table 10.10.2, nor are they capable of calculating percentages of marginal totals. Special provision is made for such percentages, but is limited to expressing all values of each wafer as percentages of the first row or column or of the top left corner value.

The above instructions can be used to produce tables of types (a), (b) and (c) of Section 10.6, relating to hierarchical data. There is no limit to the number of levels in the hierarchy.

Post compute statements can be used to compute standard deviations or standard errors. Thus for a variable X (CTX for count)

COMPUTE SQ_X = X*X

POST COMPUTE ST_DEV_X =

$$\text{SQRT}((SQ\_X - X*X/CTX)/(CTX - 1))$$

will give standard deviations for each group of X's included in a table. Such standard deviations and standard errors are, of course, only strictly relevant to random samples.

The V4 version also makes provision for the computation of medians, quantiles, maxima and minima and their incorporation in any desired manner in the printed tables.   Details need not be given here.

Printing facilities

A major aim of TPL is to provide printed tables of a high standard suitable for direct photographic reproduction.   Names can be in upper and lower case, and when variate names are followed by level names they are overprinted to give them greater prominence.   A further important enhancement in the V4 version is the ability to produce an output tape for direct processing by a photocomposition machine.   The resulting tables appear to be manually typeset without the time and expense of that approach.

The specification of a table determines the form in which it is to be printed. The labelling of the wafers, rows and columns depends on the variate and level names (up to 30 characters) specified in the code book and in the COMPUTE, DEFINE and POST COMPUTE statements.   The formats of the printed values are similarly defined by means of " masks " ; these indicate, for each variate, the required accuracy (number of decimal places) and permit appended dollar and percent signs, and subdivision of numbers by commas.   The stub width, i.e. the space allotted to row labels, and the column width (the same for all columns) can be specified by the user or set by default.   A row or column label too long for the allotted space will be subdivided automatically in a sensible manner (with optional additional guidance by the user on suitable division points for long words) and will be printed on two or more lines.

A problem in the labelling of tables that always presents difficulty is that the variate names of classifying variates as well as their level names may be required.   In the simple case of a two-factor cross-classified table the name of the row factor can best appear at the head of the column of the corresponding level names, and that of the column factor as an extended heading above all the columns.   For cross-classified and composite tables with three or more factors in the rows there does not appear to be any general system which can be automatically applied and which will give both clarity and compact presentation.

Often, however, classifying variate names are unnecessary, as the level names sufficiently indicate the variates.   Thus, for example, there may be no need for " Regions " to supplement a list of regional names.   TPL adopts this solution.   If a print name (in addition to the short name) is specified for a variate it is printed, if not, not.   Similarly with level names : if not specified the levels are numbered from 1, and printed as short variate name = level number.   There is also automatic indentation of row names to distinguish different levels of nested and crossed classifications.   Thus TPL can be relied on to give clear and unambiguous labelling for all items in a table.   This labelling can be revised, if thought necessary, without retabulation by a supplementary program, PCL (Print Control Language).   This program also provides facilities for alteration of the format of printed values, addition of footnotes relating the particular values, alteration or deletion of values, specification of column widths of varying size, etc.

This ingenious labelling system may be contrasted with that adopted in other programs. RGSP ducks the problem of variate names by giving the classifying variates and the variate tabulated in the heading, and prints the level names of all but the last two factors on separate lines, distinguishing them by one, two or three asterisks. Segmentation of level names of the last two factors has to be specified by the user in his name lists ; this has some advantages, but undoubtedly requires more attention to detail. GENSTAT prints the level names of the row factors in separate columns headed by the corresponding variate names. This gives a clear layout, but even with names limited to eight characters requires considerable space for a table containing several row factors. SPSS makes no serious attempt to label tables tidily or coherently.

10.12 SPSS (Statistical Package for the Social Sciences)

SPSS is a self-contained program package designed for the analysis of investigational surveys, particularly in the social sciences. It provides only limited facilities for table formation, but includes sub-programs for correlation and regression analysis, non-orthogonal analysis of variance and covariance, discriminant functions, factor analysis, canonical correlation, and Guttman scaling. Only some salient points need be mentioned here.

(a) *SPSS data files*

A basic feature of the package, which makes it easy to use, is the self-contained system of data files. When the original data are read they can be stored on an SPSS data file in binary (unformatted) form, together with the variate and level names and any derived variates it is wished to preserve. Such files form a basis for further analysis and can be modified and up-dated in various ways. This contrasts with TPL, which reads the data for the variates defined by the codebook directly from the original file, and with RGSP, in which the standard Fortran file-handling facilities are required for incorporation of derived variates and for re-writing a data file in binary form.

SPSS data files can only contain data for one type of unit (" case "). Consequently hierarchical data cannot be accommodated on a single file ; only tables of types (a) and (b) of Table 10.6 can be produced directly, and then only by creating separate files for the data for the two levels. In the 1975 version of the program (Nie *et al.*, 1975) a function AGGREGATE has been added which enables totals, means, etc. over sets of units belonging to the lower level of a two-level hierarchy to be calculated from the lower-level file. These are written to a subsidiary data file which can then be merged with the data file of either level of the hierarchy. However, if merging is required for both levels two separate files have to be produced in separate runs ; that for the lower level must contain repetitions of the function values, as the merging process requires one-to-one correspondence between units of files that are to be merged. This somewhat clumsy procedure enables tables of types (d) and (e) of Table 10.6 to be formed, but it is still impossible to form tables of types (c) and (f). Tables of type (c) are certainly very commonly required.

There appears to be no means of referring to the value of a variate for the preceding unit in a file, and consequently a file cannot be checked for sequence and reference errors ; such checks may be required even for non-hierarchical data when there is more than one card per unit. Arithmetical and logical tests of the types performed by VALID in RGSP can be made on unit values by suitable derived variate instructions, and the units in error can be located by a WRITE CASES instruction ; the manual, however, does not tell the user how to set about this.

(b)   Table formation

Tables of counts classified by a single factor, or cross-classified by two or more factors, can be formed. A two-factor table is printed with its marginal totals, and if requested with interleaved percentages on row or column totals, or on the grand total, or indeed on all three, but the print of the actual counts cannot be suppressed. A table with more than two factors is printed as a series of sub-tables classified by the last two factors. Thus an A $\times$ B $\times$ C table lacks the B $\times$ C margin. If required this must be separately specified, or obtained by addition on a desk calculator. Any empty row or column of a table or sub-table is omitted from the print. This makes comparison of sub-tables more difficult, and impairs legibility generally.

In common with other similar packages, the printing of tabular matter pays scant regard to compact presentation or economy of paper. Unless paging is suppressed (an option introduced in the 1975 version) each table or sub-table is printed on a separate page.

The discussion in the manual of the ways in which inferences can be drawn from tabular matter is heavily biased towards contingency tables. There are many examples for which the results could be presented much more compactly and intelligibly and in all necessary detail as tables of means with one less factor. Indeed in the earlier versions of the package (Nie *et al.*, 1970) tables of means could only be printed in hierarchical form (BREAKDOWN), including totals, means, standard deviations, variances and counts for each entry, with empty cells omitted. Thus an A $\times$ B $\times$ C table would contain an entry for the grand total and entries for the A levels, the A $\times$ B levels, and the A $\times$ B $\times$ C levels. To obtain a table of means, say, in cross-classified form with margins necessitated hand copying and either desk calculation of the missing marginal values or supplementary formation of B $\times$ C and C $\times$ A tables. The alternative CROSSBREAK now prints tables of means in the same layout as tables of counts ; if associated counts, totals and/or standard deviations are also required these are interleaved.

There are no facilities for storing tables or for performing arithmetical operations on them before they are printed, nor, in 1975, for making entries for several variates in the same table.   As mentioned below (Section 10.14), this last was a serious inconvenience in some types of questionnaire survey.

A curious, and to my mind totally unnecessary feature of the sub-program for forming tables is that the user is permitted to omit the limits of the classifying variates, with the consequence that considerably more computer time is required for the tabulations.   Moreover the limits have to be specified afresh every time a new set of tables is formed, thereby considerably increasing the possibility of error.   There is an example in the manual of a two-factor table for which one of the variates has limits of 0 and 7 ; this is broken down on the next page by sex into two sub-tables for which the lower limit is incorrectly specified as 1, with the consequence that nearly half the data are excluded.

In Section 9.15 attention was drawn to the dangers of the uncritical use of χ^2 when examining contingency tables for association.   SPSS not only provides χ^2 on request for any two-factor table, but also makes available no less than nine other " measures of association ".   Although, judging from the references given in the manual to recent statistical textbooks used by social scientists, these measures are still in current use, they do not, I think, have any merit, and only serve to confuse the beginner.*   Worse still, if a two-factor table is broken down by one or more further factors, separate values of χ^2, etc., are appended to each of the resultant sub-tables independently.

(c)   Correlation and regression

There is a set of sub-programs all of which deal with aspects of the fitting of linear models by least squares, of which the most important is that for multiple regression.   This will fit the parameters included in the equation one by one, or in groups, in a defined order, and can also be used in a stepwise manner, the parameter selected at each step being the one that gives the greatest reduction in the remaining residual sum of squares, with various stopping criteria.   The program lacks the convenient specification of additive linear models for tabular matter provided by GLIM : for such models dummy variates have to be specified by derived variate instructions.   It can also produce very voluminous output.

The main weakness of the regression sub-program is that it makes no provision for the iteration and weighting required for fitting non-linear models. Consequently, contingency tables and percentage data cannot be properly analysed.   Apparently, also, the matrix inversion subroutine cannot cope with redundant parameters.

To supplement the regression sub-program a sub-program for the analysis of variance and covariance has been added.   This avoids the need to specify dummy variates when fitting parameters to tables.   It is modelled on the form

---

* A comment as early as 1925 by R. A. Fisher in *Statistical Methods for Research Workers* on one of these measures (η) reads: " As a descriptive statistic the utility of the correlation ratio is extremely limited."   Incidentally the values given by SPSS for η are in fact those for η^2.   One may surmise that this error would not have remained undetected if the coefficient were of practical use.

that has become standard for multifactorial experiments. The simplicity and informativeness of this form depend on the orthogonality of the experimental design. For such designs the mean squares for the various effects and interactions are unaltered by the inclusion or omission of other orthogonal effects ; when, as in survey data, orthogonality is lost, the analysis of variance becomes more complicated, as explained in Section 9.7.

To cope with this various alternative forms are provided. The most useful is probably the one that fits first the group of all the main effects, then all the two-factor interactions, and then if required the three-factor interactions, etc. The total sum of squares for each group is presented, and also the sum of squares and mean square for each member of the group after elimination of the other members of that group and all members of preceding groups. For factors at more than two levels this has the advantage of providing the relevant sums of squares for each component directly ; these can only be obtained by programs such as GLIM or the SPSS regression program by doing separate analyses with each component omitted in turn. There is, however, no mechanism for including only interactions between factors which show substantial main effects. As we have seen in the examples of Chapter 9, this is a very useful technique which avoids introducing interactions that are almost certainly of no consequence and merely distort the results.

An analysis of variance does not in itself provide any estimates of the magnitude of the effects. A supplementary so-called multiple classification analysis (MCA) provides estimates of the main effects (a) unadjusted, (b) adjusted for other factors, (c) adjusted also for the covariates. It does not, however, provide similar estimates for the interactions, and it is not stated whether the estimates provided for the main effects are those obtained before or after interactions have been fitted. Indeed there is the ominous warning : " If there is strong interaction between factors the MCA scores become meaningless." As was shown in Section 9.10, they should not be if properly adjusted.

There are no facilities in the analysis of variance and covariance sub-program for separating out linear contrasts for factors at more than two levels. Facilities for this are provided in a one-way analysis sub-program, but this only covers models containing a single factor, without covariates.

There is no need to discuss the sub-programs for correlation here. Suffice it to say that inspection of the correlation matrix (which is sensibly printed at the head of the regression analysis) is often a useful preliminary to regression analysis, as it directs attention to highly correlated variates ; such correlations may sometimes be profitably reduced by replacing the offending variates by linear functions of them, e.g. maximum and minimum daily temperature may be replaced by their mean and difference. Partial correlation is, however, rarely helpful.

Any program for regression or analysis of variance is necessarily based on the variance–covariance matrix. This plus the means of the variates and the number of observations gives all the necessary information for a least-squares fit. For total and partial correlation only the correlation matrix is required. To derive the variance–covariance matrix from the correlation matrix the latter must be supplemented by the values of the means. These common requirements are recognized by SPSS, which will output either a correlation matrix or a variance–covariance matrix so that it can be used in further analyses involving the same set of variates. There is, however, no proper filing system for such matrices ; they have to be output to cards in character form and incorporated as part of the data input of the next job. This contrasts with GENSTAT, which recognizes the variance–covariance matrix, together with the means and number of observations, as a single data structure which can be named and written to a file, and with the interactive version of SPSS (described below).

GLIM will not accept a variance–covariance matrix and requires that all the data must be present in core ; it is thus not suitable for regression analysis on large amounts of quantitative data. It can cope with similar analyses on qualitative data ; as we have seen, the necessary information for these can be completely summarized in a contingency table or a table of percentages and the numbers on which they are based.

A valuable feature of the SPSS regression program is the ability to create a subsidiary file of standardized residuals ; these can be merged with the original data file and form the basis of further analyses, or can be plotted against any other chosen variate by a plotting sub-program. There are also some facilities for plotting residuals included in the regression program itself.

(d)   The interactive version (SCSS)

The current demand for interactive packages, particularly for surveys with small amounts of data, and for student use, has led to the development of an interactive version of SPSS. The 1978 version offers similar facilities for data file management, tabulation, and correlation and regression analysis as SPSS, but is structurally very different.

SCSS is a conversational system with a well-designed system of prompts which avoids excessive verbosity, while providing guidance when this is required. There are also excellent facilities for correcting errors in the commands, whether perceived at the time they are made, or at a later stage.

SCSS analyses are based on a master file and a work file. A master file can be created directly from a raw data file by SCSS, or can be formed from an SPSS file by an SPSS command. In an SCSS master file the data are held in transposed form, i.e. variable by variable instead of case by case. The transposition takes time, but once done enables the data for the variables required for a particular tabulation or other analysis to be accessed much more quickly, thus permitting speedy step-by-step interactive work.

When there is more than one card record per case, sequence and reference checks can be made before the actual transposition : any case which fails these checks is discarded. " Wild codes " can be counted, or cases containing more than a defined number of wild codes can be reported and tagged, and if desired excluded in the subsequent transposition.

The cases included in the master file can be restricted to those for which defined variables have specific values, and a systematic or random sample of the complete or restricted data can be taken. Univariate frequency distributions, etc. can be examined before the actual transposition.

Once constructed, a master file is protected from further alteration and can be used by several investigators (or a set of students) simultaneously. Each investigator has his own work file, which may be retained for further sessions. Material assembled on a work file, e.g. new or revised variables, may be amalgamated with that on the master file by creating a new master file. The original master file is retained.

SCSS handles the construction of multifactorial tables much more sensibly than SPSS. If tables of counts are required, a master table classified by all the factors of a specified set is formed (but not printed). Tables classified by any selection of these factors in any order can then be printed by DISPLAY requests. This is similar to the provisions for printing margins from a larger table in RGSP. Additionally, sub-tables for single levels of one or more of the factors can be printed ; factor levels can also be combined (" collapsed ") or omitted.

For tables of means, totals, and standard deviations, the procedure is similar. A set of master tables is constructed for each of the quantitative variables included in the list.

As in SPSS, the basic unit of printing is the two factor sub-table, with margins. The printed layout is clear and, except for column level names, compact. For tables of counts the default printing is as percentages of the row totals, as in Table 5.20.c.

Master tables cannot be preserved. When a new table or set of tables is specified the previous set is deleted. As the whole of a table or set of tables is held in the working store there are obvious limitations on the number of factors that can be included, but the user is informed if this limitation is exceeded and can suitably revise his instructions.

Output can be sent to the terminal or to a line-printer (via a file) or to both simultaneously, but when both outputs are in operation the shorter of the two required line lengths will be adopted for both.

For correlation and regression a master correlation matrix of a defined set of variables is first formed. Any such matrix can be stored on the user's work file and the whole or parts of it recovered for further work. Partial correlations and regressions can be produced. The regression procedure provides for true stepwise regression, forward inclusion and backward inclusion, in addition to forced entry and removal. The residuals can be plotted in various ways, and the residuals or predicted values or both can be stored for further analysis.

Whether the provision of such elaborate stepwise procedures is really advisable is perhaps doubtful. They certainly tend to inhibit thought and are liable to induce an uncritical approach in investigative work by the inexperienced.

10.13 GENSTAT

GENSTAT is a general statistical program primarily designed for the analysis of experimental and observational data of the types encountered in scientific research. It contains very powerful facilities for the analysis of designed experiments, provisions for multiple regression analysis and the fitting of generalized linear models similar to those of GLIM, matrix operations, multivariate and cluster analysis, and graphical presentation.

The various sections form a single integrated program with its own problem-oriented control language (Type 2 of Section 10.2). This is basically similar to, but much more sophisticated and much wider in scope than that of Part 2 of RGSP. Whereas RGSP only operates on cross-tabulated tables, the GENSTAT language contains instructions for operating on a wide variety of structures of the types required in general statistical analysis. Specific directives are provided for commonly required standard procedures, such as the analysis of experiments and the fitting of generalized linear models, while the full power of the language is available to the expert for specifying novel types of analysis simply and concisely. Newly developed procedures which are likely to be of general interest can be incorporated in a macro library (essentially the equivalent of a subroutine library) held on a GENSTAT file, and can thus be made available to a wide circle of users.

GENSTAT is not primarily designed for tabulation of survey data, but can be used quite effectively for non-hierarchical surveys, subject to certain limitations on the form in which the data are presented, and lack of convenient facilities for validation and editing. For small surveys of an investigative nature for which both tabulation and more advanced statistical procedures are required, it offers the advantages of a single integrated package which provides good presentation of tabular matter and really sound and flexible procedures for regression, generalized linear models, etc.

The basic GENSTAT table structure is the cross-classified table, with or without margins. The facilities for operating on such tables are broadly similar to those provided by the RGSP processing instructions, but lack some of the conveniences which have been found particularly useful for survey work. The printing facilities are also similar, but with the limitation that variate and level names are restricted to eight characters, and that all the values in a composite table are printed to the same number of decimal places. Basic cross-tabulated tables of counts, totals, and associated counts can be formed directly from the data. These tables (and the corresponding tables of means) can be printed directly, or can be modified and combined before printing.

In common with other GENSTAT structures, tables can be filed temporarily in a backing store for later use in the same program, or permanently for recall in later GENSTAT programs. Thus the advantages resulting from table files in RGSP accrue also to GENSTAT. The GENSTAT filing system is more complex than that of RGSP. All the structures associated with the tables that are filed, including names and headings, are retained. Consequently there is no need for respecification when the tables are recalled.

The basic GENSTAT procedure for the formation of tables differs from that of most survey programs. Instead of reading the data for one unit at a time, calculating the derived variates and making entries in the tables for that unit before proceeding to the next, the whole of the required data is read into a " data matrix " of which the rows represent units and the columns variates. Derived variates are then formed by CALCULATE instructions, their values being stored in additional columns of the matrix, and the required tables are generated by a set of TABULATE instructions.

This is the obvious procedure for the analysis of experimental data, as the amount of data from a single experiment is usually fairly small. For a survey of any size, however, an inordinately large matrix would be required to accommodate the whole of the data for the basic and derived variates. To avoid this, a sequential form of READ, which reads a defined number of units at a time, is provided, with a corresponding sequential form of TABULATE. A loop, which includes the required CALCULATE and TABULATE instructions and a test on an indicator provided by READ for the end of the file, has then to be written by the user.

This procedure is not directly applicable to hierarchical surveys. Hierarchical surveys can be analysed, but, as the manual says, " the procedure is by no means straightforward " and is not described.

The GENSTAT language, although very powerful, will, I suspect, present difficulties to the inexperienced programmer, particularly if he only uses the program occasionally. It differs markedly from most general high-level languages, in particular Fortran, so that elementary knowledge of Fortran is no great help, and conversely. The preliminary declarations also require some care, and are not particularly convenient when many tables have to be formed. Even the lists of level names for factors must have separate names assigned to them, which then have to be repeated when the factors are declared.

10.14   Assessment of the relative merits of different program packages

The proliferation of survey program packages has led to various attempts to assess their relative merits. These range from catalogues of such packages with indications of the ground they attempt to cover, to actual trials on particular problems.

One such catalogue, summarizing responses to an elaborate questionnaire prepared by the International Association for Statistical Computing (Francis, 1979), covers 45 internationally available packages and reproduces exhibits displayed at the I.A.S.C. 1977 conference. It provides a useful summary of the facilities offered and the machines for which versions are available, but inevitably gives little information on technical quality or ease of use, and none on machine efficiency. These can only be assessed by detailed study of the program manuals, backed by experience of actual users, and by comparative tests on typical jobs.

Francis, Sherman and Heiberger (1976) give an account of a test to assess the ability of various packages to produce correct and well laid-out tables. This was conducted under the aegis of the Committee on the Evaluation of Statistical Program Packages of the American Statistical Association. The test was co-operative. The developers of thirteen packages, twelve of whom furnished results, were asked to prepare three " rather conventional, but not simple " tables, typical of tables being produced by the World Fertility Survey, from card data on the number of children and other particulars relating to 985 women. The tables were :

(1) A two-way table of counts (age of mother by number of children) with marginal values for each age-group of mean number of children, standard deviation and proportion male ;

(2) A four-factor table of counts, expressed as percentages of the row totals, together with the actual row totals ;

(3) A two-factor table of means and standard deviations interleaved by columns.

It was requested that the tables be produced as nearly as possible in the specified forms so as to be suitable for photo-copying " with a minimum of cutting and pasting ". Tables 1 and 3 are composite. The printing of the component parts in the required positions therefore presents difficulties for programs which are only designed to produce cross tabulations. The component parts of these tables are, however, simple, involving only one- and two-factor tables of counts, means and proportions, and standard deviations applicable to random samples. Table 2 is a straightforward table of percentage counts, which one might expect any survey program to be capable of producing in acceptable form.

Attributes contributing to the power of the packages to produce the required tables (4 items) and to the simplicity of the language (5 items) were assessed by the three authors on scales of 1 to 5. The itemized scores for power, and the aggregate scores for simplicity, are shown in Table 10.14. Those for the four programs described above and for two others, BMDP and COCENTS, are shown separately, with two sets of means for the remaining six. Apart from SPSS, BMDP is probably the statistical package most widely used in British universities. COCENTS differs from the other tabulation packages included in the test in that it is designed for and available on a wide variety of small machines. For this reason and because of its tabulating capabilities it was adopted by the World Fertility Survey for producing the primary tables incorporated in the local reports. To make it easier to use, a preprocessor was written which translates a straightforward specification of requirements into the COCENTS form.

GENSTAT, RGSP and TPL all obtained full marks for tabulating power, and COCENTS is also outstanding, whereas SPSS and BMDP were judged to be seriously deficient. The total scores (8, 11, 13, 14) of the other four S

TABLE 10.14—AN EVALUATION OF STATISTICAL PROGRAM PACKAGES

	RGSP	TPL	SPSS	GENSTAT	BMDP	COCENTS	Others	
Tabulating power	V	T	S-V	S	S	T	T(2)	S,S-V(4)
Arithmetic ability	5	5	4	5	4	4	4·5	4·2
Placement of of numbers on page	5	5	2	5	2	5	3·5	2·2
Labelling	5	5	3	5	3	5	3·5	2·5
Visual impact	5	5	2	5	2	5	3·5	2·5
Total (4 items)	20	20	11	20	11	19	15·0	11·5
Simplicity of language								
Total (5 items)	9	25	15	16	17	5	17·5	16·5

T = tabulation package   Scores per item
S = general statistical package   1 = poor
V = survey package   5 = excellent

and S-V packages are similar to those of SPSS and BMDP, and are all somewhat lower than those of the other two tabulation packages (15, 15).   Nor
was this mainly due to difficulties with tables 1 and 3.   Not one of the S and
S-V packages other than GENSTAT was able to produce table 2 in the specified form.   Having regard to the importance of good tabular presentation of
results, not only in surveys but also in much general statistical work, it is
surprising that the provisions made for it in many general statistical packages
are so poor.

These ratings for power give a fair assessment of the ability of the different
programs to produce in acceptable form the tables chosen for the test.   The
chosen examples do not, however, by any means cover all tabulating needs.
For other types of table the power ratings would be different ; TPL, in particular, has much more limited facilities for table manipulation than RGSP and
GENSTAT.   Nor are any elaborate derived-variate computations required
in this test.

The ratings for simplicity of language are necessarily more subjective.
Considerable weight was given to the intelligibility (" readability ") , to those
unacquainted with the packages, of the instructions for forming the test
tables.   On this criterion it is not surprising that RGSP obtained low marks :
no program that uses numbers instead of names for variates can be readable in
this sense ; nor does the relegation of the derived-variate instructions to
the Fortran section of Part 1 help matters.

The real objective that the authors had in mind in their assessment of
simplicity was, I think, ease of use.   Here it is important to distinguish between
ease of learning, and ease of use once familiar.   To the beginner and occasional

user, ease of learning, which is helped by verbal terminology, is the dominant factor. Readability also facilitates checking, but the regular user may well attach greater weight to the ability to express frequently required operations concisely. To take one example : TPL contains no repetitive features. Consequently if a set of similar tables each including standard deviations is required, the formulae of Section 10.11 have to be spelt out with different variate names for each table, which is both laborious and increases the risk of error : in the above test the divisor n instead of $n - 1$ was used for table 3.

In discussing the utility of the ratings of Table 10.14 for selecting the most suitable package for different purposes it is stated that " in a research or exploratory environment the emphasis would be on simplicity, but when publication-ready tables are needed the emphasis would be on power ". The types of power required in the two environments are somewhat different, and the need for simplicity is greater in a research environment because for many research workers survey analysis is only a small part of their activities, and because also most research is associated with teaching in one form or another. For beginners it is important to have a package that is easy to use for straightforward jobs, but power there must be ; the inability of SPSS and other similar programs to produce a compact version of table 2 above is a sad example, at an elementary level, of the poor computer facilities that are available at many computer centres for even a cursory examination of survey results.

Another method of comparison, which requires less organization and is in many ways more realistic, is to take a survey which has already been analysed on one package, and re-analyse it, or typical parts of it, on another package.

A comparison of this type between RGSP and SPSS on a questionnaire survey of the career expectations of students at British universities is reported in Yates (1975), using the results from one university for the test.

Some of the questions on this survey required answers to multiple non-exclusive alternatives of up to 20 items (Yes/No, or in some cases a graded score). The RGSP facilities enabled the answers to these questions to be assembled in single well-annotated tables, whereas in SPSS each alternative required a separate table, with the consequent need for hand copying and typing. In the SPSS analysis the respondents were grouped into four broad subject classes ; in RGSP an additional cross-classification by sex could be added without any additional consumption of paper.* As expected, many sex differences were revealed. The RGSP table filing system might also have been useful for comparisons between universities or groups of them : in the SPSS analysis only the pooled results for the whole survey were summarized.

10.15 Dangers of easy-to-use statistical packages

" It is easy to make mistakes with a package that is easy to use . . . . The mind

---

* With breakdown by sex the 1970 version of SPSS would have required 15 times as much paper as RGSP for these tables ; 30 times as much paper was required for the one-way frequency distributions.

boggles at the thought of the number of published reports and papers containing incorrect statistics because they were easy to obtain." These comments were included in the report on the tabulating capabilities of different packages described in Section 10.14. If this is true of the very simple computations required in that task by programmers who were presumably well versed in the use of their package and out to show what it could do, the mind indeed boggles at the thought of the misuse of the more advanced statistical procedures included in " easy-to-use " general statistical packages, when inferential deductions rather than factually correct tables are the objective.

There are two factors which contribute to the misuse. The programs themselves are often ill-adapted to cope with commonly occurring requirements ; indeed it is very apparent from the manuals that many of them were designed by programmers who were themselves ill-acquainted with these requirements and lacking the necessary statistical knowledge and experience. Secondly, users with even less statistical understanding apply them to their data in the hope that hidden truths will thereby be revealed.

An elementary example of such misuse, emanating from Duke University, North Carolina, is provided by a note on some results obtained from a survey of contraceptive use in rural Iran (Aghajanian and Mehryar, 1979). The first five columns of Table 10.15 give the mean numbers of live births (" cumulative fertility ") of some 980 women classified by age of woman and number of child deaths before age five. At first glance these results, which are those shown in the note, might be taken to indicate that the cumulative fertility was strongly influenced by the woman's experience of child losses. Indeed, quoted references to previous studies suggest that such inferences have been drawn from

TABLE 10.15—IRAN FERTILITY SURVEY : MEAN NUMBERS OF LIVE BIRTHS

Child deaths	Age of Mother					Hypothetical sub-population
	15–19	20–24	25–29	30–34	35+	
0	0·90	2·29	3·21	4·72	4·90	4·1
1	2·00	3·62	5·07	5·92	6·80	4·8
2	—	3·66	5·13	6·93	7·43	6·0
3	—	—	7·28	7·34	7·19	7·3
4+	—	—	—	8·35	8·45	8·3

tabulations of this type. A simple hypothetical example shows, however, that apparent effects of this magnitude can emerge if such influence is negligible. The observed mortality rate in this survey for children of mothers past reproductive age was 0·33. If a mortality rate of this magnitude is assumed irrespective of family size, the probabilities of 0, 1, 2, 3, 4+ deaths in families of three, six and nine children will be :

Deaths:	0	1	2	3	4+
3 children	0·30	0·44	0·22	0·04	—
6 ,,	0·09	0·26	0·33	0·22	0·10
9 ,,	0·03	0·12	0·23	0·27	0·35

In a hypothetical sub-population of completed families containing equal numbers of families of 3, 6 and 9 children the mean family size of families with no deaths will be

$$(3 \times 0\cdot30 + 6 \times 0\cdot09 + 9 \times 0\cdot03)/(0\cdot30 + 0\cdot09 + 0\cdot03) = 4\cdot1,$$

etc. The full figures are shown in the last column of Table 10.15. They have a similar range to the actual figures for 35+ families.

The authors recognized that the results of Table 10.15 " do not eliminate the influence of increased fertility on child mortality ; the more children born, the greater the possibility of child loss ". In an attempt to eliminate this they made a multiple classification analysis of the SPSS type on the percentages of families using contraceptives, taking as factors numbers of deaths, mother's age, family income, and father's education, and claiming that " in view of the close correlation in this population of fertility with maternal age, taking account of the effect of the latter on the other variables largely eliminates the effect of varying fertility".

This is an illusion. Leaving aside income and education, the procedure is equivalent to replacing numbers of children in Table 10.15 by percentage use of contraceptives, and fitting an additive linear model for row and column effects. From their reported results it emerges that the fitted values for " contraceptives ever used " for the 35+ column are 45, 39, 32, 31, 24 per cent. Similar figures might well be obtained if use of contraceptives were entirely uninfluenced by child deaths. Thus in our hypothetical sub-population, if contraceptives were ever used in one-half of the three-child, one-third of the six-child, and one-quarter of the nine-child families—a not extreme assumption if contraceptive practice is a factor in the control of numbers of live births—a similar calculation to that above gives percentage values of 45, 41, 36, 30, 27 per cent.

There is thus no justification for the conclusion in the summary that " both the cumulative fertility level and contraceptive use are strongly influenced by the individual woman's child loss experience ". The authors' earlier well-founded doubts on the interpretation of Table 10.15 were apparently dispelled by the results provided by a statistical procedure they did not fully understand. But its use will undoubtedly impress the unsophisticated reader and lend weight to their conclusions. This and the ready availability of the procedure via SPSS may well result in others with even less understanding following their example.

10.16 Contributory sources of confusion

Intelligent application of easy-to-use statistical packages is not made easier for the statistically unsophisticated by the assorted collection of " statistics ", many of them irrelevant or misleading, that are either optionally or always appended to the results ; nor by the ill-chosen and trivial examples and unsound statistical advice presented in the manuals ; nor again by the selection of programs that are included.

The numerous measures of association that can be appended to contingency tables by SPSS have already been mentioned. BMDP outdoes even SPSS in this respect : in an example in the manual 29 such measures relating to a 2×2 table, with their significance levels, which of course differ widely, are presented.

Again, in the example of Section 10.15 the authors blind their readers with science (and I suspect themselves) by appending a coefficient β to each set of parameter values. I wonder how many statisticians could say what these β represent. Actually β^2 is the ratio of the *sum* of squares accounted for by the parameter set, after fitting the other sets, to the *total* (not residual) sum of squares, i.e. a sort of bastard correlation ratio.* The authors of the SPSS manual, however, do not appear to be aware of this, and define it as the standardized partial regression coefficient on a pseudo-variable formed from the set of multiple classification scores, a device suggested in Section 9.12 for defining generalized linear $\times$ linear components of associated *interactions*.

In regression analysis we are treated to R, R^2, and "adjusted" R^2 ; † also, in addition to the regression coefficients and their standard errors, to so-called "standardized" regression coefficients, i.e. the coefficients that would be obtained if the variates were rescaled to unit variance. Advantages claimed in the manual for the use of the standardized coefficients are (a) that in a multiple regression they are more "comparable", (b) that the constant term in the regression equation is always zero, and can be omitted, (c) that in a simple regression of x on y the standardized coefficient is equal to the correlation coefficient. Unfortunately (b) is not true, and (c) does not generalize to partial correlations. This last is got round by renaming the standardized coefficients "part correlations" (with supporting equations). Truly excessive devotion to the correlation coefficient!

From what has been said in Section 10.12 it will be apparent that there are many features of the SPSS sub-programs that are likely to confuse the unsophisticated and exasperate the sophisticated. BMDP suffers from similar defects, which I need not dwell on here, but one gem may be mentioned. This is a program for recalculating χ^2 for a two-factor contingency table when one or more of the cell values is suspect. Moreover, not content with the rejection of specified cells, a step-by-step automatic iterative process is provided which replaces successive values until the residual χ^2 is not significant, thus obliterating any meaning the table originally had. A program for providing a table of deviations from expectation or of the individual contributions to χ^2 would be much more useful. Such a table can be generated by RGSP table operations, as is illustrated in the RGSP manual.

10.17 University computer centres

University computer centres are in a position to exercise considerable

* See footnote to Section 10.12(b) for comments on this statistic.

† Adjusted R^2 is described as a "more conservative estimate of the percent of variance explained", but is actually the proportion of the variance accounted for (the R'^2 of Section 9.7)—or would be except for an error of 1 in the degrees of freedom.

influence on the quality of survey analysis. Research workers at universities, and others with university connections, rely on their centre to provide suitable facilities for the analysis of survey and observational data obtained in the course of their research. Such research covers very varied fields, medicine, biology, psychology, economics, social science, etc. In addition, postgraduate students in statistics and other subjects are often involved in the collection and analysis of survey and observational data as part of their research projects.

Provision should be made, by the programs and packages that are made available, both for users who have some statistical skill and possibly also some acquaintance with computers, and for those who have neither ; many beginners will not only have to learn how to use computers, but may well, even if they have had some training in statistics, look for guidance from the programs on the types of analysis that are likely to prove profitable. The procedures provided must therefore not only be statistically sound but should also be selected so as to be generally useful, and should be so designed that the results are not likely to mislead those with little statistical knowledge.

A good set of statistical programs can do much to educate. Conversely a poor package, particularly if it is easy to use, will give rise to much bad statistics. A package which provides answers automatically, without requiring due consideration by the user, is particularly dangerous. Step-wise regression is a case in point.

A computer centre, pressed to provide facilities for statistical analysis, is obviously attracted by a widely used and apparently comprehensive package. If one is available with a version which will run on the local computer, it may well be acquired without any critical assessment of its statistical quality. Once installed, efforts are likely to be made to popularize it and to gloss over its defects.

This has indeed happened with SPSS in the United Kingdom. As I wrote some years ago, in the paper referred to in Section 10.14 (Yates, 1975) :

> The current acceptance of SPSS as an adequate program for survey analysis is yet another example of the poor quality of many widely used statistical programs, on which I and others have commented in the past. In the main I think it is the statisticians who must be held responsible for this sad state of affairs, but in the present instance university computer centres have also played a part. There has indeed been a considerable campaign by some centres to popularize SPSS. Criticism by statisticians and suggestions for improvement have fallen on deaf ears.

> Fortunately, disquiet is now spreading, and the need for alternative programs for survey analysis is beginning to be recognized. It is up to statisticians to supply them, and to write them in such a manner that they can be readily transferred to different systems, but computer centres must cooperate in making them generally available.

The climate is now changing. GLIM and GENSTAT are available and widely used at many British university computer centres. RGSP is also attracting attention because of its much better tabulation facilities, and its ability to handle multiple punching and hierarchical data. The RGSP interface with GLIM and GENSTAT provides associated facilities for more advanced statistical analysis of better quality than is possible with SPSS.

RGSP in its present form, however, cannot be regarded as ideal for the student and for the occasional user who is not computer-oriented. To encourage its use by beginners, provision should be made for running straightforward jobs without the need for Fortran. The first steps to this end have already been taken, and an addition is now available which permits specification of arithmetical derived-variate instructions and data formats to be added to Part 1 compiler specifications, and automatically generates and runs the corresponding execution program. Hierarchical data at two levels are covered by this addition.

A further improvement which would bring RGSP more into line with the current style of program packages is the optional use of short names instead of numbers for variates in the specification of requirements. Although in surveys involving many variates it is often simpler to work from a numbered list of full names, provided these are printed on all output, short names are undoubtedly attractive for small jobs, and enable specifications to be more readily checked. This and various other improvements are planned.

At present RGSP lacks any system of data-file management other than by do-it-yourself Fortran, which becomes more tiresome if variates are denoted by names. A temporary solution of this problem would be the provision of an interface which would permit RGSP to read data, including variate and level names, from SPSS files. This would have the additional advantage that those who have embarked on the analysis of a survey using SPSS and then find that they require some facilities only available in RGSP would be able to avail themselves of these without repeating all the basic work.

10.18   Separate programs or a complete package?

The defects of many widely-used statistical packages raise the question of whether the attempt to provide for the real needs of even a single group of workers such as social scientists in a single package is the right one. One alternative is to provide independent programs for different statistical operations, designed in such a manner that the results furnished by one program can be readily used as data for others.

This approach is in fact widely adopted for survey analysis. Of the programs we have considered in detail both RGSP and TPL rely on interfaces with independently written programs to provide facilities for advanced statistical analysis, and TPL expects a reasonably clean data file. Parts 1 and 2 of RGSP are also written as independent programs, with an interface via RGSP table files. Because numbers are used for variates, these files have a particularly simple structure, which is defined in the manual, and for good measure supplementary subroutines are provided for reading any required tables and their associated particulars from a table file, and for writing tables obtained from another source to a new table file.   The first of these and the GLIM system of sub-files enabled an interface for transmitting numerical values contained in tables to GLIM and GENSTAT to be written very speedily ; a Part 1 standard subroutine was also written for transmitting tabular matter direct from Part 1.   Note also that the provision in Part 2 of RGSP for writing new table files enables Part 2 to interface with itself ; in other words, new tables constructed by Part 2 can be further manipulated without repeating all the operations required to form them.

Separate programs, by their very nature, can be relatively simply used in conjunction with other existing programs, provided that results and related particulars provided by one program are available in some suitable file form for use in other programs.   Indeed I suggested, in the paper referred to above (Yates, 1975), that Part 2 of RGSP might usefully be interfaced with SPSS if it were possible to create files of SPSS tables.

To be attractive for small jobs, and particularly to occasional users, separate programs or program " modules " must be arranged so that they can be run in sequence with minimal specification.   There are some technical problems here, but they should be resolvable, and will be made easier if consistent standards for interfacing are adopted.

Some attempt at standardization of the conventions adopted for user specifications is also required.   Thus, for example, if separate programs are used for editing data files and for tabulation, different sets of conventions for writing derived variate instructions have to be learnt ; those for grouping and re-coding in particular may differ radically.   There are not at present any agreed conventions, even on the use of symbols such as : ; $ etc., or for specifying data formats.

10.19   Programs for validation, imputation, and standard errors

For large surveys of the census type, separate programs are commonly used for validation and editing.   The CONCOR editing system, for example, used by the World Fertility Survey, provides much the same facilities for validation as those in RGSP, and in addition some limited facilities for " hot-deck " imputation.

Hot-deck imputation essentially consists of imputing for a field of the current record the value recorded in the same field of some other record (based on a match of certain specified common characteristics) which has passed all the relevant tests.   The method is therefore broadly equivalent to replacing a rejected value of a variate by the value that would be obtained by regressing the variate on the specified characteristics, with the addition of a random component dependent on the distribution of the residuals at the appropriate point on the regression surface.   This preserves the distribution of the data as represented by the records which passed the edit tests.

If the main cause of rejection is due to random errors of measurement or random lack of response and the matching variables are well chosen, hot-deck imputation will not distort the data ; thus in anthropometric surveys, for example, in which the main sources of error are usually attributable to errors of measurement and recording, imputation may well provide a satisfactory method of cleaning the data without complete rejection of units which contain one or more obvious errors.

If, however, non-response to a question is associated with some important attribute of the individual concerned, imputation may do little to improve matters and may give a false impression of the reliability of the results.   Thus in a survey of educational attainments and subsequent careers, information on the latter is particularly likely to be lacking on those who take posts abroad. If, for example, career information is available on 90 per cent. of the individuals in the sample, and 12 per cent. of these individuals are found to be employed abroad, the apparent " brain drain " will be 12 per cent., but if three-quarters of the remaining 10 per cent. are in fact abroad the true brain drain will be 18·3 per cent.   If there is no association between the recorded educational attainments and absence abroad, an estimate of about 12 per cent. will still be obtained after imputation ; if there is some association, say with a Ph.D degree, and this is specified as a matching variable in the imputation, the estimate will be somewhat increased but may still be well below the true value.

For large-scale censuses full automation of editing and imputation is clearly advantageous.   Fellegi and Holt (1976) have developed procedures for use in Canadian censuses based on the following principles:

(1) The data in each record should be made to satisfy all tests by changing the minimum number of items.

(2) As far as possible the frequency structure of the data file should be maintained.

(3) Imputation rules should be derivable from the corresponding edit rules without explicit specification, thus facilitating trials of different sets of edit rules.

The first function is particularly valuable for the identification of occasional gross errors of measurement when tests are made on a set of closely correlated quantitative variates.   An example where this would have been of great use is provided by an anthropometric survey involving some 30 measurements on some 4000 children (Healy, D′, 1952).   Errors which were discovered when the computations were well advanced led to a thorough and very laborious check based on associated measurements, using punched card machines. " A far from negligible " proportion of gross errors (0·5 per cent. of all measurements, 1 per cent. of the heights), mostly of 5 or 10 cm, were detected.   (In this survey verification of queried values was possible from photographs ; practically every queried value was found to be in error.)   Correction of these errors gave graphs of height against age, for example, which were markedly more regular ; in fact the disturbances in the original results due to gross errors were equivalent to the loss of about one-third of the observations.

. Note that the important step here was the detection of the gross errors ; imputation would have served almost equally as well as did the substitution of the correct values from the photographs. Rejection of all records containing one or more gross errors, on the other hand, would have resulted in the loss of about 15 per cent. of the observations.

Separate programs are also often used for computation of the sampling errors of means and other estimates derived from multi-stage and cluster samples. A good example is provided by CLUSTERS, developed by the World Fertility Survey. This provides estimates of the sampling errors of means and other ratio estimates applicable to sub-classes (domains of study) of hierarchically structured samples, and to contrasts between pairs of such domains ; this automates and generalizes the RGSP procedure for means outlined in Section 10.10(e). Stratification of primary (top level) sampling units can be included in the specification ; if this is done, differences between strata will be excluded from the estimates of variability. Such exclusion is correct for population estimates, and should provide reasonable approximations for domains that contain data in about the same proportion from all primary units in a stratum, as such domains will not cut across strata in the sense described in Section 7.6.1.

10.20   Where do we go from here ?

That the present facilities for survey analysis at many British universities are inadequate will not, I think, be disputed. The question that has to be decided is what should be our policy for the future.

When considering this we must bear in mind that the computer environment is changing very rapidly, and will continue to do so. Thanks to the silicon chip and large-scale integrated circuits, low-cost but very powerful mini- and micro-computers are now available, which can perform many of the functions previously relegated to a large central installation. Programmable desk calculators are steadily increasing in power and versatility. Terminals are becoming more " intelligent ". Facilities for graphical display on video terminals are well developed. At the other end of the scale, large computer installations are being linked together in networks, using high-speed data links, so that jobs submitted to one installation can be run on another, and data banks held at one location can be accessed from anywhere in the network.

A further point to bear in mind is that many under-developed countries look to Britain for suitable software to meet their needs. Easy-to-use programs developed for unsophisticated users in this country will be equally valuable in under-developed countries, provided versions can be supplied which will run on the equipment available there.

The first requirement, therefore, is that all programs intended for wide circulation must be so organized that versions for different types of machine can readily be generated, i.e. in the current jargon, they must be " portable ". This is not as simple as it sounds, even for large machines, as anyone who has faced the task will well know. Mini-computers and other small machines raise additional problems.

If any co-ordinated action is to be taken, the first question to be decided is whether to aim for a " complete " package or to adopt the alternative of assembling a library of independent but compatible component programs, making use of already existing programs (possibly with some modifications) where these are judged suitable.

The latter course is likely to provide a much more flexible approach and will give scope for much more individual initiative. Component programs can be modified or replaced to take advantage of improved techniques, and further programs can be added to cater for special requirements, such as the analysis of longitudinal surveys. New techniques can be incorporated on a trial basis at particular centres, and if proved to be of value can then be more widely circulated. Groups of individuals who are really expert in particular aspects can then work largely independently, restrained only by suitable provisions for interfacing and agreed standards for user specifications.

A suitable subdivision for survey analysis might be :

(1) Validation, editing and file management (including hierarchical data).
(2) Formation, manipulation, printing and filing tables.
(3) Regression and linear and generalized linear models.
(4) Graphical representation of tabular matter.
(5) Sampling errors applicable to hierarchical data and to ratio and regression estimates, etc.

The content of (2) could be much as that contained in the present RGSP, and that of (3) in GLIM and GENSTAT, possibly with some additional facilities particularly needed in the survey context. So far as I know, (4) has been as yet little explored.

Other multivariate techniques are, I think, rarely required in survey analysis, though they are of importance for other types of observational data, and programs for them should be able to use data files created by (1). (3) should be capable of operating on variance–covariance matrices, and of filing such matrices. Names should be transmissible between programs. Provision should be made for sequential running where this is likely to be required.

The main requirement in universities and technical colleges is for programs that can be effectively and easily used by beginners and the statistically unsophisticated, but which are sufficiently powerful and versatile to provide for more advanced needs as experience is gained. The required versatility can only be attained if needs are properly identified. To do this we must draw on the experience of research workers actively engaged in the analysis of survey material of varied types, and of teachers who are concerned with the quality of the analytical tools that are placed in the hands of their students.

This is not to say that we should not learn what is to be learnt from developments in other countries, but such developments should be assessed critically. We should be prepared to think independently, and not follow the whim of every latest fashion. Procedures such as step-wise regression, which involve automatically applied tests of significance, should be avoided ; the examples given in Chapter 9 illustrate that it is not always desirable to exclude a parameter which fails to attain formal significance.   Path analysis, a simple example of which was given in Section 9.17, is another procedure which requires critical judgment and is particularly apt to be misleading if automatically applied.

What provisions should be made for interactive working?   Such working becomes more attractive as terminals become more intelligent and improve in other ways.   It will undoubtedly be useful in selecting good graphical presentation, and indeed in examining tabular matter by graphical methods ; probably also on occasion when fitting models.   I doubt if it is of much value when studying tabular matter, which usually requires quiet contemplation rather than a hasty glance.   For correcting errors in data files, the standard text-editing facilities provided on most machines are likely to do all that is required.

The degree of skill and expertise required for an enterprise of this nature should not be underestimated.   Unless we can elicit the active collaboration of experienced statisticians, nothing worth while is likely to emerge.   That such collaboration is possible, given a few enthusiastic leaders, has been demonstrated in other fields, and indeed in statistics : GLIM was sponsored by the Royal Statistical Society, and both GENSTAT and GLIM have greatly benefited from the interchange of ideas between different groups of workers.

It should also be recognized that a great deal of routine work at various levels is involved if a software package is to be given wide circulation on many different types of computer.   I need not elaborate here.   Suffice it to say that if a high standard of excellence is to be attained and speedy implementation is regarded as of major importance—which I think it is—financial support in the initial stages is required.

One further consequence of increasing automation which affects all those concerned with data analysis is the resultant information " explosion ", or as it might be better termed the information flood, by which we are all in danger of being overwhelmed.   Much information that was previously safely tucked away in card-index files and the like is now available on computer files, and as such invites analysis.   Automatic recording instruments are another prolific source of accessible numerical material.   And by Parkinson's law, or a variant of it, that which invites analysis will get analysed.   I will not comment here beyond quoting a passage from a discussion paper on statistical teaching (Yates and Healy, 1964) :

Computers are good servants but bad masters.   There has been plenty of statistical nonsense produced on desk calculators, but this will be nothing compared with the flood that will emerge from computers if they are not wisely used and firmly controlled.

Effective programs, especially those for the analysis of survey data, will do something to mitigate such a flood.

03 47 43 73 86	36 96 47 36 61	46 98 63 71 62	33 26 16 80 45
97 74 24 67 62	42 81 14 57 20	42 53 32 37 32	27 07 36 07 51
16 76 62 27 66	56 50 26 71 07	32 90 79 78 53	13 55 38 58 59
12 56 85 99 26	96 96 68 27 31	05 03 72 93 15	57 12 10 14 21
55 59 56 35 64	38 54 82 46 22	31 62 43 09 90	06 18 44 32 53
16 22 77 94 39	49 54 43 54 82	17 37 93 23 78	87 35 20 96 43
84 42 17 53 31	57 24 55 06 88	77 04 74 47 67	21 76 33 50 25
63 01 63 78 59	16 95 55 67 19	98 10 50 71 75	12 86 73 58 07
33 21 12 34 29	78 64 56 07 82	52 42 07 44 38	15 51 00 13 42
57 60 86 32 44	09 47 27 96 54	49 17 46 09 62	90 52 84 77 27
18 18 07 92 46	44 17 16 58 09	79 83 86 19 62	06 76 50 03 10
26 62 38 97 75	84 16 07 44 99	83 11 46 32 24	20 14 85 88 45
23 42 40 64 74	82 97 77 77 81	07 45 32 14 08	32 98 94 07 72
52 36 28 19 95	50 92 26 11 97	00 56 76 31 38	80 22 02 53 53
37 85 94 35 12	83 39 50 08 30	42 34 07 96 88	54 42 06 87 98
70 29 17 12 13	40 33 20 38 26	13 89 51 03 74	17 76 37 13 04
56 62 18 37 35	96 83 50 87 75	97 12 25 93 47	70 33 24 03 54
99 49 57 22 77	88 42 95 45 72	16 64 36 16 00	04 43 18 66 79
16 08 15 04 72	33 27 14 34 09	45 59 34 68 49	12 72 07 34 45
31 16 93 32 43	50 27 89 87 19	20 15 37 00 49	52 85 66 60 44
68 34 30 13 70	55 74 30 77 40	44 22 78 84 26	04 33 46 09 52
74 57 25 65 76	59 29 97 68 60	71 91 38 67 54	13 58 18 24 76
27 42 37 86 53	48 55 90 65 72	96 57 69 36 10	96 46 92 42 45
00 39 68 29 61	66 37 32 20 30	77 84 57 03 29	10 45 65 04 26
29 94 98 94 24	68 49 69 10 82	53 75 91 93 30	34 25 20 57 27
16 90 82 66 59	83 62 64 11 12	67 19 00 71 74	60 47 21 29 68
11 27 94 75 06	06 09 19 74 66	02 94 37 34 02	76 70 90 30 86
35 24 10 16 20	33 32 51 26 38	79 78 45 04 91	16 92 53 56 16
38 23 16 86 38	42 38 97 01 50	87 75 66 81 41	40 01 74 91 62
31 96 25 91 47	96 44 33 49 13	34 86 82 53 91	00 52 43 48 85
66 67 40 67 14	64 05 71 95 86	11 05 65 09 68	76 83 20 37 90
14 90 84 45 11	75 73 88 05 90	52 27 41 14 86	22 98 12 22 08
68 05 51 18 00	33 96 02 75 19	07 60 62 93 55	59 33 82 43 90
20 46 78 73 90	97 51 40 14 02	04 02 33 31 08	39 54 16 49 36
64 19 58 97 79	15 06 15 93 20	01 90 10 75 06	40 78 78 89 62
05 26 93 70 60	22 35 85 15 13	92 03 51 59 77	59 56 78 06 83
07 97 10 88 23	09 98 42 99 64	61 71 62 99 15	06 51 29 16 93
68 71 86 85 85	54 87 66 47 54	73 32 08 11 12	44 95 92 63 16
26 99 61 65 53	58 37 78 80 70	42 10 50 67 42	32 17 55 85 74
14 65 52 68 75	87 59 36 22 41	26 78 63 06 55	13 08 27 01 50

This table forms part of a larger table of random numbers given in *Statistical Tables for Biological, Agricultural and Medical Research* by R. A. Fisher and F. Yates, Oliver & Boyd, Edinburgh (6th edition, 1963), and is reproduced by kind permission of the authors and the publishers.

414

TABLE A2—THE NORMAL DISTRIBUTION

Probability of obtaining deviations (positive or negative) greater than given multiples of the standard deviation

Deviation z/σ	Probability P	Deviation z/σ	Probability P	Deviation $z\sigma$	Probability P
0·0	1·0000	1·0	·3173	2·0	·0455
0·1	·9203	1·1	·2713	2·1	·0357
0·2	·8415	1·2	·2301	2·2	·0278
0·3	·7642	1·3	·1936	2·3	·0214
0·4	·6892	1·4	·1615	2·4	·0164
0·5	·6171	1·5	·1336	2·5	·0124
0·6	·5485	1·6	·1096	2·6	·0093
0·7	·4839	1·7	·0891	2·7	·0069
0·8	·4237	1·8	·0719	2·8	·0051
0·9	·3681	1·9	·0574	2·9	·0037
1·0	·3173	2·0	·0455	3·0	·0027

BIBLIOGRAPHY ON SAMPLING

The bibliography to the First Edition was drawn up by Mr. D. R. Read, and is reprinted here without change. It was based on a bibliography prepared by the Food and Agricultural Organisation of the United Nations. Additional references arranged as in the original bibliography will be found on pages 429 ff. The sections of the supplementary bibliographies are distinguished by dashes.

The papers have been classified under the following heads:—

(A) Theory and methods.
(B) Machine methods.
(C) Population censuses.
(D) Sociology, nutrition, health, etc.
(E) Opinion surveys and market research.
(F) Economics: surveys of industry, censuses of production, labour force, etc.
(G) Agricultural economics and farm practice.
(H) Crop estimation and forecasting, etc.
(I) Forestry and land utilization surveys.
(J) Estimation of wild populations.

Since a single paper does not necessarily deal with only one subject, the subject classification must be taken as approximate only. A certain amount of general theory, for example, will be found in papers primarily dealing with special applications. In some instances where the original paper could not be consulted the classification has been made from the title and journal. Papers by the same author may be found under more than one heading.

BOOKS

General

BAEHNE, G. W. (1935). "Practical applications of the punched card method in colleges and universities." New York: Columbia University Press.

BLANKENSHIP, A. (1943). "How to conduct consumer and opinion research" (2nd. edn., 1945). New York: Harpers.

CANTRIL, H. (1944). "Gauging public opinion." Princeton University Press.

CHURCHMAN, C. W., ACKOFF, R. L., and WAX, M. (1947). "Measurement of consumer interest." Philadelphia: University of Philadelphia Press.

FISHER, R. A. (1925). "Statistical methods for research workers" (10th edn., 1946). Edinburgh: Oliver & Boyd.

—— (1935). "The design of experiments" (4th edn., 1947). Edinburgh: Oliver & Boyd.

HARTKEMEIER, H. P. (1942). "Principles of punch-card machine operation.' New York: Thomas Y. Crowell.

KENDALL, M. G. (1943), and STUART, A. "The advanced theory of statistics," Vol. I, 1958. Vol. II, 1961. London: Griffin.

PEATMAN, J. G. (1947). "Descriptive and sampling statistics." New York: Harpers.

RHODES, E. C. (1933). "Elementary statistical methods" (8th edn., 1948). London: Routledge.

SCHUMACHER, F. X., and CHAPMAN, R. A. (1942). " Sampling methods in forestry and range management." Duke University, Durham, North Carolina.
SNEDECOR, G. (1937). " Statistical methods " (4th edn., 1946). The Collegiate Press, Ames, Iowa.
THIONET, P. (1946). " Méthodes statistiques modernes des Administrations Fédérales aux Etats-Unis." Paris : Hermann et Cie.
YULE, G. U. (1911), and KENDALL, M. G. " An introduction to the theory of statistics " (14th edn., 1950, 1958). London : Griffin.

Reports on Surveys, etc.

BOWLEY, A. L. (1930-1935). " New survey of London life and labour." Vols. I–IX. London : P. S. King.
JONES, D. CARADOG (1934). " The social survey of Merseyside." London : Hodder & Stoughton.
ROWNTREE, B. SEEBOHM (1901). " Poverty : a study of town life " (4th edn., 1908). London : Macmillan.
TOUT, H. (1939). " The standard of living in Bristol : A preliminary report of the work of the University of Bristol Social Survey." Bristol : Arrowsmith.
U.S. BUREAU OF CENSUS (1947). " A chapter in population sampling." U.S. Govt. Printing Office, Washington, D.C.

Tables

FISHER, R. A., and YATES, F. (1938). " Statistical tables for biological, agricultural and medical research " (3rd edn., 1948). Edinburgh : Oliver & Boyd.

PAPERS

A. THEORY AND METHODS

ALTSCHUL, H. (1913). " Studie über die Methode der Stichprobenerhebung." *Arch. Rass.- u. Ges. Biol.*, **10**, 110–152.
ANDERSON, P. H. (1942). " Distributions in stratified sampling." *Ann. Math. Statist.*, **13**, 42–52.
ARMITAGE, P. (1947). " A comparison of stratified with unrestricted random sampling from a finite population." *Biometrika*, **34**, 273–280.
BERNERT, E. H. (1945). *See* HAGOOD, M. J.
BOSE, C. (1943). " Note on the sampling error in the method of double sampling." *Sankhya*, **6**, 329–330.
BOWLEY, A. L. (1926). " Measurement of the precision attained in sampling." *Bull. Inst. Int. Statist.*, **22**, (1), 1–62.
COCHRAN, W. G. (1939). " The use of analysis of variance in enumeration by sampling." *J. Amer. Statist. Ass.*, **34**, 492–510.
—— —— (1942). " Sampling theory when the sampling units are of unequal sizes." *J. Amer. Statist. Ass.*, **37**, 199–212.
—— —— (1946). " Relative accuracy of systematic and stratified random samples for a certain class of populations." *Ann. Math. Statist.*, **17**, 164–177.
—— —— (1948). " Recent developments in sampling theory in the United States " (Abstract). *Econometrica*, **16**, 71–72.
CORNELL, F. G. (1947). " A stratified random sample of a small finite population." *J. Amer. Statist. Ass.*, **42**, 523–532.
CORNFIELD, J. (1942). " On certain biases in samples of human population." *J. Amer. Statist. Ass.*, **37**, 63–68.
—— (1944). " On samples from finite populations." *J. Amer. Statist. Ass.*, **39**, 236–239.
COX, G. M. (1935). *See* SNEDECOR, G. W.
CRAIG, A. T. (1939). " On the mathematics of the representative method of sampling." *Ann. Math. Statist.*, **10**, 26–34.

417

DEMING, W. E. and STEPHAN, F. F. (1940). "On a least squares adjustment of a sampled frequency table when the expected marginal totals are known." *Ann. Math. Statist.*, **11**, 427–444.

—— —— (1945). "On training in sampling." *J. Amer. Statist. Ass.*, **40**, 307–316.

FRANKEL, L. R. (1939). *See* STOCK, J.S.

GALVANI, L. (1929). *See* GINI, C.

GANGULI, M. (1941). "A note on nested sampling." *Sankhya*, **5**, 449–452.

GHOSH, B. (1947). "Bias introduced by changing the system of stratification." *Calcutta Statist. Ass. Bull.*, No. 1, 43–45.

—— (1947). "Double sampling with many auxiliary variates." *Calcutta Statist. Ass. Bull.*, No. 2, 91–93.

GINI, C. and GALVANI, L. (1929). "Di una applicazione del metodo rappresentativo all' ultimo censimento italiano della popolazione" (1° dicembre 1921). *Annali di Statistica*, VI, **4**, 1–107.

GRAVELL, W. (1923). "Die Not der Statistik und die representative Methode." *Allg. Statist. Archiv*, **13**, 345–346.

—— —— (?). "Die representative Methode." *Deutsch. Statist. Zent.*, **15**, 5.

HAGOOD, M. J., and BERNERT, E. H. (1945). "Component indexes as a basis for stratification." *J. Amer. Statist. Ass.*, **40**, 330–341.

HANSEN, M. H. and HURWITZ, W. N. (1943). "On the theory of sampling from finite populations." *Ann. Math. Statist.*, **14**, 333–362.

HENDRICKS, W. A. (1948). "Mathematics of sampling." Virginia Agric. Exp. Sta., Special Bulletin.

HURWITZ, W. N. (1943). *See* HANSEN, M. H.

JENSEN, A. (1923). "Arbejdsbesparende Metoder i Statistikken." *Nordisk Statist. Tidskrift*, Stockholm, **2**, 409–434.

—— (1926). "The application of the representative method." *Bull. Inst. Int. Statist.*, **22**, 58–60.

—— —— (1926). "Report on the representative method in statistics." *Bull. Inst. Int. Statist.*, **22**, 359–380.

—— —— (1926). "The representative method in practice." *Bull. Inst. Int. Statist.*, **22**, 381–439.

—— —— (1928). "Purposive selection." *J. R. Statist. Soc.*, **91**, 541–547.

KIAER, A. N. (1915). "Sur les méthodes représentatives ou typologiques appliquées à la Statistique." *Bull. Inst. Int. Statist.*, **11**, 180–214.

—— —— (1917). "Sur les méthodes représentatives ou typologiques appliquées à la Statistique." *Bull. Inst. Int. Statist.*, **13**, 66–101.

LUCHT, J. (1922). "Die representative Methode in der Statistik." *Zeits. preuss. statist. Landesamts*, **62**, 122–141.

MADOW, W. & L. (1944). "On the theory of systematic sampling." *Ann. Math. Statist.*, **15**, 1–24.

—— (1946). "Systematic sampling and its relation to other sampling designs." *J. Amer. Statist. Ass.*, **41**, 204–217.

MAHALANOBIS, P. C. (1944). "On large scale sample surveys." *Roy. Soc. Phil. Trans.*, B, **231**, 329–451.

—— (1946). "Recent experiments in statistical sampling in the Indian Statistical Institute." *J. R. Statist. Soc.*, **109**, 325–378.

MARCH, L. (1926). "Observations sur la méthode représentative et sur le projet de rapport relatif à cette méthode." *Bull. Inst. Int. Statist.*, **22**, 444–451.

MAYET, P. (1918). "Stichprobenerhebung in der Zwischenzeit zwischen grossen Vollzählungen längerer Periodizität." *Bull. Inst. Int. Statist.*, **14**, 258–276.

NEYMAN, J. (1934). "On the two different aspects of the representative method : the method of stratified sampling and the method of purposive selection." *J. R. Statist. Soc.*, **97**, 558–606.

—— (1938). "Contributions to the theory of sampling human populations." *J. Amer. Statist. Ass.*, **33**, 101–116.

NYBOLLE, H. C. (1923). "On Middelfejlen ved partielle Undersogelser." *Nordisk Statistik Tidskrift*, **2**, 435–437.

PIETRA, G. (1944). "La statistique méthodologique italienne de 1939 à 1942." *Rev. Inst. Int. Statist.*, 36–45.

SETTY, S. P. W. (1941). "An experimental study of some of the methods of representative sampling " (Abstract). *Sankhya*, **5**, 237.

SMITH, H. F. (1947). " Standard errors of means in sampling surveys with two-stage sampling." *J. R. Statist. Soc.*, **110**, 257-259.

SNEDECOR, G. W. (1934). " The method of expected numbers for tables of multiple classification with disproportionate sub-class numbers." *J. Amer. Statist Ass.*, **29**, 389-393.

—— and COX, G. M. (1935). " Disproportionate sub-class numbers in tables of multiple classification." Iowa Agric. Exp. Sta. Res. Bull. 180.

STEPHAN, F. F. (1940). *See* DEMING, W. E.

—— (1948). " History of the uses of modern sampling procedures." *J. Amer. Statist. Ass.*, **43**, 12-39.

STEVENS, W. L. (1948). " Statistical analysis of a non-orthogonal tri-factorial experiment." *Biometrika*, **35**, 346-367.

STOCK, J. S. and FRANKEL, L. R. (1939). " The allocation of sampling among several strata." *Ann. Math. Statist.*, **10**, 288-293.

STUART, C. A. V. (1926). " Note sur l'application de la méthode représentative." *Bull. Inst. Int. Statist.*, **22**, 440-443.

SUKHATME, P. V. (1935). " Contribution to the theory of the representative method." *J. R. Statist. Soc., Suppl.*, **2**, 253-268.

TORNQVIST, L. (1948). " An attempt to analyse the problem of an economical production of statistical data." *Nordisk Tidsskrift for Teknisk Okonomi*, **37**, 265-274.

YATES, F. (1934). " The analysis of multiple classifications with unequal numbers in the different classes." *J. Amer. Statist. Ass.*, **29**, 51-66.

—— —— (1935). " Some examples of biased sampling." *Ann. Eugen.*, **6**, 202-213.

—— —— (1946). " A review of recent statistical developments in sampling and sampling surveys." *J. R. Statist. Soc.*, **109**, 12-43.

—— —— (1948). " The influence of agricultural research statistics on the development of sampling theory " (Abstract). *Econometrica*, **16**, 69-71.

—— —— (1948). " Systematic sampling." *Roy. Soc. Phil. Trans.*, A, **241**, 345-377.

B. MACHINE METHODS

BENJAMIN, K. (1945). " An I.B.M. technique for the computation of ΣX^2 and ΣXY." *Psychometrika*, **10**, 61-67.

—— —— (1947). " Problems of multiple-punching with Hollerith machines." *J. Amer. Statist. Ass.*, **42**, 46-71.

BERKSON, J. (1941). " A punch card designed to contain written data and coding." *J. Amer. Statist. Ass.*, **36**, 535-538.

BLACK, B. J. and OLDS, E. B. (1946). " A punched card method for presenting, analysing and comparing many series of statistics for areas." *J. Amer. Statist. Ass.*, **41**, 347-355.

CARVER, H. C. (1934). " Punched card systems and statistics." *Ann. Math. Statist.*, **5**, 153-160.

COMRIE, L. J. (1933). " The Hollerith and Powers tabulating machines." (Printed for private circulation).

—— ——, HEY, G. B. and HUDSON, H. G. (1937). " The application of Hollerith equipment to an agricultural investigation." *J. R. Statist. Soc., Suppl.*, **4**, 210-224.

—— —— (1946). " Application of commercial calculating machines to scientific computing." *Math. Tables and other Aids to Computation*, **2**, 149-159.

DEMING, W. E., TEPPING, B. J. and GEOFFREY, L. (1942). " Errors in card punching." *J. Amer. Statist. Ass.*, **37**, 525-536.

DUNN, H. L. and TOWNSEND, L. (1935). " Application of punched card methods to hospital statistics." *J. Amer. Statist. Ass., Suppl.*, **30**, 244-248.

GEOFFREY, L. (1942). *See* DEMING, W. E.

HARTLEY, H. O. (1946). " The application of some commercial calculating machines to certain statistical calculations." *J. R. Statist. Soc., Suppl.*, **8**, 154-183.

HEY, G. B. (1937). *See* COMRIE, L. J.

HUDSON, H. G. (1937). *See* COMRIE, L. J.

JOLLIFFE, E. T. (1941). " Fundamental principles in tabulating machine methods of statistical analysis." *J. Exp. Educ.*, **9**, 254-274.

KEMPTHORNE, O. (1946). " The use of punched card systems for the analysis of survey data, with special reference to the analysis of the National Farm Survey." *J. R. Statist. Soc.*, **109**, 284–295.

KIMBALL, E. (?). " A fundamental punched card method for technical computations." Bureau of the Census, Washington, D.C. (Undated).

—— —— (?). " A method of technical computations by punched card equipment." Bureau of the Census, Washington, D.C. (Undated).

MANDEVILLE, J. P. (1946). " Improvements in methods of census and survey analysis." *J. R. Statist. Soc.*, **109**, 111–129.

McKAY, A. T. (1934). " A new method of handling statistical data." *J. R. Statist. Soc., Suppl.*, **1**, 62–75.

OLDS, E. B. (1946). *See* BLACK, B. J.

TEPPING, B. J. (1942). *See* DEMING, W. E.

TODD, H. A. C. (1941). " A note on systematic coding for card sorting systems." *J. R. Statist. Soc., Suppl.*, **7**, 151–154.

TOOPS, H. A. (1938). " Hollerith Coding." Ohio College Ass. Bull. No. 110, 2269–2274.

TOWNSEND, L. (1935). *See* DUNN, H. L.

VICKERY, C. W. (1938). " Punched card technique for the correction of bias in sampling." *J. Amer. Statist. Ass.*, **33**, 552–556.

—— (1939). " On drawing a random sample from a set of punched cards.' *J. R. Statist. Soc., Suppl.*, **6**, 62–66.

WISHART, J. (1947). " Statistical aspects of demobilisation in the Royal Navy." *J. R. Statist. Soc.*, **110**, 27–50.

C. POPULATION CENSUSES

BLYTHE, R. H. (1947). *See* JESSEN, R. J.

COATS, R. H. (1931). " Enumeration and sampling in the field of the census." *J. Amer. Statist. Ass.*, **26**, 270–284.

DEMING, W. E. (1940). " Sampling problems of the 1940 Census." Cowles Commission for research : Economic Report of 6th Ann. Res. Conf. on Economics and Statistics, Colorado Springs.

—— (1940). *See* STEPHAN, F. F.

—— —— and GEOFFREY, L. (1941). " On sample inspection in the processing of census returns." *J. Amer. Statist. Ass.*, **36**, 351–360.

—— —— and STEPHAN, F. F. (1941). " On the interpretation of censuses as samples." *J. Amer. Statist. Ass.*, **36**, 45–49.

—— —— (1941). *See* STEPHAN, F. F.

—— —— (1947). *See* JESSEN, R. J.

GEOFFREY, L. (1941). *See* DEMING, W. E.

GEORGE, R. F. (1936). " A sample investigation of the 1931 Population Census with reference to earners and non-earners." *J. R. Statist. Soc.*, **99**, 147–161.

GINI, C. (1928). " Une application de la méthode représentative aux matériaux du dernier recensement de la population italienne." *Bull. Inst. Int. Statist.*, **23**, (2), 198–215.

HANSEN, M. H. (1940). *See* STEPHAN, F. F.

—— —— (1941). *See* STEPHAN, F. F.

—— —— (1944). " Census to sample population growth." *Domestic Commerce*, **32**, 6.

—— —— (1944). *See* HAUSER, P. M.

—— —— and HURWITZ, W. N. (1946). " Sampling methods applied to census work." U.S. Bureau of Census.

HAUSER, P. M. (1941). " The use of sampling in the census." *J. Amer. Statist. Ass.*, **36**, 369–375.

—— —— (1942). " Proposed annual sample census of population." *J. Amer. Statist. Ass.*, **37**, 81–88.

—— —— and HANSEN, M. H. (1944). " Sample surveys in census work." U.S. Bureau of Census, Dept. of Commerce.

HURWITZ, W. N. (1946). *See* HANSEN, M. H.

JESSEN, R. J., BLYTHE, R. H., KEMPTHORNE, O. and DEMING, W. E. (1947). " On a population sample for Greece." *J. Amer. Statist. Ass.*, **42**, 357–384.

KAMEDA, T. (1930). " Application of the method of sampling to the first Japanese population census." *Bull. Inst. Int. Statist.*, **25**, (2), 121–132.

KEMPTHORNE, O. (1947). See JESSEN, R. J.

STEPHAN, F. F., DEMING, W. E. and HANSEN, M. H. (1940). " The sampling procedure of the 1940 population census." *J. Amer. Statist. Ass.*, **35**, 615–630.

—— , —— , —— , (1941). " On the sampling methods in the 1940 population Census." U.S. Dept. of Commerce Census.

—— —— (1941). *See* DEMING, W. E.

U.S. BUREAU OF CENSUS (1943). " A revised sample for current surveys." Bureau of Census.

D. SOCIOLOGY, NUTRITION, HEALTH, ETC.

ANONYMOUS (1941). " Cost of living of the working classes." *J. R. Statist. Soc.*, **104**, 53–58.

BOSE, A. N. (1941). " Some problems of field operations in labour enquiries." *Sankhya*, **5**, 229–230.

BOWLEY, A. L. (1913). " Working class households in Reading." *J. R. Statist. Soc.*, **76**, 672–701.

—— (1936). " The application of sampling to economic and sociological problems." *J. Amer. Statist. Ass.*, **31**, 474–480.

BOX, K. and THOMAS, G. (1944). " The Wartime Social Survey." *J. R. Statist. Soc.*, **107**, 151–189.

CORMICK, T. C. (1937). " Sampling theory in sociological research." *Social Forces*, **16**, 67–74.

CRUICKSHANK, E. W. H. (1947). " A survey of diets in the maternity wards of Scottish hospitals." *Proc. Nutrition Soc.*, **5**, 149–159.

DEMING, W. E. (1943). *See* HANSEN, M. H.

—— —— (1943). *See* TEPPING, B. J.

—— —— (1944). " On errors in surveys." *Amer. Social Rev.*, **9**, 359–369.

GHOSH, A. (1946). *See* MAHALANOBIS, P. C.

GOODMAN, R. (1947). " Sampling for the 1947 survey of consumer finances." *J. Amer. Statist. Ass.*, **42**, 439–448.

HANSEN, M. H. and HURWITZ, W. N. (1942). " Relative efficiencies of various sampling units in population inquiries." *J. Amer. Statist. Ass.*, **37**, 89–94.

—— —— and DEMING, W. E. (1943). " On some census aids to sampling." *J. Amer. Statist. Ass.*, **38**, 353–357.

—— —— and HURWITZ, W. N. (1946). " The problems of non-response in sample surveys." *J. Amer. Statist. Ass.*, **41**, 517–529.

HILTON, J. (1928). " Some further enquiries by sample." *J. R. Statist. Soc.*, **91**, 519–540.

HURWITZ, W. N. (1942). *See* HANSEN, M. H.

—— —— (1943). *See* TEPPING, B. J.

—— —— (1946). *See* HANSEN, M. H.

KISER, C. V. (1934). " Pitfalls in sampling for population study." *J. Amer. Statist. Ass.*, **29**, 250–256.

MAHALANOBIS, P. C., MUKHERJEA, R. and GHOSH, A. (1946). " A sample survey of after-effects of the Bengal famine of 1943." *Sankhya*, **7**, 337–400.

MASSEY, P. (1942). " The expenditure of 1,360 British middle-class households in 1938–39." *J. R. Statist. Soc.*, **105**, 159–196.

MATHEN, K. K. (1948). " Studies on the sampling procedure for a general health survey." *Calcutta Statist. Ass. Bull.* No. 3, 106–113.

McNEMAR, Q. (1940). " Sampling in psychological research." *Psychol. Bull.*, **37**, 331–365.

MUKHERJEA, R. (1946). *See* MAHALANOBIS, P. C.

NEYMAN, J. (1933). " An outline of the theory and practice of representative method applied in social research." Inst. Social Problems, Warsaw. (Polish with an English summary).

NUTRITION SOCIETY (1945). " Budgeting and dietary surveys of families and individuals." (Abstracts of series of 13 papers read at the 18th and 21st meetings of the Society, 15 Feb. 1944 and 20 May 1944.) *Proc. Nutrition Soc.*, **3**, 1–52 and 110–154.

PALMER, G. L. (1943). " Factors in the variability of response in enumerative surveys." *J. Amer. Statist. Ass.*, **38**, 143–152.

PARTEW, M. (1937). *See* SCHOENBERG, E.

PIEKALKIEWICZ, J. (1934). " Rapport sur les recherches concernant la structure de la population ouvrière en Pologne selon la méthode représentative." Inst. Social Problems, Warsaw.

PIETRA, G. (1926). " A particular case of non-representative sampling." *J. Amer. Statist. Ass.*, **21**, 330–332.

SCHOENBERG, E. and PARTEW, M. (1937). " Methods and problems of sampling presented by the urban study of consumer purchases." *J. Amer. Statist. Ass.*, **32**, 311–322.

SNEDECOR, G. W. (1939). " Design of sampling experiments in the social sciences." *J. Farm. Econ.*, **21**, 846–855.

STEPHAN, F. F. (1936). " Practical problems of sampling procedure." *Amer. Social Rev.*, **1**, 569–580.

—— —— (1939). " Representative sampling in large scale surveys." *J. Amer. Statist. Ass.*, **34**, 343–352.

TEPPING, B. J., HURWITZ, W. H. and DEMING, W. E. (1943). " On the efficiency of deep stratification in block sampling." *J. Amer. Statist. Ass.*, **38**, 93–100.

THOMAS, G. (1944). *See* BOX, K.

U.S. PUBLIC HEALTH SERVICE (1938). " The National Health Survey, preliminary reports. Significance, scope and method." U.S. Public Health Service, Washington, D.C.

WOODBURY, R. M. (1940). " Methods of family living studies." Int. Labour Office, Series N (Statistics), No. 23.

E.   OPINION SURVEYS AND MARKET RESEARCH

BEVIS, J. C. (1945). " Management of field staffs in the opinion research field." *J. Amer. Statist. Ass.*, **40**, 245–246.

BLANKENSHIP, A. (1942). " Psychological difficulties in measuring consumer preference." *J. Marketing*, **6**, 66–75.

CROSSLEY, A. M. (1941). " Theory and application of representative sampling as applied to marketing." *J. Marketing*, **5**, 456–461.

ECKLER, A. R. and STAUDT, E. P. (1943). " Marketing and sampling uses of population and housing data." *J. Amer. Statist. Ass.*, **38**, 87–92.

GALLUP, G. (1938). " Government and the sampling referendum." *J. Amer. Statist. Ass.*, **33**, 131–142.

HANSEN, M. H. (1944). *See* HAUSER, P. M.

—— —— and HAUSER, P. M. (1945). " Area sampling—some principles of sampling design." *Pub. Opin. Quart.*, **9**, 183–193.

HAUSER, P. M. and HANSEN, M. H. (1944). " On sampling in market surveys." *J. Marketing*, **9**, 26–31.

—— —— (1945). *See* HANSEN, M. H.

HILGARD, E. R. and PAYNE, S. L. (1944). " Those not at home : riddle for pollsters." *Pub. Opin. Quart.*, **8**, 254–261.

McPEAK, W. (1945). " Problems of field management in army opinion research." *J. Amer. Statist. Ass.*, **40**, 247–248.

NORDIN, J. A. (1944). " Determining sample size." *J. Amer. Statist. Ass.*, **39**, 497–506.

PAYNE, S. L. (1944). *See* HILGARD, E. R.

ROPER, E. (1940). " Sampling public opinion." *J. Amer. Statist. Ass.*, **35**, 325–334.

SILVEY, R. J. E. (1944). " Methods of listener research employed by the British Broadcasting Corporation." *J. R. Statist. Soc.*, **107**, 190–230.

STAUDT, E. P. (1943). *See* ECKLER, A. R.

STEPHAN, F. F. (1941). " Stratification in representative sampling." *J. Marketing*, **6**, 38–46.

WILKS, S. S. (1940). " Representative sampling and poll reliability." *Pub. Opin Quart.*, **4**, 261–269.

F. ECONOMICS : SURVEYS OF INDUSTRY, CENSUSES OF PRODUCTION, LABOUR FORCE, ETC.

BECKNELL, H. E. (1944). "Organising statistical work on a functional basis." *J. Amer. Statist. Ass.*, **39**, 297–302.

DEDRICK, C. A. and HANSEN, M. H. (1938). "Final report on total and partial unemployment." Vol. IV. The enumerative check census. U.S. Govt. Printing Office.

ECKLER, A. R. (1945). "The revised census series of current employment estimates." *J. Amer. Statist. Ass.*, **40**, 187–196.

—— —— (1945). "Management of field work and collection of statistics of the labor force." *J. Amer. Statist. Ass.*, **40**, 249–250.

FRANKEL, L. R. and STOCK, J. S. (1942). "On the sample survey of unemployment." *J. Amer. Statist. Ass.*, **37**, 77–80.

FRIEDMAN, M. (1936). *See* KNEELAND, H.

GEARY, R. C. (1925). "Methods of sampling applied to Irish statistics." *J. Statist. Social Inquiry Soc. Ireland*, **15**, 61–80.

GURNEY, M. (1946). *See* HANSEN, M. H.

HANSEN, M. H. (1938). *See* DEDRICK, C. A.

—— and HURWITZ, W. N. (1944). "A new sample of the population : sampling principles introduced into the Bureau's monthly reports on the labor force." U.S. Bureau of Census.

—— ——, —— —— and GURNEY, M. (1946). "Problems and methods of the sample survey of business." *J. Amer. Statist. Ass.*, **41**, 173–189.

HURWITZ, W. N. (1944). *See* HANSEN, M. H.

—— —— (1946). *See* HANSEN, M. H.

KEYFITZ, N. (1945). "The sampling approach to economic data." *Canadian J. Econ. Polit. Sci.*, **11**, 467–477.

KNEELAND, H., SCHOENBERG, E. H., and FRIEDMAN, M. (1936). "Plans for a study of the consumption of goods and services by American families." *J. Amer. Statist. Ass.*, **31**, 135–141.

SCHOENBERG, E. H. (1936). *See* KNEELAND, H.

STOCK, J. S. (1942). *See* FRANKEL, L. R.

G. AGRICULTURAL ECONOMICS AND FARM PRACTICE

ANDERSON, O. (1934). "Description of the Bulgarian agricultural sample census." Bulletin de Statistique, No. 8, Dir. Gen. de Statistique de Bulgarie. (Summary in French).

BECKER, J. A. and HARLAN, C. L. (1939). "Developments in crop and livestock reporting since 1920." *J. Farm. Econ.*, **21**, 799–827.

BENEDICT, M. R. (1939). "Development of agricultural statistics in the Bureau of the Census." *J. Farm. Econ.*, **21**, 735–760.

BEYLEVELD, A. J. (1929). "Determination of a precise indication of change in crop acreage." *J. Amer. Statist. Ass.*, **24**, 405–411.

BOYD, D. A. (1944). *See* YATES, F.

DIMITROFF, ST. (1931). "Essai d'élaboration des matériaux du recensement agricole bulgare du 31 Décembre 1926 par l'application de la méthode représentative." *Bull. Inst. Int. Statist.*, **26** (2), 799–817.

FINKER, A. L., MORGAN, J. J. and MONROE, R. J. (1943). "Methods of estimating farm employment from sample data in North Carolina." N. Carolina Agric. Exp. Sta, Tech. Bull. 75.

GOODSELL, W. D., JESSEN, R. J., and WILCOX, W. W. (1940). "Procedures which increase the usefulness of farm management research." *J. Farm. Econ.*, **22**, 753–761.

HARLAN, C. L. (1939). *See* BECKER, J. A.

HENDRICKS, W. A. (1944). "The relative efficiencies of groups of farms as sampling units." *J. Amer. Statist. Ass.*, **39**, 367–376.

HOLMES, I. (1939). "Results of four methods of sampling individual farms." *J. Farm. Econ.*, **21**, 365–374.

HOLMES, I. (1939). "Research in sample farm census methodology. Part 1. Comparative statistical efficiency of sampling units smaller than the minor Civil Division for estimating year to year change." U.S. Dept. Agric., Agric. Marketing Service.

—— —— (1943). "Some sampling uses of data from the census of agriculture." *J. Amer. Statist. Ass.*, **38**, 78–86.

—— —— (1944). "Value of farm products by colour and tenure of farm operator." (Includes description of sample and notes on precision of sample estimates). U.S. Bureau of Census.

HOPKINS, J. A. (1942). "Statistical comparisons of record-keeping farms and a random sample of Iowa farms for 1939." Iowa Agric. Exp. Sta. Res. Bull. 308.

HOUSEMAN, E. E. (1944). *See* JESSEN, R. J.

—— —— (?). "Some developments in sampling and the design of a general purpose sample for enumerative surveys in B.A.E." U.S. Dept. Agric., Bur. Agric. Econ.

JESSEN, R. J. (1939). "An experiment in the design of agricultural surveys." *J. Farm. Econ.*, **21**, 856–863.

—— —— (1940). *See* GOODSELL, W. D.

—— —— (1942). "Statistical investigation of a sample survey for obtaining farm facts." Iowa Agric. Exp. Sta. Res. Bull. 304.

—— —— (1943). *See* STRAND, N. V.

—— —— and HOUSEMAN, E. E. (1944). "Statistical investigations of farm sample surveys taken in Iowa, Florida, and California." Iowa Agric. Exp. Sta. Res. Bull. 329.

—— —— (1945). *See* KING, A. J.

—— —— (1947). "The master sample project and its use in agricultural economics." *J. Farm. Econ.*, **29**, 531–540.

KING, A. J. and SIMPSON, G. D. (1940). "New developments in agricultural sampling." *J. Farm. Econ.*, **22**, 341–349.

—— —— and McCARTY, D. E. (1941). "Application of sampling to agricultural statistics with emphasis on stratified samples." *J. Marketing*, **5**, 462–474.

—— —— (1942). *See* SNEDECOR, G. W.

—— —— and JESSEN, R. J. (1945). "The master sample of agriculture." *J. Amer. Statist. Ass.*, **40**, 38–56.

MATHISON, I. (1944). *See* YATES, F.

McCARTY, D. E. (1941). *See* KING, A. J.

MINISTRY OF AGRICULTURE (1946). "National Farm Survey of England and Wales." H.M.S.O.

MONROE, R. J. (1943). *See* FINKNER, A. L.

MORGAN, J. J. (1943). *See* FINKNER, A. L.

NAGASAWA, R. (1930). "The method of statistical investigation concerning agricultural production in Japan." *Bull. Inst. Int. Statist.*, **25** (2), 149–178.

SABIN, A. R. (1940). "A new technique for the estimation of changes in farm employment." (No. 1 of a series of analyses of sample farm data). U.S. Dept. Agric., Agric. Marketing Service.

SARLE, C. F. (1938). "Methods in sample census research." *J. Farm. Econ.*, **20**, 669–672.

—— —— (1939). "Development of partial and sample census methods." *J. Farm. Econ.*, **21**, 356–364.

—— —— (1939). "Future improvements in agricultural statistics." *J. Farm. Econ.*, **21**, 838–845.

—— —— (1940). "The possibilities and limitations of objective sampling in strengthening agricultural statistics." *Econometrica*, **8**, 45–61.

SCHUTZ, H. H. (1937). *See* SHEPARD, J. B.

SHEPARD, J. B., and SCHUTZ, H. H. (1937). "Selection of areas for sample agricultural enumerations." *J. Farm. Econ.*, **19**, 454–469.

SIMPSON, G. D. (1940). *See* KING, A. J.

SNEDECOR, G. W. and KING, A. J. (1942). "Recent developments in sampling for agricultural statistics." *J. Amer. Statist. Ass.*, **37**, 95–102.

—— —— (1947). "An experiment in the collection of morbidity and mortality data on farm animals." *Proc. U.S. Livestock Sanitary Ass.* (51st meeting), 218–225.

SOCIAL SCIENCE RESEARCH COUNCIL (1937). "The census of agriculture." Bull. 40. 230, Park Ave., N.Y.

STRAND, N. V. and JESSEN, R. J. (1943). " Some investigations on the suitability of the township as a unit for sampling Iowa agriculture." Iowa Agric. Exp. Sta. Res. Bull. 315.

TAEUBER, C. (1944). *See* TOLLEY, H. R.

TOLLEY, H. R. and TAEUBER, C. (1944). " Wartime developments in agricultural statistics." *J. Amer. Statist. Ass.,* **39,** 411-427.

U.S. BUREAU OF AGRICULTURAL ECONOMICS (1936). " Proceedings of conferences on statistical methods of sampling agricultural data." Bur. Agric. Econ., Washington, D.C.

WILCOX, W. W. (1940). *See* GOODSELL, W. D.

YATES, F. (1943). " Methods and purposes of agricultural surveys." *J. R. Soc. Arts,* **91,** 367-379.

—— ——, BOYD, D. A. and MATHISON, I. (1944). " The manuring of farm crops ; some results of a survey of fertilizer practice in England." *Emp. J. Exp. Agric.,* **12,** 163-176.

H. CROP ESTIMATION, FORECASTING, ETC.

BARNARD, M. M. (1936). " An examination of the sampling observations on wheat of the crop–weather scheme." *J. Agric. Sci.,* **26,** 456-487.

BREWBAKER, H. E. and BUSH, H. L. (1942). " Pre-harvest estimate of yield and sugar percentage based on random sampling technique." Ann. Amer. Soc. Sugar Beet Tech.

BUCKER, M. I. (?). *See* PETERS, J. H.

BUSH, H. L. (1942). *See* BREWBAKER, H. E.

CLAPHAM, A. R. (1929). " The estimation of yield in cereal crops by sampling methods." *J. Agric. Sci.,* **19,** 214-235.

—— —— (1929). *See* WISHART, J.

—— —— (1931). " Studies in sampling technique : cereal experiments. I. Field technique." *J. Agric. Sci.,* **21,** 366-371.

—— —— (1931). " Studies in sampling technique. III. Results and discussion." *J. Agric. Sci.,* **21,** 376-390.

COCHRAN, W. G. and WATSON, D. J. (1936). " An experiment on observer's bias in the selection of shoot heights." *Emp. J. Exp. Agric.,* **4,** 69-76.

—— —— (1938). *See* IRWIN, J. O.

—— —— (1940). " The estimation of the yields of cereal experiments by sampling for the ratio of grain to total produce." *J. Agric. Sci.,* **30,** 262-275.

DHANNALAL (1940). *See* KALAMKAR, R. J.

EDGAR, J. L. and DE WET, A. F. (1935). " An experiment in sampling technique for size and colour of apples : Bramley's Seedling on Rootstock No. 11, 1934 crop." Rep. E. Malling Res. Sta., 130-4.

—— —— (1938). *See* HOBLYN, T. N.

FEDERER, W. T. (1946). *See* HOUSEMAN, E. E.

GHOSH, B. (1947). " Crop estimation in India : a brief review of the sampling methods." Calcutta Statist. Ass. Bull., No. 1, 5-12.

HEATH, O. V. S. (1934). " Sampling and growth observations in plant development studies in cotton." Report of Proceedings of Empire Cotton Growing Corporation : 2nd conference on cotton growing problems ; 96-110.

HENDRICKS, W. A. (1942). " Theoretical aspects of the use of the crop meter." No. 2 of a series of analyses of sample farm data. U.S. Dept. Agric., Agric. Marketing Service.

HOBLYN, T. N. and EDGAR, J. L. (1938). " Experiments in sampling techniques. II. Size and colour of Allington Pippin, 1936 crop." Rep. E. Malling Res. Sta. for 1937, 168-172.

HOUSEMAN, E. E., WEBER, C. R., and FEDERER, W. T. (1946). " Pre-harvest sampling of soybeans for yield and quality." Iowa Agric. Exp. Sta. Res. Bull. 341.

HUBBACK, J. A. (1927). " Sampling for rice yield in Bihar and Orissa." Imp. Agric. Res. Inst., Pusa, Bull. 166. (Reprinted 1946 : *Sankhya,* **7,** 281-294).

HUDSON, H. G. (1939). " Population studies with wheat. I. Sampling." *J. Agric. Sci.,* **29,** 76-110.

IMMER, F. R. (1932). " A study in sampling technique with sugar beets." *J. Agric. Res.,* **44,** 633-647.

IRWIN, J. O., COCHRAN, W. G., and WISHART, J. (1938). " Crop estimation and its relation to agricultural meteorology." *J. R. Statist. Soc., Suppl.*, **5**, 1–45.

JEBE, E. H. (1940). *See* KING, A. J.

KALAMKAR, R. J. (1932). " A study in sampling techniques with wheat." *J. Agric. Sci.*, **22**, 783–796.

—— —— and DHANNALAL (1940). " Sampling studies in cotton varietal trial." *Sankhya*, **4**, 567–576.

—— —— (1941). " A note on crop-cutting experiments on cotton." *Sankhya*, **5**, 345–348.

—— —— (1944). *See* PANSE, V. G.

—— —— (1944). *See* PANSE, V. G.

—— —— (1945). *See* PANSE, V. G.

KING, A. J. and JEBE, E. H. (1940). " An experiment in pre-harvest sampling of wheat fields." Iowa Agric. Exp. Sta. Res. Bull. 273.

—— ——, McCARTY, D. E. and McPEAK, M. (1942). " An objective method of sampling wheat fields to estimate production and quality of wheat." U.S. Dept. Agric. Tech. Bull. 814.

LEGATT, C. W. (1941). " A study of the relative efficiency of seed sampling methods." *Can. J. Res.*, **19**, 156–162.

MAHALANOBIS, P. C. (1940). " A sample survey of the acreage under jute in Bengal." *Sankhya*, **4**, 511–530.

—— —— (1945). " Report on the Bihar crop survey : Rabi season, 1943–4." *Sankhya*, **7**, 29–106.

—— —— (1946). " Sample surveys of crop yields in India." *Sankhya*, **7**, 269–280.

MALLIK, A. K., SATAKOPAN, V. and RAO, S. G. (1945). " A study of the estimation of the yield of wheat by sampling." *Indian J. Agric. Sci.*, **15**, 219–225.

McCANDLISS, D. A. (1941). " Objective sampling in estimating southern crops." *J. Farm. Econ.*, **23**, 246–255.

McCARTY, D. E. (1942). *See* KING, A. J.

McPEAK, M. (1942). *See* KING, A. J.

PANSE, V. G. (1938). " Preliminary studies on sampling in field experiments." *Sankhya* **4**, 139–148.

—— —— and KALAMKAR, R. J. (1944). " Forecasting and estimation of crop yields." *Current Science*, **13**, 120–124.

—— —— and KALAMKAR, R. J. (1944). " A further note on the estimation of crop yields." *Current Science*, **13**, 223–225.

—— —— and KALAMKAR, R. J. (1945). " A large-scale yield survey on cotton." *Current Science*, **14**, 287–291.

—— —— (1946). " Plot size in yield surveys on cotton." *Current Science*, **15**, 218–219.

PEARCE, S. C. (1943). " An investigation into means of reducing the labour needed for recording crops." Rep. E. Malling Res. Sta. for 1942, 36–40.

—— —— (1944). " Sampling methods for the measurement of fruit crops." *J. R. Statist. Soc.*, **107**, 117–126.

PETERS, J. H. and BUCKER, M. I. (?). " The 1940 section survey of crop acreages in Indiana and Iowa." U.S. Dept. Agric., Bur. Agric. Econ.

RAO, S. G. (1945). *See* MALLIK, A. K.

SARLE, C. F. (1932). " Adequacy and reliability of crop yields estimates." U.S. Dept. Agric. Tech. Bull. 311.

SATAKOPAN, V. (1945). *See* MALLIK, A. K.

SREENIVASAN, P. S. (1943). " Studies on the estimation of growth and yield of jowar by sampling." *Indian J. Agric. Sci.*, **13**, 399–412.

SUKHATME, P. V. (1945). " Random sampling for estimating rice yields in Madras Province." *Indian J. Agric. Sci.*, **15**, 308–318.

—— —— (1946). " Bias in the use of small-size plots in sample surveys for yield." *Nature*, **157**, 630.

—— —— (1946). " Size of sampling unit in yield surveys." *Nature*, **158**, 345.

—— —— (1947). " The problem of plot size in the large scale yield surveys." *J Amer. Statist. Ass.*, **42**, 297–310.

—— —— (1947). " Use of small size plots in yield surveys." *Nature*, **160**, 542.

WATSON, D. J. (1934). *See* YATES, F.

—— —— (1936). *See* COCHRAN, W. G.

—— —— (1937). " The estimation of leaf areas." *J. Agric. Sci*, **27**, 474–483.

WEBER, C. R. (1946). *See* HOUSEMAN, E. E.

WET, A. F. DE (1935). *See* EDGAR, J. L.

WISHART, J. and CLAPHAM, A. R. (1929). " A study in sampling technique. The effect of artificial fertilizers on the yield of potatoes." *J. Agric. Sci.*, **19**, 600–618.

—— —— (1938). *See* IRWIN, J. O.

YATES, F. and WATSON, D. J. (1934). " Observer's bias in sampling observations on wheat." *Emp. J. Exp. Agric.*, **2**, 174–177.

—— —— and ZACOPANAY, I. (1935). " The estimation of the efficiency of sampling, with special reference to sampling for yield in cerea experiments." *J. Agric. Sci.*, **25**, 545–577.

—— —— (1935). " The place of quantitative measurements on plant growth in agricultural meteorology and crop forecasting." Conference of Empire Meteorologists, London, 169–172.

—— —— (1936). " Crop estimation and forecasting : indications of the sampling observations on wheat." *J. Min. Agric.*, **43**, 156–162.

—— —— (1936). " Applications of the sampling technique to crop estimation and forecasting." Manchester Statist. Soc.

ZACOPANAY, I. (1935). *See* YATES, F.

I. FORESTRY AND LAND UTILIZATION SURVEYS

BICKERSTAFF, A. (1947). " One-fifth acre versus one-tenth acre plots in sampling immature stands." Dominion Forest Service, Canada. Silvicultural Research Note 83.

—— —— (1947). " Sampling efficiency of line plot survey on Riding Mountain research area." Dominion Forest Service, Canada. Silvicultural Research Note 84.

BLYTHE, R. H. (1945). " The economics of sample size applied to the scaling of saw logs." *Biometrics*, **1**, 67–70.

FINNEY, D. J. (1949). " Random and systematic sampling in timber surveys." *Forestry*, **22**, 64–99.

FOGH, I. F. (1943). " Sampling methods in log scaling." *For. Chron.*, **19**, 127–138.

GEVORKIANTZ, S. R. (1934). *See* MUDGETT, B. D.

—— —— (1939). *See* GIRARD, J. W.

GIRARD, J. W. and GEVORKIANTZ, S. R. (1939). " Timber cruising." U.S. Dept. Agric. Forest Service, Washington.

HASEL, A. A. (1937). " Arrangement of cruise plots to permit a valid estimate of sampling error." Forest and Range Exp. Sta.

—— —— (1938). " Sampling error in timber surveys." *J. Agric. Res.*, **57**, 713–736.

—— —— (1941). " Estimation of vegetation type areas by linear measurement." *J. For.*, **39**, 34–40.

—— —— (1942). " Estimation of volume in timber stands by strip sampling." *Ann. Math. Statist.*, **13**, 179–206.

HURWITZ, W. N. (1943). " Working plan for annual census of lumber produced in 1943." U.S. Dept. Agric. Forest Service.

LEXEN, B. (1941). " The application of sampling to log scaling." *J. For.*, **39**, 624–631.

LOOMIS, R. D. (1946). " Accuracy in timber estimating." *For. Chron.*, **22**, 201–202.

MUDGETT, B. D. and GEVORKIANTZ, S. R. (1934). " Reliability of forest surveys." *J. Amer. Statist. Ass.*, **29**, 257–281.

MULLOY, G. A. (1946). *See* ROBERTSON, W. M,

OSBORNE, J. G. (1942). " Sampling errors of systematic and random surveys of cover-type areas." *J. Amer. Statist. Ass.*, **37**, 256–264.

PECHANEC, J. (1941). " Sampling error in range surveys of sagebrush grass vegetation." *J. For.*, **39**, 52–54.

PROUDFOOT, M. J. (1942). " Sampling with transverse traverse lines." *J. Amer. Statist. Ass.*, **37**, 265–270.

ROBERTSON, W. M. and MULLOY, G. A. (1946). " Sample plot methods." Dominion Forest Service, Ottawa.

THOMPSON, A, P, (1945). " A sampling approach to New Zealand timber cruising problems." *N.Z. J. For.*, **5**, 103–117.

J. ESTIMATION OF WILD POPULATIONS

BEALL, G. (1939). " Methods of estimating the population of insects in a field."
  Biometrika, **30**, 422–439.

COCHRAN, W. G. (1938). " The information supplied by the sampling results." *Ann.
  Appl. Biol.*, **25**, 383–389. (Appendix to LADELL, R. S., " Field experiments on
  the control of wireworms." *Ann. Appl. Biol.*, **25**, 341–382.)

DELURY, D. B. (1947). " On the estimation of biological populations." *Biometrics*,
  3, 145–167.

FINNEY, D. J. (1941). " Wireworm populations and their effect on crops." *Ann.
  Appl. Biol.*, **28**, 282–295.

—— —— (1942). *See* YATES, F.

—— —— (1946). " Field sampling for the estimation of wireworm populations."
  Biometrics, **2**, 1–7.

GREENSLADE, R. M. and PEARCE, S. C. (1940). " Field sampling for the comparison
  of infestations of strawberry crops by the aphis *Capitophorus fragariae Theob.*"
  Pomology and Hort. Sci., **17**, 308–317.

HOEL, P. G. (1943). " The accuracy of sampling methods in ecology." *Ann. Math.
  Statist.*, **14**, 289–300.

JACKSON, C. H. N. (1939). " The analysis of an animal population." *J. Anim. Ecol.*,
  8, 238–246.

JONES, E. W. (1937). " Practical field methods of sampling soil for wireworms."
  J. Agric. Res., **54**, 123–134.

MEYERS, M. T. and PATCH, L. H. (1937). " A statistical study of sampling in field
  surveys of the fall population of the European Corn Borer." *J. Agric. Res.*, **55**,
  849–872.

MINISTRY OF AGRICULTURE (1944). " Wireworms and food production." A wireworm
  survey of England and Wales. 1939–1942. Bull. 128. H.M.S.O.

PATCH, L. H. (1937). *See* MEYERS, M. T.

PEARCE, S. C. (1940). *See* GREENSLADE, R. M.

WADLEY, F. M. (1945). " An application of the Poisson series to some problems of
  enumerations." *J. Amer. Statist. Ass.*, **40**, 85–92.

YATES, F. and FINNEY, D. J. (1942). " Statistical problems in field sampling for
  wireworms." *Ann. Appl. Biol.*, **29**, 156–167.

SUPPLEMENTARY BIBLIOGRAPHY
TO SECOND EDITION

This Supplementary Bibliography has been prepared by Mr. D. H. Rees with the assistance of a bibliography on sampling compiled by Dr. Tore Dalenius of the Central Bureau of Statistics, Stockholm. For the most part it contains papers published since the appearance of the first edition. References which were inadvertently omitted from the first edition are distinguished by an asterisk. The subject classification followed is that given on page 416 with the exception that an additional section

(K') Industrial applications

has been added.

BOOKS

General

CASEY, R. S. and PERRY, J. W. (1951). "Punched cards : their applications to science and industry." New York : Reinhold Publishing Corporation. (London : Chapman and Hall.)

COCHRAN, W. G. (1953). "Sampling techniques." New York : Wiley.

DEMING, W. E. (1950). "Some theory of sampling." New York : Wiley.

FINNEY, D. J. (1952). "Statistical method in biological assay." London : Griffin.

GALLUP, G. (1948). "A guide to public opinion polls." Princeton : University Press.

JONES, D. CARADOG (1949). "Social surveys." London : Hutchinson.

MATERN, B. (1947). "Metoder att Uppskatta Noggrannheten vid Linje- och Provyte-taxering." Stockholm : State Forestry Research Institute.

PARTEN, M. B. (1950). "Surveys, polls and samples : practical procedures." New York : Harpers.

PAYNE, S. L. (1951). "The art of asking questions." New Jersey : Princeton University Press. (London : Oxford University Press.)

QUENOUILLE, M. H. (1952). "Associated measurements." London : Butterworths Scientific Publications.

VANCE, L. L. (1950). "Scientific method for auditing : applications of statistical sampling theory to auditing procedure." Berkeley : University of California Press. (London : C.U.P.)

Reports on Surveys, etc.

JOINT COMMITTEE OF THE ROYAL COLLEGE OF OBSTETRICIANS AND GYNAECOLOGISTS AND THE POPULATION INVESTIGATION COMMITTEE (1948). "Maternity in Great Britain : A survey of social and economic aspects of pregnancy and childbirth." London : Geoffrey Cumberlege.

M'GONIGLE, G. C. M. and KIRBY, J. (1936). "Poverty and public health." London : Gollancz.

NUFFIELD FOUNDATION (1947). "Old people." London : Geoffrey Cumberlege.

UNITED NATIONS (1947-51). "Reports of the Sub-Commission on Statistical Sampling." E/CN. 3/37, 52, 83, 114, 140. New York.

PAPERS

A'. THEORY AND METHODS

ANDERSON, O. (1949). "Die Grundprobleme der Stichprobenmethode. I." *MittBl. Math. Statist.*, **1**, 37–52.

—— —— (1949). "Die Grundprobleme der Stichprobenmethode. II." *MittBl. Math. Statist.*, **1**, 81–95.

—— —— (1950). "Die Grundprobleme der Stichprobenmethode. III." *MittBl. Math. Statist.*, **2**, 1–16.

BIRNBAUM, Z. W. and SIRKEN, M. G. (1949). "Non-response and repeated call backs in sampling surveys" (Abstract). *Ann. Math. Statist.*, **20**, 136.

BOSE, C. (1951). "Some further results on errors in double sampling technique." *Sankhya*, **11**, 191–194.

BUCKLAND, W. R. (1951). "A review of the literature of systematic sampling." *J. R. Statist. Soc. B.*, **13**, 208–215.

COCHRAN, W. G. (1947). "Recent developments in sampling theory in the United States." *Proc. Int. Statist. Confs.*, **3**, A, 40–66.

DALENIUS, T. (1950). "The problem of optimum stratification." *Skand. Aktuar. Tidskr.*, **33**, 203–213.

—— —— (1952). "Eine einfache geometrische Veranschaulichung der Theorie des geschichteten Stichprobenverfahrens." *MittBl. Math. Statist.*, **4**, 121–128.

DAS, A. C. (1950). "Two dimensional systematic sampling and the associated stratified and random sampling." *Sankhya*, **10**, 95–108.

DOERING, C. R. (1947). *See* LOMBARD, H. L.

DYKE, G. V. and PATTERSON, H. D. (1952). "Analysis of factorial arrangements when the data are proportions." *Biometrics*, **8**, 1–12.

EVANS, W. D. (1951). "On stratification and optimum allocations." *J. Amer. Statist. Ass.*, **46**, 95–104.

GEARY, R. C. (1950). "Most efficient sample sizes for the two stage sampling process in the case of the limited universe." *Bull. Inst. Int. Statist.*, **32**, 228–239.

GHOSH, B. (1949). "Interpenetrating (networks of) samples." *Calcutta Statist. Ass. Bull.*, **2**, 108–119.

GODAMBE, V. P. (1951). "On two stage sampling." *J. R. Statist. Soc. B.*, **13**, 216–218.

GOODMAN, R. and KISH, L. (1950). "Controlled selection—a technique in probability sampling." *J. Amer. Statist. Ass.*, **45**, 350–372.

HANSEN, M. H. and HURWITZ, W. N. (1949). "On the determination of optimum probabilities in sampling." *Ann. Math. Statist.*, **20**, 426–432.

—— ——, HURWITZ, W. N., MARKS, E. S. and MAULDIN, W. P. (1951). "Response errors in surveys." *J. Amer. Statist. Ass.*, **46**, 147–190.

HENDRICKS, W. A. (1951). "Variance components as a tool for the analysis of sample data." *Biometrics*, **7**, 97–101.

HURWITZ, W. N. (1949). *See* HANSEN, M. H.

—— —— (1951). *See* HANSEN, M. H.

KELLERER, H. (1949). "Elementare Ausführungen zur Theorie und Technik des Stichprobenverfahrens. I." *MittBl. Math. Statist.*, **1**, 96–114.

—— —— (1949). "Elementare Ausführungen zur Theorie und Technik des Stichprobenverfahrens. II." *MittBl. Math. Statist.*, **1**, 203–218.

—— —— (1950). "Elementare Ausführungen zur Theorie und Technik des Stichprobenverfahrens. III." *MittBl. Math. Statist.*, **2**, 36–49.

KEYFITZ, N. (1951). "Sampling with probabilities proportional to size : adjustment for changes in the probabilities." *J. Amer. Statist. Ass.*, **46**, 105–109.

KING, A. J. (1952). *See* ROBSON, D. S.

KISH, L. (1950). *See* GOODMAN, R.

KRIESBERG, M. (1952). *See* VOIGHT, R. B.

LOMBARD, H. L. and DOERING, C. R. (1947). "Treatment of the four-fold table by partial correlation as it relates to Public Health problems." *Biometrics*, **3**, 123–128.

MADOW, W. G. (1948). "On the limiting distributions of estimates based on samples from finite universes." *Ann. Math. Statist.*, **19**, 535–545.

—— —— (1949). "On the theory of systematic sampling. II." *Ann. Math. Statist.*, **20**, 333–354.

MALONEY, C. J. (1949). " Precision of estimates from samples selected under marginal restrictions " (Abstract). *Ann. Math. Statist.*, **20**, 137.

MARCUSE, S. (1949). " Optimum allocation and variance components in nested sampling with an application to chemical analysis." *Biometrics*, **5**, 189–206.

MARKS, E. S. (1951). *See* HANSEN, M. H.

MATTHAI, A. (1951). " Estimation of parameters from incomplete data with application to design of sample surveys." *Sankhya*, **11**, 145–152.

MAULDIN, W. P. (1951). *See* HANSEN, M. H.

MIDZUNO, H. (1950). "A survey method using two kinds of surveys." *Ann. Inst. Statist. Math.* (Japan), **1**.

—— (1950). "An outline of the theory of sampling systems." *Ann. Inst. Statist. Math.* (Japan), **1**.

MOKASHI, V. K. (1950). "A note on interpenetrating samples." *J. Ind. Soc. Agric. Statist.*, **2**, 189–195.

NANDI, H. K. (1948) " Choosing a random sample." *Calcutta Statist. Ass. Bull.*, No. 4, 143–152.

NARAIN, R. D. (1951). " On sampling without replacement with varying probabilities." *J. Ind. Soc. Agric. Statist.*, **3**, 169–174.

—— (1952). *See* SUKHATME, P. V.

PATTERSON, H. D. (1950). " Sampling on successive occasions with partial replacements of units." *J. R. Statist. Soc. B.*, **12**, 241–255.

—— —— (1952). *See* DYKE, G. V.

QUENOUILLE, M. H. (1949). " Problems in plane sampling." *Ann. Math. Statist.*, **20**, 355–375.

—— —— (1950). "An application of least squares to family diet surveys." *Econometrica*, **18**, 27–44.

—— —— (1950). " Computational devices in the application of least squares." *J. R. Statist. Soc. B.*, **12**, 256–272.

ROBSON, D. S. and KING, A. J. (1952). " Multiple sampling of attributes." *J. Amer. Statist. Ass.*, **47**, 203–215.

SEAL, K. C. (1951). " On errors of estimates in various types of double sampling procedure." *Sankhya*, **11**, 125–144.

SENG, Y. P. (1951). " Historical survey of the development of sampling theories and practice." *J. R. Statist. Soc. A.*, **114**, 214–231.

SETH, G. R. (1952). *See* SUKHATME, P. V.

SILBER, J. (1948). " Multiple sampling for variables." *Ann. Math. Statist.*, **19**, 246–257.

SIRKEN, M. G. (1949). *See* BIRNBAUM, Z. W.

SUKHATME, P. V. and SETH, G. R. (1952). " Non-sampling errors in surveys." *J. Ind. Soc. Agric. Statist.*, **4**, 5–41.

—— and NARAIN, R. D. (1952). " Sampling with replacement." *J. Ind. Soc. Agric. Statist.*, **4**, 42–49.

TUKEY, J. W. (1950). "Some sampling simplified." *J. Amer. Statist. Ass.*, **45**, 501–519.

VOIGHT, R. B. and KRIESBERG, M. (1952). " Some principles of processing census and survey data." *J. Amer. Statist. Ass.*, **47**, 222–231.

WEIBULL, M. (1951). " The regression problem involving non-random variates in the case of a stratified sample from normal parent populations with varying regression coefficients." *Skand. Aktuar. Tidskr.*, **34**, 53–71.

YATES, F. (1947). " The influence of agricultural research statistics on the development of sampling theory." *Proc. Int. Statist. Confs.*, **3**, A, 27–39.

B'. MACHINE METHODS

CASTONE, G. F. and DYE, W. S. (1949). "A simplified punch card method of determining sums of squares and sums of products." *Psychometrika*, **14**, 243–250.

DYE, W. S. (1949). *See* CASTONE, G. F.

HAMMERSLEY, J. M. (1952). " The computation of sums of squares and products on a desk calculator." *Biometrics*, **8**, 156–168.

KING, G. W. (1949). "A method of plotting on standard IBM equipment." *Math. Tables and other Aids to Computation*, **3**, 352–355.

LESTER, A. M. (1949). "The edge marking of statistical cards." *J. Amer. Statist. Ass.*, **44**, 293–294.

OPLER, A. (1951). "Monte Carlo matrix calculation with punched card machines." *Math. Tables and other Aids to Computation*, **5**, 115–120.

RAFFERTY, J. A. (1951). *See* VOTAW, D. F. (Jr.).

VOTAW, D. F. (Jr.) and RAFFERTY, J. A. (1951). "High speed sampling." *Math. Tables and other Aids to Computation*, **5**, 1–8.

C'. POPULATION CENSUSES

CHEVRY, G. (1949). "Control of a general census by means of an area sampling method." *J. Amer. Statist. Ass.*, **44**, 373–379.

DEDRICK, C. L. (1948). "Some problems of the 1950 census of the Americas." *Estadistica*, **6**, 354–359.

DEMING, W. E. (1949). *See* SEKAR, C. C.

GENERAL REGISTER OFFICE (1952). "Census 1951, Great Britain, one per cent. sample tables. Part I. Ages and marital condition, occupations, industries, housing of private households." H.M.S.O., London.

HAUSER, P. M. (1950). "Some aspects of methodological research in the 1950 census." *Pub. Opin. Quart.*, **14**, 5–13.

KEYFITZ, N. (1953). "A factorial arrangement of comparisons of family size." *Amer. J. Sociol.* (in the Press)

KOOP, J. C. (1951). "Notes on the estimation of gross and net reproduction rates by methods of statistical sampling." *Biometrics*, **7**, 155–166.

LAWRENCE, N. (1949). *See* SHRYOCK, H. S. (Jr.).

MARKS, E. S. and MAULDIN, W. P. (1950). "Response errors in census research." *J. Amer. Statist. Ass.*, **45**, 424–438.

MAULDIN, W. P. (1950). *See* MARKS, E. S.

ROYAL COMMISSION ON POPULATION (1949). "Report on an enquiry into family limitation and its influence on human fertility during the past fifty years." By E. LEWIS-FANING. Papers of the Royal Comm. Pop. Vol. I. H.M.S.O., London.

—— —— —— (1950). "Reports and selected papers of the Statistics Committee." Papers of the Royal Comm. Pop. Vol. II. H.M.S.O., London.

SCHMITT, R. C. (1952). "Short-cut methods of estimating county populations." *J. Amer. Statist. Ass.*, **47**, 232–238.

SEKAR, C. C. and DEMING, W. E. (1949). "On a method of estimating birth and death rates and the extent of registration." *J. Amer. Statist. Ass.*, **44**, 101–115.

SENG, Y. P. (1949). "Practical problems in sampling for social and demographic inquiries in undeveloped countries." *Population Studies*, **3**, 170–191.

SHAUL, J. R. H. (1952). "Sampling surveys in Central Africa." *J. Amer. Statist. Ass.*, **47**, 239–253.

SHRYOCK, H. S. (Jr.) and LAWRENCE, N. (1949). "The current status of state and local population estimates in the Census Bureau." *J. Amer. Statist. Ass.*, **44**, 157–173.

STEINER, P. O. (1951). "A source of bias in one of the samples of the 1950 census." *J. Amer. Statist. Ass.*, **46**, 110–113.

SWOBODA, W. (1951). "Eine Frage zur Stichprobenweisen Vorauswertung der Volkszählung 1950." *MittBl. Math. Statist.*, **3**, 135–139.

UNITED NATIONS (1949). "Population census methods." Population Studies No. 4. New York.

* U.S. BUREAU OF THE CENSUS (1945). "Notes on precision of sample estimates : technical notes on the formulas used to evaluate the precision of data." U.S. Govt. Printing Office, Washington, D.C.

D'. SOCIOLOGY, NUTRITION, HEALTH, ETC.

BHATTACHARYA, B. (1951). *See* MITRA, K. N.

BIRNBAUM, Z. W. and SIRKEN, M. G. (1950). "Bias due to non-availability in sampling surveys." *J. Amer. Statist. Ass.*, **45**, 98–111.

BOOKER, H. S. and DAVID, S. T. (1952). "Differences in results obtained by experienced and inexperienced interviewers." *J. R. Statist. Soc. A.*, **115**, 232–257.

BRANSBY, E. R., DAUBNEY, C. G. and KING, J. (1948). " Comparison of results obtained by different methods of individual dietary survey." *Brit. J. Nutrit.*, **2**, 89–110.

BRESARD, M. (1950). " Mobilité sociale et dimension de la famille." *Population* 1950, 533–566.

CANNELL, C. F. (1950). *See* GOODMAN, R.

CHIAN, C. L. (1951). " On design of mass medical surveys." *Hum. Biol.*, **23**, 242–271.

COCHRAN, W. G. (1951). " Modern methods in the sampling of human populations : general principles in the selection of a sample." *Amer. J. Publ. Hlth.*, **41**, 647–653.

CORLETT, T. (1950). *See* GRAY, P. S.

DALY, J. F. (1949). *See* JESSEN, R. J.

DAUBNEY, C. G. (1948). *See* BRANSBY, E. R.

DAVID, S. T. (1952). *See* BOOKER, H. S.

DAWN, C. S. (1951). *See* MITRA, K. N.

DEMING, W. E. (1949). *See* JESSEN, R. J.

—— —— (1950). *See* HANSEN, M. H.

DEY, K. (1951). *See* MITRA, K. N.

DIVISIA, F. J. (1950). "Aspects de la technique des sondages statistiques dans le domaine social." *Bull. Inst. Int. Statist.*, **32**, 240–244.

DURBIN, J. and STUART, A. (1951). " Differences in response rates of experienced and inexperienced interviewers." *J. R. Statist. Soc. A.*, **114**, 163–206.

GALVANI, L. (1952). " I concetti fondamentali del metodo rappresentativo." *I Problemi del Servizio Sociale*, Anno VII, n. 3.

GAYEN, A. K. (1951). *See* MITRA, K. N.

GIRARD, A. (1950). " Une enquête sur les besoins des familles." *Population* 1950, 713–732.

—— —— (1950). *See* STOETZEL, J.

GOODMAN, R. and CANNELL, C. F. (1950). " Sampling errors and components of interview costs in relation to sample design." Survey Research Center, Univ. of Michigan.

—— —— and MACCOBY, E. E. (1950). " Sampling methods and sampling errors in surveys of consumer finances." Survey Research Center, Univ. of Michigan.

GRAY, P. S. and CORLETT, T. (1950). " Sampling for the Social Survey." *J. R. Statist. Soc. A.*, **113**, 150–206.

HANSEN, M. H. (1947). " Sampling of human populations." *Proc. Int. Statist. Confs.*, **3**, 113–128.

—— —— and DEMING, W. E. (1950). " On an important limitation to the use of data from samples." *Bull. Inst. Int. Statist.*, **32**, 214–219.

HEALY, M. J. R. (1952). " Some statistical aspects of anthropometry." *J. R. Statist. Soc. B*, **14**, 164–184.

INDIAN STATISTICAL INSTITUTE (1949). " Report on the diet and health survey of middle class families in Bombay city." Ind. Statist. Inst., Bombay, India.

JESSEN, R. J., KEMPTHORNE, O., DALY, J. F. and DEMING, W. E. (1949). " Observations on the 1946 Elections in Greece." *Amer. Sociological Rev.*, **14**, 11–16.

KATONA, G. (1950). " Financial surveys among consumers." Survey Research Center, Univ. of Michigan.

KELLERER, H. V. (1950). " Stichprobenverfahren in der amtlichen deutschen Statistik seit 1946." *Bull. Inst. Int. Statist.*, **32**, 245–255.

KEMPTHORNE, O. (1949). *See* JESSEN, R. J.

KING, J. (1948). *See* BRANSBY, E. R.

KISH, L. (1949). "A procedure for objective respondent selection within the household." *J. Amer. Statist. Ass.*, **44**, 380-387.

LIKERT, R. (1950). " The sample interview survey." Survey Research Center, Univ. of Michigan.

LIVI, L. (1950). " Sur la mesure de la mobilité sociale : résultats d'un sondage effectué sur la population italienne." *Population* 1950, 65–76.

MACCOBY, E. E. (1950). *See* GOODMAN, R.

MAHALANOBIS, P. C., MAJUMDAR, D. N. and RAO, C. R. (1949). "Anthropometric survey of the United Provinces, 1941 : a statistical study." *Sankhya*, **9**, 93–324.

MAJUMDAR, D. N. (1949). *See* MAHALANOBIS, P. C.

METZNER, C. A. (1950). "An application of scaling to questionnaire construction." *J. Amer. Statist. Ass.*, **45**, 112–118.

MITRA, K. N., BHATTACHARYA, B., DEY, K., DAWN, C. S., OBADIAH, M. and GAYEN, A. K. (1951). "A study of recent trend in infantile mortality rates in Calcutta by longitudinal survey." *Sankhya*, 11, 167–182.

MORRELL, A. J. H. (1950). "The estimation of age-standardized ratio by sampling methods." *The Incorporated Statistician*, 1, 17–20.

MOSER, C. A. (1949). "The use of sampling in Great Britain." *J. Amer. Statist. Ass.*, 44, 231–259.

MOSS, L. (1950). "The Government Social Survey." *Operational Research Quarterly*, 1, 55–65.

NANDI, H. K. (1949). "A critique of United Provinces anthropometric survey, 1951 : a statistical study." *Calcutta Statist. Ass. Bull.*, 2, 95–107.

—— —— (1950). "Indian national sample survey." *Calcutta Statist. Ass. Bull.*, 3, 11–20.

*NORTHRUP, M. S. (1943). *See* WEBB, J. N.

NUTRITION SOCIETY (1950). "The Irish national nutrition survey." *Brit. J. Nutrit.*, 4, 269–295.

OBADIAH, M. (1951). *See* MITRA, K. N.

*PAYNE, S. L. (1943). *See* WEBB, J. N.

POLITZ, A. and SIMMONS, W. (1949). "An attempt to get the 'not at homes' into the sample without callbacks." *J. Amer. Statist. Ass.*, 44, 9–31.

RADVANYI, L. (1951). "Measurement of the effectiveness of basic education." The Social Sciences, Mexico.

—— —— (1952). "Ten years of sample surveying in Mexico." The Social Sciences, Mexico.

RAO, C. R. (1949). *See* MAHALANOBIS, P. C.

ROGOFF, N. (1950). "Les recherches américaines sur la mobilité sociale." *Population* 1950, 669–688.

SCHULZ, T. (1952). "Ten years of family surveys." *Oxford Univ. Inst. Stat. Bull.*, 14, 83–95.

SIMMONS, W. (1949). *See* POLITZ, A.

SIRKEN, M. G. (1950). *See* BIRNBAUM, Z. W.

SOCIAL SURVEY (1950). "The register of electors as a sampling 'frame'," by P. G. GRAY, T. CORLETT and P. FRANKLAND. Central Office of Information, London.

STOETZEL, J. and GIRARD, A. (1950). "Une enquête nationale sur le niveau intellectuel des enfants d'âge scolaire." *Population* 1950, 567–576.

STUART, A. (1951). *See* DURBIN, J.

*U.S. DEPARTMENT OF STATE (1946). "Report of the Allied Mission to observe the Greek Election." Dept. of State Publn. 2522. U.S. Govt. Printing Office, Washington, D.C.

*WEBB, J. N., NORTHRUP, M. S. and PAYNE, S. L. (1943). "Practical applications of theoretical sampling methods." *J. Amer. Statist. Ass.*, 38, 69–77.

WEST BENGAL STATE STATISTICAL BUREAU (1951). "Report on the sample survey for estimating the socio-economic characteristics of displaced persons migrating from Eastern Pakistan to the State of West Bengal." Alipore, West Bengal.

E'. OPINION SURVEYS AND MARKET RESEARCH

*CASSADY, R. (1945). "Statistical sampling techniques and marketing research." *J. Marketing*, 9, 317–341.

DURANT, H. and GREGORY, W. (1951). "Behind the Gallup Poll." *News Chronicle*, London.

GHOSH, M. N. (1947). "Survey of public opinion." *Calcutta Statist. Ass. Bull.*, No. 1, 13–18.

GREGORY, W. (1951). *See* DURANT, H.

GREVILLE, T. N. E. (1949). "Opinion polls and sample surveys." *Estadistica*, 7, 92–93.

HOCHSTIM, J. R. (1947). *See* STOCK, J. S.

KELLERER, H. (1952). "Marktforschung—ein Anwendungsgebiet des Stichprobenverfahrens." *MittBl. Math. Statist.*, 4, 209–230.

MOSER, C. A. (1952). "Quota sampling." *J. R. Statist. Soc. A.*, 115, 411–423.

SOCIAL SCIENCE RESEARCH COUNCIL (1949). "The pre-election polls of 1948." Bull. 60, 230 Park Ave., New York.

STOCK, J. S. and HOCHSTIM, J. R. (1947). "Commercial uses of sampling." *Proc. Int. Statist. Confs.*, **3**, A, 129–143.

◆WATSON, A. N. (1942). "Use of small area census data in marketing analysis." *J. Marketing*, **6**, 42–47.

F'. ECONOMICS: SURVEYS OF INDUSTRY, CENSUSES OF PRODUCTION, LABOUR FORCE, ETC.

BLYTHE, R. H. (1951). *See* ROSANDER, A. C.

DURBIN, J. (1950). *See* STONE, R.

JOHNSON, D. E. (1951). *See* ROSANDER, A. C.

KEYFITZ, N. and ROBINSON, H. L. (1949). "The Canadian sample for labour force and other population data." *Population Studies*, **2**, 427–443.

ROBINSON, H. L. (1949). *See* KEYFITZ, N.

ROSANDER, A. C., BLYTHE, R. H. and JOHNSON, D. E. (1951). "Sampling 1949 Corporation income tax returns." *J. Amer. Statist. Ass.*, **46**, 233–241.

STONE, R., UTTING, J. E. G. and DURBIN, J. (1950). "The use of sampling methods in national income statistics and social accounting." *Accounting Research*, **7**, 333–356.

UTTING, J. E. G. (1950). *See* STONE, R.

G'. AGRICULTURAL ECONOMICS AND FARM PRACTICE

ANDERSON, O. (1949). "Uber die repräsentative Methode und deren Anwendung auf die Aufarbeitung der Ergebnisse der bulgarischen landwirtsch. Betriebszahlung vom 31. Dezember 1926." Bayer. Statist. Landesamt, München.

BOYD, D. A. (1951). *See* YATES, F.

CHURCH, B. M. (1952). "Recent trends in fertilizer practice in England and Wales. Part 1. The National Position." *Emp. J. Exp. Agric.*, **20**, 249–256.

CRUMP, S. L. (1948). *See* NORDSKOG, A. W.

JEBE, E. H. (1952). "Estimation for subsampling designs employing the county as a primary sampling unit." *J. Amer. Statist. Ass.*, **47**, 49–70.

JESSEN, R. J. (1949). "Some inadequacies of the federal censuses of agriculture." *J. Amer. Statist. Ass.*, **44**, 279–292.

MADOW, L. H. (1950). "On the use of the county as the primary sampling unit for state estimates." *J. Amer. Statist. Ass.*, **45**, 30–47.

MAHALANOBIS, P. C. (1949). "Statistical tools in resource appraisal and utilization." *Proc. U. N. Sci. Conf. on the Conservation and Utilization of Resources*, **1**, 196–200.

NORDSKOG, A. W. and CRUMP, S. L. (1948). "Systematic and random sampling for estimating egg production in poultry." *Biometrics*, **4**, 223–233.

STRECKER, H. (1952). "Anwendung von Stichprobenverfahren zur Erfassung der Schweinebestände in Bayern." *MittBl. Math. Statist.*, **4**, 258–267.

YATES, F. (1949). "The place of experimental investigations in the planning of resource utilization." *Proc. U. N. Sci. Conf. on the Conservation and Utilization of Resources* **1**, 192–196.

—— (1950). "Agriculture, sampling and operational research." *Bull. Inst. Int. Statist.*, **32**, 220–227.

—— and BOYD, D. A. (1951). "The Survey of Fertilizer Practice: an example of operational research in agriculture." *Brit. Agric. Bull.*, **4**, 206–209.

H'. CROP ESTIMATION, FORECASTING, ETC.

BLASER, R. E. (1948). *See* RIGNEY, J. A.

BOSE, P. K. (1948). "Crop estimation and its relation to agricultural meteorology." *Calcutta Statist. Ass. Bull.* No. 3, 114–119.

BOYD, D. A. and DYKE, G. V. (1950). "Maincrop potato growing in England and Wales." *N.A.A.S. Quarterly Review*, No. 10, 47–57.

CAMERON, J. M. (1951). "The use of components of variance in preparing schedule for sampling of baled wool." *Biometrics*, **7**, 83–96.

DYKE, G. V. (1950). *See* BOYD, D. A.

—— —— and SIMPSON, E. P. (1952). "A collaborative investigation into methods of chemical analysis of ground pyrethrum flowers." *Appl. Statist.*, **1**, 95–105.

*IRWIN, J. O. (1929). "Crop forecasting and the use of meteorological data in its improvement." Conf. Emp. Meteorologists, 1929. H.M.S.O., London.

KOSHAL, R. S. (1947). *See* SUKHATME, P. V.

MAHALANOBIS, P. C. (1950). "Cost and accuracy of results in sampling and complete enumeration." *Bull. Inst. Int. Statist.*, **32**, (2), 210–213.

*McVAY, F. E. (1947). "Sampling methods applied to estimating numbers of commercial orchards in a commercial peach area." *J. Amer. Statist.*, **42**, 533–540.

*PANSE, V. G. and SAHASRABUDHE, V. B. (1943). "A rapid method of sampling for fibre weight determination in cotton." *Indian J. Genet.*, **3**, 28–44.

—— —— and SUKHATME, P. V. (1948). "Crop surveys in India—I." *J. Ind. Soc. Agric. Statist.*, **1**, 34–58.

—— —— (1951). *See* SUKHATME, P. V.

RIGNEY, J. A. and BLASER, R. E. (1948). "Sampling Alyce clover for chemical analyses." *Biometrics*, **4**, 234–239.

*SAHASRABUDHE, V. B. (1943). *See* PANSE, V. G.

SIMPSON, E. P. (1952). *See* DYKE, G. V.

SMITH, H. FAIRFIELD (1948). "A sampling survey of tappings on small holdings, 1939-40." *J. Rubber Res. Inst. Malaya*, **12**, 79–125.

SUKHATME, P. V. (1947). "Report on random sample survey for estimating the outturn of paddy in Central Province and Berar 1945-46." Ind. Coun. Agric. Res., India.

—— —— (1947). "Report on a random sample survey for estimating the outturn of paddy in Madras 1945-46." Ind. Coun. Agric. Res., India.

—— —— and KOSHAL, R. S. (1947). "Report of the scheme for crop cutting : experimental survey on paddy in the Bombay Province 1945-46." Imperial (now Indian) Coun. Agric. Res., India.

—— —— (1948). *See* PANSE, V. G.

—— —— (1950). "Efficiency of subsampling designs in yield surveys." *J. Ind. Soc. Agric. Statist.*, **2**, 212–228.

—— —— and PANSE, V. G. (1951). "Crop surveys in India—II." *J. Ind. Soc. Agric. Statist.*, **3**, 97–168.

TAYLOR, J. (1951). "The estimation of fruit size of cherries by sampling methods." Rep. East Malling Res. Sta., 1950, 93–99.

WEST BENGAL STATE STATISTICAL BUREAU (1950). "Report on the experimental sample survey of autumn crops, 1949, namely jute and paddy, in the districts of Burway and Hooghly." Alipore, West Bengal.

I′. FORESTRY AND LAND UTILIZATION SURVEYS

BECKER, M. E. (1950). "Forest survey procedures for area and volume determination." *J. For.*, **48**, 465–469.

BICKFORD, C. A. (1952). "The sampling design used in the forest survey of the North East." *J. For.*, **50**, 290–293.

*FINNEY, D. J. (1947). "Volume estimation of standing timber by sampling." *Forestry*, **21**, 179–203.

—— —— (1949). "The efficiency of enumerations. 1. Volume estimation of standing timber by sampling. 2. Random and systematic sampling in timber surveys." Bull. No. 146, For. Res. Inst., Dehra Dun, India.

—— —— and PALCA, H. (1949). "The elimination of bias due to edge effects in forest sampling." *Forestry*, **23**, 31–47.

—— —— (1950). "Investigations in random and systematic sampling." *The Incorporated Statistician*, **1**, 15–20.

—— —— (1950). "An example of periodic variation in forest sampling." *Forestry* **23**, 96–111.

JOHNSON, F. A. (1949). " Statistical aspects of timber volume sampling in the Pacific North West." *J. For.*, **47**, 292–295.
—— —— (1950). " Estimating forest areas and volumes for large tracts." *J. For.*, **48**, 340–342.
MOUNTAIN, H. S. (1949). " Determining the solid wood volume of four foot pulpwood stacks." *J. For.*, **47**, 627–631.
PALCA, H. (1949). *See* FINNEY, D. J.

J'. ESTIMATION OF WILD POPULATIONS

ANSCOMBE, F. J. (1948). " On estimating the populations of aphids in a potato field." *Ann. Appl. Biol.*, **35**, 567–571.
—— —— (1950). " Soil sampling for potato root eelworm cysts." *Ann. Appl. Biol.*, **37**, 286–295.
BAILEY, N. T. J. (1951). " On estimating the size of mobile populations from recapture data." *Biometrika*, **38**, 293–306.
CHAPMAN, D. G. (1951). " Some properties of the hyper-geometric distribution with applications to zoological sample censuses." *Univ. Calif. Publ. Statist.*, **1**, 131–159.
CHITTY, D. (1951). *See* LESLIE, P. H.
GRUNDY, P. M. (1951). " The expected frequencies in a sample of an animal population in which the abundance of species are log normally distributed. I." *Biometrika*, **38**, 427–434.
LESLIE, P. H. and CHITTY, D. (1951). " The estimation of population parameters from data obtained by means of the capture-recapture method : 1. The maximum likelihood equation for estimating the death rate." *Biometrika*, **38**, 269–292.
MORAN, P. A. P. (1951). "A mathematical theory of animal trapping." *Biometrika*, **38**, 307–311.
—— —— (1952). " The estimation of death-rates from capture-mark-recapture sampling." *Biometrika*, **39**, 181–188.
OAKLAND, G. B. (1950). "An application of sequential analysis to whitefish sampling." *Biometrics*, **6**, 59–67.
WILLIAMS, C. B. (1950). " The application of the logarithmic series to the frequency of occurrence of plant species in quadrats." *J. Ecol.*, **38**, 107–138.

K'. INDUSTRIAL APPLICATIONS

BRITISH ELECTRICAL AND ALLIED INDUSTRIES RESEARCH ASSOCIATION (1948). " Some aspects of sampling as applied to electricity supply," by P. SCHILLER. Technical report K/T 118, London.
—— —— —— —— —— (1948). "A large scale sampling of domestic consumers," by P. SCHILLER. Technical report K/T 125a, London.
BRITISH ELECTRICITY AUTHORITY (1950). "A further analysis of the returns of the British Electrical Research Association sampling survey of domestic consumers." Utilisation Res. Rep. No. 4, London.
INESON, J. L. (1951). " The applications of statistical methods to large scale surveys in connection with the fixing of electricity tariffs." *J. Inst. Municipal Treasurers and Accountants*, **7**, 115–130.
JOWITT, G. H. (1952). " The accuracy of systematic sampling from conveyor belts." *Appl. Statist.*, **1**, 50–59.
MINISTRY OF WORKS (1950). " Productivity in house building : a pilot sample survey in the South, East and West of England and in South Wales, August, 1947–October, 1948." H.M.S.O., London.
PEARSON, E. S. (1934). " Problems of industrial sampling." *J. R. Statist. Soc. Suppl.*, **1**, 107–151.
WHITWELL, J. C. (1951). " Estimating precision of textile instruments." *Biometrics*, **7**, 102–112.

SUPPLEMENTARY BIBLIOGRAPHY
TO THIRD EDITION

This Supplementary Bibliography has been prepared by Mr. H. R. Simpson. For the most part it contains papers published since the appearance of the second edition. References which were inadvertently omitted from the second edition are distinguished by an asterisk. Cross-references have been omitted to save space. Only papers in English or with English summaries have been included.

We received very considerable help from Mr. J. A. King, Librarian of the Royal Statistical Society, for which we are most grateful.

BOOKS

General

CANNING, R. G. (1956). " Electronic data processing for business and industry." New York: Wiley. (London: Chapman & Hall.)

DALENIUS, T. (1957). " Sampling in Sweden: Contributions to the method and theories of sample survey practice." Stockholm: Almqvist and Wiksell.

HANSEN, M. H., HURWITZ, W. N. and MADOW, W. G. (1953). " Sample survey methods and theory." Vol. I. " Methods and applications." Vol. II. " Theory." New York and London: Wiley.

*HARTKEMEIER, H. P. (1952). " Punch-card methods." Dubuque, Iowa: Brown.

HENDRICKS, W. A. (1956). " The mathematical theory of sampling." New Brunswick: Scarecrow Press. (London: Bailey Bros. & Swinfen.)

HYMAN, H. H. (1955). " Survey design and analysis: principles, cases and procedures." Free Press, Illinois.

MOSER, C. A. (1958). " Survey methods in social investigation." London: Heinemann.

SANDERSON, F. H. (1954). " Methods of crop forecasting." Cambridge, Mass.: Harvard University Press.

SOCIAL SURVEY (1956). " Some useful data when sampling the population of England and Wales." London.

SUKHATME, P. V. (1953). " Sampling theory of surveys with applications." New Delhi: The Indian Society of Agricultural Statistics. (Ames, Iowa: The Iowa State College Press.)

*UNITED NATIONS (1949). " Population census methods." Population Studies, No. 4, ST/SOA/Series A. New York: United Nations. London: H.M.S.O.

Reports on Surveys, etc.

CARR SAUNDERS, A. M., CARADOG JONES, D. and MOSER, C. A. (1958). " A survey of social conditions in England and Wales as illustrated by statistics." Oxford: Clarendon Press.

GLASS, D. V. and GREBENIK, E. (1954). " The trend and pattern of fertility in Great Britain: A report on the family census of 1946." Part I (report); Part II (tables). London: H.M.S.O.

INSTITUTE OF PRACTITIONERS IN ADVERTISING (1956). " A national survey of readership of newspapers and magazines on a continuing basis." London.

LOGAN, W. P. D. and BROOKE, E. M. (1957). " The survey of sickness 1943 to 1952." London: H.M.S.O.

LONDON TRANSPORT EXECUTIVE (1956). " London travel survey, 1954." London.

UNITED STATES DEPARTMENT OF HEALTH, EDUCATION AND WELFARE (1957–). " Health statistics from the U.S. National Health Survey." Washington.

*UNITED NATIONS, Statistical Office of (1952–). " Statistical papers." New York.

A″. THEORY AND METHODS

BASU, D. (1954). " On the optimum character of some estimators used in multistage sampling problems." *Sankhya*, 13, 363–368.

BELSON, W. A. (1959). " Matching and prediction on the principle of biological classification." *Appl. Statist.*, 8, 65–75.

BRANKO, B. (1958). " A formula for the upper bound of relative variance, in a simple case of two-stage sampling." *Statist. Rev.*, Belgrade, 9, 298–301.

BROOKS, E. M. (1953). " Planning and operating sample surveys." *Estadist.*, 11, 63–71.

BROOKS, S. H. (1955). " The estimation of an optimum subsampling number." *J. Amer. statist. Ass.*, 50, 398–415.

CANSADO, E. (1957). " Sampling without replacement from finite populations." *Trab. estadist.*, 8, 3–12.

CHAKRAVARTI, I. M. (1954). " On the problem of planning a multistage survey for multiple correlated characters." *Sankhya*, 14, 211–216.

COCHRAN, W. G., MOSTELLER, F. and TUKEY, J. W. (1954). " Principles of sampling." *J. Amer. statist. Ass.*, 49, 13–35.

*DALENIUS, T. (1952). " The problem of optimum stratification in a special type of design." *Skand. AktuarTidskr.*, 35, 61–70.

———— (1953). " The economics of one-stage stratified sampling." *Sankhya*, 12, 351–356.

———— (1953). " The multivariate sampling problem." *Skand. AktuarTidskr.*, 36, 92–122.

*———— and GURNEY, M. (1951). " The problem of optimum stratification—II." *Skand. AktuarTidskr.*, 34, 133–148.

———— and HODGES, J. L. (1957). " The choice of stratification points." *Skand. AktuarTidskr.*, 40, 198–203.

————, ———— (1959). " Minimum variance stratification." *J. Amer. statist. Ass.*, 54, 88–101.

DEMING, W. E. (1953). " On the distinction between enumerative and analytic surveys." *J. Amer. statist. Ass.*, 48, 244–255.

———— (1953). " On a probability mechanism to attain an economic balance between the resultant error of response and the bias of non-response." *J. Amer. statist. Ass.*, 48, 743–772.

———— (1956). " On simplifications of sampling design through replication with equal probabilities and without stages." *J. Amer. statist. Ass.*, 51, 24–53.

———— and GLASSER, G. J. (1959). " On the problem of matching lists by samples." *J. Amer. statist. Ass.*, 54, 403–415.

DOUGLAS, J. W. B. and BLOMFIELD, J. M. (1956). " The reliability of longitudinal surveys." *Milbank mem. Fd. Quart.*, 34, 227–252.

DURBIN, J. (1953) " Some results in sampling theory when the units are selected with unequal probabilities " *J. R. statist. Soc. B*, 15, 262–269.

———— (1954). " Non-response and call-backs in surveys." *Bull. Inst. int. Statist.*, 34, 3, 3–17.

———— (1958). " Sampling theory for estimates based on fewer individuals than the number selected." *Bull. Inst. int. Statist.*, 36, 3, 113–119.

———— and STUART, A. (1954). " Callbacks and clustering in sample surveys: an experimental study." *J. R. statist. Soc. A*, 117, 387–428.

————, ———— (1954). " An experimental comparison between coders." *J. Marketing*, 19, 54–66.

ECIMOVIC, J. P. (1956). " Three-stage sampling with varying probabilities of selection." *J. Ind. Soc. agric. Statist.*, 8, 14–44.

ECKLER, A. R. (1955). " Rotation sampling." *Ann. math. Statist.*, 26, 664–685.

———— and PRITZKER, L. (1951). " Measuring the accuracy of enumerative surveys." *Bull. Inst. int. Statist.*, 33, 4, 7–24.

EKMAN, G. (1959). " An approximation useful in univariate stratification." *Ann. math. Statist.*, 30, 219–229.

GHOSH, B. (1957). " Enumerational errors in surveys." *Bull. Calcutta statist. Ass.*, 7, 50–59.

GODAMBE, V. P. (1955). " A unified theory of sampling from finite populations." *J. R. statist. Soc. B*, 17, 269–278.

GRUNDY, P. M. (1954). " A method of sampling with probability exactly proportional to size." *J.R. statist. Soc. B*, 16, 236–238.

*HAYASHI, C. (1952). "On the prediction of phenomena from qualitative data and the quantification of qualitative data from the mathematico-statistical point of view." *Ann. Inst. statist. Math.*, Tokyo, **3**, 70–98.

*HORVITZ, D. G. and THOMPSON, D. J. (1952). "A generalization of sampling without replacement from a finite universe." *J. Amer. statist. Ass.*, **47**, 663–685.

KATZ, L. (1953). "Confidence intervals for the number showing a certain characteristic in a population when sampling is without replacement." *J. Amer. statist. Ass.*, **48**, 256–261.

KEYFITZ, N. (1957). "Estimates of sampling variance where two units are selected from each stratum." *J. Amer. statist. Ass.*, **52**, 503–510.

KITAGAWA, T. (1955). "Some contributions to the design of sample surveys." *Sankhya*, **14**, 317–362.

——— (1956). "Some contributions to the design of sample surveys. Part IV. Exact sampling theories and analysis of variance schemes associated with designs of sample surveys." *Sankhya*, **17**, 1–36.

*MAHALANOBIS, P. C. (1952). "Some aspects of the design of sample surveys." *Sankhya*, **12**, 1–7.

MAITLAND, D. W. (1958). "Some sample survey techniques." *Bull. statist. Soc. N.S.W.*, No. 21, 26–34.

MASUYAMA, M. (1957). "Ratio estimates in line-grid sampling." *Bull. math. Statist.*, **7**, 73–76.

MATTHAI, A. (1954). "On selecting random numbers for large-scale sampling." *Sankhya*, **13**, 257–260.

MICKEY, M. R. (1959). "Some finite population unbiased ratio and regression estimators." *J. Amer. statist. Ass.*, **54**, 594–612.

*MIDZUNO, H. (1952). "On the sampling system with probability proportional to sum of sizes." *Ann. Inst. statist. Math.*, Tokyo, **3**, 99–107.

MILNE, A. (1959). "The centric systematic area-sample treated as a random sample." *Biometrics*, **15**, 270–297.

MOKASHI, V. K. (1954). "Efficiency of stratification of sub-sampling designs for the ratio method of estimation." *J. Ind. Soc. agric. Statist.*, **6**, 77–82.

MOORE, P. G. (1957). "Sampling techniques and some applications." *J. Inst. Actuar. Stud. Soc.*, **14**, 111–128.

MURTHY, M. N. and NANJAMMA, N. S. (1959). "Almost unbiased ratio estimates based on interpenetrating sub-sample estimates." *Sankhya*, **21**, 381–392.

NANJAMMA, N. S., MURTHY, M. N. and SETHI, V. K. (1959). "Some sampling systems providing unbiased ratio estimators." *Sankhya*, **21**, 299–314.

NARAIN, R. D. (1953). "On the recurrence formula in sampling on successive occasions." *J. Ind. Soc. agric. Statist.*, **5**, 96–99.

NORDBOTTEN, S. (1954). "On the determination of an optimal sample size." *Skand. AktuarTidskr.*, **37**, 60–64.

——— (1956). "Allocation in stratified sampling by means of linear programming." *Skand. AktuarTidskr.*, **39**, 1–6.

——— (1957). "On errors and optimal allocation in a census." *Skand. AktuarTidskr.*, **40**, 1–10.

PATTERSON, H. D. (1954). "The errors of lattice sampling." *J.R. statist. Soc. B*, **16**, 140–149.

PRICE, O. O. and McKEON, A. J. (1953). "A comparison of serial number digit sampling with systematic and random sampling." *Appl. Statist.*, **2**, 39–43.

RAJ, D. (1954). "Ratio estimation in sampling with equal and unequal probabilities." *J. Ind. Soc. agric. Statist.*, **6**, 127–138.

——— (1956). "Some estimators in sampling with varying probabilities without replacement." *J. Amer. statist. Ass.*, **51**, 269–284.

——— (1956). "On the method of overlapping maps in sample surveys." *Sankhya*, **17**, 89–98.

——— (1956). "A note on the determination of optimum probabilities in sampling without replacement." *Sankhya*, **17**, 197–200.

——— (1957). "On estimating parametric functions in stratified sampling designs." *Sankhya*, **17**, 361–366.

— ——— (1958). "On the relative accuracy of some sampling techniques." *J. Amer. statist. Ass.*, **53**, 98–101.

RANGARAJAN, R. (1957). "A note on two-stage sampling." *Sankhya*, 17, 373–376.

RAO, J. N. K. and CHAWLA, H. K. (1956). "Efficiency of stratification in sub-sampling designs for the ratio method of estimation with varying probabilities of selection." *J. Ind. Soc. agric. Statist.*, 8, 91–101.

RIDDERSTROM, S. (1955). "On ratio estimates in simple random sampling with some practical applications." *Skand. AktuarTidskr.*, 38, 135–162.

ROSHWALB, I. (1953). "Effect of weighting by card-duplication on the efficiency of survey results." *J. Amer. statist. Ass.*, 48, 773–777.

ROY, J. (1957). "A note on estimation of variance components in multistage sampling with varying probabilities." *Sankhya*, 17, 367–372.

SAITO, K. (1957). "Some results in the theory of sampling on successive occasions with partial replacement of units." *Rep. statist. Appl. Res.* (JUSE), 4, 15–22.

SEAL, K. C. (1953). "On certain extended cases of double sampling." *Sankhya*, 12, 357–362.

SEN, A. R. (1953). "On the estimate of the variance in sampling with varying probabilities." *J. Ind. Soc. agric. Statist.*, 5, 119–127.

—— —— (1954). "A comparison of stratified two-stage sampling systems." *J. Amer. statist. Ass.*, 49, 539–558.

—— —— (1955). "A simple design in sampling with varying probabilities." *J. Ind. Soc. agric. Statist.*, 7, 57–69.

SINGH, D. (1954). "On efficiency of the sampling with varying probabilities without replacement." *J. Ind. Soc. agric. Statist.*, 6, 48–57.

—— —— (1956). "On efficiency of cluster sampling." *J. Ind. Soc. agric. Statist.*, 8, 45–55.

*SKELLAM, J. G. (1949). "The distribution of the moment statistics of samples drawn without replacement from a finite population." *J. R. statist. Soc.* B, 11, 291–296.

SOM, R. K. (1959). "Self-weighting sample design with an equal number of ultimate stage units in each of the selected penultimate stage units." *Bull. Calcutta statist. Ass.*, 9, 59–66.

STEVENS, W. L. (1958). "Sampling without replacement with probability proportional to size." *J. R. statist. Soc.* B, 20, 393–397.

STUART, A. (1954). "A simple presentation of optimum sampling results." *J. R. statist. Soc.* B, 16, 239–241.

SUKHATME, P. V. (1953). "Measurement of the observational errors in surveys." *Rev. Inst. int. statist.*, 20, 121–134.

TIKKIWAL, B. D. (1953). "Optimum allocation in successive sampling." *J. Ind. Soc. agric. Statist.*, 5, 100–102.

—— —— (1956). "A further contribution to the theory of univariate sampling on successive occasions." *J. Ind. Soc. agric. Statist.*, 8, 84–90.

TULSE, R. (1957). "Sampling for variables with a very skew distribution." *Appl. Statist.*, 6, 40–44.

*UNITED NATIONS ECONOMIC AND SOCIAL COUNCIL—SUB-COMMISSION ON STATISTICAL SAMPLING (1952). "Report to the Statistical Commission on the fifth session of the Sub-Commission on Statistical Sampling held from 19 to 31 December, 1951." *Sankhya*, 12, 165–204.

YAMAMOTO, S. (1955). "On the theory of sampling with probabilities proportional to given values." *Ann. Inst. statist. Math.*, Tokyo, 7, 25–38.

YATES, F. (1953). "The work of the United Nations Sub-Commission on Statistical Sampling." *Sankhya*, 12, 305–306.

—— —— and GRUNDY, P. M. (1953). "Selection without replacement from within strata, with probability proportional to size." *J. R. statist. Soc.* B, 15, 253–261.

ZARKOVIC, S. S. (1956). "Note on the history of sampling methods in Russia." *J. R. statist. Soc.* A, 119, 336–338.

—— —— (1958). "On some consequences of the heterogeneity of units in sampling with varying probabilities and replacement." *Statist. Rev.*, Belgrade, 8, 45–47.

—— —— (1958). "On some problems of sampling work in under-developed countries." *Statist. Rev.*, Belgrade, 9, 1–11.

B″. MACHINE METHODS

*AYERS, J. D. and STANLEY, J. P. (1952). " The rolling totals method of computing sums, sums of squares, and sums of cross-products." *Psychometrika*, **17**, 305–310.

*BOTHWELL, R. D. and LEVINE, D. B. (1952). " Document sensing in large-scale enumerative surveys." *Estadist.*, **10**, 710–720.

CHURCH, B. M. and LIPTON, S. (1956). " The use of an electronic computer in the estimation of sampling errors in a nutritional survey." *Brit. J. Nutrit.*, **10**, 27–32.

CONKLIN, M. R. and GRETTON, O. C. (1958). " Some experience with electronic computers in processing the 1954 census of manufacturers." *Amer. Statist.*, **12**, 19–23.

DOUGLAS, A. S. (1959). " Techniques for the recording of, and reference to data in a computer." *Comp. J.*, **2**, 1–9.

FRY, T. C. (1956). " The automatic computer in industry." *J. Amer. statist. Ass.*, **51**, 565–575.

HEALY, M. J. R. (1956). " The electronic computer at Rothamsted." *Rep. Rothamst. exp. Sta.*, 229–235.

McGEE, W. C. (1959). " Generalization: Key to successful electronic data processing." *J. Assoc. Comp. Mach.*, **6**, 1–23.

McPHERSON, J. L. (1953). " Large-scale electronic tabulating equipment in use at the U.S. Bureau of the Census." *Estadist.*, **11**, 830–839.

MERRIMAN, J. H. H. *et al.* (1957). " A discussion on the use of electronic data processing equipment." *J. R. statist. Soc. A*, **120**, 291–307.

PODDER, K. C. (1954). " On the punched card method in smoothing for age bias in census returns." *Sankhya*, **13**, 261–266.

WARD, D. H. (1959). " The use of edge-punched cards in statistical computation." *Appl. Statist.*, **8**, 104–113.

YATES, F. and REES, D. H. (1958). " The use of an electronic computer in research statistics: four years' experience." *Comp. J.*, **1**, 49–58.

ZIOLA, R. (1953). " Mark sensing in the Canadian census records." *Estadist.*, **11**, 783–791.

C″. POPULATION CENSUSES

BANCROFT, G. (1954). " Special uses of the current population survey mechanism." *Estadist.*, **12**, 198–206.

BELLAMY, J. (1953). " A note on occupation statistics in British censuses." *Population Stud.*, **6**, 306–308.

BENE, L. (1958). " Complete enumeration and sampling surveys in the population census." *Demográfia*, **1**, 161–181.

BENJAMIN, B. (1955). " Quality of response in census taking." *Population Stud.*, **8**, 288–293.

BOYARSKY, A. Y. (1958). " An experiment in the theory of a census with control rounds." *Rev. Inst. int. statist.*, **26**, 48–55.

COALE, A. J. (1955). " The population of the United States in 1950 classified by age, sex and color—a revision of census figures." *J. Amer. statist. Ass.*, **50**, 16–54.

*COX, P. R. (1952). " Estimating the future population." *Appl. Statist.*, **1**, 82–94.

CROSETTI, A. H. and SCHMITT, R. C. (1956). " A method of estimating the intercensal population of counties." *J. Amer. statist. Ass.*, **51**, 587–590.

ECKLER, A. R. (1953). " Extent and character of errors in the 1950 census." *Amer. Statist.*, **7**, 15–20.

EL-BADRY, M. A. and STEPHAN, F. F. (1955). " On adjusting sample tabulations to census counts." *J. Amer. statist. Ass.*, **50**, 738–762.

GREBENIK, E. (1955). " The sources and nature of statistical information in special fields of statistics. Population and vital statistics." *J. R. statist. Soc. A*, **118**, 452–462.

HAJNAL, J. (1955). " The prospects for population forecasts." *J. Amer. statist. Ass.*, **50**, 309–322.

HANSEN, H. (1957). " Effect of the new design for the current population survey." *Estadist.*, **15**, 418–421.

HANSEN, M. H., HURWITZ, W. N., NISSELSON, H. and STEINBERG, J. (1955). " The redesign of the census current population survey." *J. Amer. statist. Ass.*, **50**, 701–719.

MARKS, E. S., MAULDIN, W. P. and NISSELSON, H. (1953). " The post-enumeration survey of the 1950 Census: a case history in survey design." *J. Amer. statist. Ass.*, **48**, 220–243.

PROTHERO, R. M. (1956). " The population census of Northern Nigeria 1952: problems and results." *Population Stud.*, **10**, 166–183.

SCHNEIDER, J. R. L. (1954). " Note on the accuracy of local population estimates." *Population Stud.*, **8**, 148–150.

SHIELDS, B. F. (1957–58). " An analysis of the Irish census of distribution, 1951." *J. Statist. Social Inquiry Soc.*, **20**, 118–135.

SHRYOCK, H. S. (1954). " Accuracy of population projections for the United States." *Estadist.*, **12**, 587–598.

SILCOCK, H. (1954). " Precision in population estimates." *Population Stud.*, **8**, 140–147.
—— —— (1954). " Note on Mr. Schneider's analysis." *Population Stud.*, **8**, 151.

STEVENS, W. L. (1957). " Bias of the census method." *J. Inst. Actuar. Stud. Soc.*, **14**, 192–198.

SUDAN GOVERNMENT MINISTRY FOR SOCIAL AFFAIRS (1955). " The 1953 pilot population census for the first population census in Sudan." Khartoum, Sudan.
—— —— —— —— —— (1956). " First interim report." Khartoum, Sudan.
—— —— —— —— —— (1956). " Supplement to interim reports." Khartoum, Sudan.

*U.S. BUREAU OF THE CENSUS (1952). " Concepts of small areas under the U.S. census of 1950, and program for tabulation and publication of the data." *Estadist.*, **10**, 254–267.

WINDLE, C. (1959). " The accuracy of census literacy statistics in Iran." *J. Amer. statist. Ass.*, **54**, 578–581.

ŽARKOVIC, S. S. (1955). " Sampling methods in the Yugoslav 1953 census of population." *J. Amer. statist. Ass.*, **50**, 720–737.
—— —— (1956). " Some remarks on coverage checks in population censuses." *Population Stud.*, **9**, 271–275.

D″. SOCIOLOGY, NUTRITION, HEALTH, ETC.

BAUER, T. J. et al. (1954). " Do persons lost to long term observation have the same experience as persons observed ? " *J. Amer. statist. Ass.*, **49**, 36–50.

BELLOC, N. B. (1954). " Validation of morbidity survey data by comparison with hospital records." *J. Amer. statist. Ass.*, **49**, 832–846.

CARTWRIGHT, A. (1957). " The effect of obtaining information from different informants on a family morbidity inquiry." *Appl. Statist.*, **6**, 18–25.
—— —— (1959). " Some problems in the collection and analysis of morbidity data obtained from sample surveys." *Milbank mem. Fd. Quart.*, **37**, 33–48.
—— —— (1959). " The families and individuals who did not cooperate in a sample survey." *Milbank mem. Fd. Quart.*, **37**, 347–368.

COHEN, S. E. and LIPSTEIN, B. (1954). " Response errors in the collection of wage statistics by mail questionnaire." *J. Amer. statist. Ass.*, **49**, 240–250.

COLE, D. E. (1956). " Field work in sample surveys of household income and expenditure." *Appl. Statist.*, **5**, 49–61.

*CORLETT, T. (1952). " A use for the jury qualification in sample design." *Appl. Statist.*, **1**, 34–36.

DEMING, W. E. (1954). " On the presentation of the results of sample surveys as legal evidence." *J. Amer. statist. Ass.*, **49**, 814–825.

DEMING, J. (1957). " Application of the Gompertz curve to the observed pattern of growth in length of 48 individual boys and girls during the adolescent cycle of growth." *Hum. Biol.*, **29**, 83–122.

DOWNHAM, J. S. (1954). " Social class in sample surveys." *Inc. Statist.*, **5**, 17–38.

GAITO, J. and GIFFORD, E. C. (1958). " Components of variance in anthropometry." *Hum. Biol.*, **30**, 120–127.

GALES, K. (1957). " Discriminant functions of socio-economic class." *Appl. Statist.*, **6**, 123–132.

HARRIS, F. F. (1954). "The use of sampling methods for ascertaining total morbidity in the Canadian sickness survey, 1950–1." *Bull. World Hlth. Org.*, **11**, 25–50.

HAUSER, P. M. (1954). "The use of sampling for vital registration and vital statistics." *Bull. World Hlth. Org.*, **11**, 5–24.

HAYASHI, C. (1957). "Note on sampling from a sociometric pattern." *Ann. Inst. statist. Math.*, Tokyo, **9**, 49–52.

HEALY, M. J. R. (1958). "Variations within individuals in human biology." *Hum. Biol.*, **30**, 210–218.

*HEWITT, D. and STEWART, A. (1952). "The Oxford Child Health Survey: a study of the influence of social and genetic factors on infant weight." *Hum. Biol.*, **24**, 309–319.

HILL, T. P., KLEIN, L. R. and STRAW, K. H. (1955). "The Savings Survey 1953: response rates and reliability of data." *Bull. Oxf. Univ. Inst. Statist.*, **17**, 89–126.

HOCKING, W. S. (1958). "A method of forecasting the future composition of Great Britain by marital status." *Population Stud.*, **12**, 131–148.

INDIAN STATISTICAL INSTITUTE (1953). "Reports on the Indian National Sample Survey. General report No. 1. First round: October 1950–March 1951." *Sankhya*, **13**, 47–214.

———————— (1954). "Report on the Indian National Sample Survey No. 5. Technical paper on some aspects of the development of the sample design." *Sankhya*, **14**, 265–316.

KEMSLEY, W. F. F. (1959). "Designing a budget survey." *Appl. Statist.*, **8**, 114–123.

KISH, L. and LANSING, J. B. (1954). "Response errors in estimating the value of homes." *J. Amer. statist. Ass.*, **49**, 520–538.

—— and HESS, I. (1958). "On non-coverage of sample dwellings." *J. Amer. statist. Ass.*, **53**, 509–524.

*KLEIN, L. R. (1951). "Estimating patterns of savings behaviour from sample survey data." *Econometrica*, **19**, 438–454.

—— (1955). "Patterns of savings: the surveys of 1953 and 1954." *Bull. Oxf. Univ. Inst. Statist.*, **17**, 173–214.

*—— —— and MORGAN, J. N. (1951). "Results of alternative statistical treatments of sample survey data." *J. Amer. statist. Ass.*, **46**, 442–460.

—— and VENDOME, P. (1957). "Sampling errors in the savings surveys." *Bull. Oxf. Univ. Inst. Statist.*, **19**, 85–95.

LIEBERMAN, M. D. (1958). "Philippine statistical program development and the survey of households." *J. Amer. statist. Ass.*, **53**, 78–88.

*LITTELL, A. S. (1952). "Estimation of the *T*-year survival rate from follow-up studies over a limited period of time." *Hum. Biol.*, **24**, 87–116.

LYDALL, H. F. (1954). "The methods of the Savings Survey." *Bull. Oxf. Univ. Inst. Statist.*, **16**, 197–244.

MANDEL, B. J. (1953). "Sampling the federal old-age and survivors' insurance records." *J. Amer. statist. Ass.*, **48**, 462–475.

MARSHALL, J. T. (1953). "Utilisation of census data in national health activities." *Estadist.*, **11**, 455–462.

*MERRELL, M. (1952). "The family as a unit in public health research." *Hum. Biol.*, **24**, 1–11.

MOSER, C. A. (1955). "Recent developments in the sampling of human populations in Great Britain." *J. Amer. statist. Ass.*, **50**, 1195–1214.

NISSELSON, H. and WOOLSEY, T. D. (1959). "Some problems of the household interview design for the National Health Survey." *J. Amer. statist. Ass.*, **54**, 69–87.

PEAKER, G. F. (1953). "A sampling design used by the Ministry of Education." *J. R. statist. Soc.* A, **116**, 140–165.

PETT, L. B. and OGILVIE, G. F. (1956). "The Canadian weight-height survey." *Hum. Biol.*, **28**, 177–188.

POTI, S. J., RAMAN, M. V., BISWAS, S. and CHAKRABORTY, B. (1959). "A pilot health survey in West Bengal—1955." *Sankhya*, **21**, 141–204.

*SILCOCK, H. (1952). "Estimating by sample the size and age-sex structure of a population." *Population Stud.*, **6**, 55–68.

SOM, R. K. and DAS, N. C. (1959). "On recall lapse in infant death recording." *Sankhya*, **21**, 205–208.

TANNER, J. C. (1957). "The sampling of road traffic." *Appl. Statist.*, **6**, 161–170.

U.S. Bureau of the Census (1957). " Description of the sample design of the U.S. National Health Survey." *Estadist.*, **15**, 428–431.
Utting, J. E. G. and Cole, D. (1953). " Sample surveys for the social accounts of the household sector." *Bull. Oxf. Univ. Inst. Statist.*, **15**, 1–23.
Vinski, I. (1958). " Methods of national wealth estimation." *Statist. Rev.*, Belgrade, **8**, 137–148.
World Health Organisation (1959). " Immunological and haematological surveys. Report of a Study Group." Tech. Rep. No. 181. Geneva.

E". OPINION SURVEYS AND MARKET RESEARCH

Abrams, M. A. (1957). " Problems of media research: the readership survey of the Institute of Practitioners in Advertising." *Statist. Neerlandica*, **11**, 63–76.
—————— (1958). " Technical problems in the I.P.A. readership survey." *Inc. Statist.*, **8**, 55–66.
*Ackoff, R. L. and Pritzker, L. (1951). " The methodology of survey research." *Int. J. Opin. Attitude Res.*, **5**, 313–334.
Aoyama, H. (1954). " On the interviewing bias." *Ann. Inst. statist. Math.*, Tokyo, **5**, 73–76.
Brunk, M. E. and Federer, W. T. (1953). " Experimental designs and probability sampling in marketing research." *J. Amer. statist. Ass.*, **48**, 440–452.
Brownlee, K. A. (1957). " A note on the effects of non-response on surveys." *J. Amer. statist. Ass.*, **52**, 29–32.
Cauter, T. (1956). " Some aspects of classification data in market research." *Inc. Statist.*, **6**, 133–144.
Dalenius, T. (1956). " The Survey Research Centre of the Central Bureau of Statistics, Sweden." *Sankhya*, **17**, 225–244.
Du Plessis, J. C. (1956). " On the results of the South African business opinion survey." *S. Afr. J. Econ.*, **22**, 221–228.
Durant, H. (1954). " The Gallup poll and some of its problems." *Inc. Statist.*, **5**, 101–112.
Edwards, F. (1953). " Aspects of random sampling for a commercial survey." *Inc. Statist.*, **4**, 9–26.
El-Badry, M. A. (1956). " A sampling procedure for mailed questionnaires." *J. Amer. statist. Ass.*, **51**, 209–227.
Ferber, R. (1956). " The effect of respondent ignorance on survey results." *J. Amer. statist. Ass.*, **51**, 576–586.
Gales, K. and Kendall, M. G. (1957). " An inquiry concerning interviewer variability." *J. R. statist. Soc. A*, **120**, 121–147.
Gray, P. G. (1955). " The memory factor in social surveys." *J. Amer. statist. Ass.*, **50**, 344–363.
—————— (1956). " Examples of interviewer variability taken from two sample surveys." *Appl. Statist.*, **5**, 73–85.
—————— (1957). " A sample survey with both a postal and an interview stage." *Appl. Statist.*, **6**, 139–153.
Moser, C. A. and Stuart, A. (1953). " An experimental study of quota sampling." *J. R. statist. Soc. A*, **116**, 349–405.
Shankleman, E. (1955). " Measuring the readership of newspapers and magazines." *Appl. Statist.*, **4**, 183–194.
Slonim, M. J. (1957). " Sampling in a nutshell." *J. Amer. statist. Ass.*, **52**, 143–161.
Sobol, M. G. (1959). " Panel mortality and panel bias." *J. Amer. statist. Ass.*, **54**, 52–68.
Thrift, H. J. (1959). " The Newspaper Society regional readership survey." *Inc. Statist.*, **9**, 115–137.
Van Den Berg, N. and Verburgh, C. (1956). " The accuracy of forecasts in the S. African business opinion surveys." *S. Afr. J. Econ.*, **24**, 37–62.
—————, —————— (1956). " A reply by the authors of ' the accuracy of forecasts in the S. African business opinion surveys '." *J. Econ.*, **22**, 229–230.
*Wadsworth, R. N. (1952). " The experience of a user of a consumer panel." *Appl. Statist.*, **1**, 169–178.
Williams, N. (1953). " The use of samples in auditing." *Appl. Statist.*, **2**, 180–183.

F". ECONOMICS: SURVEYS OF INDUSTRY, CENSUSES OF PRODUCTION, LABOUR FORCE, ETC.

CLARK, D. A. (1953). "The census of distribution." *Appl. Statist.*, **2**, 1–12.
DOMINION BUREAU OF STATISTICS, CANADA (1956). "Sources, procedures of compilation, and types of current industrial statistics in Canada." *Estadist.*, **14**, 432–441
GEARY, R. C. and FORECAST, K. G. (1955). "The use of census of industrial production material for the estimation of productivity." *Rev. Inst. int. Statist.*, **23**, 6–19.
GEORGE, C. O. (1954). "Industrial censuses in the United States." *Inc. Statist.*, **5**, 59–82.
HENDERSON, P. D. (1954). "Retrospect and prospect: the economic survey, 1954." *Bull. Oxf. Univ. Inst. Statist.*, **16**, 137–178.
MAURICE, R. J. (1956). "Industrial censuses in Great Britain and the United States." *Appl. Statist.*, **5**, 177–194.
*ORGANIZATION FOR EUROPEAN ECONOMIC COOPERATION (1951). "Industrial censuses in Western Countries: report of a group of experts." Paris. London: H.M.S.O.
———— ———— (1955). "Industrial censuses in the United States." Paris.
U.S. BUREAU OF THE CENSUS (1955). "Concepts and methods used in the current labour force statistics prepared by the U.S. Bureau of the Census (1954)." *Estadist.*, **13**, 280–290.

G". AGRICULTURAL ECONOMICS AND FARM PRACTICE

BEHRMANN, H. I. (1954). "Sampling technique in an economic survey of sugar cane production." *S. Afr. J. Econ.*, **22**, 326–336.
BURROWES, W. D. (1953). "The problem of exceptional samples in agricultural sample surveys." *J. R. statist. Soc.* A, **116**, 175–176.
BURROWES, W. D. (1957). "The problem of exceptional samples in agricultural sample surveys." *Estadist.*, **15**, 601–604.
CHURCH, B. M. (1954). "Problems of sample allocation and estimation in an agricultural survey." *J. R. statist. Soc.* B, **16**, 223–235.
EBLING, W. H. (1953). "Some problems of tabulation in agricultural mail sampling." *Estadist.*, **11**, 103–111.
HENDRICKS, W. A. (1953). "Some specific problems in collecting agricultural data." *Estadist.*, **11**, 48–55.
HOUSEMAN, E. E. (1957). "Sample design for the survey of farm operators' 1955 expenditures." *Estadist.*, **15**, 591–600.
LEECH, F. B. et al. (1955). "Methods of milk production: some results of a survey in four areas of England and Wales." *J. agric. Sci.*, **46**, 78–89.
*MIDZUNO, H. (1952). "Report of the survey design for agricultural production estimates in Ryuku Islands." *Ann. Inst. statist. Math.*, Tokyo, **3**, 109–121.
PALCA, H. (1953). "An experiment in the sampling of agricultural returns." *Appl. statist.*, **2**, 152–159.
———— (1955). "Some sampling problems in agriculture." *Inc. statist.*, **6**, 18–33.
SELLERS, K. C. and LEECH, F. B. (1955). "Survey of losses associated with pregnancy and parturition in Yorkshire sheep." *J. agric. Sci.*, **46**, 90–96.
SLATER, J. K. W. and MANBY, T. C. D. (1955). "Pilot survey of the mechanical conditions and fuel consumptions of agricultural tractors." *J. agric. Sci.*, **45**, 264–276.
TAYLOR, W. B. and CLEMENT, D. V. P. (1956). "The New Zealand agricultural sample survey." *J. R. statist. Soc.* A, **119**, 409–424.

H". CROP ESTIMATION, FORECASTING, ETC.

FREEMAN, G. H. (1958). "A comparison of methods of measuring leaf areas in the field." *Rep. E. Malling Res. Sta. for 1957*, 83–86.
———— and BOLAS, B. D. (1956). "A method for the rapid determination of leaf areas in the field." *Rep. E. Malling Res. Sta. for 1955*, 104–107.
HEMINGWAY, R. G. (1955). "Soil-sampling errors and advisory analyses." *J. agric. Sci.*, **46**, 1–8.

HENDRICKS, W. A. (1957). " Research on sample survey procedures and crop estimating methods." *Estadist.*, **15**, 346–355.

JOLLY, G. M. and HOLLAND, D. A. (1958). " Sampling methods for the measurement of extension growth of apple trees." *Rep. E. Malling Res. Sta. for 1957*, 87–90.

KELLY, B. W. (1958). " Objective methods for forecasting Florida citrus production." *Estadist.*, **16**, 56–64.

MASUYAMA, M. (1954). " Analysis of the 1939 model sample survey results from the viewpoint of integral geometry." *Sankhya*, **13**, 229–234.

—— —— (1954). " Mathematical note on area sampling." *Sankhya*, **13**, 241–242.

—— —— and SENGUPTA, J. M. (1955). " On a bias in a crop-cutting experiment (Application of integral geometry to areal sampling problems—Part V)." *Sankhya*, **15**, 373–376.

MOKASHI, V. K. (1953). " Investigations on sampling for estimation of crop acreages —I." *J. Ind. Soc. agric. Statist.*, **5**, 128–143.

—— —— (1954). " Investigations on sampling for estimation of crop acreages—II." *J. Ind. Soc. agric. Statist.*, **6**, 115–126.

OWEN, P. C. (1957). " Rapid estimation of the areas of the leaves of crop plants." *Nature, London*, **180**, 611.

RUBIN, S. S. and DANILEVSKAJA, O. M. (1957). " The measurement of leaf area in fruit trees." *Bot. Zurnal*, **42**, 728–730.

SENGUPTA, J. M. (1954). " Some experiments with different types of area sampling for winter paddy in Giridih, Bihar, 1945." *Sankhya*, **13**, 235–240.

STEVENS, W. L. (1955). " Estimation of the Brazilian coffee harvest by sampling survey." *J. Amer. statist. Ass.*, **50**, 775–787.

U.S. BUREAU OF AGRICULTURAL ECONOMICS (1953). " Sampling methods for agricultural estimating and forecasting, and elements to be considered in their adoption." *Estadist.*, **11**, 72–90.

—— —— —— —— (1953). " Use of check data in crop estimation." *Estadist.*, **11**, 91–102.

I''. FORESTRY AND LAND UTILIZATION SURVEYS

FINNEY, D. J. (1953). " The estimation of error in the systematic sampling of forests." *J. Ind. Soc. agric. Statist.*, **5**, 6–16.

MASUYAMA, M. (1953). " A rapid method of estimating basal area in timber survey— an application of integral geometry to areal sampling problems." *Sankhya*, **12**, 291–302.

MATUSITA, K. et al. (1955). " Some problems of sampling in a forest survey." *Ann. Inst. statist. Math.*, Tokyo, **7**, 1–23.

MOKASHI, V. K. (1954). " Efficiency of sampling methods in forest surveys." *J. Ind. Soc. agric. Statist.*, **6**, 101–114.

RAO, J. N. K. (1957). " Double ratio estimate in forest surveys." *J. Ind. Soc. agric. Statist.*, **9**, 191–204.

J''. ESTIMATION OF WILD POPULATIONS

DARROCH, J. N. (1958). " The multiple-recapture census. I. Estimation of a closed population." *Biometrika*, **45**, 343–359.

—— —— (1959). " The multiple-recapture census. II. Estimation when there is immigration or death." *Biometrika*, **46**, 336–351.

ELTON, C. S. and MILLER, R. S. (1954). " The ecological survey of animal communities: with a practical system of classifying habitats by structural characters." *J. Ecol.*, **42**, 460–496.

HAMMERSLEY, J. M. (1953). " Capture-recapture analysis." *Biometrika*, **40**, 265–278.

LESLIE, P. H., CHITTY, D. and CHITTY, H. (1953). " The estimation of population parameters from data obtained by means of the capture-recapture method. III. An example of the practical applications of the method." *Biometrika*, **40**, 137–169.

K″. INDUSTRIAL APPLICATIONS

ASHFORD, J. R. (1958). " The design of a long-term sampling programme to measure the hazard associated with an industrial environment." *J. R. statist. Soc.* A, 121, 333–347.

BILLETER, E. P. (1956). " Optimum design in mixed sampling plans " *Rev. Inst. int. Statist.*, 24, 73–76.

CHAMPERNOWNE, D. G. (1953). " The economics of sequential sampling procedures for defectives." *Appl. Statist.*, 2, 118–130.

HAMAKER, H. C. (1958). " Some basic principles of sampling inspection by attributes." *Appl. Statist.*, 7, 149–159.

*HEBDEN, J. and JOWETT, G. H. (1952). " The accuracy of sampling coal." *Appl. Statist.*, 1, 179–191.

*JOWETT, G. H. (1952). " The accuracy of systematic sampling from conveyor belts." *Appl. Statist.*, 1, 50–59.

KETTMAN, G. (1958). " A comparison of the leading sampling plans and a new German acceptance sampling system." *E. Q. Bull. No.* 3, 47–74.

MIKAMI, M. (1956). " On a multi-level sampling inspection plan for continuous production." *Bull. math. Statist.*, 7, 1–10.

TIPPETT, L. H. C. (1954). " Statistics in research and management in the cotton industry." *Inc. Statist.*, 5, 147–159.

—— —— (1958). " A guide to acceptance sampling." *Appl. Statist.*, 7, 133–148.

FOURTH EDITION

This list contains references not included in the bibliographies to the earlier editions.

For books and computer program manuals the dates given are to the editions consulted:

BOOKS AND PAPERS

AGHAJANIAN, A. and MEHRYAR, A. H. (1979). " Effect of child mortality on contraceptive use in rural Iran." *Bull. Eugen. Soc.*, **11**, 14–18.

BARTLETT, M. S. (1935). " Contingency table interactions." *Suppl. J.R. statist. Soc.*, **2**, 248–252.

CHURCH, B. M. and LEWIS, D. A. (1977). " Fertilizer use on farm crops in England and Wales: Information from the Survey of Fertilizer Practice, 1942–1976." *Outl. Agric.*, **9**, 186–193.

—— —— and WEBBER, J. (1971). " Fertiliser practice in England and Wales: a new series of surveys." *J. Sci. Fd Agric.*, **22**, 1–7.

COCHRAN. W. G. (1977). *Sampling Techniques*. 3rd edn. New York: Wiley.

DALENIUS, T. E. (1977). " Bibliography on non-sampling errors in surveys." *Int. Stat. Rev.*, **45**, 71–89, 181–197, 303–317.

—— —— (1978). " Information privacy and statistics, a topical bibliography." U.S. Dept. of Commerce, Bureau of the Census.

FELLEGI, I. P. and HOLT, D. (1976). " A systematic approach to automatic edit and imputation." *J. Amer. statist. Assoc.*, **71**, 353, 17–35.

FRANCIS, I. (1979). " A Comparative Review of Statistical Software." The International Association for Statistical Computing, Voorburg, Netherlands.

—— ——, SHERMAN, S. P. and HEIBERGER, R. M. (1976). " Languages and programs for tabulating data from surveys." *Proceedings, Computer Science and Statistics: Ninth Annual Symposium on the Interface, Cambridge, USA.*

HIGGINS, J. E. and KOCH, G. G. (1977). " Variable selection and generalized chi-square analysis of categorical data applied to a large cross-sectional occupational health survey." *Int. Stat. Rev.*, **45**, 51–62.

KENDALL, M. G. and STUART, A. *The Advanced Theory of Statistics*. Vol. I, 4th edn, 1977. Vol. II, 4th edn, 1979. London and High Wycombe: Griffin.

KISH, L. (1965). *Survey Sampling*. New York: Wiley.

KOCH, G. G., TOLLEY, D. and FREEMAN, J. L. (1976). " An application of the clumped binomial model to the analysis of clustered attribute data." *Biometrics*, **32**, 337–354.

LEECH, F. B., DAVIS, M. E., MACRAE, W. D. and WITHERS, F. W. (1960). *Disease, wastage and husbandry in the British dairy herd: Report of a National Survey in 1957–58·* London: Ministry of Agriculture, Fisheries & Food.

WARNER, S. L. (1965). " Randomized response: A survey technique for eliminating evasive answer bias. *J. Amer. statist. Assoc.*, **60**, 63–69.

WHITTAKER, J. and AITKIN, M. (1978). " A flexible strategy for fitting complex log-linear models." *Biometrics*, **34**, 487–495.

YATES, F. (1947). " The analysis of contingency tables with groupings based on quantitative characters." *Biometrika*, **35**, 176–181.

—— —— (1970). *Experimental Design: Selected Papers* London: Griffin.

—— —— (1973). " The analysis of surveys on computers—features of the Rothamsted Survey Program." *Appl. Statist.*, **22**, 161–171.

—— —— (1975). " The design of computer programs for survey analysis—a contrast between the Rothamsted General Survey Program (RGSP) and SPSS." *Biometrics*, **31**, 573–584.

—— —— and HEALY, M. J. R. (1964). " How should we reform the teaching of statistics ?." *J.R. statist. Soc. A*, **127**, 199–210.

—— —— and SIMPSON, H. R. (1960). " A general programme for the analysis of surveys." *Comp. J.*, **3**, 136–140.

—— ——, —— —— (1961). " The analysis of surveys: processing and printing of the basic tables." *Comp. J.*, **4**, 20–24.

COMPUTER PROGRAM MANUALS

BMDP. W. J. Dixon *et al.*, Health Sciences Computing Facility, U.C.L.A., University of California Press, 1977.

CLUSTERS. Vijay Verma and Mick Pearce, I.S.I. World Fertility Survey, London, 1977.

CONCOR. Editing System by the Latin-American Centre of Demography (CELADE), I.S.I. World Fertility Survey, London, 1979.

GENSTAT. J. A. Nelder *et al.*, Rothamsted Experimental Station, 1977.

GLIM. R. J. Baker and J. A. Nelder, Numerical Algorithm Group, Oxford, 1978.

RGSP. F. Yates, Rothamsted Experimental Station, Part 1, 1975; Part 2, 1980.

SCSS. N. H. Nie *et al.*, SPSS Inc., McGraw-Hill, 1978.

SPSS. N. H. Nie *et al.*, SPSS Inc., McGraw-Hill, 1970, 1975.

TPL. R. C. Mendelssohn *et al.*, U.S. Dept. of Labor, Bureau of Labor Statistics, 1978.

INDEX

When a subject is dealt with in the whole of a section, only the first page of the section is entered. Relevant examples which occur at the end of such sections are not separately indexed.

For technical aspects of the application of computers to survey analysis see supplementary index.

Accuracy, relative, 31, 135, 256; definition, 256.
Addressograph list, 85.
Administrative areas, sampling by, 37, 79.
Administrative organization for surveys, 108.
Advertising, 83.
Advisory Economists, 62.
Aerial photographs, 33, 43, 77, 78, 89.
Aghajanian, A., 404.
Agricultural Meteorological Committee, 13.
Agriculture, frames for, 84–91.
Aitkin, M., 357.
Allied Mission for Observing Greek Election, 74.
Alternative estimates, 135.
Analysis and presentation of results, planning of, 117; step-by-step, 118; critical, 126–34, 321–69; sampling for, 121; overall control of accuracy, 119–21; adjustment for defects in the sample, 124.
Analysis of variance, 205; interpretation of, 286–90, 331; applications of, 206, 240 259, 263, 281, 288, 298, 331; examples of, 208, 210, 211, 227, 229, 261, 283, 290, 293, 332, 334.
Animal populations, 45.
Anthropometric surveys, 410.
Area sampling, 71–83, 86–91; see also systematic sample from areas.
Areas, measurement by sampling, see point and line sampling.
Areas, selection with equal probability, 86.
Attenuation, 360.
Attributes, see qualitative variates.
Automatic recording, 110.

Balanced differences, 232.
Balanced sample, 40, 164, 222.
Bartlett, M. S., 358.
Bias, 9–17; permissible, 17; estimation of, 35, 246, 247; in selection, 9–15, 69, 81, 84, 87, 92, 93, 124, 155, 246; in systematic samples, 10, 233, 236–7; in demarcation of units, 15, 154; in eye estimates, 92, 153, 155, 223; in estimation, 16, 35–7, 75, 81, 135, 152, 164; in estimate of error, 189; relative precision of biased and unbiased estimates, 249.
Biological sampling, 51, 244.
Blanks (on cards), 116.
Blocks, city, 72, 75.

Box, K., 58.
Boyd, D. A., 85, 360.
British Institute of Public Opinion, 186.
Bureau of Agricultural Economics, 77.
Bureau of the Census, 77.
Byssinosis, 349.

Calcutta Institute of Statistics, 87.
Calibration, see eye estimates.
Canada, population census, 337; labour force survey, 48, 271.
Cards, 114; alternatives to, 114; multiple punching, 115; format (fixed and free), 116; card-type numbers, 117; unit references, 117.
Career expectations of university students, 403.
Causal relations, 126, 178, 321, 365.
Cells, 42.
Census of Woodlands—1938–9, 16, 153, 221, 270, 307.
Census of Woodlands—1942, 45, 49, 87, 106, 234, 251.
Central Office of Information, 82.
Change, see successive occasions.
Checks, on field work, 112; on analysis, 119–21; by comparison with complete returns, 29.
Child losses, influence on future births (Iran), 404.
Child nutrition, 321–4.
Chi-squared, 22, 191, 349, 358, 395, 406; formulae for, 354–6; examples, 191, 356–9.
Cluster sample, 21; errors, 216; see also multi-stage.
Cochran, W. G., 290.
Coding of data, 110, 114–16; literal, 115; missing, 116.
Coefficient of variation, 174.
Commercial undertakings, 82.
Comparability, 53.
Complete census, 3; combination with sample, 50.
Composite sampling scheme, 49.
Compulsory returns, 63.
Computers, uses of, 114–21, 129, 134.
Consumer preferences, 83.
Contingency coefficient, 354.
Contingency tables, 353; chi-squared for, 354; other measures of association, 356, 395, 406.

451

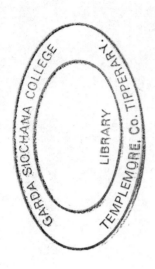